Ludwig Erik Aguilar
Biomaterial Science

Also of Interest

Materials for Medical Application
Heimann, 2020
ISBN 978-3-11-061919-5, e-ISBN 978-3-11-061924-9

Biomaterials and Engineering for Implantology.
In Medicine and Dentistry
Oshida, Miyazaki, 2022
ISBN 978-3-11-074011-0, e-ISBN 978-3-11-074013-4

Medical Physics.
Models and Technologies in Cancer Research
Bajek, Tylkowski (Eds.), 2021
ISBN 978-3-11-066229-0, e-ISBN 978-3-11-066230-6

Smart Polymers.
Principles and Applications
García JM, García FC, Reglero Ruiz, Vallejos, 2022
ISBN 978-1-5015-2240-6, e-ISBN 978-1-5015-2246-8

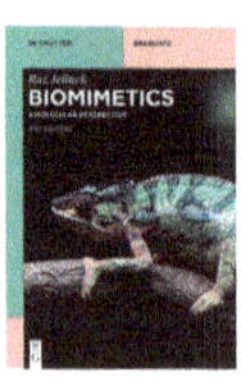

Biomimetics.
A Molecular Perspective
Jelinek, 2021
ISBN 978-3-11-070944-5, e-ISBN 978-3-11-070949-0

Ludwig Erik Aguilar

Biomaterial Science

Anatomy and Physiology Aspects

DE GRUYTER

Author
Dr. Ludwig Erik Aguilar
Department of Bionanosystem Engineering
Jeonbuk National University
Deokjin-gu
Jeonju-si 54896
Republic of Korea

Disclaimer: The figures for this book were created with BioRender.com.

ISBN 978-3-11-065535-3
e-ISBN (PDF) 978-3-11-065537-7
e-ISBN (EPUB) 978-3-11-065568-1

Library of Congress Control Number: 2022934438

Bibliographic information published by the Deutsche Nationalbibliothek
The Deutsche Nationalbibliothek lists this publication in the Deutsche Nationalbibliografie; detailed bibliographic data are available on the Internet at http://dnb.dnb.de.

Cover image: John Paul "Pejie" Abia
Typesetting: Integra Software Services Pvt. Ltd.
Printing and binding: CPI books GmbH, Leck

www.degruyter.com

To my firstborn Son, Søren; may you grow up with a sense of wonder and dedicate your life to the pursuit of truth and happiness.

About the authors

Ludwig Erik Aguilar, Ph.D. grew up in the Philippines where he studied BSc. in Nursing at Bulacan State University. Afterwards, he completed his MSc. & Ph.D. degrees in Bionanosystem Engineering at Jeonbuk National University, South Korea. His main area of research was in design and synthesis of nanobiomaterials (e.g. nanoparticles, nanofibers, and nanocoatings) for drug delivery and medical device applications whereas he published more than 20 academic papers in peer-reviewed journals.

His clinical and materials background gave him insights on how biomaterials can be both safe and effective if used inside the body. He worked as an Assistant Research Professor at Jeonbuk National University, where he was also a lecturer of anatomy & physiology at the graduate school of engineering. During this time, he realized the need to create abasic understanding for students of non-clinical background to learn about the human body if they aspire to become a biomedical engineer or biomaterialist. As of now, he works as a Senior R&D Engineer at Novo Nordisk A/S, researching novel drug delivery devices and contributed as well to stem cell delivery research.

https://doi.org/10.1515/9783110655377-202

Preface

When someone wants to venture to the field “biomaterials,” one should have a working knowledge of the human body to complement their knowledge in chemistry or engineering. To be effective in designing and engineering materials that could be used in biological systems, it is crucial to be able to have a balanced understanding of biology, chemistry, and engineering. Biomaterials field is a complex and highly collaborative one. It is an amalgamation of the basic sciences and how it is translated to useful biomedical devices, drug delivery formulations, and artificial substitutes to impaired organs. This book tries to find a perfect balance of necessary parts of information in human anatomy and physiology and materials engineering.

https://doi.org/10.1515/9783110655377-203

Goals

Anatomy and physiology are the studies of body parts and functions in relation to human beings. Biomaterials are synthetic or natural nonindigenous materials that are used in biological systems to augment, treat, and cure various physiological functions. Here we elucidate different biomaterials used in the anatomical systems and how those materials react to the human body and vice versa. We will also explore established and future prospective of biomaterial types/design considerations/characterization and synthesis, to guide students and academics alike in understanding the relations of material science and the human body. We will explain complex body functions in a level that non-clinicians can understand and create a synergy with all related sciences involved in biomaterials.

To explain important concepts in anatomy and physiology in relation to biomaterials engineering.

To focus on important physiological concepts in relation to biomaterials.

https://doi.org/10.1515/9783110655377-204

Contents

Chapter 1
Importance of human anatomy and physiology in biomaterial science

The study of the human body and its related functions is called anatomy and physiology. Anatomy refers to the internal and external structures of the body and their physical relationships, whereas physiology refers to the study of the functions of those structures [1–3]. This field is mostly studied in life sciences and is the foundational knowledge of clinicians such as medical doctors, nurses, and physical therapists. It is one of the most difficult subjects in life sciences as it requires large memorization of terms, processes, and functions of individual body parts and systems. But understanding how the human body works is vital in becoming a biomaterialist as the development of biomaterials should utilize the anatomical structure and physiological functions of the human body. A civil engineer who is making a bridge must first know the lay of the land to create an efficient construction plan to build a bridge that will last for a long time; similarly, a biomaterial scientist/engineer should also know the basics of the human body and how it functions. In biomaterial science, the ability to identify each part of the body in a systematic and coordinated way is crucial when designing a device, drug delivery system, or prosthesis that is intended for the human body. Rational designs based on the knowledge of anatomy will give rise to an effective treatment and intervention. If a biomaterial engineer or scientist cannot distinguish the difference between the different structures and functions in the body, then it might endanger the lives of the patients by having created or designing materials that are not suitable for each target organ or tissue.

The human body consists of 14 systems that work in unison to maintain a biological balance called "homeostasis." If one system fails or malfunctions, it could lead to a cascade of effects and would create imbalances in the body, and accumulation of imbalances is called a disease. The very nature of the human body is complex and until now being studied upon to further understand more of its fundamental physiological process. With more in-depth understanding, this helps to cure and treat diseases that normally would permanently impair or even be detrimental to a person. Right now, the human body is now being integrated with multiple non-native materials to treat various anatomical and physiological maladies, with modern medicine utilizing novel biomaterials to aid in treating or curing these diseases (as shown in Figure 1.1). New methods of tissue regeneration, delivering drugs, and development of medical devices are now being revolutionized by the advancement of biomaterial science and creating a better world in the process.

https://doi.org/10.1515/9783110655377-001

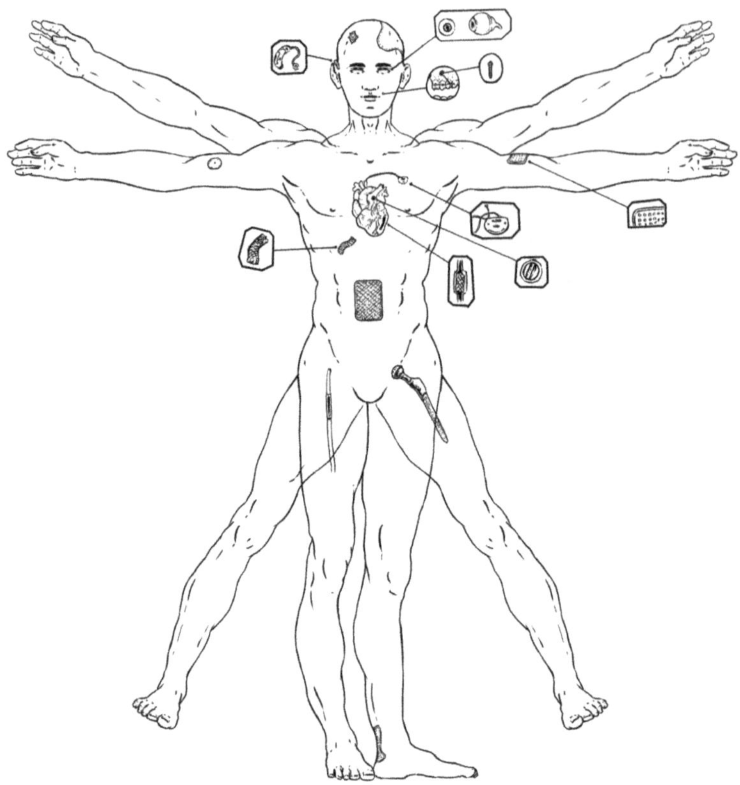

Figure 1.1: A creative illustration of biomaterials used in biomedical devices, prosthesis, drug delivery systems, and tissue-engineered materials. It shows how many systems can be applied into the body. It varies in complexity and functions.

1.1 Biomaterials

Materials that are used inside or outside the body to augment or facilitate normal physiological functions are called biomaterials. It is an umbrella term used to classify materials having properties that are useful for the human body [4–6]. A biomaterial must be compatible to the human body by allowing the body to tolerate its presence without the material causing any harm to the cells or the surrounding tissue. This is called biocompatibility and has been a general requirement for all biomaterials [4, 7–9]. There are different ways to measure biocompatibility of a material depending on the application [9, 10]. A standard called ISO-10993 is one of the benchmarks for testing a biomaterial for potential commercial use. Without passing this standard, a material would not be classified suitable for biological use or called as a biomaterial. With the advancement of the field, bioactivity or adding functionality to a biomaterial is now being explored. This is to create favorable therapeutic and biological effects in the body to either enhance or compliment a treatment, endow newer functionalities to a medical device, and create smarter drug delivery systems. In contrast to conventional biomaterials which aim to be tolerated by the body, these newer class of biomaterials aim to be more functional and smarter by providing new material properties to otherwise inert materials.

Furthermore, biomaterials are the cornerstone of biomedical technology, and both fields go hand in hand in the development of subfields within them. This chapter discusses the basic and necessary information about human anatomy and physiology for a biomaterial scientist to design and test materials intended for use in different systems in the body. A lean discussion on anatomy and physiology is provided to systems where biomaterials are currently being applied. An introduction to examples of biomaterial concepts is also included to set the scene for what relevant technologies are currently being used in the field.

1.2 Biomaterials for the integumentary system

The skin is composed of several layers, which are divided into three parts: epidermis, dermis, and hypodermis (see Figure 1.2) [11–14]. There are specialized structures found in these layers and these are sweat and sebaceous glands, hair follicle, capillaries, fat, and connective tissue (collagen) [15, 16]. The most superficial layer is called epidermis, which is composed of dead skin cells that provide waterproof protection and the primary barrier from pathogens. The next layer is the dermis where specialized structures can be found, e.g., the duct of the sebaceous gland or sweat gland, sensory nerves to enable us to have tactile sensation on our skin, and the dermis is also where new skin cells are constantly regenerating. The hypodermis is where most of the blood vessels and connective tissues made up of collagen matrix can be found. Specialized immune cells can also be found in the skin constantly patrolling for allergens, or any pathogens that might have crossed the skin barrier.

When certain pathologies are involved, it might affect the ability of the skin to regenerate and provide the primary barrier from pathogens. Examples of which are when there is chronic inflammation created by diabetes [17, 18] or different kinds of burns (chemical, radiative, or fire related) can lead to dysfunction of the normal physiological function of the skin. Naturally, the skin can heal itself given enough time, where scarring is left on the skin after a wound or injury has healed [19–21]. Scars are a natural part of the healing process and usually fade and become paler over time, although they never completely disappear. Biomaterial science can assist in this healing process in ways where it can prevent initial infection and provide a conducive environment for the skin tissue to optimize its healing process [22]. Examples of contemporary research on wound healing materials are polymeric systems of different iterations, e.g., hydrogel and nano- and microfibrous materials

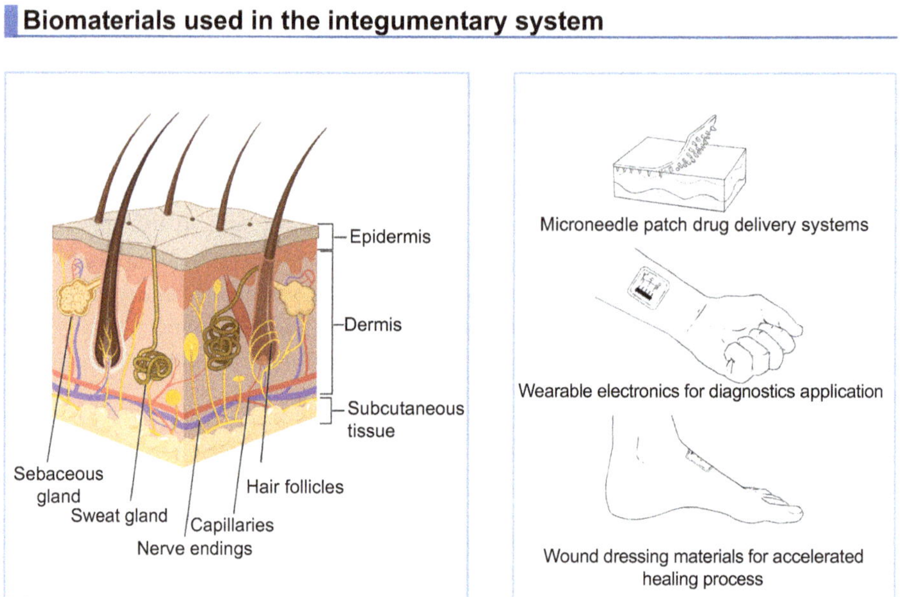

Figure 1.2: (Left panel) The gross anatomy of the skin which is the biggest organ in the integumentary system and the whole body. (Right panel) The examples of biomaterials used in this system encompass from drug delivery system through transdermal patches (polymeric microneedles), wearable electronics (polymeric adhesives with integrated circuits), and wound dressing materials (polymers with antibacterial and healing agents).

that release antibacterial agents such as metallic nanoparticles made up of silver, gold, or copper, and lactam-containing molecules (Figure 1.2) [22–24]. The idea is to absorb the wound exudate, while facilitating oxygenation and reducing the pathogens around the wound site. These factors assist the skin to recuperate faster compared to unassisted wounds.

The skin could also be used as an entry point for therapeutics to the body, as it is done via transdermal patches with microneedles (Figure 1.2) [25–27]. Microneedle technology is a promising way to deliver therapeutics that need to be released in a controlled manner [26, 28]. The needles used in these patches are in millimeter to microscale in length and could be hollow or solid in form. The drugs are then absorbed in the body via the dermal capillaries and the interstitial spaces of the dermis. Examples of which are female contraceptive patches that deliver hormones (progesterone) and nicotine patches for people who are trying to quit smoking [29, 30]. Smart patches that react with external and internal stimuli such as glucose levels [31, 32], heat, and ultrasound are also being developed to achieve an on-demand delivery of therapeutics [33].

The skin is a dynamic organ and can serve as noninvasive site for monitoring physiological conditions. An emerging technology of wearable sensors uses flexible electronics made up of composites of carbon-based materials (Figure 1.2) [34], various metallic nanoparticles, and polymers to receive biochemical cues and translate it to electrochemical signal [35, 36]. For example, sweat from the skin can be used as a biochemical sample to analyze and diagnose some patients with cystic fibrosis, as these patients have increased chloride concentrations on their sweat [37, 38]. Sweat can also be a source of energy which can also be used to power electronic devices as it could be an anchor point for energy harvesting. This is the domain of electrical and biomaterial engineering that focuses on biosensors and microfluidic technologies. These technologies can be combined with novel biomaterial manufacturing techniques that will be discussed in detail at the later parts of this book.

1.3 Biomaterials for the musculoskeletal system

The skeletal system supports the human body by providing structural support and by protecting vital organs, e.g., the brain, the heart, and lungs. It also serves as the anchoring point for the muscles to provide the body's wide range of motion. The skeletal system is comprised of rigid bone material of hierarchical structure constantly created by osteoblast and destroyed by osteoclast cells in a process called bone remodeling. The skeletal system has an array of multiple bone structures in the macrolevel, and different types of the bone are seen depending on the type of load and function it should provide. Long bones are high load carrying and primarily involved in providing structural and mechanical support to the muscular system, and the marrow is also seen inside long bones such as the femur and it provides other functions such as regulating formation of blood cells called hematopoiesis (see Figure 1.3, left panel). Stem cells can also be found being produced in the marrow called bone marrow stem cells (BMSC). Flat bones provide protection to the cavities in the body. Most prominent part of this is the cranial bone or the skull which houses the brain and other sensory organs. Flat bones act like plate armoring to provide protection to highly sensitive organs of the body. Complex bones provide articulation and usually found in the joints of the limbs such as in hands (metacarpals) and feet (metatarsals), and the spinal column is also considered as a complex bone structure. Bone is regarded as a complex, highly organized, and specialized tissue that is composed of nearly 70% inorganic constituents (primarily, hydroxyapatite (HA), $Ca_{10}(PO_4)_6(OH)_2$), nearly 20% organic constituents and 10% water. Among many organic constituents, the type I collagens are most abundant (nearly 90%), which anchors HA crystals and enhances cellular attachment due to its abundant RGD (Arg–Gly–Asp tripeptide sequence) residues. The smallest structure on the bone is the minerals being secreted by the osteoblast cells. It is a combination of calcium and phosphate at 1:1.25 ratio, respectively. This mineral is called HA and the mineralization of this helps to bring about the rigid structure of the bone [39]. When we look closely on the microstructure of the bone, we can see that it has porosity, and this porous structure is called

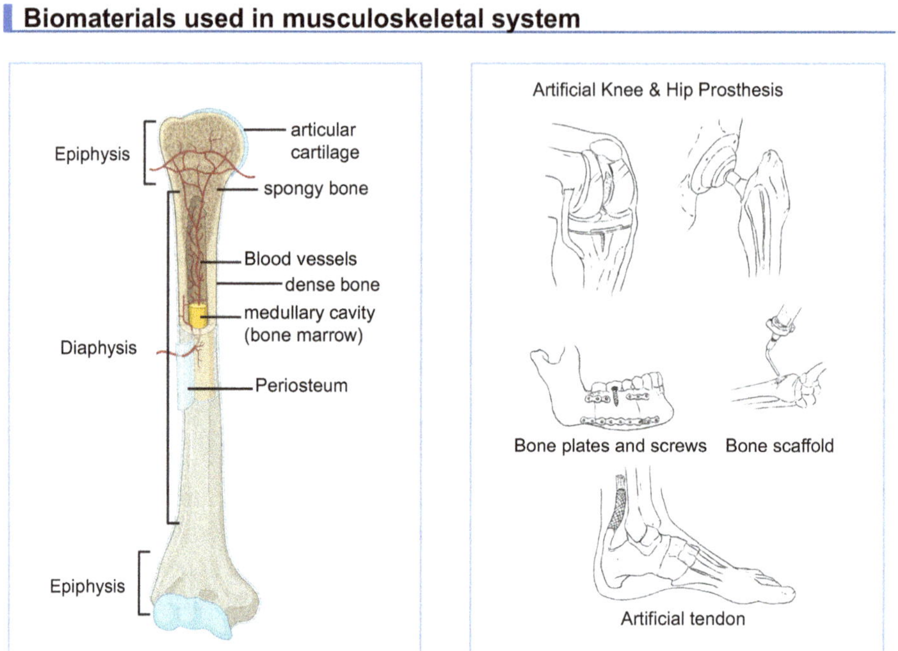

Figure 1.3: (Left panel) The gross anatomy of a long bone exhibiting the intrinsic structure that allows it to give structural support and mobility to the body and create the very important physiological function of hematopoiesis (red blood cell generation). (Right panel) The various examples of bone tissue prosthesis (metal and ceramics) that restores the ability of the damaged bone to provide structural support and mobility to the body.

trabeculae or spongy bone. People who have osteoporosis had increased bone porosity giving it weaker structural integrity, thus making it prone for fractures.

The most common application of biomaterials are in orthodontics when there is a need to increase the bone volume and density so that screws can be anchored on the jawbone. (see Figure 1.3, right panel). It is also noted that metal alloys, usually titanium or stainless steel, can replace or support healing bones in events of fractures. These metal implants may be bioinert and does not cause any reaction such as inflammation or hemotoxicity and cytotoxicity [40]. They also must be corrosive resistant so that metallic ions would not be able to diffuse into the body and cause damage [41]. New generation of metallic implants can also deliver therapeutics such as antibiotics to attenuate bacterial colonization [41–43]. Strategies such as adding nontherapeutic agents such as silver and copper nanoparticles on the surface of the implant can also attenuate bacterial colonization [44]. Bioactivity is also highly valuable in metallic implants that seek bone regenerative properties. Addition of HA [45–47] or beta-tricalciumphosphate is also a contemporary strategy to increase osteoinduction on the implantation site [48].

When people suffer from bone injuries, biomaterial science can help in guiding the growth of bone tissue so that it can regain its structural integrity. When creating biomaterials for bone tissue applications should mimic the native environment of the skeletal tissue (Figure 1.3, right panel). Ideally, synthetic bone tissue scaffolds are expected to mimic both the structures and functions of natural bone besides having good mechanical properties and biocompatibility. Therefore, it becomes necessary and relevant to take know-how about physicochemical architecture of native bone along with pertinent biomechanical and biochemical properties to gain an insight into choosing and harnessing the best biomaterial types that are used in the fabrication of bone tissue scaffolds. In order to remodel the damaged bone, effective biomaterials should be able to induce the following: osteoinduction must be first initiated on the scaffold, which allows the recruitment of immature cells and the stimulation of these cells to become preosteoblasts. This is followed by osteoconduction, whereas the recruited cells after developing into preosteoblast further proliferate in number in the presence of the scaffold. Lastly, the scaffold must facilitate osteointegration where it becomes part of the native hard tissue environment slowly being replaced by the native mature osteocytes, thus repairing the defective bone [49, 50]. The scaffold material should also exhibit biocompatibility by virtue of noninflammatory and noncytotoxic effects. Leachable materials are generally avoided in creating bioscaffold for any application and this also applies to hard tissue engineering. The degraded product of the scaffold should also be easily eliminated by the body. While the whole process of bone remodeling is happening, the scaffold must also provide adequate nutrition and gas transport onto the cells [51]. That is why highly porous materials are good candidates in tissue engineering, whether it is for soft or hard tissue applications. Porosity facilitates efficient nutrient and gas transport within the scaffold by increasing the surface area, thus making diffusion relatively easy [52, 53].

Connective tissues that facilitate the attachment of muscles and bones to each other are called ligaments and tendons. Ligaments are bands of elastic collagenous tissue that wraps around the joints of the bones [54]. They connect bone to bone and provide support and limit the range of motion of the joint, while tendons connect muscles to bones and are also made up of elastic collagenous tissue. When these tissues are damaged, healing may take some time and would typically require surgical intervention. However, in cases of severe damage, e.g., sport injuries or vehicular accidents, total replacement of new tendons or ligaments is required to repair the mobility of the affected limb. If a limb's range of motion is not restored, the muscles will undergo irreversible atrophy (degeneration) leading to muscle mass loss and permanent disability [55, 56]. Here, biomaterials are used to recreate these collagenous elastic tissues using both synthetic material and a combination of synthetic and natural polymers (see Figure 1.3, right panel). Polymeric materials made into hydrogels [57, 58] or nonwoven fibers [59] can be implanted to replace the damaged connective tissue with the hope that it will perform at par with a healthy tendon or ligament. However, finding the right balance of elasticity and biocompatibility remains a challenge as pure polymeric materials are prone for degradation overtime. Here strategies such as incorporating chondrocytes (collagen-forming cells) together with the biomaterial is employed to reinforce the implant with collagen and gradually replace the elastic tissue and facilitate full repair of the tendon or ligament [60–62].

Another kind of connective tissue in the musculoskeletal system is cartilages. It is a flexible connective tissue that keeps joint motion and provides cushioning support to bones against impact and friction from physical activities. There are three types of cartilages in the body, namely, hyaline, fibrous, and elastic cartilages. Hyaline cartilage is predominant in the nose, ear, trachea, and smaller respiratory tubes, while elastic cartilages are found in the external ear flaps and larynx. Fibrous cartilages are found in the synovial joints of the musculoskeletal system. They are essential in providing mechanical support in load-bearing joints such as the hip and knee. It cushions the ends of the bone from excessive friction and impact to ensure smooth articulation. However, several pathologies can affect cartilages such as osteoarthritis or rheumatoid arthritis. This happens when the cartilage in the synovial joint is degenerated due to old age or autoimmune response and could cause disfigurement and chronic pain during movement. Cartilages have limited repair capabilities as the chondrocytes are bound by a structure called lacunae, which limits their ability to migrate to damaged areas in the joint to produce more collagen. When this happens, artificial cartilages can be made to replace the damaged joint and provide support in the joint relieving it from pain and restoring its mobility [63–66]. When it is too advanced and the cartilage is beyond repair, metallic prosthesis is implanted to replace the degenerated joints. These are common occurrences in the hip and knee joint of people who had either osteo- or rheumatoid arthritis. Hip or knee replacement is done by surgically removing the diseased joint and replacing it with a metallic prosthesis (see Figure 1.3, right panel) [65, 67, 68]. This metallic prosthesis is made up of either titanium or stainless steel stem with a ceramic or polymer end to facilitate smooth articulation of the joint [69, 70]. These prostheses are either cemented or uncemented depending on the implant surface. Porous implants will let bones to grow into its surface, thereby removing the need for a bone cement for anchorage while smooth surfaced implants typically require the use of bone cements to keep it in place. Surface modification of metallic implants is also an emerging field of biomaterials due to the rise of implant bacterial infection post implantation. This is one of the leading causes of implant failure and it is very detrimental to the patient as it could cause septic shock if left untreated. Furthermore, the need for another surgical intervention would put severe medical and financial burden to the healthcare system and to the patient. Therefore, antibacterial coatings are placed on the surfaces of the prosthesis to avoid acute infection that could rise post-surgery. New coating methods, besides the traditional dip coating, are being developed (e.g., electropolymerization coating and electrospraying) to create a surface that is hostile to bacterial colonization while keeping the biocompatibility of the implant [71–73]. Addition of antibacterial agents such as silver nanoparticles and antibacterial drugs to the surfaces of these implants is also being done to reduce the infection rate related to the use of metallic prosthesis [74–76].

1.4 Biomaterials for the cardiovascular system

The cardiovascular system can be considered as the primary system that is responsible for transporting nutrients, gasses, and biomolecules needed for sustaining homeostasis. The main organ of this system is the heart and on the macro-level is a type of involuntary muscle that contracts at a specific rate (80–120 beats per minute) and can increase or decrease depending on the physical activity, emotional status, stress level, or physiological conditions of the body. Its normal rhythm is also called as sinus rhythm, whereas a fast beating of the heart is called tachycardia and slow beating is called bradycardia. The central nervous system (CNS) (hypothalamus) and hormonal response (e.g., adrenaline and epinephrine) also influence the rate of contraction of the heart. The heart anatomy is made up of four chambers as shown in Figure 1.4. The heart is also made up of different layers called pericardium (outermost), myocardium (middle), and endocardium (innermost). The pericardium, also called pericardial sac, is a double-walled sac containing the heart and the roots of the great vessels. It has two layers: an outer layer made up of strong connective tissue (fibrous pericardium), and an inner layer made up of serous membrane (serous pericardium). It protects the heart from infection, blunt trauma, and lubricates the heart motion by using the pericardial fluid. The myocardium is the contractile part of the heart. It houses the contractile nodes responsible for the electrical stimulation needed to maintain a sinus rhythm while the endocardium provides protection to the valves and heart chambers.

Circulation of the blood inside the heart is discussed in the following, First, deoxygenated blood coming from the superior (blood coming from the upper extremities and head) and inferior (blood coming from lower extremities) vena cava enters the right atrium, the blood then passes through the first heart valve called the tricuspid valve, the tricuspid valve is made up of cartilaginous tissue that allows the one way passage of blood from the atrium to the ventricle. The tricuspid valve is opened when the right atrium contracts forcing the blood to pass through it and being pulled by chordae tendineae that are anchored in the inner walls of the right ventricle. The chordae tendineae (tendinous cords) are tendon-resembling fibrous cords of connective tissue that connect the papillary muscles to the tricuspid valve (right side) and the mitral (left side) valve in the heart. When the right ventricle is then filled, it then contracts to close the tricuspid valve and pushes the deoxygenated blood to the pulmonary artery passing to the pulmonary valve to enter the lungs. While in the lungs, the blood undergoes gas exchange of carbon dioxide (product of metabolism) and oxygen from the regular inspiration of air from the environment. The oxygenated blood is then circulated to the pulmonary vein and fills the left atrium of the heart. From the left atrium, the blood is then pumped into the left ventricle passing through the mitral valve. The left ventricle then pushes the blood through the aortic valve to undergo systemic circulation. To describe the gross anatomy of the heart, we can notice that there is clear separation from the left and right sides of the heart. The septal wall ensures that the oxygenated and deoxygenated blood does not mix. Also, the left ventricular wall is much thicker than the right ventricular wall because the left ventricle needs to provide the enough pressure to maintain proper circulation all throughout the body, whereas the right side of the heart only needs to provide blood circulation toward the lungs for oxygenation and back to the heart. The normal blood pressure on a healthy adult is around 120/80 mmHg. This number indicates the systolic (heart contraction) and diastolic (heart relaxation) pressure of blood (Figure 1.4, left panel).

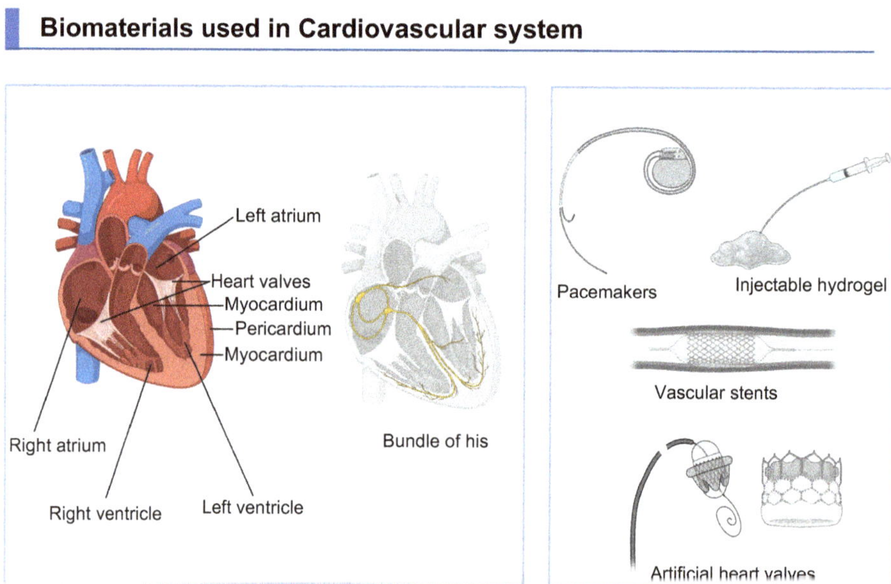

Figure 1.4: (Left panel) The gross anatomy of the heart that shows the four heart chambers, valves, muscle layers, and bundle of His which is a part of the electrical system of the heart. (Right panel) The biomedical devices and materials that are applied in the cardiovascular system are pacemakers (metal probe) which restore the normal heart rhythm, injectable cell and non-cell-laden hydrogel to recuperate the damage heart muscles, and artificial heart valves (metal, polymeric, and bioderived).

The blood is now then circulated throughout the body by the blood vessels (arteries, veins, and capillaries); the difference between the two is that arteries mostly carry oxygenated blood (except for the pulmonary artery) and the veins transport deoxygenated blood (except for the pulmonary vein) while the capillaries are the smallest unit where diffusion and exchange of nutrients and waste from cellular metabolic processes happen. Blood vessels have the capability to increase and decrease in diameter depending on the temperature, pressure, or biomolecular response that the body experiences and this is called vasoconstriction and vasodilation. When a person experiences low blood pressure caused by a trauma, the body will maintain homeostatic pressure by constricting the blood vessels from its periphery (limbs) and prioritizing blood flow by dilating the vessels supplying the vital organs to ensure that the person receives adequate blood supply in times of stresses. This is achieved because the blood vessels are made up of three layers called tunica intima (inner), tunica media (middle), and tunica adventitia (outer). However, structurally, there are differences between an artery and a vein. An artery has thicker tunica media, while a vein contains valves to ensure that there is no backflow of deoxygenated blood that could cause fluid retention on the limbs. The capillaries are also structurally different as it is only made up of single layer of endothelial cells to facilitate ease of diffusive transport of nutrients, gas, and metabolic waste from the tissue.

Several pathological conditions could affect the circulatory system, and one of the most common is myocardial infarction [77] or when the heart muscle is impaired due to severe lack of oxygenation (ischemia) [78, 79] caused by reduction of blood flow to the heart muscle. The heart is supplied by blood from the coronary arteries, and pathological conditions may arise when these arteries are blocked or constricted, reducing blood flow to the myocardium. This is called coronary artery disease [80] usually predicated by build-up of plaques (atherosclerosis) [81, 82] and loss of elasticity (arteriosclerosis) [81, 83] of the arterial wall. Plaque build-ups can be caused by lipid accumulation or deposition inside the vessel lining and would trigger a cascade of inflammation and immune responses. When the vessel is clogged by these plaques, the blood is unable to reach the myocardium creating a condition called hypoxia. This limits the oxygenation and nutrient uptake of the heart leading to what we call an infarct. The infarcted heart is then compromised and unable to pump blood to the systemic circulation, and severe cases would eventually lead to death. In non-severe cases, the heart is then impaired permanently and unable to supply adequate blood to the systemic circulation and would lead to a condition called chronic heart failure [84, 85], where the heart function is severely reduced from normal values. The heart should have a normal ejection fraction of above 50%, meaning that specific blood volume must be ejected from the left ventricle to supply the systemic circulation. Chronic heart failure is diagnosed when the ejection fraction of the heart is less than 40%. This is a serious condition due to its permanence and deteriorating characteristics; the heart muscle has a very low turn rate and does not heal effectively after a severe injury caused by myocardial infarction. This leads to a need in creating new treatments that could help the myocardium remodel and/or assist the heart in providing the adequate ejection fraction needed for the individual to maintain a normal lifestyle. Other structural abnormalities can occur in the heart where the valves are unable to close properly due to prolapse, due to the degeneration of the chordae tendineae (heart strings) [86, 87]. This condition requires total replacement of the compromised valve with an artificial one [88, 89]. Artificial heart valves are a one-way valve implanted into a person's heart can be separated into three broad classes: mechanical heart valves, bioprosthetic tissue valves, and engineered tissue valves. Biomaterial science provides new treatment strategies, from growth factor-laden injectable hydrogels, to cell therapy using stem cells, to pacemakers, and to eventually creating a whole new heart from scratch using decellularized scaffold or 3D bioprinting [90]. All these new strategies will be further discussed at the later chapters of this book.

There are also other conditions that could affect the circulation of blood in the body. Structural abnormalities in the vessels called aneurysms [91, 92] can also be detrimental to an individual if left untreated. Aneurysms are caused by weakened vessel walls, leading the arterial wall to bulge or get distended. The most common causes of aneurysms are atherosclerosis [81] and high blood pressure. Deep wounds and infections can also lead to an aneurysm [93]. It is very common to see aneurysms on the cerebral arterial vessels due to chronic high blood pressure, and when an aneurysm rupture due to a sudden surge on arterial wall pressure, it could lead to internal bleeding called hemorrhagic stroke and, in severe cases, sudden death. Other pathologies such as arterial clogging due to plaques are also a common cause of circulatory impairment; as discussed earlier, these plaques accumulate inside the vessel lining reducing the blood flow to the tissues. Clinical interventions using biomaterials are being used to address these arterial diseases, and one of the leading strategies when it comes to athero- and arteriosclerosis [94, 95] is using vascular stents. Stents are made up of metallic meshes that are deployed inside the coronary artery using a catheter to create patency on the occluded area, in order to restore blood flow to the infarcted tissues. Stents are also used in aneurysm treatment to ensure a laminar blood flow on the affected vessel. Stents can also be deployed on other arterial vessel such as the femoral artery to restore blood flow on the limbs. This intervention is called angioplasty [96, 97] or surgical restoration of the vessel.

Historically, angioplasty uses an inflatable balloon attached on the proximal end of a catheter to dilate the occluded vessel and restore blood flow (Figure 1.4, right panel). However, re-occlusion happens over time (restenosis) [98, 99] creating a need to create a material or device that could hold the patency of the vessel for a significant amount of time. Modern angioplasty uses these metallic stents to maintain patency inside the vessel. These stents are tubular meshes made up of biomedical-grade alloys such as stainless steel, chromium–cobalt, and nickel–titanium (nitinol). Self-expandable stents are made up of nitinol, while balloon-expandable stents are usually made up of stainless steel and chromium–cobalt alloys. However, stents are made up of metallic materials that attract platelets and could induce proliferation of endothelial cells on the implantation site [100, 101; hence, newer generation of stents incorporate antiproliferative and antiplatelet drugs to reduce the occurrence of restenosis over time. Polymeric stents are also being developed to ensure that there will be no need to remove the stent when the life span of the device is done. Polymeric stents are designed to be mechanically robust while having the capability to degrade inside the vessel [102, 103]. Biodegradable alloys are also being developed as a potential stent material [104, 105]; an example of which is magnesium alloy [106, 107]. The advantages of biodegradable stents are obvious; however, creating a balance between degradation and mechanical stability remains a challenge in the field. Combination of metallic and polymeric materials is also quite promising as the metallic stents are covered with a polymeric membrane [108–110] that acts as a physical barrier from endothelial cell proliferation toward the lumen of the stent, and these polymeric membranes can also be a platform for delivery of antiproliferative and antiplatelet drugs [111–115].

1.5 Biomaterials for the nervous system

The nervous system is a complex network that coordinates and regulates functions in the body [116, 117]. It has a level of organization that coordinates the body's multiple and complex functions. It is divided into two systems called the central nervous system (CNS) and the peripheral nervous system (PNS) (Figure 1.5, left panel). The CNS is comprised of the brain and the spinal cord, whereas the PNS is composed of nerves and ganglia outside the brain and the spinal cord. PNS can be further divided into sensory and motor division. The sensory (afferent) division picks up various stimuli from the environment, whereas the motor (efferent) division sends direction from the brain toward the muscles and glands. The motor division can further be divided into somatic (voluntary) nervous system and autonomic (involuntary nervous system). The somatic nervous system controls all voluntary movements' musculoskeletal movements, while the autonomic nervous system controls all involuntary processes such as cardiac and respiratory rhythm, gland stimulation, and peristalsis (gastrointestinal (GI) movement). The autonomic nervous system can be further classified into sympathetic and parasympathetic nervous systems. The sympathetic nervous system is responsible for "fight or flight" response of the body, while the parasympathetic nervous system counteracts this response by inhibiting the body from overworking and restores the body's relaxation state.

The human nervous system may be the most complex and intricate biological system known to man with all its amazing capabilities; however, the function of the nervous system can be simplified into these three processes: sensory input, integration, and output. We experience various extrinsic and intrinsic stimuli, and the nervous system is responsible in coordinating and processing those stimuli so homeostasis can be achieved. Sensory nerves accept the sensory input and send those signal (chemical or electrical) to the PNS or CNS. The signal is relayed to the spinal cord and communicated to the brain where signal integration happens, and the brain decides what kind of output must be done in response to the stimuli. For example, if we touch a hot surface, the nerves on our skin experiences that heat and sends an electrical potential signal to the spinal cord and by reflex the signal travels back to the motor

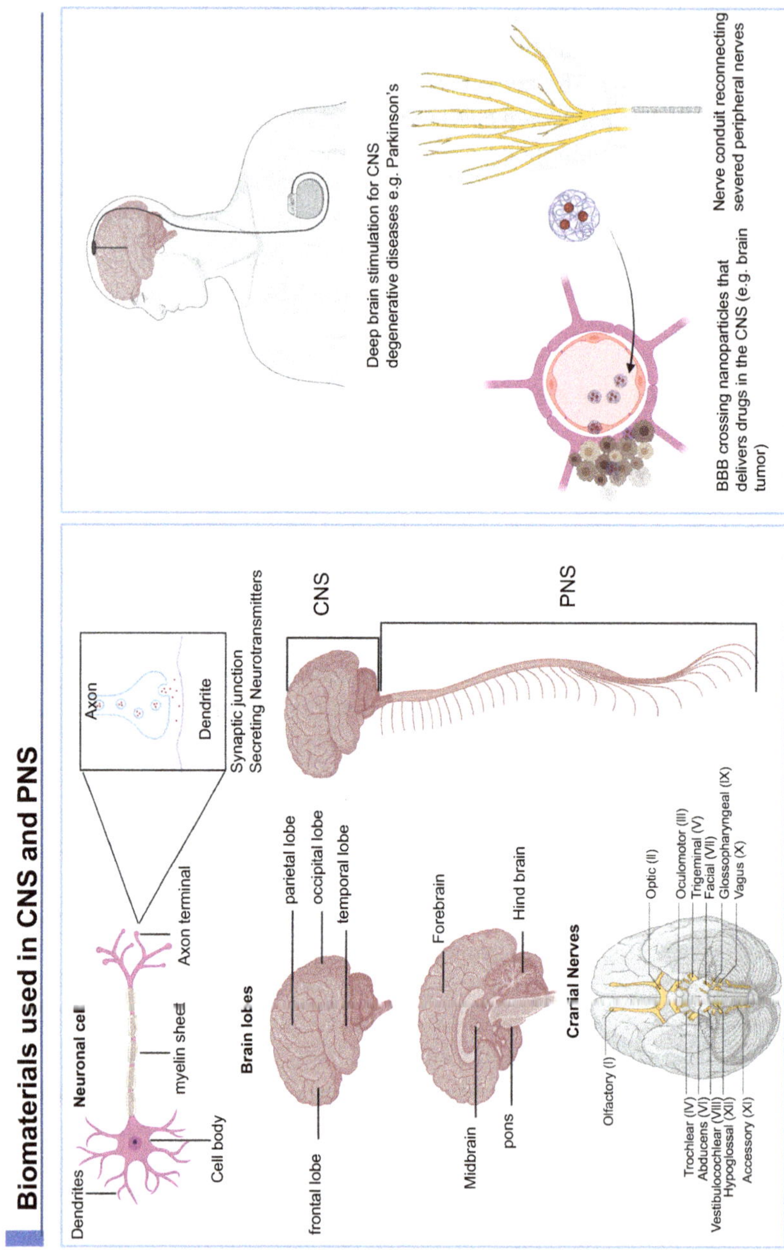

Figure 1.5: (Left panel) The basic parts of a neuronal cell and how it communicates with neighboring neurons using neurotransmitters. The gross anatomy of the central and peripheral nervous systems (CNS and PNS), the four lobes of the brain (cerebrum) and different cranial nerves. The examples of biomaterials that are applied in the CNS and PNS. Deep brain stimulation probe (metal) for Parkinson's disease, nanoparticles that can traverse the blood brain barrier, and nerve conduits (polymers with growth factors) for reconnecting damaged peripheral motor nerves.

neurons and orders the muscle on our arms to flinch to remove the hand from the heat source avoiding damage. This is called the reflex arch mechanism, when the signal is only processed in the spinal cord. Other example of sensory stimuli is when light is sensed through our eyes via photoreceptors and is processed in our brain, most specifically in the occipital lobe. The light that we see is converted to a recognizable image in the brain giving us a sense of vision. That vision can then trigger other processes in the brain such as when we feel hunger, lust, and extreme emotional stress. All these complex processes are sent to the brain by what we call the cranial nerves. All in all, the human body has 12 cranial nerves and they are responsible in controlling different processes in the body.

Table 1.1: Cranial nerve description.

Cranial nerve	Function	Sensory/ motor/ autonomic	Ref.
I. Olfactory	Smell	Yes/no/no	[118]
II. Optic	Vision	Yes/no/no	[119]
III. Oculomotor	Eye movement	No/yes/yes	[120]
IV. Trochlear	Eye inner rotation	No/yes/no	[121]
V. Trigeminal	Eye/forehead/maxilla/mandible movement	Yes/yes/no	[122]
VI. Abducens	Eye movement lateral	No/yes/no	[123]
VII. Facial	Taste and ear sensation/facial expression/nose and mouth glands	Yes/yes/yes	[124]
VIII. Vestibulocochlea	Balance and hearing	Yes/no/no	[125]r
IX. Glossopharyngeal	Taste/tongue and ear sensation/swallowing/ parotid gland secretion	Yes/yes/yes	[126]
X. Vagus	Stomach and ear sensation/taste/pharynx and larynx movement/respiratory and digestive tract glands secretion/cardiac muscle stimulation	Yes/yes/yes	[127]
XI. Accessory	Neck and shoulder movement	No/yes/no	[128]
XII. Hypoglossal	Tongue movement	No/yes/no	[129]

The cerebrum is the uppermost part of the brain anatomy and can be divided into left and right hemispheres and the frontal, parietal, temporal, and occipital lobes. The mid-brain is where the thalamus, hypothalamus, pons, and the pituitary gland are located. It also connects the left and right sides of the brain together. The spinal cord is connected to the brain in a section called the brainstem and runs through the spinal canal down to the vertebral column. Nerves are then spread outward from the spinal column to communicate with the rest of the body.

The most basic unit of the nervous system is called neurons. Neurons are cells responsible for signal communication via electrochemical signaling to other nerves, glands, or muscle. A neuron is comprised of dendrites, cell body, and axon with myelin sheet cover [130]. The dendrites and axons are structures that connect neurons to other neurons, and in those connection, there is a micro-distanced gap called a synapse. Neurons communicate via electrochemical signaling via these synapses. The biochemicals that are being sent to and from synapses of neurons are called neurotransmitters [131, 132]. Different neurotransmitters can affect different processes in the body and an imbalance of it can cause physiological and psychological disorders. One example of which is Parkinson's disease, when the part of the brain has a reduction in a neurotransmitter called dopamine [133, 134]. Dopaminergic neurons are deteriorated causing the body to have physiological disorder that affects the person's voluntary movement. Most neurological disorders are progressive, meaning that over time, an individual's condition will only be worsened. This is because once a neuron is degenerated or injured via ischemia, trauma, or other endogenous and exogenous factors, it does not have the capability to regenerate. However, there are cells responsible in maintaining the neuronal cells and they are called glial cells [135–137]. Glial cells maintain homeostasis within the nervous system by creating myelin to cover the axons of the neurons to facilitate faster signaling time. There are also different types of glial cells in the CNS: ependymal cells, astrocytes, oligodendrocytes, and microglia. While in the PNS, Schwann cells and satellite cells are responsible. Ependymal cells line the spinal cord and ventricles of the brain and responsible in creating cerebrospinal fluid (CSF). Astrocytes are star-shaped cells that help maintain a neuron's microenvironment. They regulate the levels of neurotransmitter around synapses, controlling the levels of action potential ions like potassium and calcium and providing metabolic support. Oligodendrocytes provide support to axons of neurons in the CNS, especially those that are elongated within the brain by producing the myelin, which is wrapped around axons to insulate and provide increased transmission of electrochemical signals. Microglia are the brain's immune cells, protecting it against injury and pathogens. Schwann cells myelinate neurons in the PNS and satellite cells surround neurons in the sensory, sympathetic and parasympathetic ganglia and help regulate the chemical environment. The brain must be selective to what kind of substances it allows to cross the neuronal tissue from the systemic circulation. The brain is protected by the blood–brain barrier (BBB), a highly selective semipermeable membrane made up of tightly packed endothelial cells. This barrier is highly selective and only allows to select micro- and macromolecules that are needed in maintaining brain homeostasis such as glucose, organic anions, water, and some amino acids, and the entry of these substances are regulated via active (receptor or channel mediated) or passive (diffusion) transport. It also protects the brain from pathogens such as bacterial, viral, and spores. It allows diffusion of waste from the brain to the blood and the CSF. Knowing this, it becomes very difficult to target the brain with therapeutics needed to correct some diseases. Hence, biomaterial science is employed to allow the crossing of large molecular therapeutics in the brain by using nanoparticles to exploit the active transport mechanism in the BBB [138]. A detailed discussion of this strategy will be discussed in the later chapter of this book.

When a nerve is injured, it has a very low capability to repair itself; hence, there is a need to advance the field of nerve tissue engineering using biomaterials as a strategy to assist the nerves to regain its function. For example, if a person had a spinal trauma, it would lead to paralysis because the signals from the brain to the muscle is severed. New advancement in the field of nerve tissue engineering helps in spinal injuries by creating biomaterials for guiding axonal regrowth to facilitate nerve regeneration [139]. Synthetic and natural-based polymers can be used as an artificial scaffold for nerves to attach and guide their growth, thus reconnecting severed neuronal connection due to injury [140–143]. Key design elements for polymeric neuronal scaffolds are biodegradability, electrical conductivity, high porosity, and with an adequate mechanical strength. It should also facilitate cell adhesion and directional growth for the neuronal and glial cells, so that it can ensure that the severed nerve is properly connected and would not form a teratoma (tumor) (Figure 1.5, right panel).

Degenerative diseases in the CNS can also be alleviated with the use of biomaterials, most specifically new therapy options that use stem cells to replace the degenerated cells within the brain [144–146]. Stem cell therapy is an emerging option to treat Parkinson's, Alzheimer's, and Huntington's diseases. Stem cells are harvested and differentiated to neuronal progenitor cells and implanted directly to the interior of the brain to replace the degenerated neurons of the affected individual. However, these cells must survive during the implantation and should evade immune cells to avoid rejection, that is why in a landmark trial, dopaminergic progenitor neurons are encapsulated in gelatin microspheres in hopes to increase their viability and enhance their engraftment post implantation. Although this strategy has been effective in delivering the cells to the target site in the brain, engraftment and survivability still have been lacking to create significant therapeutic effects to the patients [147]. Hence, there is still a pressing need to improve on these biomaterials to enhance cell viability and engraftment to the local implantation site (Figure 1.5, right panel).

There is also an emerging field of brain–machine interfaces that will allow patients who had been suffered from degenerative disease such as amyotrophic lateral sclerosis (ALS) also known as Lou Gehrig's disease [148]. Due to the nature of this condition, people who are suffering with ALS will become paralyzed and unable to communicate as they lose their control over their muscles. That is why new brain–machine interfaces made up of soft electrodes can be implanted directly to the brain of the patient to allow them to communicate or send signal directly to an external communicator device. However, this field is relatively at its infancy and only shows the viability and proof of concept. Elon Musk's new company Neuralink exhibited that this concept can be done in vivo and shows relatively successful results as they are able to implant a pig with their Neuralink device and safely remove it without causing detrimental effects to the test subject. Although their results are not substantiated with data, it can be said that there is a growing interest in the field (Figure 1.5, right panel).

One Food and Drug Association approved the use of this brain–machine interface in the field of deep brain stimulation [149–151], where a device is implanted in specific areas deep in the brain of the patient and sends electrical or light impulses to stimulate the specific areas of the brain. This intervention has been approved to be used in Parkinson's disease, obsessive compulsive disorder, and epilepsy. However, there are certain risks for this kind of devices, as the implantation procedure is highly invasive and could cause hemorrhagic trauma (bleeding) to the brain of the patient [152]. Thus, this requires newer breed of electrodes that have key design characteristics. Besides being able to conduct electrical impulses, it should also be biocompatible or bioinert, and thin in size to avoid damage during implantation.

1.6 Biomaterials for the digestive system

The digestive system is made up of GI tract and other accessory organs responsible for ingestion of food, secretion of digestive enzymes, digestion and absorption of nutrients and liquid, and elimination of food wastes (Figure 1.6, left panel) [153, 154]. This system is essential in providing the body its energy, nutritional, and hydration requirements to keep on living [155]. The digestive system can be imagined as a long tube with multiple functionalities as the tube progresses. It all starts with the mouth, where the initial stages of digestion occur. Food is first mechanically and enzymatically broken down in the mouth with a movement called maceration, and the enzyme amylase that can be found in the saliva initially breaks down carbohydrates present in the food. The parotid glands secrete the saliva, and the jaw, teeth, and tongue are responsible for macerating the food to smaller pieces. Once the food is properly macerated, it is now then swallowed and enters the esophageal tract. The esophagus is a long muscular tube made up of three smooth muscle layers. The movement of the food material along the esophagus is highly controlled by these muscle layers and this is what we call peristalsis, which is to avoid the food from being regurgitated and ensuring that the food only travels toward the stomach. Before the food enters the stomach, the esophagus ends with the lower esophageal sphincter. This sphincter acts like a gate that divides the esophageal tract and the stomach and it ensures that stomach contents do not go back toward the esophagus. This is very important as stomach contents are very acidic due to the hydrochloric acid (HCl) that it produces. If the lower esophageal sphincter is compromised, it will allow the reflux of the gastric contents and damage the cells that line the esophagus and this may lead to

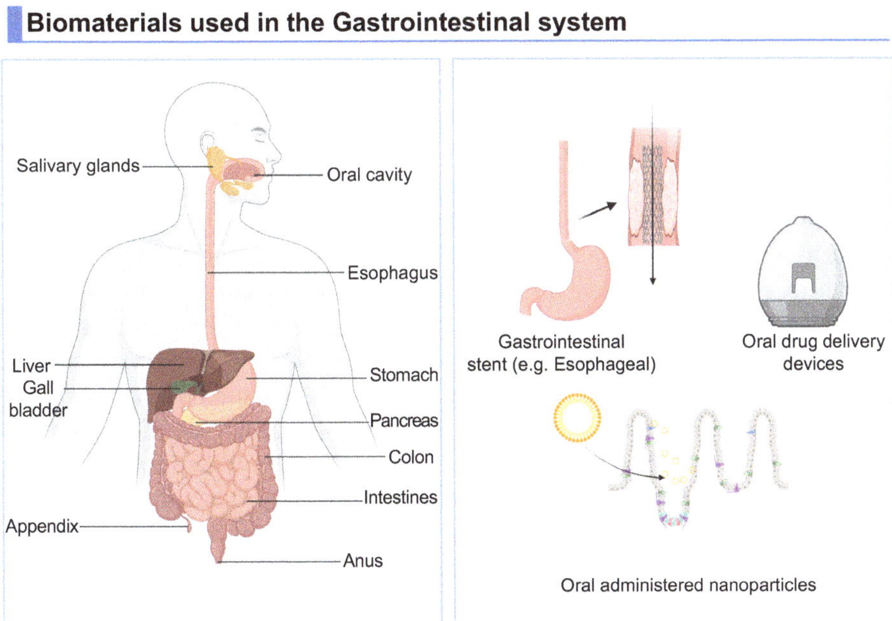

Figure 1.6: (Left panel) The gross anatomy of the gastrointestinal (GI) system including accessory organs (e.g., liver, pancreas, and the gall bladder). The GI is a long canal where digestion of food occurs. This digestion of food matter gives the body the energy and raw materials to maintain homeostasis. (Right panel) The examples of biomedical devices (esophageal stents – metallic) (oral drug delivery devices – metal and polymers) and nanoparticle drug delivery systems (polymeric).

esophageal cancer, if not treated immediately. Once the food has entered the stomach, further enzymatic and mechanical digestion happens with the addition of HCl to speed up the breakdown of the food to its basic components. HCl is produced by specialized cells called the parietal cells that are found at the mucosal layer of the stomach [156]. The stomach's anatomical surface has multiple folds called the rugae, and this allows the stomach to expand once it is filled with the food material. Stomach expansion will also trigger its involuntary movement to churn the food material into smaller pieces and facilitate gastric emptying. The beginning of absorption starts at the stomach where some of the nutrients that are digested traverse the stomach lining toward the systemic circulation via projections in the stomach called the villi and microvilli, which are made up of epithelial cells that are tightly packed together so that only small-sized solubilized molecules can pass through. Once the food has been solubilized, it can now then pass the cells that line this epithelial layer and enter the systemic circulation via capillary diffusion. Carbohydrates are broken down into simple sugars, while proteins are turned into smaller amino acids during this phase of digestion. The simple sugars (glucose) are then used by the body for energy while the amino acids are utilized for other metabolic processes such as in making cell proteins and other biomolecules needed to maintain the body. Once the food has been churned and digested in the stomach, it then passes through the pyloric sphincter via gastric emptying and enters the first segment of the intestines called the duodenum. At this point, the food material has been turned into what is called chyme for ease of absorption. The intestine is also lined with epithelial cells that will only allow absorption of small molecules, such as simple sugars, amino acids, and fatty acids. The increase of blood glucose from the initial absorption in the stomach then triggers the pancreas to start producing insulin. Insulin is needed for glucose to enter the cell membrane; diabetes is a condition where an individual cannot produce insulin (type I) or can produce insulin, which we call insulin tolerance (type II) [157]. The passage of the chyme to the duodenum also triggers the liver to produce bile. Bile is needed to digest fat that is included in the chyme. It breaks down the fat molecules to smaller fatty acids so that it can be absorbed in the intestines and be utilized by the body for energy and other vital body functions. Fatty acids are needed to make certain hormones such as testosterone, adrenaline, and estrogen to name a few. The liver also plays an important role in filtering toxins and metabolize drugs that are present in the blood coming from the digestive tract before entering the systemic circulation [158, 159]. It can also turn extra glucose that is not utilized and store it in the liver as glycogen, and with the help of insulin convert it to fatty acid which are then stored in the adipose tissue as fat. It can also convert fat storage into glucose via the process called gluconeogenesis.

After being added with extra digestive juices from the liver, the chyme will now then pass through the small intestine called the jejunum and ileum where further absorption takes place. The chyme will still be churned into small constituents via enzyme and mechanical processes (peristalsis) as it passes through the small intestines [160]. The undigested material will now then pass through the end part of the small intestine and enters the first part of the large intestine (colon) called the cecum, where the appendix is also located. There is very little nutrient absorption happening in the colon as it is mainly responsible in reabsorbing water from the undigested material that passed through the intestines. At this point, the undigested material will traverse the colon via ascending, transverse, and descending colon gradually losing water content and turning into fecal matter. It stays there until the parasympathetic nervous system kicks in and triggers bowel elimination and expels the fecal matter through the rectum. The rectum is the end part of the alimentary tract and is terminated by the rectal sphincter.

As mentioned in the earlier paragraph, there are diseases linked to the malfunction of the GI tract where biomaterials can be utilized. First, nonvascular stents are deployed when there is an occlusion on the GI tract, whether it may be from the esophagus, duodenum, or bile duct [161–163]. These stents are made up of metal alloys such as stainless steel or self-expandable nitinol, and like what vascular stents do, they create patency for the food material, or bile to pass through (Figure 1.6, right panel). These occlusions are mainly made up of malignant tumors (cancer). As the cancer progresses, the tumor size increases and creates a blockade on the alimentary tract. A patient who suffers from this will manifest dysphagia (difficulty in swallowing) in esophageal cancer, and extreme body weight loss in cases of duodenal or gall bladder cancer. This is due to the reduction of nutritional absorption of body because the food material can be stuck or blockaded on some parts of the alimentary tract. The deployment of these alimentary stents will create a patency on the GI tract allowing the passage and absorption of food material to the rest of the system. However, this creates a positive feedback loop as the cancer tumor can rapidly progress due to restoration of normal nutritional uptake of the patient [164]. This will render the alimentary stents to be occluded again by the tumor; however, further intervention will make it difficult as the metallic mesh of the stent is incorporated by the tumor tissue. Like vascular stents, polymeric mesh covers can alleviate this re-occlusion scenario by creating a physical barrier. These polymeric covers are made up of elastomers and can be loaded with anticancer drugs such as paclitaxel [165, 166] to inhibit the growth of the tumor and prolong the life span of the stent. Current challenges for development of these stents are providing a balance between mechanical stability and giving a sustained or on-demand release profile of therapeutics over a significant period. Also, proper anchorage must be ensured to avoid migration of the stent to proximal parts of the alimentary tract due to peristaltic movement [167–169].

A current field where biomaterials are being utilized is the oral delivery of peptides using various nanomaterials and stimuli-responsive materials. Oral delivery of large molecules like therapeutic peptides such as insulin has been one of the goals of the pharmaceutical industry [170–172]. Oral delivery route for medicine is one of the easiest ways of delivering medicine with high tolerance, adherence, and effectiveness. Insulin is an essential hormone that is secreted by the pancreas to assist in the metabolism of sugars by making the cells in the body permeable to glucose [173–175]. People who suffer from type 1 diabetes cannot produce insulin because their pancreas has been compromised by their own immune system, while an individual who has type 2 diabetes still has the capability to produce insulin; however, their body cannot develop a resistance to it because of prolonged high intake of high caloric food in their diet. Insulin is a peptide with 50 amino acids long, and has been discovered and isolated by Dr. Banting back in the early 1900s. He first tried to deliver insulin to diabetic patients via oral route but its effectivity is not seen due to the fact that peptides cannot survive the harsh conditions of the GI tract, especially the acidic environment in the stomach. Since then, insulin has been delivered via the subcutaneous route to increase its therapeutic effect and bioavailability. However, this creates a burden to the patient who needs to inject themselves on a daily or weekly basis. Also, there are side effects of frequent subcutaneous injections such as injection site hypertrophy [176]. That is why, there are new emerging techniques in biomaterial science that aims to deliver sensitive payloads such as insulin to the body via the oral route. This is done by either encapsulating or conjugating insulin with nanoparticles [177–179] to protect it from degradation and exploit active mechanisms that can be found in the GI system. This is done via two ways: there are specialized cells in the GI system called the Peyer's patches. These are specialized immune cells that protect the system from bacteria and other pathogens. They are specifically targeted so that large molecules that are encapsulated or conjugated by nanoparticles can be absorbed via phagocytosis [180, 181] and enter the systemic circulation. Other strategies use increased mucoretention by adding mucoadhesive materials [182]. This strategy increases the likelihood of being absorbed by the epithelial cells by making the nanoparticles stick to the surface of the epithelial mucosa and be absorbed via phagocytosis or receptor-mediated cytosis. Surface charge-modified particles (cationic) can also be used to ensure that the nanoparticles are easily absorbed by the epithelium of the digestive system (Figure 1.6, right panel). A much more mechanical approach is being used to deliver peptides using oral devices. These devices facilitate tiny peptide injections in the stomach or intestine to facilitate high bioavailability. One example of which is the RANI pill that is made to inject the intestine with small-chain peptides using expanding stimuli-responsive polymers. Another example is the SOMA device that is shaped like a leopard tortoise shell, to ensure self-righting orientation, and using a spring-loaded mechanism, it injects insulin to the sublayer of the gastric epithelium (Figure 1.6, right panel). Detailed discussion of these topics will be further elaborated at the drug delivery section of this book.

As mentioned earlier, the pancreas and the liver are the two main accessory organs of the digestive system. However, if one of them fails to function, the life of an individual is highly affected and can lead to detrimental effects. As the early example suggests, if a person has type 1 diabetes, then the pancreas fails to produce insulin and the person is unable to utilize the glucose for energy, leading to various complications. Even though insulin can be administered exogenously to the body, its strict regimen like glucose monitoring and accurate dosing is very limiting to a lot of people living with type 1 diabetes. More importantly, the risk of hyperglycemia (increased) and hypoglycemia (decreased) (blood glucose) is always a looming threat for a person undergoing insulin treatment. Hence, strategies such as insulin pumps with glucose monitor loop [183, 184] are being developed to simultaneously monitor the blood glucose level of an individual and accurately deliver the required insulin depending on the person's current need. As promising as this technology is, it requires high maintenance, such as replacement of the needle probes that measure and deliver insulin every 2 weeks and replenishing the insulin payload of the device once all of it is consumed. That is why tissue engineering using insulin-producing cells (pancreatic beta-cells) together with biomaterials are being developed so that we can replace the dysfunctional pancreas of a type 1 diabetic patient with an engineered one [184, 185]. This is all in the hope that the engineered pancreas will produce insulin and sense blood glucose levels in situ, removing the need for an external device that requires weekly maintenance. An example of this is what the company Procyon is currently developing. They created a small implantable device that encapsulates beta-cells to produce insulin. The key design element of this implantable device is that it allows the release of insulin and waste materials generated by the beta-cells, while shielding it from the body's immune system. However, the main challenge is to ensure that the encapsulated beta-cells receive their required oxygenation and energy requirements while keeping them away from the radar of immune cells. The device should also not trigger creation of fibrotic tissue around it as it will render to isolate the device and make ineffective.

Liver failure is a detrimental condition and would require immediate medical attention. If left untreated, the person may die within days. This is due to the inability of the liver to detoxify the blood causing multiple organ failure. There are multiple causes of liver failure; it could be viral, or diet induced [186]. Hepatitis virus strains can damage the liver by infecting the hepatocytes (liver cells) and cause inflammation and scarring. This condition is called liver cirrhosis [187, 188]. Too much fat and sugar in a diet can also cause the liver to fail by making it store too much fat around the hepatic tissue. This is called hepatic steatosis [189]. Alcoholism can also cause the liver to fail due to being overworked, called alcoholic hepatitis. All these conditions, if left untreated, may require liver transplant in a person. However, due to organ shortage, a person might not live long enough to receive a liver transplant. Hence, biomaterials can be used as a solution for this problem. Hepatic cells can be grown outside the body using the person's own stem cell or hepatic cell. A scaffold may be 3D printed, or a decellularized scaffold from an exogenous donor may be used to grow an engineered liver. However, it will still take a significant amount of research and development to create a fully functional liver in vitro as the technology of 3D bioprinting, bioreactors, and stem cell differentiation techniques must be streamlined altogether.

1.7 Biomaterials for the reproductive system

The reproductive system is responsible primarily for sexual reproduction, and there are vast differences between the male and female genitalia when it comes to anatomical features [190]. The male genitalia are responsible for creating the sperm (testis), which has the role to fertilize the egg (ovaries) inside the female genitalia. The act of reproduction is called sexual intercourse or coitus, and a successful one can bring about another person after 9 months of gestation. A male produces millions of sperms per ejaculate while a female usually releases a solitary egg per month. These reproductive cells are called gametes and are the only haploid cells in an individual's body. Haploid means they only have half of the 46 chromosomes that a human has. Once a sperm meets an egg is called fertilization, and the 23 chromosomes for each gamete match up and bring about a zygote that contains 46 chromosomes. These chromosomes are found in the nucleus of the newly formed zygote and have all the genetic information of the baby and what kind of traits it will inherit from the parents.

The male reproductive organ is made up of these parts: the scrotum which houses the testis, and the penis comprising the glans penis which is covered by a layer of skin called the foreskin, and the shaft or the body of the penis and collectively they are called the external genitalia (Figure 1.7, left panel). The scrotum is outside the body because the sperm and sperm production are highly sensitive to body temperature. The scrotum has the capability to contract and relax depending on the temperature of the surroundings. It contracts when the environment is cold and relaxes when it is too hot. This is a way to protect the testis from extreme temperatures and maintain the health of the sperm. The testis is responsible for production of sperm via a process called spermatogenesis. The testis is connected to the rest of the genitalia via a series of ducts called the efferent ductules that connect to the epididymis and finally the vas deferens. The vas deferens is the one being ligated or cut-off during a procedure called vasectomy [191, 192]. It is a form of male contraception so that the sperm cannot travel from the testis and be ejaculated by the penis during intercourse. But during a normal ejaculation, the sperm is carried along the vas deferens, and additional liquid is added from a gland called the prostate. The urethra senses the sperm and fluid mixture, and it sends signals to your spinal cord, which in turn sends signals to the muscles at the base of your penis. This creates a strong contraction that forces the semen out of the penis as you climax. In order to have a successful coitus, the penis must be able to stand erect to facilitate penetration to the woman's vagina. An erection is achieved when the body sexually aroused from visual or olfactory cues (pheromones). This leads to the penis to be filled with blood as the heart diverts blood flow to the arteries of the penis and fills a specialized tissue called corpus cavernosa. Some men will experience erectile dysfunction that will render them unable to perform sexual acts. However, biomaterials can solve this by creating a device that is called penile prosthesis (Figure 1.7, right panel) [193], which allows artificial erection by pumping it with a fluid from a reservoir that is under the scrotal sac and inflating a tube that is implanted on the penile shaft. Other non-inflatable iterations of this penile prosthesis called semirigid implants [194–196] are also available in the market. This solution is more reliable compared to oral erectile dysfunction medication such as phosphodiesterase type 5 inhibitors [197, 198] and intracavernosal vasoactive agents [199, 200] because many men become refractory to pharmaceutical treatment or unfit to take those medicine due to underlying cardiovascular conditions.

The women genitalia are made up of these parts: the external genitalia are comprised of the vulva, the labia (majora and minora), the clitoris, and the urethra. The internal female genitalia are made up of the vaginal canal, the cervix, the uterus, the fallopian tube, and the ovaries (Figure 1.7, left panel). The female egg is developed and released from the ovaries each month and travels the fallopian tube

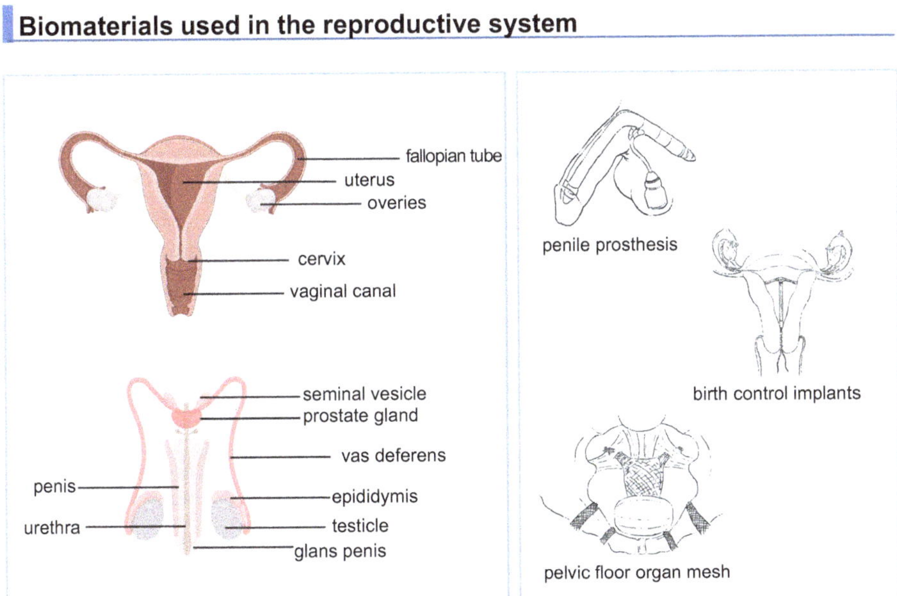

Figure 1.7: (Left panel) The gross anatomy of the male and female reproductive organs responsible for procreation and various hormonal functions. (Right panel) The biomaterials used in this area are penile prosthesis (polymers) to restore the ability to have an erection, birth control implants (intrauterine device – metallic and polymeric loaded with hormones), and pelvic floor meshes to provide support in cases of uterine prolapse.

waiting to be fertilized. This is called the menstrual cycle. The menstrual cycle can be divided into two phases: first is the follicular phase and second is the luteal phase. The follicular phase starts when the follicles inside the ovaries start to develop into an ovum. This is regulated by the follicle-stimulating hormone that comes from the gonadotropic cells of the anterior pituitary gland, and regulates the development, growth, pubertal maturation, and reproductive processes of the body. Other hormones are also involved in this process, and they are estrogen, and luteinizing hormone, and peak secretion of these hormone triggers what we call ovulation. Once ovulation occurs, the follicle that contained the egg transforms into something called a corpus luteum and begins to produce progesterone as well as estrogen afterward, the luteal phase begins. Progesterone levels peak about halfway through this phase and hormonal changes of the luteal phase are associated with common premenstrual symptoms that many people experience, such as mood changes, headaches, acne, bloating, and breast tenderness. If an egg is fertilized, progesterone from the corpus luteum supports the early pregnancy by preparing the endometrial lining of the uterus for implantation. If the ovum is not fertilized, the corpus luteum will start to break down between 9 and 11 days after ovulation. This results in a drop in estrogen and progesterone levels, which causes menstruation. The luteal phase typically lasts about 14 days, but between 9 and 16 days is common. All in all, the menstrual cycle (follicular and luteal phases) lasts on average for 28–30 days.

Women who seek not to be pregnant are taking hormones that will disrupt these processes. Oral contraceptives are very effective in disrupting this process; however, it must be taken daily, and dosing should be strictly observed. That's why implantable hormonal contraceptives were developed to facilitate the release of hormonal contraceptives over a period without the need for daily dosing. These matchstick-sized implants are made up of polymers and allow the controlled release of progesterone, which thickens the cervical mucus and hinders sperm motility inside the female genitalia. Other nonhormonal options are also available by using an intrauterine device (IUD) implant (Figure 1.7, right panel). This device is coated with copper and works by not allowing the sperm to fertilize the egg by creating toxic environment for the sperm and the egg. It may also make it harder for a fertilized egg to implant in the uterus. An IUD coated with progestin works in a similar way, but also thickens the cervical mucus and thins the uterine lining.

The female genitalia can also suffer from a condition called uterine prolapse [201, 202]. This is when the muscle and ligaments that support the uterus becomes weak and are no longer able to support the uterus. In some cases, the uterus can protrude from the vaginal opening. This may be caused by several reasons. It could be a complication from pregnancy, chronic straining, old age, genetic predisposition, and history of pelvic surgery [203]. It is a debilitating condition and would require surgical intervention to reinforce the pelvic floor muscles that support the internal structures of the female genitalia. This is done by adding a polymeric mesh [204–206] to provide additional mechanical support to the pelvic floor muscles. However, there is a recent scandal that shocked the medical industry because of the material used in commercialized meshes that create severe rejection response [207, 208]. Some cases of rejection were so severe that women who underwent this surgery were paralyzed due to nerve damage and some lead to death because of sepsis (blood toxicity). Surgeons were unable to completely remove the mesh support in cases of rejection because the polymer mesh was incorporated into the local tissue (Figure 1.7, right panel). This is a stark reminder that even though the field of biomaterials can be of help, careful and rigorous experimentation must be done to ensure the safety of these biomaterials.

1.8 Biomaterials for the urinary system

The urinary or renal system is responsible for regulating the blood volume, pressure, and pH, and eliminating waste in a form of fluids, controlling the level of electrolytes and metabolites in the body [209, 210]. It is composed of the kidneys, ureters, the bladder, and the urethra (Figure 1.8, left panel). Its primary function is to produce and eliminate urine from the body, and by doing so it can perform the functions stated above. How it creates urine is a process by filtering 1.25 L/min of blood from systemic circulation in the kidneys. The most basic unit of the kidneys is called the nephron and its primary function is to regulate the concentration of water and solutes like sodium and other electrolytes by filtering the blood and reabsorbing what is needed and excreting the wastes through urination. The nephron has a structure called the Bowman's capsule and it filters the blood from the circulatory system into the tubules. Filtration occurs by virtue of osmotic and pressure gradients between the semipermeable membrane of the Bowman's capsule and the glomerulus (network of capillaries). Water, small molecules, and ions can easily pass through the filtration membrane. However, larger molecules such as proteins and blood cells are prevented from passing through the filtration membrane. The filtrate collected travels to the ureters and is collected inside the bladder, and once the bladder is full, it will send a signal to the parasympathetic nervous system to initiate the relaxation of the sphincters in the bladders to release the urine. The urine then travels to the urethra in a process called urination. The length of the urethra is different between male and female. The urethra is also used by the male reproductive system as a pathway for ejaculates, while the female urethra is solely

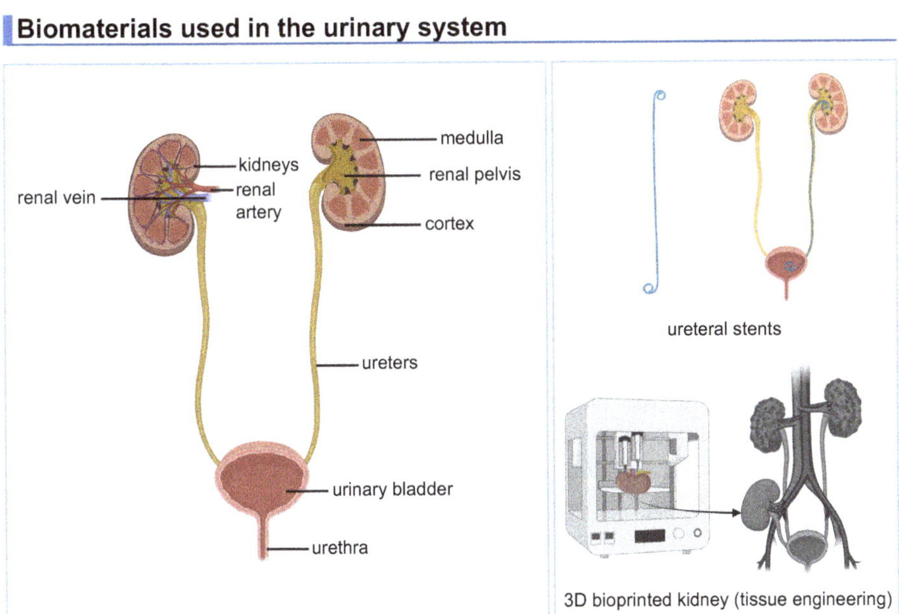

Figure 1.8: (Left panel) The gross anatomy of the urinary system (kidneys, ureters, urethra, and bladder) responsible for filtering the blood and elimination of waste through the excretion of urine. (Right panel) In cases of blockage of flow of urine from the kidneys to the bladder through the ureters, a polymeric stent is used. In more advanced applications, where kidneys must be replaced, a tissue-engineered kidney can be used (nonclinical and early research stage).

used for urination. The rate of urination and amount of output are also regulated by hormones; as an example, frequent urination can cause the blood volume to be reduced and consequently also reduce the blood pressure, and the antidiuretic hormone can reduce the glomerular filtration rate, thus maintaining normal blood volume and pressure. Another example is aldosterone; it helps the kidneys manage water, sodium, and potassium and too much of it can cause the body to hold on to water and sodium and increase blood volume.

In cases of kidney failure, this normal filtration process gets compromised, and the kidneys are unable to eliminate waste in the bloodstream. This can be due to various reasons, such as severe dehydration, toxin exposure, adverse certain medications, and underlying chronic disease such as diabetes. There are various kinds of renal failure and they can be classified as acute (short term) or chronic (long term). Nevertheless, the underlying disease pathology is the inability of the kidneys to filter the blood of toxins. When severe kidney failure happens, the patient must undergo dialysis [211, 212], where a machine filters the blood outside the body and return after filtering it with solutes that are toxic. This is done on a regular basis and the frequency depends on the severity of the kidney failure. This treatment although effective is not a long-term solution because a dialysis treatment must be done in a clinical setting. A group in University of California, San Francisco, working on "The Kidney Project" is trying to solve this problem by replacing the damaged kidney with an artificial implantable device that can filter the blood of toxins [213, 214]. The technology is based on a hemofilter that uses silicon nanotechnology to produce a highly efficient and compact membrane, which relies on the body's blood pressure to perform filtration without needing pumps or a power supply. There are still a lot of challenges to overcome such as making sure of meaningful ultrafiltration volumes at driving pressures like capillary perfusion pressure, while avoiding biofouling during the service life of the device. Furthermore, if this technology is to be generalized to implantable continuous therapies, protein losses must be minimal. The other component of this device is the cell bioreactor which utilizes recent advances in the field of tissue engineering to grow and maintain renal tubule cells. This bioreactor must be capable of high-volume salt and water reabsorption from the ultrafiltrate while maintaining a barrier to reabsorption of toxins while also being able to regulate physiological processes such as blood pressure and vitamin D production.

Other devices that rely on biomaterials that are used in the renal system are stents [215]. They facilitate the flow of urine from the ureters to the bladder in cases of obstruction due to kidney stones [216, 217]. They are made up of strong polymeric materials like polyurethane [218] or silicone [219] and are placed by a doctor using a cystoscope (thin, flexible tube with a camera) to visualize the opening of the ureters. Once they can see this opening, the physician will thread a ureteral stent through the cystoscope and into your ureter (Figure 1.8, right panel). After the stent is in place, the cystoscope will be removed. The stent will stay inside the ureters for about 3–6 months and must be replaced after the said time. This is to avoid biofilm formation inside the lumen of the stent [220, 221]. A biofilm is a pathogenic agent caused by the aggregation of microorganisms that adhere to a surface. These adherent bacterial colonies become embedded within a slimy extracellular matrix that is composed of extracellular polymeric substances. This substance (film) provides a safe environment for these microorganisms to thrive. If left inside the body for too long, they might cause severe infection and would lead to detrimental effects such as sepsis. Significant research efforts are being poured to attenuate the formation of biofilms on biomedical devices such as the ureteral stent [221]. Biofilm formation is also a problem in other devices used in the urinary system, namely, long-term use of urinary catheters that are used to drain the bladder of a patient of urine. Urinary catheters [222] like ureteral stents are made up of flexible polymers (silicone and polyurethane). Urinary tract infection might occur from long-term use of these catheters [223]. Strategies such as adding antibacterial agents or modifying the surface charge of the polymers are being done to attenuate biofilm formation. This will not only increase the life span of the stent but also save the patient from having detrimental consequences of biofilm formation.

1.9 Biomaterials for the sensory organs (eyes and ears)

For us to make sense of our surroundings, we need organs that would make us able to hear, see, taste, smell, and feel. The five basic sensory systems are visual (eyes), auditory (ears), olfactory (nose), gustatory (tongue), and tactile (skin) [224]. These sensory organs allow us to make sense of the world around us by receiving stimuli from the environment and sending it the sensory system located in the brain to be processed. Impairment to these organs such as in the eyes or ears would make a difficult life for an individual.

The eyes are one of the most complex organs in the body and are responsible for our sense of sight. The anatomical feature of the eye can be distinguished into three layers [225]. The outer layers of the eyes are the cornea and the sclera (Figure 1.9, left panel). The cornea is a structure that refracts the incoming light to the lens and the retinal layer. It also protects the eye from damages and other pathogenic factors. The sclera is a connective tissue that layers and protects the eye from internal and external forces and retains its shape. The middle layer of the eye is composed of the iris, the ciliary body, and the choroid. The iris controls the opening of the pupil like a camera aperture and regulates the amount of light reaching the retina. This also gives an individual their eye color; the ciliary body controls how the lens are shaped so that we can focus on an object and is the site of aqueous humor production. The choroid provides oxygenation and nutrients to the outer retinal layers. The inner layer of the eye is the retina, a layered structure of neurons and epithelium that captures and processes light through cells called photoreceptors. These photoreceptors are supported by a layer of cells called the retinal pigmented epithelium (RPE) that provides structural support and nourishment. The three transparent structures surrounded by the ocular layers are called aqueous, vitreous, and lens.

Impairment on these structures can hinder the ability of the person to see and would require intervention to correct the visual pathway of the eye (Figure 1.9, right panel). The most common application of biomaterials in the eyes are with corrective contact lenses. Contact lenses are made up of clear polymeric materials that can refract and focus the incoming light to enhance the visual acuity of the wearer. Contact lenses work correcting refractive errors due to near-sightedness (myopia) or farsightedness (hyperopia) [226]. In more advance scenarios, biomaterials can be used to create artificial structures in the eye such as the cornea. Corneal damage may be due to trauma and can cause blindness and requires a complete replacement to restore sight. Cornea can be harvested from eye donors but due to shortage, artificial cornea can also be used as an alternative (known as a keratoprosthetics). Commercially available ones coming from Boston KPro (types I and II), AlphaCor, osteo-odonto-keratoprosthesis, and Fyodorov-Zuev KPro are used in the clinic to replace the cornea of a blind individual, partially restoring their eyesight.

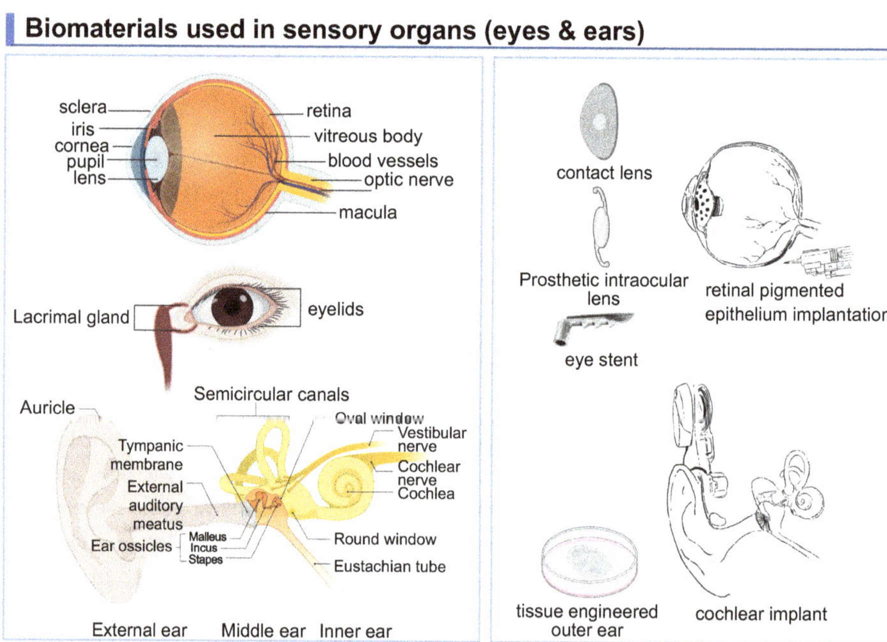

Figure 1.9: (Left panel) The gross anatomy of the eyes and ears which are responsible for vision and hearing senses, respectively. (Right panel) To alleviate cases of impairment in both senses, biomedical devices made from biomaterials (contact lenses, prosthetic intraocular lenses, lacrimal (eye) stents, artificial lab-grown external ear structure (early research level), and cochlear implants) are used.

Macular degeneration [227] can also lead to vision impairment as the photoreceptors in the retina become ischemic due to lack of nourishment and structural support. This is a condition where the RPE is degenerated and unable to support the photoreceptors. It can be classified as wet or dry macular degeneration. The difference between the two is that wet macular degeneration is characterized by overexpression of vascular growth factors that cause bleeding in the subretinal space, whereas dry macular degeneration is the accumulation of drusen (yellow deposits in the subretinal space). Wet macular degeneration is usually treated with anti-vascular endothelial growth factor injections to the eye [228–230], while the dry macular degeneration is where biomaterials can be applied [231]. RPE is first cultured on a layer of polymeric membrane and is transplanted on the subretinal space. This, in theory, will regenerate the damaged RPE layer of a dAMD patient and slow down the progression of the disease. However, it is not meant to restore the vision of the patient since photoreceptors, once damaged does not have the ability to regenerate. Humans are born with fixed number of photoreceptors in the eye and this amount does not change and only reduced in cases of degeneration due to macular damage. Future prospect of this intervention is to replace the RPE layers first, and then implant new photoreceptors from stem cell-derived sources.

The ears are responsible for our sense of hearing and can be divided into three sections: the outer ear, the middle ear, and the inner ear (Figure 1.9, left panel) [232, 233]. The outer ear captures sound from the environment. It is made up of the visible pinna (also called the auricle), the ear canal, and the outer layer of the eardrum (also called the tympanic membrane). The sound waves that enter the pinna then travels through the ear canal all the way to the tympanic membrane. This membrane is a thin layer of tissue that vibrates when it receives sound waves. These vibrations will in turn vibrate three small bones in the middle ear collectively called the ossicles. The ossicles (malleus, incus, and stapes) are the smallest bones in the body and are responsible for transmitting and amplifying the sound waves that it receives from the tympanic membrane to the cochlea. The cochlea is a snail-like structure found in the inner ear and is filled with a fluid that creates ripples when moved by the ossicles. Using these ripples, small hair-like structures called stereocilia or hair cells moved along the wave and then send neurotransmitter signals down to the auditory nerve to the brain. Different hair cells respond to different sound frequencies, and the hair cells located at the bottom part of the cochlear spiral detect high pitched sounds, while the hair cells located at the middle and upper part of the spiral detect progressively lower pitched sounds. These signals that travel through the auditory nerve is then translated by the brain as recognizable sounds.

Deafness can be caused by multiple disruption of this process; sound quality is mediated by how efficient the sound wave can travel from the outer ear all the way to the inner ear. If, for example, the tympanic membrane is ruptured, it would not be able to move the ossicles that relay the vibrations to the cochlea, or if the cochlear inner structure is degenerated then it would not be able to excite the auditory nerve to send signals to the brain. Here, biomaterials can be applied to restore the ability of a person to hear. For tympanic membrane regeneration, a scaffold from electrospun or 3D-printed polymers can be used to grow the intricate network of collagen membrane in vitro [234–236]. The resulting membrane can then be directly transplanted to the patient, reducing the need to use the patient's own tissue for reconstruction surgery.

In cases of cochlear deafness [237], an implant may be introduced to the patient to stimulate the auditory nerve so that the patient may hear [238, 239]. This implant is placed at the back of the ear and is made up of two components: the outer electrical component receives and processes the sound waves through a microphone and then translates that sound waves to an electrical signal (Figure 1.9, right panel). That electrical signal is then relayed to the internal part of the implant which is located under the skin. The signal travels all the way through an electrode array that is threaded inside the cochlea. Unlike hearing aids that only amplify incoming sound waves, a cochlear implant directly stimulates the auditory nerves, thus providing an artificial pathway for hearing.

1.10 Conclusion

After knowing the different anatomical processes that govern the body, we can now analyze what key design considerations must be done to create an effective biomaterial. The most common applications of biomaterials are with tissue engineering, medical device implants and prosthesis, and drug delivery systems. Tissue engineering is a strategy to grow a piece of the body on a temporary scaffold to regenerate, replace, or induce the growth of local tissue in the implantation site. Here polymers work best as we can fine-tune their mechanical and degradation properties. A fine balance between mechanical properties, degradation rate, biocompatibility, and nourishment transport must be considered when choosing the right material for tissue engineering. The key application of effective biomaterial selection for medical devices and prosthesis is to improve the capabilities of the said device. This is done by choosing a more biocompatible or bioinert material that can interface with the tissue of choice, creating a lighter material or energy-efficient material, having the capability to avoid bacterial colonization during implantation, or to serve as a platform for potential drug delivery. Finally, biomaterials are used to increase the pharmacokinetics and pharmacodynamics of drugs. Drugs can be packaged in nanoparticles and microparticles, polymeric gels and fibers, and microneedles to increase the bioavailability, create a controlled or on-demand release properties, target specific cells, tissues, or organs, and overcome physiological barriers. The following chapter will discuss the different kinds of biomaterials, identifying major biomaterial classifications and sources. It will discuss the properties and special effects of different materials with respect to the human body.

References

[1] Blanchard S. 3 – Anatomy and physiology. In: Enderle JD, Blanchard SM, Bronzino JD, editors. Introduction to Biomedical Engineering (Second Edition). Boston: Academic Press; 2005. p. 73–125.

[2] Alasmari WA. Medical students' feedback of applying the virtual dissection table (anatomage) in learning anatomy: A cross-sectional descriptive study. Advances in Medical Education and Practice. 2021;12:1303–07.

[3] Dempsey AMK, Lone M, Nolan YM, Hunt E. Universal design for learning in anatomy education of healthcare students: A scoping review. Anatomical Science Education. 2021.

[4] Heness G, Ben NB. Biomaterials-classifications and behavior of different types of biomaterials. Materials Forum 2013.

[5] Parida P, Behera A, Mishra SC. Classification of Biomaterials used in Medicine. 2012.

[6] Davis J. Overview of biomaterials and their use in medical devices. Handbook of Materials for Medical Devices. 2003:1–12.

[7] Williams DF. Regulatory biocompatibility requirements for biomaterials used in regenerative medicine. Journal of Materials Science: Materials in Medicine. 2015;26(2):89.

[8] Williams D. Biomaterials and biocompatibility. Medical Progress through Technology. 1976;4(1–2):31–42.

[9] Huzum B, Puha B, Necoara RM, Gheorghevici S, Puha G, Filip A, et al. Biocompatibility assessment of biomaterials used in orthopedic devices: An overview. Experimental and Therapeutic Medicine. 2021;22(5):1–9.

[10] Shang T, Zhang JY, Klonoff DC. The FDA Pilot Accreditation Scheme for Conformity: Will It Pertain to Cybersecurity of Diabetes Devices?. Los Angeles, CA: SAGE Publications Sage CA; 2021.

[11] Kolarsick PA, Kolarsick MA, Goodwin C. Anatomy and physiology of the skin. Journal of the Dermatology Nurses' Association. 2011;3(4):203–13.

[12] Schommer NN, Gallo RL. Structure and function of the human skin microbiome. Trends in Microbiology. 2013;21(12):660–68.

[13] Proksch E, Brandner JM, Jensen JM. The skin: An indispensable barrier. Experimental Dermatology. 2008;17(12):1063–72.

[14] Kalafa EG, Hixona K, Kadakia P, Dunn A, Sell S. 9.1. 1 Skin anatomy and physiology. Electrospun materials for tissue engineering and biomedical applications. Research, Design and Commercialization. 2017:179.

[15] Graham HK, Eckersley A, Ozols M, Mellody KT, Sherratt MJ. Human Skin: Composition, Structure and Visualisation Methods. Skin Biophysics: Springer; 2019. p. 1–18.

[16] Caon M. Integumentary System. Examination Questions and Answers in Basic Anatomy and Physiology: Springer; 2020. p. 157–72.

[17] Van Hattem S, Bootsma AH, Thio HB. Skin manifestations of diabetes. Cleveland Clinic Journal of Medicine. 2008;75(11):772–74.

[18] Sibbald RG, Landolt SJ, Toth D. Skin and diabetes. Endocrinology and Metabolism Clinics. 1996;25(2):463–72.

[19] Eming SA, Martin P, Tomic-Canic M. Wound repair and regeneration: Mechanisms, signaling, and translation. Science Translational Medicine. 2014;6(265):265sr6–sr6.

[20] Akita S. Wound Repair and Regeneration: Mechanisms, Signaling: Multidisciplinary Digital Publishing Institute; 2019.

[21] Takeo M, Lee W, Ito M. Wound healing and skin regeneration. Cold Spring Harbor Perspectives in Medicine. 2015;5(1):a023267.

[22] Wang S, Urban MW. Self-healing polymers. Nature Reviews Materials. 2020;5(8):562–83.

[23] Guo B, Dong R, Liang Y, Li M. Haemostatic materials for wound healing applications. Nature Reviews Chemistry. 2021:1–19.

[24] Mihai MM, Dima MB, Dima B, Holban AM. Nanomaterials for wound healing and infection control. Materials. 2019;12(13):2176.

[25] Zahoor I, Singh S, Behl T, Sharma N, Naved T, Subramaniyan V, et al. Emergence of microneedles as a potential therapeutics in diabetes mellitus. Environmental Science and Pollution Research. 2021:1–21.

[26] Agrawal S, Gandhi SN, Gurgar P, Saraswathy N. Microneedles: An advancement to transdermal drug delivery system approach. Journal of Applied Pharmaceutical Science. 2020;10(3):149–59.

[27] Kochhar JS, Lim WXS, Zou S, Foo WY, Pan J, Kang L. Microneedle integrated transdermal patch for fast onset and sustained delivery of lidocaine. Molecular Pharmaceutics. 2013;10 (11):4272–80.

[28] Chang H, Chew SW, Zheng M, Lio DCS, Wiraja C, Mei Y, et al. Cryomicroneedles for transdermal cell delivery. Nature Biomedical Engineering. 2021:1–11.

[29] Mofidfar M, Prausnitz MR. Electrospun transdermal patch for contraceptive hormone delivery. Current Drug Delivery. 2019;16(6):577–83.

[30] Galzote RM, Rafie S, Teal R, Mody SK. Transdermal delivery of combined hormonal contraception: A review of the current literature. International Journal of Women's Health. 2017;9:315.

[31] Tamada JA, Bohannon NJ, Potts RO. Measurement of glucose in diabetic subjects using noninvasive transdermal extraction. Nature Medicine. 1995;1(11):1198–201.

[32] Lipani L, Dupont BG, Doungmene F, Marken F, Tyrrell RM, Guy RH, et al. Non-invasive, transdermal, path-selective and specific glucose monitoring via a graphene-based platform. Nature Nanotechnology. 2018;13(6):504–11.

[33] Makvandi P, Jamaledin R, Chen G, Baghbantaraghdari Z, Zare EN, Di Natale C, et al. Stimuli-responsive transdermal microneedle patches. Materials Today. 2021.

[34] Li Y, Ai Q, Mao L, Guo J, Gong T, Lin Y, et al. Hybrid strategy of graphene/carbon nanotube hierarchical networks for highly sensitive, flexible wearable strain sensors. Scientific Reports. 2021;11(1):1–9.

[35] Gong S, Schwalb W, Wang Y, Chen Y, Tang Y, Si J, et al. A wearable and highly sensitive pressure sensor with ultrathin gold nanowires. Nature Communications. 2014;5(1):1–8.

[36] Yoon S, Kim YJ, Lee YR, Lee N-E, Won Y, Gandla S, et al. Highly stretchable metal-polymer hybrid conductors for wearable and self-cleaning sensors. NPG Asia Materials. 2021;13(1):1 14.

[37] Choi D-H, Thaxton A, Cheol Jeong I, Kim K, PR S, GR C, et al,. Sweat test for cystic fibrosis: Wearable sweat sensor vs. standard laboratory test. Journal of Cystic Fibrosis. 2018;17(4): e35–e8.

[38] Kaya T, Liu G, Ho J, Yelamarthi K, Miller K, Edwards J, et al. Wearable sweat sensors: Background and current trends. Electroanalysis. 2019;31(3):411–21.

[39] Palmer LC, Newcomb CJ, Kaltz SR, Spoerke ED, Stupp SI. Biomimetic systems for hydroxyapatite mineralization inspired by bone and enamel. Chemical Reviews. 2008;108 (11):4754–83.

[40] Sahoo N, Anand S, Bhardwaj J, Sachdeva V, Sapru B. Bone response to stainless steel and titanium bone plates: An experimental study on animals. Medical Journal Armed Forces India. 1994;50(1):10–14.

[41] Casaña-Ruiz MD, Bellot-Arcís C, Paredes-Gallardo V, García-Sanz V, Almerich-Silla JM, Montiel-Company JM. Risk factors for orthodontic mini-implants in skeletal anchorage biological stability: A systematic literature review and meta-analysis. Scientific Reports. 2020;10(1):1–10.

[42] Wu S, Xu J, Zou L, Luo S, Yao R, Zheng B, et al. Long-lasting renewable antibacterial porous polymeric coatings enable titanium biomaterials to prevent and treat peri-implant infection. Nature Communications. 2021;12(1):1–14.
[43] Lohberger B, Stuendl N, Glaenzer D, Rinner B, Donohue N, Lichtenegger HC, et al. CoCrMo surface modifications affect biocompatibility, adhesion, and inflammation in human osteoblasts. Scientific Reports. 2020;10(1):1–8.
[44] Zegan G, Cărăuşu E, Golovcencu L, Sodor Botezatu A, Cernei ER, Anistoroaei D. Nanoparticles deposition on mini-implants for osseo-integration and antibacterial properties improvement. Revista de Chimie. 2017;68(12):2929–31.
[45] Funao H, Nagai S, Sasaki A, Hoshikawa T, Tsuji T, Okada Y, et al. A novel hydroxyapatite film coated with ionic silver via inositol hexaphosphate chelation prevents implant-associated infection. Scientific Reports. 2016;6(1):1–9.
[46] Ong JL, Chan DC. Hydroxyapatite and their use as coatings in dental implants: A review. Critical Reviews™ in Biomedical Engineering. 2000;28:5&6.
[47] Tian Q, Lin J, Rivera-Castaneda L, Tsanhani A, Dunn ZS, Rodriguez A, et al. Nano-to-submicron hydroxyapatite coatings for magnesium-based bioresorbable implants–deposition, characterization, degradation, mechanical properties, and cytocompatibility. Scientific Reports. 2019;9(1):1–27.
[48] Lewallen EA, Riester SM, Bonin CA, Kremers HM, Dudakovic A, Kakar S, et al. Biological strategies for improved osseointegration and osteoinduction of porous metal orthopedic implants. Tissue Engineering Part B: Reviews. 2015;21(2):218–30.
[49] Albrektsson T, Johansson C. Osteoinduction, osteoconduction and osseointegration. European Spine Journal. 2001;10(2):S96–S101.
[50] Barradas A, Yuan H, van Blitterswijk CA, Habibovic P. Osteoinductive biomaterials: Current knowledge of properties, experimental models and biological mechanisms. European Cells and Materials. 2011;21(407):29.
[51] Chang -C-C, Zhou S, Li Q. Structural Optimization for Transport Efficiency in Porous Tissue Scaffolds.
[52] Petersen A, Princ A, Korus G, Ellinghaus A, Leemhuis H, Herrera A, et al. A biomaterial with a channel-like pore architecture induces endochondral healing of bone defects. Nature Communications. 2018;9(1):1–16.
[53] Wang X, Nie Z, Chang J, Lu ML, Kang Y. Multiple channels with interconnected pores in a bioceramic scaffold promote bone tissue formation. Scientific Reports. 2021;11(1):1–15.
[54] Harley BJ, Bergman JW. Tendon and ligament anatomy, biology, and biomechanics. Oncology and Basic Science. 2008:463–75.
[55] Järvinen M, Kannus P. Quadriceps Muscle Atrophy and Long-term Recovery after knee Ligamentous Injury. Annales Chirurgiae et Gynaecologiae; 1987. p. 108–13.
[56] Kannus P, Järvinen M, Jozsa L. Thigh muscle atrophy and postoperative osteoarthrosis after knee ligament tears. Acta Chirurgica Hungarica. 1988;29(2):151–61.
[57] Liu R, Zhang S, Chen X. Injectable hydrogels for tendon and ligament tissue engineering. Journal of Tissue Engineering and Regenerative Medicine. 2020;14(9):1333–48.
[58] Liu J, Lin S, Liu X, Qin Z, Yang Y, Zang J, et al. Fatigue-resistant adhesion of hydrogels. Nature Communications. 2020;11(1):1–9.
[59] Sensini A, Massafra G, Gotti C, Zucchelli A, Cristofolini L. Tissue engineering for the insertions of tendons and ligaments: An overview of electrospun biomaterials and structures. Frontiers in Bioengineering and Biotechnology. 2021;9:98.
[60] Shao F, Liu Q, Zhu Y, Fan Z, Chen W, Liu S, et al. Targeting chondrocytes for arresting bony fusion in ankylosing spondylitis. Nature Communications. 2021;12(1):1–14.

[61] Neidlin M, Korcari A, Macheras G, Alexopoulos LG. Cue-signal-response analysis in 3d chondrocyte scaffolds with anabolic stimuli. Annals of Biomedical Engineering. 2018;46(2):345–53.
[62] Huang X, Hou Y, Zhong L, Huang D, Qian H, Karperien M, et al. Promoted chondrogenesis of cocultured chondrocytes and mesenchymal stem cells under hypoxia using in-situ forming degradable hydrogel scaffolds. Biomacromolecules. 2018;19(1):94–102.
[63] Thorvaldsson A, Stenhamre H, Gatenholm P, Walkenström P. Electrospinning of highly porous scaffolds for cartilage regeneration. Biomacromolecules. 2008;9(3):1044–49.
[64] Meng Q, Man Z, Dai L, Huang H, Zhang X, Hu X, et al. A composite scaffold of MSC affinity peptide-modified demineralized bone matrix particles and chitosan hydrogel for cartilage regeneration. Scientific Reports. 2015;5(1):1–14.
[65] Ansari M, Eshghanmalek M. Biomaterials for repair and regeneration of the cartilage tissue. Bio-Design and Manufacturing. 2019;2(1):41–49.
[66] Kuo Y-C, Ku I-N. Cartilage regeneration by novel polyethylene oxide/chitin/chitosan scaffolds. Biomacromolecules. 2008;9(10):2662–69.
[67] Fuchs A, Eberbach H, Izadpanah K, Bode G, Südkamp NP, Feucht MJ. Focal metallic inlay resurfacing prosthesis for the treatment of localized cartilage defects of the femoral condyles: A systematic review of clinical studies. Knee Surgery, Sports Traumatology, Arthroscopy. 2018;26(9):2722–32.
[68] Lecocq M, Linares J-M, Chaves-Jacob J, Coyle T, Roffino S, Eyraud M, et al. Total knee arthroplasty with a Ti6Al4V/PEEK prosthesis on an osteoarthritis rat model: Behavioral and neurophysiological analysis. Scientific Reports. 2020;10(1):1–15.
[69] Cho SM, Park J-W, Han H-S, Seok H-K, Moon M-W, Kim YC. Multifunctional composite coating as a wear-resistant layer for the bearing in total hip joint replacement. ACS Applied Materials & Interfaces. 2013;5(2):395–403.
[70] Cheng K-Y, Gopal V, McNallan M, Manivasagam G, Mathew MT. Enhanced tribocorrosion resistance of hard ceramic coated Ti-6Al-4V alloy for hip implant application: In-vitro simulation study. ACS Biomaterials Science & Engineering. 2019;5(9):4817–24.
[71] Javadi A, Solouk A, Nazarpak MH, Bagheri F. Surface engineering of titanium-based implants using electrospraying and dip coating methods. Materials Science and Engineering: C. 2019;99:620–30.
[72] Korsacilar D, Albayrak A, Malayoğlu U. Production of a bio-compatible metallic implant via electropolymetrization method by coating a synthetic hydrogel on a titanium alloy. Journal of Biomechanics. 2011;44:17.
[73] De Giglio E, Guascito M, Sabbatini L, Zambonin G. Electropolymerization of pyrrole on titanium substrates for the future development of new biocompatible surfaces. Biomaterials. 2001;22(19):2609–16.
[74] Civantos A, Martinez-Campos E, Ramos V, Elvira C, Gallardo A, Abarrategi A. Titanium coatings and surface modifications: Toward clinically useful bioactive implants. ACS Biomaterials Science & Engineering. 2017;3(7):1245–61.
[75] Xie C-M, Lu X, Wang K-F, Meng F-Z, Jiang O, Zhang H-P, et al. Silver nanoparticles and growth factors incorporated hydroxyapatite coatings on metallic implant surfaces for enhancement of osteoinductivity and antibacterial properties. ACS Applied Materials & Interfaces. 2014;6(11):8580–89.
[76] Lyndon JA, Boyd BJ, Birbilis N. Metallic implant drug/device combinations for controlled drug release in orthopaedic applications. Journal of Controlled Release. 2014;179:63–75.
[77] Thygesen K, Alpert JS, White HD. Infarction JEAAWTFftRoM. Universal definition of myocardial infarction. Journal of the American College of Cardiology. 2007;50(22):2173–95.

[78] Goldstein JA. Pathophysiology and management of right heart ischemia. Journal of the American College of Cardiology. 2002;40(5):841–53.
[79] Zipes DP. Influence of myocardial ischemia and infarction on autonomic innervation of heart. Circulation. 1990;82(4):1095–105.
[80] Libby P, Theroux P. Pathophysiology of coronary artery disease. Circulation. 2005;111 (25):3481–88.
[81] Falk E. Pathogenesis of atherosclerosis. Journal of the American College of Cardiology. 2006;47(8S):C7–C12.
[82] Ross R. Cell biology of atherosclerosis. Annual Review of Physiology. 1995;57(1):791–804.
[83] Giannoglou GD, Chatzizisis YS, Zamboulis C, Parcharidis GE, Mikhailidis DP, Louridas GE. Elevated heart rate and atherosclerosis: An overview of the pathogenetic mechanisms. International Journal of Cardiology. 2008;126(3):302–12.
[84] Morrissey RP, Czer L, Shah PK. Chronic heart failure. American Journal of Cardiovascular Drugs. 2011;11(3):153–71.
[85] Guzzetti S, Magatelli R, Borroni E, Mezzetti S. Heart rate variability in chronic heart failure. Autonomic Neuroscience. 2001;90(1–2):102–05.
[86] Cranz A, Greinacher A, Nagy E, Friederich H-C, Katus HA, Geis N, et al. Broken heartstrings – post-traumatic stress disorder and psychological burden after acute mitral regurgitation due to chordae tendineae rupture. Journal of Clinical Medicine. 2020;9(12):4048.
[87] Ferrario M, Signorini MG, Magenes G. Complexity analysis of the fetal heart rate for the identification of pathology in fetuses. Computers in cardiology, 2005: IEEE; 2005. p. 989–92.
[88] Flanagan TC, Pandit A. Living artificial heart valve alternatives: A review. European Cells and Materials. 2003;6(1):28–45.
[89] David TE. Surgical treatment of aortic valve disease. Nature Reviews Cardiology. 2013;10(7):375–86.
[90] Fioretta ES, Motta SE, Lintas V, Loerakker S, Parker KK, Baaijens FP, et al. Next-generation tissue-engineered heart valves with repair, remodelling and regeneration capacity. Nature Reviews Cardiology. 2021;18(2):92–116.
[91] Brisman JL, Song JK, Newell DW. Cerebral aneurysms. New England Journal of Medicine. 2006;355(9):928–39.
[92] Rhoton JAL. Aneurysms. Neurosurgery. 2002;51(suppl_4):S1-121–S1-58.
[93] Olijhoek JK, van der Graaf Y, Banga J-D, Algra A, Rabelink TJ, Visseren FL. The metabolic syndrome is associated with advanced vascular damage in patients with coronary heart disease, stroke, peripheral arterial disease or abdominal aortic aneurysm. European Heart Journal. 2004;25(4):342–48.
[94] White RA. Atherosclerosis and Arteriosclerosis: CRC Press; 2019.
[95] Stary HC, Chandler AB, Dinsmore RE, Fuster V, Glagov S, Insull JW, et al. A definition of advanced types of atherosclerotic lesions and a histological classification of atherosclerosis: A report from the Committee on Vascular Lesions of the Council on Arteriosclerosis, American Heart Association. Circulation. 1995;92(5):1355–74.
[96] Castaneda-Zuniga W, Formanek A, Tadavarthy M, Vlodaver Z, Edwards J, Zollikofer C, et al. The mechanism of balloon angioplasty. Radiology. 1980;135(3):565–71.
[97] Bittl JA. Advances in coronary angioplasty. New England Journal of Medicine. 1996;335 (17):1290–302.
[98] Kim MS, Dean LS. In-stent restenosis. Cardiovascular Therapeutics. 2011;29(3):190–98.
[99] Brower V. Stents and biology combination for restenosis. Nature Biotechnology. 1996;14 (4):422.

[100] Davé V, Buensuceso C, Colter D, Zhao J, Falotico R. Endothelial cell attachment and proliferation studies on modified metal stent surfaces. MRS Online Proceedings Library (OPL). 2009:1236.
[101] Inoue T, Croce K, Morooka T, Sakuma M, Node K, Simon DI. Vascular inflammation and repair: Implications for re-endothelialization, restenosis, and stent thrombosis. JACC: Cardiovascular Interventions. 2011;4(10):1057–66.
[102] Avinash K, Ramya A, Pooja B, Priya V, Naresh B. Design methodology of a balloon expandable polymeric stent. Journal of Biomedical Engineering and Medical Device. 2019;4:139.
[103] Eberhart RC, Su S-H, Nguyen KT, Zilberman M, Tang L, Nelson KD, et al. Bioresorbable polymeric stents: Current status and future promise. Journal of Biomaterials Science, Polymer Edition. 2003;14(4):299–312.
[104] Hermawan H, Dubé D, Mantovani D. Developments in metallic biodegradable stents. Acta Biomaterialia. 2010;6(5):1693–97.
[105] Hermawan H, Purnama A, Dube D, Couet J, Mantovani D. Fe–Mn alloys for metallic biodegradable stents: Degradation and cell viability studies. Acta Biomaterialia. 2010;6(5):1852–60.
[106] Kumar K, Gill R, Batra U. Challenges and opportunities for biodegradable magnesium alloy implants. Materials Technology. 2018;33(2):153–72.
[107] Ma J, Zhao N, Betts L, Zhu D. Bio-adaption between magnesium alloy stent and the blood vessel: A review. Journal of Materials Science & Technology. 2016;32(9):815–26.
[108] Strohbach A, Busch R. Polymers for cardiovascular stent coatings. International Journal of Polymer Science. 2015:2015.
[109] Sydow-Plum G, Tabrizian M. Review of stent coating strategies: Clinical insights. Materials Science and Technology. 2008;24(9):1127–43.
[110] Qi P, Yang Y, Zhao S, Wang J, Li X, Tu Q, et al. Improvement of corrosion resistance and biocompatibility of biodegradable metallic vascular stent via plasma allylamine polymerized coating. Materials & Design. 2016;96:341–49.
[111] Du R, Wang Y, Huang Y, Zhao Y, Zhang D, Du D, et al. Design and testing of hydrophobic core/hydrophilic shell nano/micro particles for drug-eluting stent coating. NPG Asia Materials. 2018;10(7):642–58.
[112] Wessely R. New drug-eluting stent concepts. Nature Reviews Cardiology. 2010;7(4):194–203.
[113] Torii S, Jinnouchi H, Sakamoto A, Kutyna M, Cornelissen A, Kuntz S, et al. Drug-eluting coronary stents: Insights from preclinical and pathology studies. Nature Reviews Cardiology. 2020;17(1):37–51.
[114] Roopmani P, Satheesh S, Raj DC, Krishnan UM. Development of dual drug eluting cardiovascular stent with ultrathin flexible poly (l-lactide-co-caprolactone) coating. ACS Biomaterials Science & Engineering. 2019;5(6):2899–915.
[115] Polyak B, Medved M, Lazareva N, Steele L, Patel T, Rai A, et al. Magnetic nanoparticle-mediated targeting of cell therapy reduces in-stent stenosis in injured arteries. Acs Nano. 2016;10(10):9559–69.
[116] Cantile C, Youssef S. Nervous system. Jubb, Kennedy & Palmer's Pathology of Domestic Animals. 2016;Volume 1:250.
[117] Campese VM, Romoff MS, Levitan D, Lane K, Massry SG. Mechanisms of autonomic nervous system dysfunction in uremia. Kidney International. 1981;20(2):246–53.
[118] Leblanc A. Olfactory Nerves (I). The Cranial Nerves: Anatomy Imaging Vascularisation. Berlin, Heidelberg: Springer Berlin Heidelberg; 1995. p. 1–16.
[119] Leblanc A. Optic Nerve (II). The Cranial Nerves: Anatomy Imaging Vascularisation. Berlin, Heidelberg: Springer Berlin Heidelberg; 1995. p. 17–32.

[120] Leblanc A. Oculomotor Nerve (III). The Cranial Nerves: Anatomy Imaging Vascularisation. Berlin, Heidelberg: Springer Berlin Heidelberg; 1995. p. 33–46.
[121] Leblanc A. Trochlear Nerve (IV). The Cranial Nerves: Anatomy Imaging Vascularisation. Berlin, Heidelberg: Springer Berlin Heidelberg; 1995. p. 47–58.
[122] Leblanc A. Trigeminal Nerve (V). The Cranial Nerves: Anatomy Imaging Vascularisation. Berlin, Heidelberg: Springer Berlin Heidelberg; 1995. p. 59–153.
[123] Leblanc A. Abducent Nerve (VI). The Cranial Nerves: Anatomy Imaging Vascularisation. Berlin, Heidelberg: Springer Berlin Heidelberg; 1995. p. 155–70.
[124] Leblanc A. Facial Nerve (VII). The Cranial Nerves: Anatomy Imaging Vascularisation. Berlin, Heidelberg: Springer Berlin Heidelberg; 1995. p. 171–210.
[125] Leblanc A. Vestibulocochlear Nerve (VIII). The Cranial Nerves: Anatomy Imaging Vascularisation. Berlin, Heidelberg: Springer Berlin Heidelberg; 1995. p. 211–26.
[126] Leblanc A. Glossopharyngeal Nerve (IX). The Cranial Nerves: Anatomy Imaging Vascularisation. Berlin, Heidelberg: Springer Berlin Heidelberg; 1995. p. 227–40.
[127] Leblanc A. Vagus Nerve (X). The Cranial Nerves: Anatomy Imaging Vascularisation. Berlin, Heidelberg: Springer Berlin Heidelberg; 1995. p. 241–50.
[128] Leblanc A. Accessory Nerve (XI). The Cranial Nerves: Anatomy Imaging Vascularisation. Berlin, Heidelberg: Springer Berlin Heidelberg; 1995. p. 251–58.
[129] Leblanc A. Hypoglossal Nerve (XII). The Cranial Nerves: Anatomy Imaging Vascularisation. Berlin, Heidelberg: Springer Berlin Heidelberg; 1995. p. 259–72.
[130] Heiligenberg W, Keller CH, Metzner W, Kawasaki M. Structure and function of neurons in the complex of the nucleus electrosensorius of the gymnotiform fish Eigenmannia: Detection and processing of electric signals in social communication. Journal of Comparative Physiology A. 1991;169(2):151–64.
[131] Lovinger DM. Communication networks in the brain: Neurons, receptors, neurotransmitters, and alcohol. Alcohol Research & Health. 2008.
[132] Simpkins CA, Simpkins AM. Neurons and Neurotransmitters. Neuroscience for Clinicians: Springer; 2013. p. 77–91.
[133] Balestrino R, Schapira A. Parkinson disease. European Journal of Neurology. 2020;27 (1):27–42.
[134] Höglinger GU, Rizk P, Muriel MP, Duyckaerts C, Oertel WH, Caille I, et al. Dopamine depletion impairs precursor cell proliferation in Parkinson disease. Nature Neuroscience. 2004;7 (7):726–35.
[135] Parpura V, Heneka MT, Montana V, Oliet SH, Schousboe A, Haydon PG, et al. Glial cells in (patho) physiology. Journal of Neurochemistry. 2012;121(1):4–27.
[136] Jessen KR. Glial cells. The International Journal of Biochemistry & Cell Biology. 2004;36 (10):1861–67.
[137] He F, Sun YE. Glial cells more than support cells?. The International Journal of Biochemistry & Cell Biology. 2007;39(4):661–65.
[138] Masserini M. Nanoparticles for brain drug delivery. International Scholarly Research Notices. 2013:2013.
[139] Chiono V, Tonda-Turo C. Trends in the design of nerve guidance channels in peripheral nerve tissue engineering. Progress in Neurobiology. 2015;131:87–104.
[140] Manzari-Tavakoli A, Tarasi R, Sedghi R, Moghimi A, Niknejad H. Fabrication of nanochitosan incorporated polypyrrole/alginate conducting scaffold for neural tissue engineering. Scientific Reports. 2020;10(1):1–10.
[141] Talebi A, Labbaf S, Atari M, Parhizkar M. Polymeric nanocomposite structures based on functionalized graphene with tunable properties for nervous tissue replacement. ACS Biomaterials Science & Engineering. 2021;7(9):4591–601.

[142] Ghasemi-Mobarakeh L, Prabhakaran MP, Morshed M, Nasr-Esfahani MH, Baharvand H, Kiani S, et al. Application of conductive polymers, scaffolds and electrical stimulation for nerve tissue engineering. Journal of Tissue Engineering and Regenerative Medicine. 2011;5 (4):e17–e35.
[143] Wang S, Cai L. Polymer gel systems for nerve repair and regeneration. Biomaterials. 2010:43–63.
[144] Zhu J, Shen Y, Chen L, Zhang HL. Biology and therapy of adult human brain stem cells. Cell Research. 2008;18(1):S113–S.
[145] Schonberg DL, Lubelski D, Miller TE, Rich JN. Brain tumor stem cells: Molecular characteristics and their impact on therapy. Molecular Aspects of Medicine. 2014;39:82–101.
[146] Holley RJ, Wood SR, Bigger BW. Delivering hematopoietic stem cell gene therapy treatments for neurological lysosomal diseases. ACS Chemical Neuroscience. 2018;10(1):18–20.
[147] Yan Y, Shin S, Jha BS, Liu Q, Sheng J, Li F, et al. Efficient and rapid derivation of primitive neural stem cells and generation of brain subtype neurons from human pluripotent stem cells. Stem Cells Translational Medicine. 2013;2(11):862–70.
[148] Kiernan MC, Vucic S, Cheah BC, Turner MR, Eisen A, Hardiman O, et al. Amyotrophic lateral sclerosis. The Lancet. 2011;377(9769):942–55.
[149] Sorkhabi MM, Benjaber M, Brown P, Denison T. Physiological artifacts and the implications for brain-machine-interface design. 2020 IEEE International Conference on Systems, Man, and Cybernetics (SMC): IEEE; 2020. p. 1498.
[150] Awan NR, Lozano A, Hamani C. Deep brain stimulation: Current and future perspectives. Neurosurgical Focus. 2009;27(1):E2.
[151] Krauss JK, Lipsman N, Aziz T, Boutet A, Brown P, Chang JW, et al. Technology of deep brain stimulation: Current status and future directions. Nature Reviews Neurology. 2021;17 (2):75–87.
[152] Jebari K, Hansson S-O. European public deliberation on brain machine interface technology: Five convergence seminars. Science and Engineering Ethics. 2013;19(3):1071–86.
[153] Lewis AJ, Southern LL. Anatomy of the Digestive System and Nutritional Physiology. Swine Nutrition: CRC Press; 2000. p. 51–84.
[154] Denbow DM. Gastrointestinal Anatomy and Physiology. Sturkie's avian physiology: Elsevier; 2015. p. 337–66.
[155] Svihus B. Function of the digestive system. Journal of Applied Poultry Research. 2014;23 (2):306–14.
[156] Engevik AC, Kaji I, Goldenring JR. The physiology of the gastric parietal cell. Physiological Reviews. 2020;100(2):573–602.
[157] Jacobson AM, De Groot M, Samson JA. The evaluation of two measures of quality of life in patients with type I and type II diabetes. Diabetes Care. 1994;17(4):267–74.
[158] Kaplan JB, Kalra A, Biggins SW. Liver Anatomy and Function. Radiation Therapy for Liver Tumors: Springer; 2017. p. 3–11.
[159] Campbell I. Liver: Functional anatomy and blood supply. Anaesthesia & Intensive Care Medicine. 2006;7(2):49–51.
[160] Diamant NE. Neuromuscular mechanisms of primary peristalsis. The American Journal of Medicine. 1997;103(5):40S–3S.
[161] Vanderburgh L, Ho C-S. Nonvascular stents. Progress in Cardiovascular Diseases. 1996;39 (2):187–200.
[162] van Boeckel PG, Sijbring A, Vleggaar FP, Siersema PD. Systematic review: Temporary stent placement for benign rupture or anastomotic leak of the oesophagus. Alimentary Pharmacology & Therapeutics. 2011;33(12):1292–301.

[163] Coté GA, Mullady DK, Jonnalagadda SS, Keswani RN, Wani SB, Hovis CE, et al. Use of a pancreatic duct stent or guidewire facilitates bile duct access with low rates of precut sphincterotomy: A randomized clinical trial. Digestive Diseases and Sciences. 2012;57 (12):3271–78.
[164] Jang JK, Song H-Y, Kim JH, Song M, Park J-H, Kim EY. Tumor overgrowth after expandable metallic stent placement: Experience in 583 patients with malignant gastroduodenal obstruction. American Journal of Roentgenology. 2011;196(6):W831–W6.
[165] Kim S-Y, Kim M, Kim M-K, Lee H, Lee DK, Lee D-H, et al. Paclitaxel-eluting nanofiber-covered self-expanding nonvascular stent for palliative chemotherapy of gastrointestinal cancer and its related stenosis. Biomedical Microdevices. 2014;16(6):897–904.
[166] Jeong D, Lee DH, Lee DK, Na K. Nonvascular drug-eluting stent coated with sodium caprate-incorporated polyurethane for the efficient penetration of paclitaxel into tumor tissue. Journal of Biomaterials Applications. 2015;29(8):1133–44.
[167] Jin Z, Chen Z, Wu K, Shen Y, Guo S. Investigation of migration-preventing tracheal stent with high dose of 5-fluorouracil or paclitaxel for local drug delivery. ACS Applied Bio Materials. 2018;1(5):1328–36.
[168] Garbey M, Salmon R, Fikfak V, Clerc CO. Esophageal stent migration: Testing few hypothesis with a simplified mathematical model. Computers in Biology and Medicine. 2016;79:259–65.
[169] Sloan MA, Hauter WE, Sloan PA, Hafner JA. Esophageal stent migration as a cause of severe upper airway obstruction. 2014.
[170] Deutel B, Greindl M, Thaurer M, Bernkop-Schnürch A. Novel insulin thiomer nanoparticles: In vivo evaluation of an oral drug delivery system. Biomacromolecules. 2008;9(1):278–85.
[171] Carino GP, Mathiowitz E. Oral insulin delivery. Advanced Drug Delivery Reviews. 1999;35 (2–3):249–57.
[172] Hubálek F, Hh R, Gram-Nielsen S, Madsen P, Nishimura E, Münzel M, et al. Molecular engineering of safe and efficacious oral basal insulin. Nature Communications. 2020;11 (1):1–9.
[173] Gerozissis K. Brain insulin: Regulation, mechanisms of action and functions. Cellular and Molecular Neurobiology. 2003;23(1):1–25.
[174] Stewart CE, Rotwein P. Growth, differentiation, and survival: Multiple physiological functions for insulin-like growth factors. Physiological Reviews. 1996;76(4):1005–26.
[175] Ralston SL. Insulin and glucose regulation. Veterinary Clinics: Equine Practice. 2002;18 (2):295–304.
[176] Seyoum B, Abdulkadir J. Systematic inspection of insulin injection sites for local complications related to incorrect injection technique. Tropical Doctor. 1996;26(4):159–61.
[177] Han X, Lu Y, Xie J, Zhang E, Zhu H, Du H, et al. Zwitterionic micelles efficiently deliver oral insulin without opening tight junctions. Nature Nanotechnology. 2020;15(7):605–14.
[178] Kim S-Y, Lee H, Cho S, Park J-W, Park J, Hwang J. Size control of chitosan capsules containing insulin for oral drug delivery via a combined process of ionic gelation with electrohydrodynamic atomization. Industrial & Engineering Chemistry Research. 2011;50 (24):13762–70.
[179] Sung H-W, Sonaje K, Liao Z-X, Hsu L-W, Chuang E-Y. pH-responsive nanoparticles shelled with chitosan for oral delivery of insulin: From mechanism to therapeutic applications. Accounts of Chemical Research. 2012;45(4):619–29.
[180] Anselmo AC, Zhang M, Kumar S, Vogus DR, Menegatti S, Helgeson ME, et al. Elasticity of nanoparticles influences their blood circulation, phagocytosis, endocytosis, and targeting. ACS Nano. 2015;9(3):3169–77.
[181] Richards DM, Endres RG. The mechanism of phagocytosis: Two stages of engulfment. Biophysical Journal. 2014;107(7):1542–53.

[182] Mumuni MA, Kenechukwu FC, Ofokansi KC, Attama AA, Díaz DD. Insulin-loaded mucoadhesive nanoparticles based on mucin-chitosan complexes for oral delivery and diabetes treatment. Carbohydrate Polymers. 2020;229:115506.
[183] Aye T, Block J, Buckingham B. Toward closing the loop: An update on insulin pumps and continuous glucose monitoring systems. Endocrinology and Metabolism Clinics of North America. 2010;39(3):609–24.
[184] Fuchs J, Hovorka R. Closed-loop control in insulin pumps for type-1 diabetes mellitus: Safety and efficacy. Expert Review of Medical Devices. 2020;17(7):707–20.
[185] Shalitin S, Phillip M. Closing the loop: Combining insulin pumps and glucose sensors in children with type 1 diabetes mellitus. Pediatric Diabetes. 2006;7:45–49.
[186] Lee WM. Acute liver failure. New England Journal of Medicine. 1993;329(25):1862–72.
[187] Ferrell L. Liver pathology: Cirrhosis, hepatitis, and primary liver tumors. Update and diagnostic problems. Modern Pathology. 2000;13(6):679–704.
[188] Tsochatzis EA, Bosch J, Burroughs AK. Liver cirrhosis. The Lancet. 2014;383(9930):1749–61.
[189] Siegelman ES, Rosen MA. Imaging of hepatic steatosis. Seminars in Liver Disease: Copyright© 2001 by Thieme Medical Publishers, Inc., 333 Seventh Avenue, New . . .; 2001. p. 071–80.
[190] Jones RE, Lopez KH. Human Reproductive Biology: Academic Press; 2013.
[191] Weiske WH. Vasectomy. Andrologia. 2001;33(3):125–34.
[192] Schwingl PJ, Guess HA. Safety and effectiveness of vasectomy. Fertility and Sterility. 2000;73 (5):923–36.
[193] Sadeghi-Nejad H. Penile prosthesis surgery: A review of prosthetic devices and associated complications. The Journal of Sexual Medicine. 2007;4(2):296–309.
[194] Mulcahy JJ, Austoni E, Barada JH, Choi HK, Hellstrom WJ, Krishnamurti S, et al. The penile implant for erectile dysfunction. The Journal of Sexual Medicine. 2004;1(1):98–109.
[195] Verze P, Capece M, Califano G, La Rocca R. Two-piece inflatable and semi-rigid penile implants: An effective alternative?. International Journal of Impotence Research. 2020;32 (1):24–29.
[196] Mirheydar HS, Palazzi KL, Parsons JK, Chang D, Hsieh TC. Hospital-based trends in penile prosthetic surgery. The Journal of Sexual Medicine. 2015;12(4):1092–98.
[197] Rosen RC, Kostis JB. Overview of phosphodiesterase 5 inhibition in erectile dysfunction. The American Journal of Cardiology. 2003;92(9):9–18.
[198] Carson CC, Lue TF. Phosphodiesterase type 5 inhibitors for erectile dysfunction. BJU International. 2005;96(3):257–80.
[199] Bella AJ, Brock GB. Intracavernous pharmacotherapy for erectile dysfunction. Endocrine. 2004;23(2):149–55.
[200] Bivalacqua TJ, Champion HC, Hellstrom WJ, Kadowitz PJ. Pharmacotherapy for erectile dysfunction. Trends in Pharmacological Sciences. 2000;21(12):484–89.
[201] Doshani A, Teo RE, Mayne CJ, Tincello DG. Uterine prolapse. BMJ. 2007;335(7624):819–23.
[202] Hendrix SL, Clark A, Nygaard I, Aragaki A, Barnabei V, McTiernan A. Pelvic organ prolapse in the Women's Health Initiative: Gravity and gravidity. American Journal of Obstetrics and Gynecology. 2002;186(6):1160–66.
[203] Mäkinen J, Söderström K-O, Kiilholma P, Hirvonen T. Histological changes in the vaginal connective tissue of patients with and without uterine prolapse. Archives of Gynecology. 1986;239(1):17–20.
[204] Mancuso E, Downey C, Doxford-Hook E, Bryant MG, Culmer P. The use of polymeric meshes for pelvic organ prolapse: Current concepts, challenges, and future perspectives. Journal of Biomedical Materials Research Part B: Applied Biomaterials. 2020;108(3):771–89.

[205] Edwards SL, Werkmeister JA, Rosamilia A, Ramshaw JA, White JF, Gargett CE. Characterisation of clinical and newly fabricated meshes for pelvic organ prolapse repair. Journal of the Mechanical Behavior of Biomedical Materials. 2013;23:53–61.
[206] Powell J. Uterine suspension using mesh (including sacrohysteropexy) to repair uterine prolapse: Interventional procedures guidance [IPG584]. 2017.
[207] Mangir N, Roman S, Chapple CR, MacNeil S. Complications related to use of mesh implants in surgical treatment of stress urinary incontinence and pelvic organ prolapse: Infection or inflammation?. World Journal of Urology. 2020;38(1):73–80.
[208] Falagas ME, Velakoulis S, Iavazzo C, Athanasiou S. Mesh-related infections after pelvic organ prolapse repair surgery. European Journal of Obstetrics & Gynecology and Reproductive Biology. 2007;134(2):147–56.
[209] Hallgrímsson B, Benediktsson H, Vize PD. Anatomy and Histology of the Human Urinary System. The Kidney: Elsevier; 2003. p. 149–64.
[210] Jones TC, Hard GC, Mohr U. Urinary System: Springer Science & Business Media; 2013.
[211] Daugirdas JT, Blake PG, Ing TS. Handbook of Dialysis: Lippincott Williams & Wilkins; 2012.
[212] Pastan S, Bailey J. Dialysis therapy. New England Journal of Medicine. 1998;338 (20):1428–37.
[213] Chevtchik N, Pinto PC, Masereeuw R, Stamatialis D. Membranes for bioartificial kidney devices. World Scientific Series in Membrane Science and Technology: Biological and Biomimetic Applications. Energy and the Environment. 2018;2.
[214] van Gelder MK, Mihaila SM, Jansen J, Wester M, Verhaar MC, Joles JA, et al. From portable dialysis to a bioengineered kidney. Expert Review of Medical Devices. 2018;15(5):323–36.
[215] Zähringer M, Pattynama P, Talen A, Sapoval M. Drug-eluting stents in renal artery stenosis. European Radiology. 2008;18(4):678–82.
[216] Portis AJ, Sundaram CP. Diagnosis and initial management of kidney stones. American Family Physician. 2001;63(7):1329.
[217] Khan SR, Pearle MS, Robertson WG, Gambaro G, Canales BK, Doizi S, et al. Kidney stones. Nature Reviews Disease Primers. 2016;2(1):1–23.
[218] Singh I, Gupta N, Hemal A, Aron M, Seth A, Dogra P. Severely encrusted polyurethane ureteral stents: Management and analysis of potential risk factors. Urology. 2001;58 (4):526–31.
[219] Schmehl JM, Harder C, Wendel HP, Claussen CD, Tepe G. Silicon carbide coating of nitinol stents to increase antithrombogenic properties and reduce nickel release. Cardiovascular Revascularization Medicine. 2008;9(4):255–62.
[220] Zumstein V, Betschart P, Albrich WC, Buhmann MT, Ren Q, Schmid H-P, et al. Biofilm formation on ureteral stents-Incidence, clinical impact, and prevention. Swiss Medical Weekly. 2017;147:0506.
[221] Arkusz K, Pasik K, Halinski A, Halinski A. Surface analysis of ureteral stent before and after implantation in the bodies of child patients. Urolithiasis. 2021;49(1):83–92.
[222] Feneley RC, Hopley IB, Wells PN. Urinary catheters: History, current status, adverse events and research agenda. Journal of Medical Engineering & Technology. 2015;39(8):459–70.
[223] Maki DG, Tambyah PA. Engineering out the risk for infection with urinary catheters. Emerging Infectious Diseases. 2001;7(2):342.
[224] Csillag A. Atlas of the Sensory Organs: Functional and Clinical Anatomy: Springer Science & Business Media; 2007.
[225] Oyster CW. The Human Eye. Sunderland, MA: Sinauer; 1999.
[226] Khoo C, Chong J, Rajan U. A 3-year study on the effect of RGP contact lenses on myopic children. Singapore Medical Journal. 1999;40(4):230–37.

[227] Lim LS, Mitchell P, Seddon JM, Holz FG, Wong TY. Age-related macular degeneration. The Lancet. 2012;379(9827):1728–38.
[228] Senra H, Ali Z, Balaskas K, Aslam T. Psychological impact of anti-VEGF treatments for wet macular degeneration – a review. Graefe's Archive for Clinical and Experimental Ophthalmology. 2016;254(10):1873–80.
[229] Adrean SD, Chaili S, Ramkumar H, Pirouz A, Grant S. Consistent long-term therapy of neovascular age-related macular degeneration managed by 50 or more anti–VEGF injections using a treat-extend-stop protocol. Ophthalmology. 2018;125(7):1047–53.
[230] Kovach JL, Schwartz SG, Flynn HW, Scott IU. Anti-VEGF treatment strategies for wet AMD. Journal of Ophthalmology. 2012:2012.
[231] Jemni-Damer N, Guedan-Duran A, Fuentes-Andion M, Serrano-Bengoechea N, Alfageme-Lopez N, Armada-Maresca F, et al. Biotechnology and biomaterial-based therapeutic strategies for age-related macular degeneration. Part I: Biomaterials-based drug delivery devices. Frontiers in Bioengineering and Biotechnology. 2020;8:1257.
[232] Møller AR. Hearing: Anatomy, Physiology, and Disorders of the Auditory System: Plural Publishing; 2012.
[233] Burnett CH. The Ear: Its Anatomy, Physiology, and Diseases: HC Lea's son & Company; 1884.
[234] Jang CH, Cho YB, Yeo M, Lee H, Min EJ, Lee BH, et al. Regeneration of chronic tympanic membrane perforation using 3D collagen with topical umbilical cord serum. International Journal of Biological Macromolecules. 2013;62:232–40.
[235] Kim J, Kim SW, Choi SJ, Lim KT, Lee JB, Seonwoo H, et al. A healing method of tympanic membrane perforations using three-dimensional porous chitosan scaffolds. Tissue Engineering Part A. 2011;17(21–22):2763–72.
[236] Kim SW, Kim J, Seonwoo H, Jang K-J, Kim YJ, Lim HJ, et al. Latent progenitor cells as potential regulators for tympanic membrane regeneration. Scientific Reports. 2015;5(1):1–8.
[237] Shepherd RK, Hardie NA. Deafness-induced changes in the auditory pathway: Implications for cochlear implants. Audiology and Neurotology. 2001;6(6):305–18.
[238] Loizou PC. Introduction to cochlear implants. IEEE Engineering in Medicine and Biology Magazine. 1999;18(1):32–42.
[239] Guo R, Ma X, Liao M, Liu Y, Hu Y, Qian X, et al. Development and application of cochlear implant-based electric-acoustic stimulation of spiral ganglion neurons. ACS Biomaterials Science & Engineering. 2019;5(12):6735–41.

Chapter 2
Classification of biomaterials

2.1 Description of different classes of biomaterials

Biomaterials is a broad field, encompassing multiple kinds of materials and processing methods. Different types of biomaterials can be classified into three classes (natural, synthetic, and composite), and selection and design considerations of a biomaterial depend on the specific applications to the body. Biomaterials can be of natural (polysaccharides, e.g., cellulose, chitin, chitosan proteins, e.g., silk, collagen, and gelatin) or synthetic (metals, ceramic, and polymers) in nature. Composite biomaterials are combinations of different types of materials to perform or enhance specific properties that require specific application.

Derivation of natural biomaterials depends on isolating it from organic sources such as animals and plants. Synthetic biomaterials are derived from inorganic sources (metal ores and minerals) excluding polymers which are derived from crude oil or other hydrocarbons. Composite biomaterials combine two or more distinct materials using various syntheses or manufacturing processes.

Natural biomaterials are known for their biocompatibility, functionality, and mechanical properties. They can exhibit great functionality due to their chemistry as they are derived from organic sources. Natural biomaterials are heavily applied in tissue engineering as they are biocompatible and do not produce toxic by-products when they degrade, cells can also easily adhere on their surface and induce proliferation and integration to the local tissue. They can also be used in drug delivery systems where drugs can be added via covalent or electrostatic bonding. Natural biomaterials can also be cleared in the body without harming the kidneys or the liver.

Synthetic biomaterials, on the other hand, possess superb mechanical properties, corrosion resistance, and stability, making them attractive in the production of prosthesis, implants, and various medical devices. They can stay inside the body for an indefinite period without causing adverse effects such as inflammation, trauma, necrosis, or fibrosis. This is called bioinertness, meaning that a material does not induce any reaction on the body. Even if they degrade a synthetic biomaterial must not create toxicity to the body by releasing ions, gasses, or toxic leachable to the system.

Composites are known for possessing the best of both worlds, as we can combine different materials to harness their unique properties. Composite biomaterials are used to combine the properties of natural and synthetic materials to achieve multifunctionality. A material that can do more than one function is highly desirable in the field of biomedical devices and drug delivery. New generation of devices are equipped with drug delivery capabilities to enhance their performance and to complement their primary function. Whereas in drug delivery, composite materials are being used by functionalizing metallic nanoparticles as a platform not just to

https://doi.org/10.1515/9783110655377-002

deliver therapeutics but also to track their fate inside the body by using conventional imaging techniques such as x-ray or CT scans or are used in adjunct therapies such as hyperthermia.

Selecting the correct material for different applications to the body is crucial in developing an effective biomedical device, drug delivery system, or engineered tissues. In this chapter, different types, properties, and usages of biomaterials are discussed.

2.2 Natural biomaterials

To create materials out of the things we see in nature has always been innate to us humans. Since the early history of humans, we are using materials that we see in nature for various applications, examples of which are sinew from animals and fibers from plants. Nature provides a wide variety of materials that have various properties that could be considered valuable to the field of biomaterials. The field has always looked into nature for inspiration to create biomaterials that are biocompatible and biodegradable, and that have robust mechanical strength, with interesting chemical properties that could be used for functionalizing with various active compounds. The growing field of nature-inspired biomaterials had given us significant advancement in medicine, from creating tissue constructs out of the body, to delivery of drugs, nature has always been generous in providing us new avenue for discovery. The use of natural materials has clear advantages not just to the body but also to the environment as these materials are widely available and self-renewing, and do not cause pollutants as they degrade in less time compared to inorganic-based materials.

The following section describes what materials are being used in the field for various biomedical applications. These are different polysaccharides, polypeptides, polyphenols, and protein-based materials that had garnered significant attention from researchers from the past decades.

2.2.1 Polysaccharides used as biomaterials

Polysaccharides are long-chain polymers of repeating monosaccharide units linked by glycosidic bonds that are either branched or linear in structure [1]. They are mostly derived from terrestrial and aquatic plants with other examples coming from animal sources, e.g., crustaceans, arthropods, and sponges. Polysaccharides are used by plants, animals, and fungi as either components of their extracellular matrices for structural support or as a form of short-term energy storage. Polysaccharides are the largest component of biomass in the world, and it is estimated that more than 90% of the mass in nature is in the form of polysaccharides, making them one of the most widely researched materials in the field [2–5] (see Figure 2.1).

Figure 2.1: Common source of biologically derived polymers (polysaccharides) that are used as biomaterials for various applications and the chemical structures of these representative biologically derived polysaccharides.

2.2.1.1 Alginates

Alginates are derived from the cell wall of seaweed or other sea flora and have a repeating subunit of a linear copolymer with homopolymeric blocks of (1→4)-linked β-D-mannuronate (M) and α-L-guluronate (G) residues, respectively, covalently linked together in different sequences or blocks [6, 7]. This sequence can be homopolymeric blocks of consecutive G-residues (G-blocks), consecutive M-residues (M-blocks), or alternating M- and G-residues (MG-blocks), and the sequence depends on species to species and determines what kind of properties it will possess. Some alginates will yield strong gel materials while others will be weak and less viscous [8–10]. The color can also be affected by the sequence of the homopolymeric blocks and can yield white, to yellow, to brown alginate. Alginates are refined mostly in brown seaweed (Phaeophyceae) [11] such as *Laminaria hyperborea*, *Laminaria digitata*, *Laminaria japonica*, *Ascophyllum nodosum*, and *Macrocystis pyrifera*. They are harvested from natural occurring sources or farmed in large-scale aquaculture to sustainably source the material. Salts such as sodium, potassium, or calcium are added to yield sodium alginate ($NaC_6H_7O_6$), potassium alginate ($KC_6H_7O_6$), and calcium alginate ($C_{12}H_{14}CaO_{12}$). These final products are traditionally used in various food, pharmaceutic, and cosmetic applications as a natural thickener or emulsifying agent [3, 12, 13].

The most common form of alginate to be used in biomaterials is in the form of hydrogel. Hydrogels are three-dimensional polymeric network with high water content [14, 15]. Alginate is cross-linked via ionic bonding using a cross-linking agent, most typically calcium cations (Ca^{2+}). The cross-linking of alginate is rapid, and the gelation time is almost instantaneous. In order to control this gelation process, buffer solutions such as sodium hexametaphosphate are used to slow down the reaction [16]. The phosphate groups compete with the carboxylate groups of alginates, retarding the reaction to a slower phase. Using a different cation source can also lead to slow down the reaction. Gelation rate is crucial in controlling gel homogeneity and rheological properties; thus, a slower, well-controlled gelation must be made in order to maximize the properties of alginate. Alginate hydrogels are being researched for their use in tissue engineering, especially as a bioscaffold for pancreatic cells, or nephrotic cells as they can be engineered to be a shear-thinning material, where their viscosity is reduced when shear forces are applied to them, making them suitable for injectable applications and 3D bioprinting, where the material must pass through a narrow needle. It is ideal to make hydrogel formulations with alginate as it readily absorbs water at a high swelling ratio. Sensitive therapeutics such as proteins or growth factors can be easily co-formulated with alginate due to its nontoxic synthesis method and ease of gelation. Growth factors such as vascular endothelial growth factors (VEGF) can be entrapped in the hydrogel network of alginate [17–19] and slowly release it to induce angiogenesis in infarcted tissues. This in turn creates a more stable release kinetics to avoid complications of VEGF treatment and increase the half-life of the growth factor. The use of alginate hydrogel means that it can be injected locally to the target tissue and increase the therapeutic effect by avoiding loss in the systemic circulation. However, even though alginate has been proven to be biocompatible in vitro and in vivo, there is still an ongoing debate on the extent of biocompatibility of this material. It has been seen that the sequence of the polymeric block chains can affect the host's immune response [20, 21] and that high M content in alginates were immunogenic and approximately 10 times more potent in inducing cytokine production compared with high G-alginates. Others report that impurities such as heavy metals, endotoxins, proteins, and polyphenolic compounds could be found in some alginates extracted from natural sources [22–24]. Therefore, there should be a multistep purification process for this nature-derived material before being marketed or tested in humans.

Other derivatives of alginates are also being used in biomedical applications by adding alkyl chains or hydrophobic polymers to the alginate backbone to bring about amphiphilic alginate [25–27]. These create a wide variety of applications such as in micro- and nanoencapsulation of peptides, and small molecules for systemic drug delivery. Examples of these long-chain alkyl chains that are added to the alginate backbone are dodecyl and octadecyl moieties connected via ester bond formation [28]. This kind of modification extends the stability of alginate in aqueous media compared to those in pure alginate form, thus providing long-term circulation time in the body. Synthetic polymers can also be conjugated with alginate and the example is the grafting of poly(ε-caprolactone) (PCL) and poly(butyl-methacrylate) that are used in drug delivery applications [29–31]. Peptides can also be added to the molecular structure of alginate to increase its cell adhesion and proliferation properties [32–34]. These peptide–alginate derivatives are done by introducing the peptide as a side chain moiety to the alginate backbone using carboiimide chemistry [35, 36]. These couple the free carboxylic groups of alginates to the amine-terminated groups of the peptide. Pure alginate lacks the ability to increase the adhesion and proliferation rate of mammalian cells; therefore, adding these peptide side chains can increase the bioactivity properties of alginate and making it more suitable for tissue engineering applications. Examples of these peptides are arginine–glycine–aspartic acid [37, 38] (RGD) that acts as an adhesion ligand for the integrin receptors of mammalian cells. Other peptide sequences such as DGEA (Asp–Gly–Glu–Ala) and YIGSR (Tyr-Ile-Gly-Ser-Arg) [16, 39] that are derived from extracellular matrix (ECM) can be added using the same method, thus mimicking the ECM microenvironment of the body and increasing the cellular activity of the seeded cells in the peptide–alginate scaffold. Alginate hydrogels cannot be metabolized inside the body as humans lack the enzyme needed to degrade it (alginase). However, it can be dissolved to its oligomeric form by the release of divalent ions that cross-link the gel. But in order to clear alginate in the system via renal system, its molecular weight must be reduced [40–42], and this is done via partial oxidation of alginate with sodium periodate. The periodate oxidation can cleave the polymeric chain by altering the chain conformation of the carbon–carbon bond to an open-chain adduct. This leaves the alginate backbone weak from hydrolysis and therefore can be cleared from the body.

2.2.1.2 Agarose

Agarose is a seaweed-derived polysaccharide and linear polymer made up of repeating units of agarobiose, which is a disaccharide of D-galactose and 3,6-anhydro-L-galactopyranose [43]. It exhibits similar gelation properties of alginate, with high water uptake and hydrogen bond self-gelation with or without cross-linking agent. However, its mechanical properties are highly affected when seeded with cells and therefore attracts very little attention in tissue engineering applications [44]. This is due to the interference of cell proteins with the hydrogen bonding within the polymeric network of agarose. This can be circumvented by grafting other polymeric materials such as collagen, chitosan, or cellulose to increase cell attachment and handleability [45]. Agarose also exhibits higher viscosity at lower concentrations compared to alginate which may limit its applicability as a bioink or injectable delivery [46].

2.2.1.3 Carrageenan

Carrageenan is another seaweed-derived polysaccharide made up of alternating chains of α-1,3-D-galactose and (β-1,4) 3,6-anhydro-galactose with ester sulfates [47, 48]. There are three identified types called kappa-carrageenan, iota-carrageenan, and lambda-carrageenan. K-Carrageenan forms high mechanical strength and rigid gels [49–51] and is sourced from *Kappaphycus alvarezii.* I-Carrageenan can form soft gels in the presence of calcium ions [52], like how alginate can be cross-linked, and it is primarily sourced from *Eucheuma denticulatum*, while L-carrageenan cannot form gels and are mostly used only as a congealing agent for dairy products [53]. The differences of these types of carrageenan are mostly with the numbers of ester sulfate groups present in their polymeric structure which lowers the solubility temperature of the molecule. This in turn makes K-carrageenan able to form rigid gels because it has the greatest number of ester sulfate groups [54] compared to I- and L-carrageenans which have lesser ester sulfates in their respective polymeric chains. The applications of this natural polymer range from tissue engineering via 3D bioprinting to mucoadhesive to increase the retention rate of therapeutics in the stomach.

2.2.1.4 Fucoidan

Fucoidan is an anionic sulfated polysaccharide, also derived from algal and seaweed sources [55]. It is a branched polymer that contains repeating units of poly-α-(1→3)-fucopyranoside backbone with a high degree of branching, mainly α-(1→2)-fucopyranoside single units. Fucoidan is marketed as a dietary supplement and has been reported to have diverse biological properties from being antioxidant, anti-inflammatory, antiallergic, antitumor, antiobesity, anticoagulant, antiviral, antihepatopathy, antiuropathy, and antirenalpathy activities [56, 57]. Fucoidan has also been combined with chemotherapeutic agents and as a coating for gold nanoparticles (AuNP) as a drug delivery [58–61] vector for cancer treatment and is believed to possess indirect antimicrobial activity.

2.2.1.5 Cellulose

Cellulose ($C_6H_{10}O_5$) is a polysaccharide abundant in plant matter. It is a linear polymer chain of low to high molecular weight and is linked by β-D-glucose units [62, 63]. It is the most abundant natural occurring organic polymer in the world and is used by plants for mechanical support as it is the main constituent of walls of plant cells, keeping the plant upright and rigid. This rigidity comes from the extensive inter- and intramolecular hydrogen bonds between the molecules. Cellulose can be derived in almost all terrestrial plant matter and some algal species with varying content depending on the species of the plant. Extraction is done by mechanical processing of the hard plant material into small constituents (pulp) and then digesting the molecular polymeric chain using a strong alkali solution [64]. Further purification can be done with the use of bleaching agents and hydrolyzing acids [64]. Other sources of cellulose can be biosynthesized using bacterial cultures [65, 66]. *Acetobacter xylinum* is a gram-negative bacterium that can synthesize large amounts of high-grade cellulose organized in microbial fibrils. Using the carbon and nutrient sources from bacterial culture medium, it can then polymerize these organic contents and turn it to linear β-1,4-glucan chains. Cellulose is secreted by these bacteria and then extracted and used as an alternative source. The main difference of bacterial cellulose to plant-based cellulose is that it is free from other organic impurities such as lignin and hemicelluloses that could create biocompatibility issues, thereby making it attractive for biomedical use [67–69].

Pure cellulose is not that suitable for tissue engineering and drug delivery as it is too rigid, and the human body cannot degrade it enzymatically. However, cellulosic materials had been used in the medical industry as wound dressing and disposable materials (gauze and cotton swab) [70, 71] as they are highly absorbent and bioinert. Hence, cellulosic derivatives are being made to make them useful for other applications in biomedical engineering such as in antibacterial wound dressing and dialysis filters and even in long-term implantable materials [72, 73]. Most specifically, the use of semicrystalline nanocellulose has been widely researched upon [74, 75]. The attractiveness of nanocellulose comes from it being a natural nano-sized material with various physical, chemical, and biological properties. Nanocellulose has superb longitudinal modulus due to the combination of highly ordered crystalline domain and amorphous regions which affects its mechanical properties [76, 77]. Therefore, it is used as a reinforcement material for tissue engineering to strengthen mechanically weak but highly functional biomaterials. It is reported that Young's modulus ranges from 100 to 200 GPa, making it stronger than Kevlar. It can also increase the conductivity of the material due to its crystalline structure which can be highly favorable in nerve tissue engineering [78]. Its chemical properties are also being used as a drug delivery vector due to its receptive surface properties. Nanocellulose has glucose units that contain free hydroxyl bunches which can be conjugated with various amine, thiol, and catechol groups containing compounds. Its surface reactivity can also be utilized to capture proteins for biosensor applications [79, 80]. Finally, its biological properties can be like alginate, where it is generally regarded as biocompatible or bioinert due to the body's inability to degrade it, and must be used in a way that its concentration does not hamper normal physiological processes.

2.2.1.6 Pectin

Pectin is a natural heteropolysaccharide composed of (1, 4)-linked α-D-galacturonic acid residues and variety of neutral sugars such as rhamnose, galactose, and arabinose [81, 82]. It is the second most abundant organic substance in plant cell wall next to cellulose and has a similar gelation property. Its main sources are pulps and peel of fruits, e.g., citruses (orange, lemon, grapefruit, and lime) and apple pomace. Since it can also be made into a gel it is also being investigated for bioink and drug delivery applications [83–86]. When cross-linked with a divalent calcium ion, it can form a robust high elastic moduli gel suitable for mimicry of various human tissues. Gelation can also happen in the presence of acidic conditions, making it pH responsive and suitable for gastrointestinal applications. Like other biopolymers that are derived from plant materials, it has a low cell adhesion property making it less attractive for tissue engineering. However, it can be grafted onto other bioactive molecules, thereby increasing its bioactivity.

2.2.1.7 Starch

Starch ($C_6H_{10}O_5$) is a polysaccharide comprising glucose monomers joined in α-1,4-linkages [87]. The simplest form of starch is the linear polymer comprising amylose (30%) and amylopectin (70%). It is abundant in tubers and other root crops that hold a lot of water for the plant. It is hygroscopic, which means that it attracts water molecules enabling reversible hydration and creates a hydrogel. Starch is currently being used in heat-extruded 3D printing meant for edible products as it demonstrates ease in printability with high shape fidelity and resolution [88–90]. However, it does not yet translate to tissue engineering. Here, starch molecules can be grafted with other biopolymers or macromolecules to increase their applicability for more advanced biomedical applications such as tissue engineering and drug delivery [91]. It is easy to assume that the rheological flexibility of starch can be tuned to mimic different native tissues in the body, thereby making it an ideal 3D culture platform for different cell types to study differentiation and migration patterns on various topological cues.

2.2.1.8 Dextran

Dextran is a complex, branched and hydrophilic polysaccharide consisting of linear chains of α-1,6-linked D-glucopyranose residues and a result of enzymatic degradation of sucrose [92, 93]. It is produced by *Leuconostoc mesenteroides* and *Streptococcus mutans* bacterium which are lactic acid-producing bacterial species. Dextran can be found as a by-product in wine production and as a component of dental plaque that contributes to degradation of the enamel on teeth [94]. It is used as an antithrombotic agent in the clinic to reduce blood viscosity and increase the blood volume in cases of hypovolemia (reduced blood volume) [95, 96]. It has garnered attention as a drug delivery vehicle [97–99] due to its molecular structure that is rich in reactive hydroxyl groups and its superb biocompatibility toward the blood, which makes it an attractive material for various systemic deliveries of fast clearing and short half-lived drugs [100, 101]. Dextran can form nanoparticle complexes with various therapeutics such as growth factors, antibiotics, steroids, and increasing the pharmacokinetics and dynamics of the drug inside the systemic circulation of the body.

2.2.1.9 Pullulan

Pullulan is a linear water-soluble polysaccharide containing repeating units of maltotriose linked by α-1,6-glycosidic units. Its maltotriose units are linked by α-(1–4)-glycosidic bond, whereas successive maltotriose units are linked to each other by α-(1–6)-glycosidic linkages [102]. Pullulan is produced from starch by the fungus *Aureobasidium pullulans* that is abundant in marine environments. Pullulans' unique glycosidic linkages mean that it can be tailored to be either partially soluble or insoluble in water. It can be modified to become a pH-sensitive polymer by presenting reactive groups that can have different isoelectric points. Pullulan has multiple industrial applications such as a food additive, a blood plasma substitute, a flocculant, an adhesive, and a film [103]. It is one of strongest natural polymers and its mechanical properties make it similar in strength and elasticity as polystyrene.

As a biomaterial, it can be enzymatically degraded by pullulanase that hydrolyzes the α-(1–6)-glycosidic linkages and turning them to maltotriose units but not attacked by digestive enzymes of the human gut. It can form a film on the lumen of the gastrointestinal lining. This property is useful for increasing the retention rate of orally delivered drugs, as it can be released over a longer period. It can also form a strong adhesion to the mucosal lining of the body, e.g., buccal area, and it has been used as a buccal patch that can release drugs on a controlled manner [104, 105]. This solid patch and membrane made of pullulan has a strong resistance to both oxygen and water vapor penetration, thus protecting the underlying environment from factors that could degrade its drug payload. Pullulan can also be conjugated with chemotherapeutic drugs, e.g., adriamycin, 5-fluorouracil, or paclitaxel to form a nanoparticulate system that can target malignant cells in the body [106, 107]. Its biocompatibility exhibits nonimmunogenic, nontoxic, and inertness, and its degradation rate in human blood serum is relatively quicker compared to other polysaccharide biopolymers like dextran. Like other biopolymers, it can be used for tissue engineering applications as it possesses the properties required for a bioscaffold [108–110].

2.2.1.10 Chitin

Chitin is a polysaccharide chain of repeating units of *N*-acetylglucosamine which is a derivative of glucose. This linear polymer is present in the ECM of invertebrates, e.g., arthropods, mollusks, sponges, nematodes, and fungi. It gives protection and support to the body of these organisms as they lack internal structures to support their bodies that all vertebrates have [111]. Chitin can form fibers adopting three crystalline iterations, e.g., α-, β-, or γ-chitin. These fibers interact with other chitin-binding proteins to form higher order structures [112, 113]. An example of which is the structure called chitin laminae [113]. It is made up of two-dimensional sheets of α-chitin crystals with antiparallel chitin fibers in association with β-folded proteins. This structure of chitin is the primary constituent of arthropod exoskeleton and the fibers that support the extracellular matrices of sea sponges. β-Chitin and γ-chitin are found in harder biomaterials such as the squid beak and in cocoons of terrestrial insects, respectively. Chitin as a biomaterial has been extensively used in tissue engineering applications since it possesses biocompatibility and degradability [114–117]. Unlike other polysaccharide-based biomaterials, chitin can be degraded and eliminated inside the body by enzymes such as lysozyme that breaks down the linkage between the acetylated units and degrades chitosan to oligosaccharides. Chitosan also possesses antibacterial properties [118] as it can react with anions on the cell wall of a bacterium leading to the breakage of the bacterial cell wall causing its death. Chitosan can also bind to negatively charged peptides to form a complex. This is beneficial if the goal is to make a bioactive scaffold that can carry or release growth factors and other macromolecules. This property of chitosan can also be used for drug delivery applications as it can protect and prolong the stability of growth factors, peptides, and other bioactive compounds [119–122]. Chitosan can also be grafted onto other synthetic or natural polymers to make it into other products such as hydrogels [122], electrospun fibers [123], or nanoparticle formulations that have enhanced biocompatibility, bioactivity, and degradation.

2.2.1.11 Chitosan

Chitosan is a linear polysaccharide composed of random units of β-(1→4)-linked D-glucosamine (deacetylated unit) and *N*-acetyl-D-glucosamine (acetylated unit). It is derived from chitin of crustacean origins after treating it with alkaline solution such as sodium hydroxide and it is the second most abundant polysaccharide in the world after cellulose [124–126]. Chitosan possesses similar biocompatibility and degradability as chitin without the immunogenic effects. It can also bind with other anionic molecules to create complexes, thus making it a suitable chelating agent for cholesterol, fats, proteins, and metal ions [126, 127]. The cationic charge of chitosan can also interact with the negatively charged parts of the cell membrane, which can serve as an adhesion substrate for tissue engineering applications [128–130]. However, using pure chitosan as a biomaterial for tissue engineering is difficult as it has poor mechanical properties. In order to create a usable scaffold, it is usually blended with other natural or synthetic polymers to enhance its mechanical properties. The cell membrane reactivity of chitosan has also been explored to enhance permeation of oral formulated peptides as it can open the tight junction proteins of epithelial cells in the gastrointestinal tract [131–133]. Knowing this, nanoparticle formulation of chitosan conjugated with insulin, octreotide, and GLP-1 [134–137] is being explored to enhance the rate of absorption by increasing their retention time on the surface of the epithelial and mucosal lining of the gastrointestinal tract. Like chitin, chitosan can also increase the stability of sensitive bioactive compounds such as growth factors, or peptides making it an attractive material for drug delivery applications.

2.2.2 Polypeptides used as biomaterials

Polypeptides are amino acid-based materials which have primary, secondary, tertiary, and quaternary structures. Examples of which are hyaluronic acid, collagen, gelatin, and silk fibroin. These classes of materials are specially relevant to biological applications due to their superb functionality, high biocompatibility, and ease of degradation. They are widely accepted to be biocompatible as they are naturally present in biological systems, including humans. They are being mostly used for tissue-engineered applications as they possess cytocompatibility and can enhance cell proliferation both in vitro and in vivo (Figure 2.2).

2.2.2.1 Hyaluronic acid

Hyaluronic acid or hyaluronan (HA) is a heteropolymer of repeating β-1,4-D-glucuronic acid and β-1,3-*N*-acetyl-D-glucosamine units [138–140]. It can be found in the ECM of connective, epithelial, and neuronal tissues and being most abundantly in tissues subjected to mechanical loads, such as cartilages, dermis, and vocal folds. HA can be grouped into two classes: class I is a single-domain integral membrane protein and class II is a two-domain soluble/membrane-anchored protein [140]. Class I HA is ubiquitous in the body while class II can only be found in *Pasteurella multocida* bacterial species. Polymers of hyaluronic acid can range from 5.0 to 20,000 kDa in vivo and are synthesized in the body by a class of integral membrane proteins called HA synthase. These synthases are called HAS1, HAS2, and HAS3, and are responsible for the polymerization of HA via repeated addition of D-glucuronic acid and *N*-acetyl-D-glucosamine to the nascent polysaccharide as it is extruded via ABC transporter through the cell membrane into the extracellular space. It is currently being used in cosmetics and dermatological products to promote healing by increasing collagen production on the skin [141–143]. Whereas its more advanced applications are in cell tissue engineering as a bioscaffold [144–146] as it can mediate various cellular

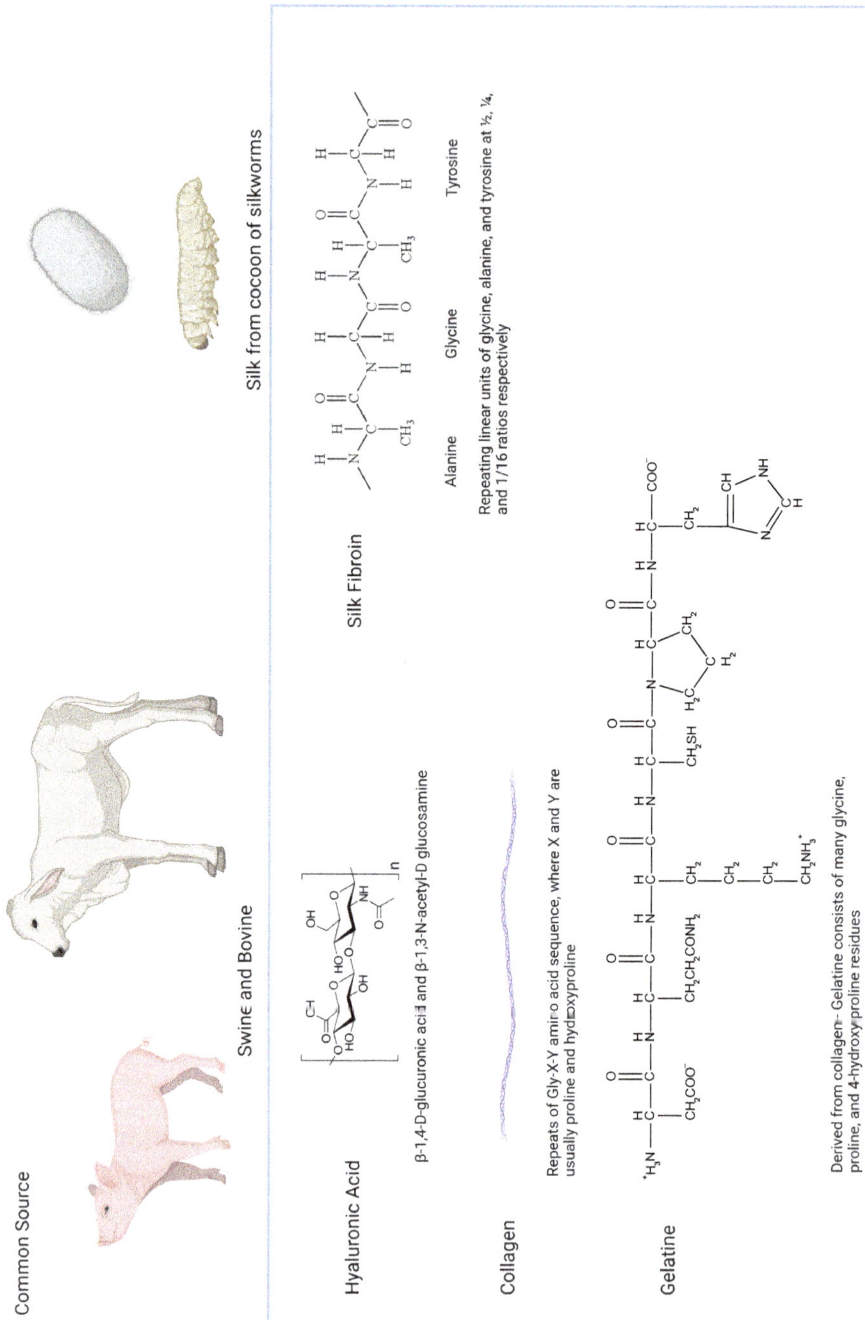

Figure 2.2: Common source of biologically derived polymers (polypeptides) that are used as biomaterials for various applications and the chemical structures of these representative biologically derived polypeptides.

responses such as cell division and migration, angiogenesis, wound healing, and tissue regeneration.

HA is highly attractive for tissue engineering because of its superb biocompatibility, biodegradability, and functionalization capabilities compared to other plant-based polymeric materials. Since HA is a component of the mammalian ECM, it can be tolerated and degraded by the human body at high concentrations without the worry of severe inflammatory response. HA is degraded in the body by the hyaluronidase enzyme family and can be eliminated after 3–5 days. Right now, extraction of HA is from animal sources or by microbial cultures [140, 147]. There are multiple ways to prepare HA as a bioscaffold [148], and it can be made into a hydrogel [149], sponge [150], or fibrous material [151, 152]. All these iterations mimic the topographical environment of ECM and are highly suitable for further increasing proliferation and migration rate of seeded cells. However, HA has very weak mechanical properties and should be chemically cross-linked to improve its handleability during processing into a scaffold [153, 154]. The sites for modification are the carboxyl groups, hydroxyl group, and $-NHCOCH_3$. Using carbodiimides or carbonyldiimidazole, the carboxyl group will be covalently modified to form amide bonds, while the hydroxyl group can be modified to ether formation, ester formation, hemiacetal formation, and oxidation, and $-NHCOCH_3$ is modified by deacetylation, amidation, hemiacetylation, and hemiacetal formation. HA can be processed using multiple fabrication techniques to form porous structures. It can be electrospun to produce fibers [155] at nano- to micro-scales or use it as a bioink in 3D bioprinting [156]. HA can also be made into a pure hydrogel scaffold and freeze drying it to sublimate the water content thus leaving a highly porous structure suited for cell proliferation [157, 158]. Current use of HA bioscaffold is in the field of cartilage repair [159], where chondrocytes (collagen-forming cells) are seeded on an HA scaffold [160]. Due to the highly biocompatible environment that HA provides, these chondrocytes secrete type II collagen on a much higher concentration compared to other natural polymeric materials such as alginate or cellulose, thus yielding a higher-quality-engineered cartilage.

2.2.2.2 Collagen

Collagen is a ubiquitous polypeptide chain containing repeats of Gly-X-Y amino acid sequence, where X and Y are usually proline and hydroxyproline, respectively [161]. All types of collagen have similar triple-helix polypeptide chains and variations highly related to their structure, where the differences are restricted to the length of the helix and the number and nature of the carbohydrates attached to the triple helix. Collagen types I, II, III, V, and XI are arranged in primary (amino acids), secondary (peptide chain), tertiary (helix structure), and quaternary (microfibril structures). These microfibrils give rise to structural support on the ECM of an organism and are abundant in fibrous tissues. Collagen type I is found in the skin, tendons, and blood vessels while type II and III are abundant in cartilage tissues and in the lining of blood vessels, respectively. Collagen is extracted from animal sources and there are three major ways to harvest it, and these are with neutral salt, acid isolation, and enzymatic solution. It can also be extracted using pepsin digestion [162] to depolymerize the peptide chains and lead to smaller molecular weight solutions. Extracted collagen have poor thermal stability and mechanical strength due to the denaturization of its assembly, it must be kept in well controlled environment or cross-linked to avoid further destabilization. Cross-linking it makes more biocompatible by reducing its antigenic properties. Animal-sourced collagen is known to induce immune response and this strategy helps in reducing the over reaction of the host's adaptive immunity.

Collagen's various types and abundance give it a wide range of application and make it one of the most studied in the fields of biomaterials. Collagen is known to increase cell activity by providing attachment points and mimicking the natural ECM environment of the human tissue making it an attractive material for various tissue engineering applications [163–165]. It can be manufactured to become various biomaterials; however, adding synthetic polymers makes it easier for other fabrication methods such as 3D printing, electrospinning, and sponge manufacturing. A wide variety of synthetic polymers (e.g., poly-ethylglycolic acid, polyurethane (PU), and PCL) can be incorporated with collagen to increase their biocompatibility and cell activity. Collagen is also an attractive material for adding in bioceramics such as hydroxyapatite (HA) for bone tissue engineering to accelerate bone regeneration in cases of defects or injuries [166, 167]. Growth factors such as VEGF can be easily encapsulated in collagen matrixes for improved revascularization of ischemic tissues. Collagen has also found its way in ophthalmic applications [168] as an eye implant and for sustained release of drugs in the subretinal space. This reduces the need for frequent eye injections that can be burdensome for retinopathy patients. Collagen is also used in other drug delivery platforms such as in nanoparticle and microparticle preparations for gene and peptide delivery [169–171].

2.2.2.3 Gelatin

Gelatin is a peptide chain polymer derived through enzymatic or thermal degradation of the helical structure of collagen [172]. Gelatin consists of many glycine, proline, and 4-hydroxy proline residues with a shorter linear chain compared to collagen. The type of gelatin obtained (type A and type B) is governed by the process of collagen hydrolysis (i.e., acidic and basic hydrolyses). Given that gelatin is derived from collagen, it has the same biological properties but with higher biocompatibility by exhibiting lower antigenicity. Type A and B gelatin are used heavily in the pharmaceutical industry for capsule manufacturing, and as a food gelling agent [173]. It is thermal sensitive and exists as a single molecule at temperatures above 35 °C as it cannot form interchain hydrogen bonds. At lower temperatures, it congeals and forms a gel with a relatively soft modulus. Gelatin is clear, flexible, and oxygen permeable, making it suitable for various biomedical applications. Gelatin is also a major component of ECM and permits the activity of cells by providing anchorage points for cell adhesion proteins of seeded cells, thus receiving increased attention in recent years in the field of tissue engineering, especially as a bioink in 3D bioprinting [174–176]. Gelatin is cross-linked with a methacryloyl to form a hydride construct called GelMA bioink [175]. The methacryloyl in the polymeric chain increases the thermal stability of gelatin, allowing it to exist as a gel even at temperatures higher than 35 °C, which is vital to ensure that the structural integrity of the printed bioscaffold remains even after implantation in the body. GelMA can be cross-linked under photoinitiated radical polymerization, meaning that UV light can be used to cross-link the hydrogel constructs made from it. This is vital to create a 3D structure in hopes of creating the intricate structures of the body. However, this requires the use of a photo-cross-linker and could cause cell toxicity if residues are left unreacted in the scaffold. Also, UV light is known to affect the cells by giving ionizing radiation that could damage the intracellular proteins and the DNA inside the nucleus. Therefore, a crucial element of using a low-intensity UV light source with highly reactive photo-cross-linker is needed. It is very important to mention that the chemical modification of gelatin by methacryloyl only uses about 5% of its total amino acid residues, thereby most of the functional moieties of the peptide chain are left unaffected. These sequences such as RGD- and matrix metalloproteinase (MMP)-degradable motifs can still retain the cell adhesion properties of gelatin. Most importantly, the degradability of GelMA is also maintained because these motifs are conserved; therefore, enzymes such as type I and II collagenases also known as MMP-1 and MMP-8, respectively, can still cleave the peptide chains of GelMA.

Gelatin can also be made into a drug delivery vehicle such as in nanoparticle formulation [177, 178]. It is widely used as a drug or gene carrier for various treatment applications (e.g., cancer treatment, immunotherapy, gene therapy, vessel pathologies, and various kinds of viral and bacterial infections). Due to gelatin's biocompatibility and biodegradability, coating quantum dots or other inorganic nanoparticles with it can also reduce their cytotoxicity [179, 180]. The preparation method of gelatin as a nanoparticle is done using nanoprecipitation, a nontoxic and nondegrading method that could keep the active agents safe and effective. The release kinetics of active agents from gelatin is mediated by either simple diffusion, erosion of gelatin, or by enzymatic cleavage of its matrix. Targeting ligands are also easily attached to the gelatin matrix, providing tissue-specific release of therapeutics. Lastly, gelatin can also be made to deliver cells for Parkinson's treatment [181, 182]. Dopaminergic progenitor cells are encapsulated in gelatin microspheres to provide protection from autonomic immune response and by providing initial proliferation anchorage points for the cells to engraft.

2.2.2.4 Keratin

Keratin is a fibrous protein that is complimentary to collagen, having both alpha (α) and beta (β) structures that have similar functions and amino acid sequences [183, 184]. Keratin is the main component of hair, scales, hooves, and horns of animals. Research of the structural proteins of keratin had classified it into two distinct groups based on their structure, function, and regulation. "Hard" keratins form ordered arrays of intermediate filaments embedded in a matrix of cysteine-rich proteins and contribute to the tough structure of epidermal structures (e.g., hair, horn, and scales), while "soft" keratins preferentially form loosely packed bundles of cytoplasmic intermediate filaments that give mechanical resilience to epithelial cells. Keratin as a biomaterial has been studied for its potential in tissue engineering as it possesses cell-binding motifs, such as leucine–aspartic acid–valine and glutamic acid–aspartic acid–serine binding residues, which are capable of supporting cellular attachment and proliferation [185, 186]. However, pure keratin possesses low elastic moduli making it brittle and uneasy to handle, and blending it with other natural or synthetic polymers can give rise to more elastic keratin that could be used in various biomedical applications [187].

2.2.2.5 Silk fibroin

Silk is an amino acid-based polymer made up of repeating linear units of glycine, alanine, and tyrosine at ½, ¼, and 1/16 ratios, respectively, with trace amounts of other amino acids at 3/16 ratio [188]. Different kinds of silk are produced by several Lepidoptera (moths and butterflies) and all arachnids (spiders) and are classified also as keratin but from invertebrate origin. They are used for cocoon formation, web building, egg protection, and transportation. The most known silk sources are from silkworms (*Bombyx mori*) and the golden orb spiders (*Nephila*). For centuries, silk from these organisms has been harvested and used for intricate tapestry and clothing. Known for their resiliency and lightness, it is one of the strongest materials known to man that are comparable to other materials such as steel or Kevlar. Since it is biocompatible and biodegradable, it has garnered a lot of attention in biomedical research as a material for tissue engineering [189, 190], drug delivery [191], and flexible electronics [192]. The active sites of silk polymeric chain can be incorporated with therapeutics and prolong its activity to reduce the need for cold storage. It can also be manufactured into other drug delivery platforms such as a microneedle patch [193] since it has a superb mechanical property when fully dehydrated. It has also demonstrated good performance as a bioscaffold for cells, and its ability to ensure initial attachment can increase the proliferation and migration of nascent cells. In order to make silk a biomaterial, it must be made into an aqueous solution. This is done by processing raw silk from silkworms and spiders by boiling it in alkaline solution (distilled water with NaOH) to remove the sericin in the silk. This process is called degumming and is the precursor of other silk-based materials such as silk cord, or non-woven mats made through electrospinning. The de-gummed silk fiber is further treated with lithium bromide salts and packed in a dialysis bag to remove residual short amino acid chains. This will produce an aqueous silk solution that can be made into various biomaterials such as sponges, hydrogels, microneedles, films, and 2D nanofibers.

Table 2.1: Summary of polysaccharides and polypeptides and their applications as a biomaterial.

Natural polymers	Source	Applications	Ref.
Alginate	Plant polysaccharide – brown and red algae	Wound dressing, tissue engineering, drug delivery	[15]
Agarose	Plant polysaccharide – red seaweed	Bacterial culture media, tissue engineering	[44]
Carrageenan	Plant polysaccharide – red seaweed	Mucoadhesive, tissue engineering	[49]
Fucoidan	Plant polysaccharide – brown seaweed	Wound dressing, drug delivery, tissue engineering	[55, 58, 60]
Cellulose	Plant polysaccharide – all terrestrial plants	Filters, wound dressing, tissue engineering, drug delivery	[73, 194]
Pectin	Plant polysaccharide – all terrestrial plants	Tissue engineering	[83]
Starch	Plant polysaccharide – tubers, root crops	Tissue engineering	[91]
Dextran	Bacterial polysaccharide	Drug delivery	[96]
Pullulan	Fungal polysaccharide	Tissue engineering, drug delivery	[107, 108]
Silk	Animal polypeptide – moths and spiders	Flexible electronics, drug delivery, tissue engineering	[189, 191, 192, 195]
Keratin	Animal polypeptide – horn, hair, scales, hoof of vertebrates	Tissue engineering	[187]
Chitin	Animal polysaccharide – arthropods, mollusks, sponges	Tissue engineering, drug delivery	[124]
Chitosan	Animal polysaccharide – crustaceans	Tissue engineering, drug delivery	[124]
Hyaluronic acid	Animal polypeptide extracted from poultry	Cosmetics, tissue engineering	[148, 149]
Collagen	Animal polypeptide extracted from bovine	Cosmetics, tissue engineering	[170, 196, 197]
Gelatin	Animal polypeptide extracted from bovine and porcine	Wound dressing, tissue engineering, drug delivery	[172]

2.2.3 Phenolics used as biomaterials

Phenolic compounds are a group of small molecules that are characterized to have at least one or more hydroxyl groups bonded directly to an aromatic hydrocarbon [198, 199]. There are different subgroups of phenolic compounds depending on their structures. These are phenolic acids, flavonoids, tannins, coumarins, lignans, quinones, stilbenes, and curcuminoids. These phenolic compounds are abundant in plant material mostly found in the intracellular endoplasmic reticulum of plants and stored in vacuoles. They are present in fruits and vegetables and are proven to have antioxidant, anti-inflammatory, and antibacterial properties which are beneficial in reducing risks of cancer and combating chronic and infectious diseases. Phenolics can be easily functionalized with other macromolecules or added to polymeric materials, that is why they are being investigated in various biomedical applications, especially in nanomedicine, biosensors, and tissue engineering (Figure 2.3).

2.2.3.1 Caffeic acid

Caffeic acid or hydroxycinnamic acid contains both phenolic and acrylic functional groups and is an intermediate in lignin synthesis in all plants. It has a strong antioxidant property but may exhibit pro-oxidant effects at large doses [200]. It is abundant in all plant matter and can be extracted at large quantities making it highly accessible. It forms a yellowish powder in pure form and is hydrophobic at neutral pH. It has been applied to various applications in drug delivery [201, 202] and as a bioactive coating for metallic implants [203, 204]. Its hydroxyl group can be conjugated to other molecules such as thiolates targeting ligands, boronic acid containing chemotherapeutics, or other amine-terminated compounds (e.g., peptides). It can also be made as a coating to iron nanoparticles, AgNPs, and AuNPs for biosensor or drug delivery applications [201, 202, 205, 206].

2.2.3.2 Curcumin

Curcumin (also known as diferuloylmethane) is a yellowish phenolic compound that is abundant in turmeric (yellow ginger) and is recognized to have variety of health benefits [207]. It is added in wound dressing materials to decrease inflammation as a result of inhibiting cyclooxygenase enzymes and protect the wound from bacterial infection [208, 209]. It can also salvage free radicals in the cells keeping it from oxidative stresses [210, 211]. In some studies, curcumin has been encapsulated in nanoparticles as the active ingredient to destroy malignant cells due to its ability to inhibit nuclear factor κB [212] that has a direct effect as an antineoplastic that could disrupt the tumors' ability to salvage damaged proteins. Nuclear factor κB is a pro-inflammatory transcription factor that modulates the expression of different proteins – such as cytokines, interleukin (IL)-1, IL-2, and interferon-γ (IFNγ) – involved in multiple cell signaling pathways associated with cancer progression and

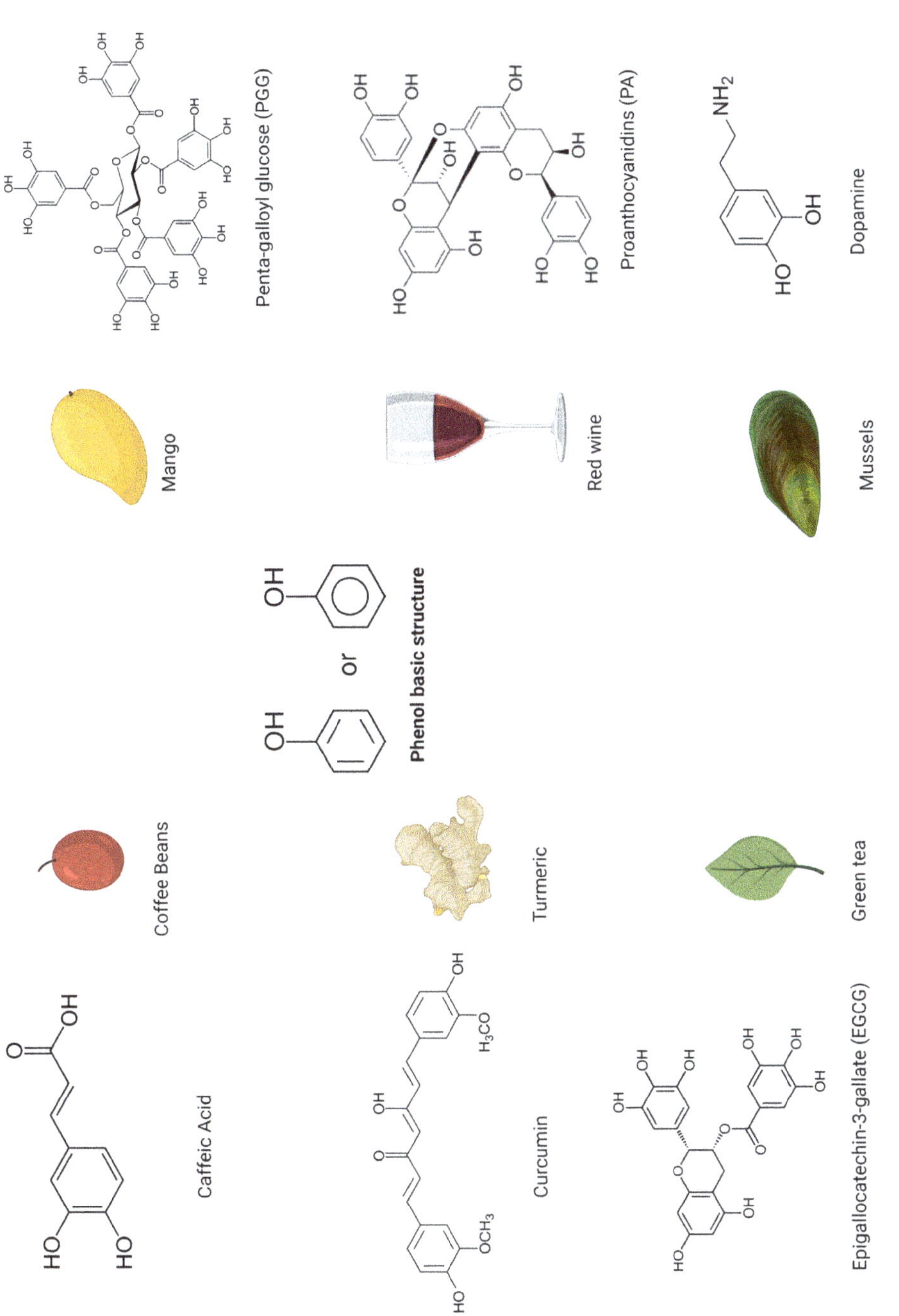

Figure 2.3: Common source of biologically derived phenolics that are used as biomaterials for various applications and the chemical structures of these representative biologically derived phenolic molecules.

inflammation. Due to the various cell signaling effects of curcumin, it can be applied to treat diseases such as rheumatoid [213] or osteoarthritis [214] to reduce pain and inflammation on the affected joints. Curcumin formulated into a nanoparticle has been reported to reduce the said symptoms of pain and inflammation with added mitigation of cartilage erosion in the joint. Curcumin has also shown to be involved in forming new and remodeling of damaged tissues [215] as it can regenerate not just cartilage but also fibroblast cells.

2.2.3.3 Epigallocatechin-3-gallate (EGCG)

Epigallocatechin-3-gallate (EGCG) is a polyphenolic compound identified as an active ingredient in green tea, accounting for 18 ~ 36% of total phenolics and 70% of total catechins [216]. It also exhibits strong antioxidant effects by salvaging reactive oxygen species in the intracellular environment, thus preventing damages in the DNA, and this reduces the risks for development of mutations that could lead to cancer. In high doses, EGCG can also enhance the effect of chemotherapeutic drugs that are used for colon cancer therapy. New nanoparticle formulations encapsulate EGCG as its main anticancer agent [217, 218]. Its anticancer properties come from its ability to halt key regulatory protein pathways such as the P13K–Akt pathway, which is an important pathway for the cell cycle, and this in turn slows down the proliferation of malignant cells in the tumor. This pathway inhibition can also reduce fibrotic tissue formation in the liver by downregulating the expression of genes involved in liver inflammation, thereby halting the progression of diseases such as hepatitis. EGCG can also increase collagen production in dentin, which is to provide enough anchorage for dental implants [219, 220]. Furthermore, since it can stimulate regenerative properties of collagen, it is also applied to arthritis treatment to increase cartilage production and prevent it from further degradation [221].

2.2.3.4 Penta-galloyl glucose (PGG)

Penta-galloyl glucose (PGG) is a polyphenol that is composed of a D-glucose molecule with five gallic acid molecules (3,4,5-trihydroxybenzoic acid). This polyphenol is one of the most potent antioxidants in the tannin group and is known for various medicinal properties such as antiviral, antibacterial, antidiabetic, anti-inflammatory, and anticancer properties [222, 223]. Compared to EGCG and caffeic acid, this polyphenol does not have pro-oxidant properties which make it attractive in nanoparticle formulation for cancer treatment. It has antiangiogenesis effect to malignant tumor [222] that could suppress the tumor's ability to metastasize and recruit nascent vessel structure. This in turn makes the tumor starve and reduce in size overtime. PGG has gallic acid functionality which has high affinity to proteins with high proline content such as collagen and elastin, which means that it can be used to target tissues such as cartilages and the liver. Also due to its high affinity to ECM components (e.g., elastin), PGG has been found to stabilize engineered heart valves [224] by increasing the valve's affinity to cell infiltration and matrix remodeling in vivo. The valves treated with PGG also exhibited improvement in flexibility in vivo compared to untreated valves, which calcified after implantation. PGG is also an attractive additive to decellularized blood vessels for vascular engraftment applications [225, 226] as it can preserve the elastic laminar integrity and prevent aneurysm formation on the graft. This is still because of the recruitment of elastin to the PGG-treated vessels, thereby making it structurally sound. Furthermore, elastin has the tendency to degrade enzymatically or calcify on vascular grafts, making it rigid and unable to adapt when there is an increase in blood pressure. PGG can increase elastin's stability by binding to it and protecting it from enzymatic degradation and calcification. It is reported that PGG can reduce susceptibility of grafts to enzymatic degradation by elastase by almost 50% [226–228].

2.2.3.5 Proanthocyanidins (PA)

Proanthocyanidins (PA) or tannins are natural polyphenolic compounds that are classified as a flavonoid [229]. They are composed of flavan-3-ol subunits and like other phenolics they have been proven to have medicinal effects such as antioxidant, anti-inflammatory, antimicrobial, antihistamine, and pain enzyme inhibition. It is highly abundant in grapes and cocoa seeds, which means it can also be found in food products such as chocolate and wine. PA could cross-link to collagen matrices making them stable from enzymatic degradation [230, 231]. Like PGG, PA exhibit high affinity to proline-rich proteins such as elastin and collagen. This affinity is translated to tissue engineering as they can delay the degradation of collagen made bioscaffolds from collagenases, while at the same time stimulate the production of nascent collagen by enhancing the activity of proline hydroxylase enzyme. Proline hydroxylation is an important step in the synthesis of collagen. More specifically, PA has been applied to dental applications to increase soft tissue integrity in the dentin area by rapidly stabilizing the demineralized dentin layer [231, 232].

2.2.3.6 Polydopamine (PD)

Polydopamine is the polymerized form of the dopamine which is a catecholamine that is found in animals. In humans, it is known for being a neurotransmitter that regulates voluntary movements and emotions [233, 234]. In other animals such as in mollusks, it is used as a material to adhere on the surface of rocks or other substrates. In recent years, it has been used as a biomaterial as it possesses functional groups that could be used for conjugating various amine, thiol, and boronic acid terminated compounds. Once, polymerized it can form a strong bond onto a substrate making it an attractive coating for metallic implants and inorganic nanoparticles [202, 235–237]. Its catechol group can reversibly form a complex with boronic acid, or other amine-terminated compounds which are present in some chemotherapeutic drugs and peptides. The reversibility of this complexation means that it can be used as a pH-responsive drug delivery platform [238–240]. The complexation happens when PD is deprotonated by an alkaline pH (above 8.0) and then the moieties such as boronic esters and amine or thiolate groups can form a covalent bonding to the hydroxyl group of PD. This reaction can then be reversed once the complexed motifs are exposed to acidic pH (below 5.0) which are present in the gastric mucosa or in a tumoral microenvironment. Examples of application of this are when PD has been electropolymerized onto a nitinol duodenal stent to form a strong coating, this served as a drug delivery platform for bortezomib (boronic acid containing chemotherapeutic) [239, 241]. The pH of the duodenum changes if a person is at a fasting or feed state, and after eating, the acidic gastric contents can trigger the protonation of bortezomib from the PD coating, thereby releasing locally directly to the tumor or absorbed in the intestinal epithelium. Nanoparticle systems are also a good platform for PD as they can be circulated in the body and functionalized with targeting ligands such as flourescein isothiocyanate albumin, or other thiolate-containing cell tracker molecules [202]. Once absorbed by the malignant cells, the tumor microenvironment can dissociate the complexation between the drug payload and PD, further reducing the side effects of the drug by allowing it to be absorbed only in the malignant tissues of the body. PD has high chemical activity, adhesion, hydrophilicity, hygroscopicity, stability, biocompatibility, and other properties that are applicable in tissue engineering. PD can increase the attachment of cells as it reacts with the adhesion molecules that the cells secrete during the initial phase of attachment in vitro. It is often used in the surface modification of bioscaffolds, especially in the peripheral nerve, articular tissue, and bone tissue regeneration as well as in stem cell differentiation [242].

PD's strong binding strength to substrates such as graphene, metals, and other reactive surfaces such as quartz crystals makes it a versatile material for biosensor applications [243–245]. Biosensors are devices that combine both biological components with a physicochemical detector. It can measure the presence of certain biochemical molecule with changes in electrical signal (e.g., impedance) when a certain biological reaction takes place. This could be from antibody–antigen interaction, protein binding, enzymatic interactions, or with cell or bacterium attachments. PD is used in this application as a highly sensitive substrate using its catechol motif as the physicochemical part of the sensor. PD can sense glucose levels, proteins, superoxide anions, and other free radicals. It can also serve as an anchorage point for certain antibodies to detect the presence of antigens for diagnostic purposes [246, 247].

Table 2.2: Summary of phenolic molecules and their applications as a biomaterial.

Phenolic name	Source	Applications	Ref.
Caffeic acid (CA)	Plant material – fruits and vegetables	Bioactive coating, drug delivery	[200, 201, 204]
Curcumin (CU)	Plant material – yellow ginger (turmeric)	Wound dressing, drug delivery	[212, 248]
Epigallocatechin-3-gallate (EGCG)	Plant material – green tea	Drug delivery	[217]
Penta-galloyl glucose (PGG)	Plant material – tea and coffee	Graft stabilizer, drug delivery	[222]
Proanthocyanidins (PA)	Plant material – cocoa beans and grape seeds	Tissue engineering	[229]
Polydopamine (PD)	Animal phenol (catecholamine) mollusks	Drug delivery, biosensor, bioactive coating	[235, 239, 243, 244, 249]

2.3 Synthetic biomaterials

Synthetic biomaterials are classified as metals, ceramics, and biodegradable and nonbiodegradable polymers. The development of these classes of material has a rich history. The first use of synthetic material to repair a body part has been recorded since ancient times, when ancient Egyptians are known to use gold to replace and fill the cavities in their teeth. But prior to the discovery of the germ theory and the subsequent development of aseptic techniques, the use of synthetic materials to repair the body has been limited. The discovery and development of synthetic biomaterials had significantly improved the field of medicine, as it broadens the catalog of materials that can be used in the body. The use of produced synthetic materials made it possible for biomedical engineers, scientists, and clinicians alike, to create devices, prosthesis, or drug delivery platforms with a wide variety of properties and reliability and not just relying on naturally occurring biocompatible means. The innate mechanical stability, bioinertness, and production reliability make synthetic biomaterials a staple in the field of biomedical engineering and medicine. This part of the chapter discusses the different kinds of synthetic biomaterials and their current use in the field of biomedicine and biomaterial science.

2.3.1 Synthetic polymers (nonbiodegradable)

Nonbiodegradable polymers are used as a biomaterial primarily in applications that require long-term use with vital properties such as mechanical and chemical stabilities

[250]. Nonbiodegradable means that it is highly resilient to enzymatic, chemical, and hydrolytic degradation inside the body. These materials can last for years and even decades without changing significantly in chemical properties and mechanical strength. Knowing this, nonbiodegradable polymers are commonly used in vascular grafts [251, 252], polymeric dental fillers [253], components of hip and knee prosthesis [250, 254], artificial heart valves [250], components of vascular and nonvascular stents, catheters, fillers and cosmetic implants, birth control implants [255], implantable meshes [256], and sutures. Polymers that are used in these applications are known to be stable when interfaced with biological tissues, which means that they are not making toxic leachable, do not react with proteins or enzymes, and do not induce inflammatory or immune response in the body. Because the polymeric chains of these synthetic polymers are very stable and are made with repeating units of nonreactive groups such as ethylene, methylene, propylene, acrylates, urethanes, and ethylene oxides to name a few.

The underlying property of these polymers is their ability to be tolerated without causing adverse effects with prolonged or repeated use inside or outside the body. Also, most of these polymers are resilient to mechanical stresses such as shear, elongation, and compression forces. Their long polymeric chains allow them to be able to withstand all applied forces inside the body without breaking or deforming making them reliable materials that could last inside the body for indefinite periods. Lastly, the polymeric structure of these materials allows them to be sterilized with conventional sterilization techniques such as gamma and UV irradiation, or gas sterilization without changing their chemical structure that could affect their physical properties. The following will discuss the different types of nondegradable polymers and their uses as a biomaterial (Figure 2.4).

2.3.1.1 Polyurethane (PU)

PU is a long-chain, high-molecular-weight polymer that is made by reacting polyol and isocyanates to form urethane links. PU is a versatile polymer and can be modified to exhibit various properties [257]. Changing the ratio of the reactants can determine if the yielded polymer will be flexible or rigid. Adding other chain extenders (e.g., diols and diamines) can also vary the behavior of PU. It could be used as an adhesive, elastomer, or foam depending on the extender or cross-linker used in the polymerization reaction. PU has found its way in a wide range of application in biomedicine due to its structural phases that exhibit both hard and soft segments, while the hard segment gives mechanical stability and the soft segment provides elastomeric properties. This unique mechanical property made PU as one of the main materials used in vascular grafts [258, 259], stent coatings [260], and catheters [261]. Its flexibility is renowned, and its superb elasticity and resistance to plastic deformation makes it an attractive polymer for applications that require elasticity requirements. An example of which is in vascular graft applications where PU exhibits tensile

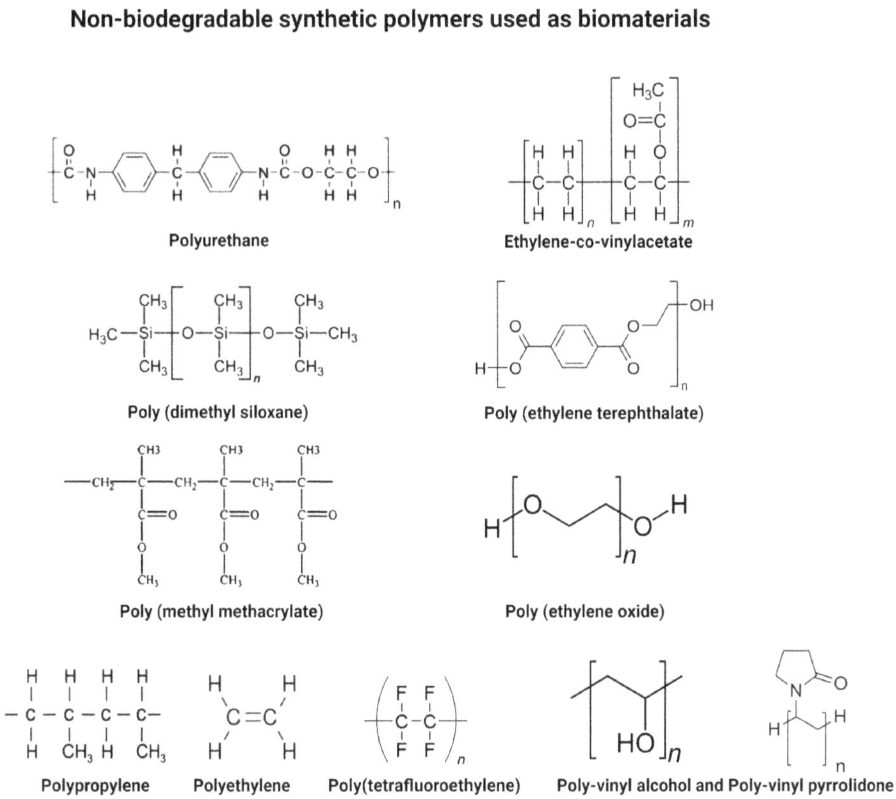

Figure 2.4: Chemical structures of nonbiodegradable synthetic polymers used as biomaterials for varied applications.

properties that can withstand the fluctuations of blood pressure without bursting or deforming. Its nonreactive hydrophobic surface makes it hemocompatible and resistant to degradation over long periods of time. Its ability to withstand high tensile strain makes it a good polymeric cover for vascular and nonvascular stents. To avoid restenosis (reocclusion of the stent), a polymeric cover might be added to the metallic mesh; however, the cover must be able to withstand the radial force that the stent is giving and the radial stress during the stent implantation. Here PU makes a good polymeric cover material for stents as it does not require a high force for it to elongate, which means that the radial force coming from the stent or the implantation balloon is enough to expand the polymeric cover without crimping or breaking [262]. As a heart valve material, PU can mimic the elastic moduli of the replaced cartilaginous valves [263, 264], as it can flutter together with the movement of the blood flow inside the heart without breaking or deforming. For balloon catheters, PU can be inflated with an assurance that it will not burst due to applied pressures.

With its mechanical stability, PU has also been used to blend with other low mechanical strength polymers to give elasticity to what may be brittle or hard polymers. For example, PU has been co-electrospun with acrylate polymers for drug delivery applications in nonvascular stents [265]. This technique synergizes both the mechanical properties of PU with the pH and thermoresponsiveness of the acrylate-based polymers. This strategy makes what may be unusable polymers for stent applications due to their mechanical weakness practically usable. PU can also be used as a drug delivery platform as it can effectively encapsulate stable drugs such as paclitaxel (antiproliferative) or heparin (anticoagulant) [266, 267]. These classes of drugs can be mixed with PU dissolved in a solvent and be made into membrane films [268], nonwoven electrospun fibers [269, 270], or injection-molded materials [271]. Passive diffusion facilitates the release of these drugs giving a sustained release over time. However, adding these drugs may affect the mechanical properties of PU negatively. But this reduction on mechanical strength and elasticity is a good trade-off for being able to release therapeutics, as it gives PU an added functionality as a biomaterial.

Surface characteristics such as wettability, hydrophilicity/hydrophobicity, bulk chemistry, surface charge and charge distribution, surface roughness, and rigidity are all important factors that could determine the behavior of the material when interfaced with a biological element. Hence, surface modification of PU is needed to alter its hydrophobic state [272]. It can be done via surface plasma treatment or UV light irradiation to modify its surface properties, in order for it to be used in other applications such as tissue engineering, it decreases the hydrophobic properties of PU by creating free hydroxy radicals on its surface, thus facilitating the adhesion of proteins or cell adhesion molecules. However, these treatments may also alter the mechanical properties of PU; thus, it must be used in applications where load or shear stresses are not abundant, but wettability is a must. PU still remains a staple in development of various biomedical products. It is one of the most successful marketed materials adapted for multiple applications (catheters, heart valves, stents, dental fillers, condoms, and bioadhesive) and it is still one of the most reliable nonbiodegradable materials that can be used inside the body.

2.3.1.2 Poly(dimethyl siloxane) (PDMS)

Poly(dimethyl siloxane) (PDMS), also known as silicone oil and its cross-linked variant silicone rubber, is one of the most used nondegradable polymers in medicine. Its polymeric structure is made up of long-chained $SiO(CH_3)_2$ units and is known for its inertness and viscoelastic properties [273]. The oil variant has flow characteristics that can be considered highly viscous that requires long flow time; however, given enough time it may cover or conform to diverse topological surfaces, which means that it can be used as a quality coating material for various highly reactive surfaces that require passivation. An example of this is in hypodermic needles, where silicone oil coating is used to protect it from oxidation and reacting with biological elements. Silicone oil is also a superb lubricant and is used to lubricate condoms and other biomedical devices such as insulin pens and syringes. Also, low-molecular-weight silicone oil is the common filler for breast implants as it can mimic the feel of native breast tissues [274].

Since PDMS can cover almost all substrates without gaps, it is also used as the main material to make channels in microfluidic chips [275, 276] and components of bio-microelectrical machines. This is also known as soft lithography where nano- and micropatterns can be made via UV light patterning of PDMS over a silicon wafer substrate [277]. These devices are used for development of organs on a chip where various drug screening assays could be made and contribute to the understanding of various biological processes without experimenting on live animals or human test subjects. With higher degree of cross-linking, PDMS is transformed into silicone rubber [278] where its increased elasticity makes it applicable as biomedical tubings, cosmetic implants, and intraocular devices. PDMS can also serve as a drug delivery platform [279] as it can be fine-tuned to be able to be a barrier while diffusing out hormones as a method for birth control [280]. This matchstick-sized implantable device is implanted via subdermal injection and can deliver levenorgestrel, a progesterone analogue, over a period of 3 months. PDMS can also be made into hydrogel preparations with varied light refractive properties. This is used as intraocular lenses or as contact lenses to correct the refraction of light that enters the eye [281]. The ability of PDMS to be bioinert is one of its attractive qualities for biomedical applications, as it can stay inside the body without causing inflammatory or immune response. However, some people have predisposed allergic reactions to silicone-based materials but, in general, silicone is still one of the staple nonbiodegradable polymers used as a biomaterial in medicine.

2.3.1.3 Poly(methyl methacrylate) (PMMA)

Poly(methyl methacrylate) (PMMA), also known as acrylic, acrylic glass, or plexiglass, is a transparent rigid thermoplastic that is lightweight with strong mechanical properties [282]. It is derived from polymerizing the photo- and chemically curable monomer methyl methacrylate. This polymer is very easy to make into various shapes as it can be cured using various molds and turned into shapes that can be used practically. PMMA transmits up to 92% of visible light (3 mm thickness) and gives a reflection of about 4% from each of its surfaces due to its refractive index, making it an attractive material for ophthalmology applications as replacements for damaged lens in cases of cataract and as a common lens for eyeglasses [283, 284]. It is relatively biocompatible, as it does not induce any inflammatory or immune response in the body. These properties make it applicable to be used in contact with tissues. Its strong and rigid mechanical properties mean that it can be used in hard tissue engineering applications. In dentistry, it is applied as a UV-curable cavity filler to fix the damaged tooth and artificial teeth used in dental prosthetics [285]; furthermore, in orthopedics it is used as a bone cement to join fractures [286, 287] and to facilitate remodeling of lost bones by filling the gaps on the defects and preventing the bone from motion and further damage. However, this bone cement is usually supplied in monomeric form (MMA), and MMA is known to be an irritant and a possible carcinogen [288]; thus, complete polymerization using dark curing must be ensured using appropriate initiators even after the removal of the light irradiation sources, as the PMMA form once completely cured is inert and biocompatible. The bioactivity of PMMA for bone tissue reconstructive procedures is also being commercialized by adding bone regenerative agents such as tricalcium phosphate and HA [289]. This allows the bone to actively regenerate on the presence of PMMA laced with these bioceramics. This is applied in cranial bone defects where the goal is to create the nascent bone tissue on the head to fully cover the cranium that protects the brain. PMMA can also be injected to the vertebral column in a procedure that is minimally invasive. This is to provide support in cases of vertebral slip disks, a condition that gives tremendous amount of pain to the patient [290, 291]. PMMA can alleviate the pressure in this scenario and restore the height of the compressed vertebral body. Here, PMMA bone cement becomes a filler between two spinal processes, stabilizing the fractured segments and alleviating the extreme pain associated with spinal compression fractures. This procedure is slowly gaining acceptance in the clinic; nevertheless, adverse effects related with chronic exposure to unpolymerized methacrylate monomer are still a pressing concern.

2.3.1.4 Poly(propylene) (PP)

Polypropylene (PP), also known as polypropene, is a methylated ethylene thermoplastic polymer with partial crystalline and nonpolar structure [292]. It is a hard and thermally and chemically stable plastic material that is one of the most common commodity plastics in the world, next only to polyethylene (PE). Although very pliable, bulk PP is not suited for high-load-bearing applications; thus, it is mostly used in fiber form to increase its load-bearing capabilities. Fiber PP is used in medical devices such as in hernia meshes [293] or transvaginal mesh [294], even as a nonabsorbable suture. Meshes made up of PP are helpful in assisting the recovery of the weakened tissue by providing adequate support from straining. PP meshes are stitched by surgeons to the torn tissue to close the hernia. PP allows manufacturers to create mesh with large pores and minimal surface area. But all hernia mesh materials come with unique advantages and disadvantages. Most obvious advantages of using PP are its inherent durability, ease of sterilization thus reducing infection risks, and pliability providing comfort post-implantation. However, recent clinical findings show extreme adverse effects with chronic exposure to PP, where adverse effects such as biological tissue tend to adhere to the surface of the PP mesh, mesh shrinkage over time, and mesh degradation [295, 296]. These scenarios are not suited for long-term treatment as in cases of adhesion. These PP mesh will be almost impossible to remove surgically. Also, PP's shrinkage after long exposure to biological tissue may cause nerve and tissue damage to the implantation site. Although PP is generally biocompatible to humans, hypersensitivity issues or other reactions to PP mesh may lead to these complications. To address these concerns, manufacturers may sometimes coat the PP mesh to create what is called a composite mesh. The coatings may contain materials designed to help the body adapt to mesh like adding omega-3-fatty acid or may provide medicines to prevent infection [297, 298].

2.3.1.5 Poly(ethylene) (PE)

PE is a lightweight durable thermoplastic made up of repeating units of ethylene (C_2H_4) polymerized in a semicrystalline manner [299]. It is the most common commercial plastic in the world and is used in variety of industrial applications. Bulk PE has low mechanical strength; however, it has high ductility and impact strength as well as low friction coefficient. Its mechanical stability and density can be enhanced if the crystallinity of the polymeric chain is increased; examples of which are HDPE (high-density PE) and medium-density PE. All these variants of PE have slightly lower thermal stability than PP which makes PE easier to form into other types of products. PE has a strong barrier property to water but high permeability to polar gases such as oxygen and carbon dioxide. It is considered as bioinert and can be implanted inside the body without causing severe adverse effects. Its low permeability to water makes it highly stable inside the body while permitting gaseous exchange in the implantation site. One variant of PE called ultra-high-molecular-weight PE (UHMWPE) [299] is a tough material with low friction coefficient and is commonly used in friction heavy applications such as in construction of articular portions of implants used for hip and knee replacements [300]. HDPE, on the other hand, can mimic hard tissue mechanical properties as it can be made rigid and impact resistant. HDPE is used for facial and cranial reconstruction surgery to replace the damaged or missing bones in the face and head of the patient [301, 302]. Surface treatments (plasma and bioactive coating) of these variants of PE can also further improve their biocompatibility and by letting the implant integrate to the local tissue, and another effective integration strategy is the creation of highly porous HDPE implants [303] that will allow the in-growth of tissue to the structure, stabilizing it and restore full mechanical properties of the damaged site. However, this strategy is indication specific and it is highly relevant to the bone tissue and soft tissue implants, while it is contraindicated in craniofacial and cosmetic surgeries. The pores are made by sintering HDPE particles into predefined anatomically relevant geometries with pore sizes ranging from 100 to 250 μm [301, 304]. In relation to being used as an implantable material for reconstructive and orthopedic surgeries, the biodegradation rate of PE is highly sought-after as it is on the matter of decades and could remain stable inside the body with the same life span as the patient. This allows the production of permanent implants that can be shaped into complex geometries with the inertness and tuneable mechanical properties. PE is also used in drug delivery implant, where UHMWPE used in hip and knee replacements is coated with various antimicrobial agents such as gentamicin to avoid post-implantation infection [305]. Another form of drug delivery using PE is with the use of HDPE balloon catheters that are coated with thrombin inhibitor to reduce thrombosis at the vessel walls [301, 306]. However, these kinds of platform rely on the surface coating of HDPE and UHMWPE, thus limiting the capability of having high drug loading and controlled release of therapeutics over time.

2.3.1.6 Poly(tetrafluoroethylene) (PTFE)

Poly(tetrafluoroethylene) (PTFE) is a fluorinated aliphatic polymer commercially known as Teflon. PTFE is an inert polymer with low dielectric constant and low friction coefficient. PTFE has a highly polarized carbon–fluorine bond along its polymeric backbone that makes the polymer highly crystalline [307, 308]. This makes PTFE hydrophobic and lipophobic, thus creating a nonreactive surface that is very resistant to adhesion to biological elements. Another iteration of PTFE is called extended PTFE (e-PTFE) [309]. This is made by extruding the PTFE polymer via anisotropic loading. Here, the microstructure of PTFE forms pores that are axially aligned along the direction of the strain, resulting in a material that has unique microporosity. This porosity created by the extension of the microstructure of PTFE is permeable to oxygen allowing it to integrate to biological tissues and provide mechanical support without biological adhesion. An example of this is in mesh [310, 311] applications to prevent intestinal adhesions by creating a physical barrier in between tissues after an abdominal surgery. Further benefit of this microstructure of e-PTFE is that it can guide the growth of hard tissues along the direction of its pores, and this is applied in the orthodontics field to allow complete regeneration of defect in the alveolar bone of the jaw [312, 313]. Furthermore, like PE polymers, PTFE-based constructs are being used in facial reconstruction and cosmetic surgeries due to their inertness [314, 315]. PTFE has a long life span and biological stability meaning that it would not create any toxic leachable that could harm the patient. PTFE can also be made into injectable particles and used to treat vesicouretic reflux [316]. Vesicoureteral reflux occurs when the flow of urine goes the wrong way. This condition is more common among infants and young children. Here, PTFE particles in paste form are injected in the bladder near the opening of the ureter. The injection creates a bulge that makes it harder for urine to flow back up the ureter. The inertness and stability of PTFE in this iteration had positive results as a long-term implant. The low coefficient of friction of PTFE makes it a suitable material for artificial large vascular grafts [317] as it repels most biological elements. This is a beneficial attribute for a vascular graft as it facilitates smooth flow dynamics by reducing the flow shear stress that the grafts are experiencing. However, in order to facilitate the integration of this kind of artificial graft to the vasculature, a certain degree of endothelialization must occur. Here, the porous structure of e-PTFE is employed to induce the ingrowth of vascular endothelial cells into the graft and create a stronger support to the damaged vessel over time. PTFE is also the material of choice for hemodialysis fistulas [318], where communications between the native artery and the vein in an extremity are surgically created for patients who need regular blood dialysis treatment. PTFE and other synthetic materials (poly(ethylene terephthalate) (PET) and PU) are used or have been used as a communication medium between the artery and the vein and are termed prosthetic hemodialysis access arteriovenous grafts. These materials' bioinertness and nonfouling properties can avoid complications such as biofilm formation that could induce sepsis. PTFE has also been investigated for hip and knee articular joint replacement prosthesis [319] since it has low friction coefficient and logically may be suitable for providing ease of gliding movements of these prosthesis; however, due to its weak mechanical properties, wear and tear occurs causing prosthesis failure and PTFE has been since obsolete for articular joint replacement applications.

2.3.1.7 Ethylene-co-vinylacetate (EVA)

Ethylene–vinyl acetate (EVA), also known as poly(ethylene-vinyl acetate), is the copolymer of ethylene and vinyl acetate at a ratio of vinyl acetate ranging from 10% to 40% while the majority being ethylene [320]. EVA polymers are known in the biomedical industry as a robust material in drug delivery applications. EVA polymers act as drug reservoir [321] through which the drug delivery is mainly governed by diffusion mechanism. As a drug delivery platform, the permeability of the polymeric network of EVA can be fine-tuned by varying the ratio of the ethylene:vinyl acetate monomers. If vinyl acetate content is increased, the crystallinity of the polymeric structure decreases, making the polymer network more permeable to water, thus increasing the rate of diffusion and release of the drug payload. This tunability of EVA makes it a versatile platform to be used in multiple drug delivery indications that require varying release rates. Examples of successfully commercialized implantable EVA-based drug delivery systems are in contraceptive implants as an intrauterine device that releases progestin [321, 322], and treatment of glaucoma as an intraocular implant placed on the lower eyelid to release pilocarpine over a 1 week period [323, 324], removing the need of daily eye drop administration. EVA polymers also exhibit hemocompatibility, allowing it to be used in drug-eluting vascular stent application [325]. However, it has been replaced by other materials such as PU, as PU exhibits better hemocompatibility and elasticity compared to EVA polymers.

2.3.1.8 Poly(ethylene terephthalate) (PET)

PET is a thermoplastic polyester derived from aromatic diacid and ethylene glycol, commercially known as Dacron®. It has multiple properties such as slow biodegradation and its fiber-forming capabilities allowing it to be manufactured into constructs that are used in vascular graft and hernia mesh applications, and its one of the most common applications is for degradable sutures [326, 327]. PET's ability to be woven into different patterns allows it to have the elasticity that is usually not allowed if PET is used as a single sheet material. Weaved PET allows it to bend and dilate during blood pressure variations that occurs frequently in the body. Since PET is mostly applied on large vessels such as the aorta and large arteries [328]. Its woven structure allows the graft to withstand the high shear forces and tensile forces that the blood flow administers on it. Furthermore, woven PET can also be manufactured to have bifurcations that allow the divergence of blood flow from a common blood vessel. However, the limitations of these artificial grafts are that they cannot dilate or constrict as seamlessly like how local blood vessels can control their diameter based on biological cues. The ability of the native blood vessels to dilate and constrict based on these natural cues highlights the shortcomings of synthetic vascular grafts. Furthermore, PET can only be applied to large vessel graft applications as small diameter vessels are more prone to the effects of thrombus formation which PET might aggravate. Another application of PET is in atrial septal defect repair [329, 330]. This is a congenital condition where holes are present in the atrial septal wall that permits the mixing of oxygenated and deoxygenated blood in the heart causing the patient to experience shortness of breath, tiredness, and reduction of oxygen perfusion to the rest of the body. This is a detrimental condition, and in advance cases, prosthesis are surgically employed to repair these holes in the heart. The prosthesis consisted of a double umbrella composed of four stainless steel arms covered with a PET (Dacron®) meshwork. This is inserted into the septal hole to form a blockade and inhibit the mixing of the blood from the left and right sides of the heart. However, after some experimental trials, arm fractures were observed in significant number of cases after 1–6 months post implantation [331, 332]. Furthermore, a high incidence of prosthesis shunting was detected, thus leading to the discontinuation of the prosthesis.

Woven PET also found its way in ligament repair applications [333, 334], as it can serve as a replacement to damaged ligaments in the body. Its woven architecture, bioinertness, and relatively strong mechanical properties make it an interesting investigational material for ligament repair. This artificial ligament is commercially known as Stryker® [335] and made by weaving ten longitudinal yarns of Dacron® (PET) and one multifilament transversal yarn to form a suturable ligament-like structure. However, this strategy is only suited for short-term replacement as after 2 years of follow-up significant number of complications have been reported such as tearing, chronic synovitis, and deep infection [336]. The material also cannot last for more than 5 years as 57.5% of recipients of Stryker® for ACL reconstruction had reported implant failure [336]. After further investigation, the external knitted shell of the implant showed compression and crushing injury near the femoral condyle, while the inner braided part showed evident abrasion damage. Thus, after years of being in the market and due to the lack of demand, this material for ligament repair has been discontinued altogether.

2.3.1.9 Poly(ethylene oxide) (PEO, PEG)

Polyethylene oxide or polyethylene glycol (PEG) is a highly biocompatible molecule with a semicrystalline property and is composed of (CH_2–CH_2O) repeating units that easily bind with water molecules via hydrogen bonding; thus, it is considered highly soluble in water [337]. However, it is classified as nonbiodegradable as the body lacks the correct enzyme to degrade the structure of the monomeric form of this polymer. PEG is one of the most used polymers for passivation of various biomaterials and medical devices [338]. The nonimmunogenic, biocompatible, and flexible nature of PEG makes it suitable for coating and grafting to varied types of biomaterials. PEG has multiple iterations such as branched, star, and comb-like macromolecules. A process called PEGylation [339, 340] is an effective process in which PEG is bonded to another molecule to increase its biocompatibility. This process is very common in drug delivery applications [341] (e.g., nanoparticles and microparticle formulations) to evade the cell and antibody-mediated immune system of the body. PEG can also hinder the protein adsorption also known as protein corona formation on particle formulations which are essential in drug delivery to maximize their circulation time and increase its pharmacokinetic and pharmacodynamic effects. In traditional pharmacology, PEGylation is co-incubated together with a target active molecule, most typically peptides, proteins, and antibody fragments [342, 343], to improve the safety and efficiency of these therapeutics. It does so by altering many physicochemical properties of these molecules such as conformation, electrostatic binding, molecular weight, solubility, and hydrophobicity. Like nano- and micro-formulated drugs, it can increase systemic retention of the therapeutic agent [344] by masking it from the body's immune system uptake and fast renal clearance. This strategy has seen recent success especially in the creation of vaccines against the novel coronavirus of 2019 [345, 346]. PEGylated lipid micelles are the platform vehicles of the mRNA vaccines that help the body create the required antibodies to fight the coronavirus. However, there is a certain unpredictability in the clearance time of PEGylated compounds that could cause toxicity to the liver and kidneys; furthermore, there are some cases that hypersensitivity to PEG can cause anaphylactic shock (severe allergic reaction) to some vaccine receivers. Hence, strategies such as changing the size (Mw), the number, the location, and the type of linkage of PEG molecule are now being investigated by multiple researchers.

PEG is used in a very utilitarian way to solve biocompatibility issues in some materials that may exhibit unwanted reactivity to the body. PEG coating on metallic implants may inhibit the protein absorption on devices that require biofouling inhibition [347, 348]. In orthodontic application, PEG coatings can serve as the grafting site for antibacterial agents such as silver nanoparticles (AgNPs) or antibacterial drugs [349, 350]. Moreover, PEG-5000 (molecular weight 5,000 kDa) can also significantly reduce the initial attachment of bacteria on the surface of metal wires used in antiadhesive orthodontic devices. Other applications of PEG coatings that can be used for devices such as PEGylated catheters in urinary and vascular applications

[351–353] can also inhibit the formation of biofilm that could induce urinary tract and intravascular catheter-associated infections.

In tissue engineering, PEG hydrogels [354] had been the most extensively studied for its benefits as a hydrated structure, resulting in typically high cytocompatibility and protein repellence to minimize inflammatory responses. While it may be counterintuitive to use PEG as a material for cell tissue engineering due to its nonadhesive properties to biological elements, including cells, incorporating cell adhesion motifs via direct conjugation to the polymeric network of PEG can be made. While PEG in itself lacks the proper rheological properties to be used in injectable delivery of hydrogel-based cell therapy, there are PEG derivatives such as injectable and star/branched PEGs that have unique potential to control the chemistry, degradability, injectability, and mechanics of PEG-based hydrogels.

2.3.1.10 Polyvinyl alcohol (PVA) and polyvinyl pyrrolidone (PVP)

Polyvinyl alcohol (PVA) is a water-soluble synthetic polymer made up of repeating units of $[CH_2CH(OH)]_n$ [355]. Although water soluble, it is considered nonbiodegradable inside the body due to the vinyl polymer having carbon–carbon bonds in its main chain and it is not expected to undergo a complete backbone chain dissociation upon oral or intravenous administration. However, its biocompatibility, low tendency for protein adhesion, and low toxicity make it an attractive material for a wide range of biomedical applications.

Pure PVA possesses certain limitations that are seen negatively in applications to the body. Its mechanical properties are very weak and brittle, and it is hard to produce a strong, workable material to be utilized in practical biomedical applications. Furthermore, although PVA in hydrogel [356] form presents a relative biocompatibility in contact with body tissues, some studies showed that it is susceptible to calcification if exposed for longer period of time to biological fluids. This is not desired if PVA is used in cartilage repair, for example, PVA hydrogels (SaluCartilage™) were clinically tested in Europe for cartilage repairs and were observed to have poor fixation and underwent rapid deswelling [357, 358]. Lastly, since PVA is a highly hydrophilic polymer, PVA hydrogels have poor stability in water. Knowing these limitations, PVA is blended with other polymers to better utilize its inherent properties [359], and this is done by copolymerization, cross-linking, or grafting with other synthetic or natural polymers. Blending with other polymers proved to be a favorable method of modifying the properties of PVA-based biomaterials. PVA has a regularly arranged hydroxyl motif that creates a crystalline region through intramolecular hydrogen bonds, and adding polymers with high affinity to these hydroxyl groups can enhance the polymer stability and would lead to a more workable material.

In this sense, the most compatible polymer to be blended with PVA is polyvinyl pyrrolidone (PVP) due to its ease in processability, biocompatibility, and nonimmunogenicity, and more importantly, it is approved by the US Food and Drug Administration as a safe polymer for pharmaceutical and biomedical applications. But like PVA, when used alone, PVP possesses weak and brittle mechanical properties that lead to certain utility limitations. However, this limitation was proved to be easily overcome by combining PVA and PVP into a copolymer (PVA/PVP biomaterial) that exhibits additional functionalities as compared with PVP or PVA alone [360]. Numerous reports demonstrating that PVA and PVP interact through intermolecular hydrogen bonding between hydroxyl groups of PVA and carbonyl ones of PVP form a thermodynamically miscible blend that allows for better swellability in the presence of biological fluids due to the pore-forming properties of PVA, and the mechanical properties of PVA/PVP biomaterials are significantly better when combined together. With the ease of processability and enhanced properties, PVA/PVP materials can be applied to a wider range of applications, such as in tissue engineering (hydrogels and solid scaffolds), local and systemic drug delivery (nanoparticles and nanofibers), and wound dressing (casted films) [360–362].

2.3.2 Synthetic polymers (biodegradable and bioresorbable)

Synthetic biodegradable polymers had been a significant factor for the development of advanced biomaterials that rely on properties that have mechanical, chemical, and biological stabilities when applied inside the body. Biodegradation is a significant property for any material that is used inside the body as it can be implanted and used without the need for subsequent removal after the material served its purpose. The material is simply broken down inside the body and the resulting compounds can be safely metabolized by the liver and eliminated by the kidneys. In practice, all synthetic polymers are biodegradable given enough time and proper conditions. Furthermore, the term "biodegradable" can be applied to polymers that undergo dispersion in vivo as a result of macromolecular degradation, but without elimination of products and subproducts by the body. This is due to the product size being too large to be eliminated by the kidneys. But to limit the scope of biodegradable materials in this chapter, we will refer to biodegradability as materials that degrade once its purpose in the body had been served. Another important term to distinguish is "bioresorbable polymers." These are types of polymers that degrade through a reduction in size and the subsequent products are resorbed in vivo, i.e., materials that can be metabolized by the liver and eliminated by the kidneys.

Synthetic biodegradable and bioresorbable polymers are essential in developing components of drug delivery platforms, tissue engineering scaffolds, and biomedical devices. Therefore, there is increased attention in designing new synthetic biodegradable polymers that could be used in medicine and biology, and in this part of the chapter, we discuss the different classes of synthetic biodegradable and bioresorbable polymers (polyesters, polyanhydrides, polyorthoesters, and polyketals) and their uses as biomaterials (Figure 2.5).

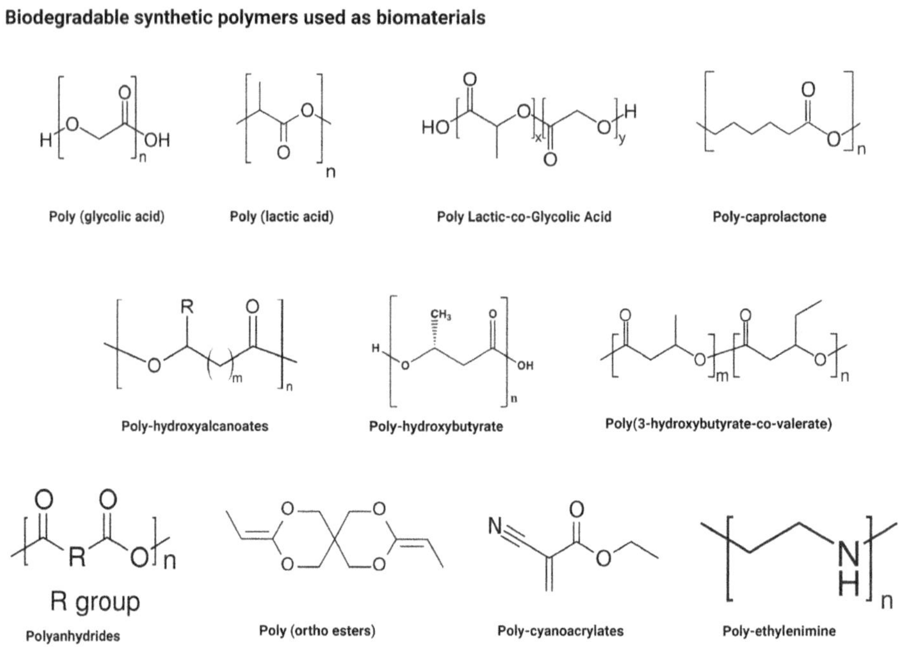

Figure 2.5: Chemical structures of biodegradable synthetic polymers used as biomaterials for varied applications.

2.3.2.1 Polyesters

2.3.2.1.1 Poly(glycolic acid) (PGA)

Poly(glycolic acid) (PGA) is an aliphatic polyester derived from fossil fuel resources. It is a bioresorbable polymer that has high mechanical strength and exhibits excellent short-term stability in ambient conditions [363]. It is known to biodegrade within approximately 12 months and degrade chemically by hydrolytic cleavage [364]. PGA is one of the first degradable synthetic polymers investigated for biomedical applications due to its proven biocompatibility and noninflammatory response in the body. However, once degraded, the resulting localized glycolic acid accumulation may potentially cause undesirable acute inflammatory response [365–367]. Its first successful application in medicine is in absorbable sutures and over time it has been utilized in more other biomedical applications such as tissue engineering for bone and nerve regeneration [368, 369], wound healing material [370], and as a bioadhesive [371]. PGA has seen the most success in facial nerve regeneration [372] getting a US FDA approval for nerve grafting; in this application, bone marrow-derived stem cells are seeded into a PGA tube construct and differentiated into adult neuronal cells. PGA has seen promising results by providing a suitable environment for the stem cell differentiation and create structures that could resemble the native facial nerve.

In practice, PGA is notorious for its hydrolytic instability as its ester backbone is highly susceptible to hydrolysis. Thus, PGA is commonly combined with more hydrolytic stable polymers such as polylactic acid (PLA) to control its degradation rate. In addition to degradation by hydrolysis, PGA is also degraded by enzymes in the body (e.g., esterase). The degraded components of PGA when hydrolyzed are ultimately converted by the body into carbon dioxide as part of metabolic byproduct and to adenosine triphosphate (ATP) which is used by the cells as energy, which means that when properly degraded, PGA will not cause any harm to the body by producing toxic compounds. Knowing this, PGA can be used for bioresorbable temporary devices as stated earlier.

2.3.2.1.2 Poly(lactic acid (PLA)

The main monomeric component of PLA is . As the name suggests, lactic acid is a natural occurring compound that is found in the body as a product of anaerobic respiration [373]. PLA chemical synthesis can be done by hydrolysis of lactonitrile by strong acids. This allows PLA to be derived from renewable resources. But the most common source of PLA is from fermented organic sources such as milk. The traditional application of lactic acid in medicine is as a supplement in intravenous fluids [374] (components of lactated Ringer's solution and Hartmann's solution). It is mostly used for fluid recovery after severe acute blood loss due to trauma, surgery, or burns. As a biomaterial, PLA behaves as a thermoplastic meaning that it can be casted into multiple geometric shapes that have robust mechanical properties. PLA can also be tuned to allow different degradation rates, and this is done by varying its density, molecular weight, and polymeric chain length to name a few. This ability of PLA to be modified to have varied degradation allows it to be applied for drug delivery applications, where fast sustained release kinetics is required. Drug delivery vehicles such as nano- and microparticles made up of PLA have also been investigated as PLA allows entrapment and/or conjugation of therapeutics into its polymeric network. Implantable drug delivery devices based on PLA have been investigated as PLA can either have bulk or surface degradation mechanics that allows either a slow or fast release of entrapped therapeutics, respectively.

The capability of being bioresorbed without causing any harm to the tissues makes PLA an ideal material if the goal of the treatment is to provide a temporary scaffold that will eventually get replaced by the nascent tissue. Here PLA has the capability to fill that niche requirement and has been investigated for short-term fixation in treating bone fracture and ligament tear [375, 376]. Stable nondegrading bone-fixing materials are not suitable for this application as they lack the compliance for the dynamic changes of skeletal tissue as it begins to recuperate. This is highlighted in pediatric orthopedics where children's skeletal tissue undergoes rapid development and bioresorbable compliance is a desired quality to have for a material. Lastly, hydrolysis products of PLA are also considered to be nonimmunogenic and biocompatible [365, 377] as they undergo the same route as PGA for excretion in the body. Here, lactic acid is converted to pyruvic acid and thus utilized by the body to create ATP or excreted by the kidneys.

2.3.2.1.3 Poly-lactic-co-glycolic acid (PLGA)

Poly-lactic-co-glycolic acid (PLGA) is a copolymer of PLA and PGA and is one of the most studied polymers in biomedical applications. The combination of PLA and PGA creates a polymer that is less susceptible to hydrolysis, thus giving it more in vivo stability over pure PLA or PGA alone. PLGA can also be modified in terms of its geometry, degradation rate, and mechanical properties to meet the requirements of a given application. This is done by varying the ratio of PLA and PGA of the copolymer chain, increasing the molecular weight, or varying the available surface area of the material. But the most significant factor is the polymer ratio, as it dictates the degradation behavior of PLGA based on the amount of glycolic acid units. PGA concentration is directly proportional to its degradation rate where degradation times of 1:1, 3:1, and 85:12 PLGA are 1–2 months, 4–5 months, and 5–6 months, respectively.

Due to the easy tunability of the degradation rate of PLGA, it can be made into various drug delivery vehicles, e.g., nano- and microparticles [378], nanofibers [379], and millimeter-sized implants [368, 380]. Various types of therapeutics can also be administered by entrapping it onto the polymeric network of PLGA, and the subsequent bulk and surface degradation mechanism facilitates the release of these therapeutics. PLGA has also been applied to tissue engineering as a degradable scaffold [368, 381] as it can promote cell adhesion and proliferation while its relatively biocompatible degradation products allow for slow reintegration of nascent tissue to the local implantation site. It has also been used to differentiate mesenchymal stem cells (MSC) to different cell types that are used in bone, tendon, and ligament regeneration. Although PLGA is considered to be biocompatible, its degradation products when not fully hydrolyzed may cause inflammatory and foreign body reactions [365]. These reactions are attributed to the acidification of the surrounding tissues in the implantation site, resulting from the accumulation of degradation products, e.g., lactic acid and glycolic acid. In order to combat this, PLGA can be made into a highly porous material to allow increased water and oxygen permeation and facilitate full hydrolyzation of lactic acid and glycolic acid to biocompatible subproducts.

2.3.2.1.4 Polycaprolactone (PCL)

PCL is a synthetic biodegradable polyester that is biocompatible and with slow degradation properties [382]. PCL is prepared by ring-opening polymerization of ε-caprolactone using a catalyst such as stannous octoate. Like other polyester materials, PCL degradation is governed by hydrolysis; however, its ester linkages are less susceptible to hydrolytic degradation compared to PGA and PLA. It takes more than 12 months for PCL to degrade significantly, and the total degradation time is up to 4 years. The degradation rate of PCL can also be fine-tuned like PGA and PLA by varying its molecular weight and increasing its available surface area for increased hydrolyzation. In terms of mechanical stability, PCL has relatively weak mechanical properties and performs poorly for load-bearing biomedical applications. In order to make PCL useful in practice, it is usually cross-linked or blended with other natural or synthetic polymers to increase not only its mechanical strength but also its bioactivity.

PCL is one of the most studied biodegradable polymers in tissue engineering, as it is proven to be biocompatible and both prolonged and short-term use of it will not elicit any adverse reactions from the body. PCL nanofibers are usually made as it mimics the ECM environment [383–385] and it promotes adhesion and proliferation of cells both in vitro and in vivo. The long-term degradability of PCL is useful in neural regeneration and bone regeneration as it can give prolonged scaffolding support. Furthermore, PCL can be easily loaded with various inorganic materials that can add other properties that are beneficial for tissue engineering, e.g., AgNPs for antibacterial effect [386, 387], HA for osteocyte activity [388], carbon-based materials for scaffold electrical conductivity [389], and iron oxides for hyperthermia cancer therapy [390]. These added properties give multifunctionality to PCL and prove that PCL is a robust platform material for multiple advance healthcare applications in the body.

PCL is applied in drug delivery applications where a prolonged release of entrapped therapeutics is required. In this application, passive diffusion and polymer erosion govern the drug release mechanism. A wide array of therapeutics can be entrapped on the PCL polymeric network, as PCL can be dissolved in organic solvents that are usually not harmful to stable drugs, e.g., chemotherapeutics and antibiotics. There are multiple iterations of drug delivery vehicles where PCL was used as the main material. It can be made into nano- and microparticle preparations and used for cancer therapy [390]. PCL's polymeric network can entrap chemotherapeutic drugs [391], which is meant to delivery at a relatively high efficiency (above 60–70%). PCL can also be grafted with targeting ligand on its surface to allow tissue-specific absorption and modified with PEG to increase its circulation time. Electrospun PCL mats are also investigated, whereas they are used for localized drug delivery in various types of medical devices, e.g., stents [392], wound dressings [393], and meshes [394].

The degradation products of PCL inside the body are mostly governed by hydrolysis, and PCL's main subproducts of caproic acid, succinic acid, valeric acid, and butyric acid have been shown to have a negative effect on cells due to localized acidification. However, this is only observed if the subproducts are not eliminated but in practice all these acids are usually further hydrolyzed to nontoxic forms and subsequently eliminated by the body.

2.3.2.2 Polyhydroxyalkanoates (PHA)

Polyhydroxyalkanoates (PHAs) are linear polymers with ester linkages synthesized by various bacteria inside their intracellular structures called inclusion bodies [395]. They behave like a thermoplastic and can be molded into various geometric shapes using traditional fabrication methods. PHA-based biopolymers are gaining popularity for biomaterial applications because of their biocompatibility and biodegradability [396, 397]. Furthermore, PHAs are easy to produce and functionalize with strong mechanical properties. The broad range of applications in biomedicine includes drug delivery systems, implants, tissue engineering scaffolds, and artificial organ constructs. PHA's varieties are synthesized in the bodies of different gram-positive and gram-negative bacterial species. They are then extracted from these bacterial cultures to form granules that can then be used for practical applications. Different varieties of PHAs are known and have slightly different chemical structure from each other. The following constituent monomers of PHA are classified on the basis of the number of carbons in the monomer as short-chain length PHAs (polyhydroxybutyrate (PHB), poly(3-hydroxyvalerate) (PHV), and poly(3-hydroxybutyrate-co-valerate) (PHBV)), medium-chain length PHAs (e.g., polyhydroxyoctanoate, polyhydroxynonanoate, polyhydroxyhexanoate, and polyhydroxyheptanoate), and long-chain length PHAs. But among these types, PHB and PHBV are used most extensively for biomaterial applications. PHAs contain reactive (hydroxyl) sites that can conjugate various peptides that can add functionality to the material. Its bulk degradation mechanism governed by hydrolysis is also a useful property for drug delivery applications.

PHAs can be functionally modified by various chemical and physical reactions (e.g., epoxidation, carboxylation, chlorination, hydroxylation, and pyrolysis) to increase the bioactivity, processability, compatibility, and biodegradability or to decrease crystallinity and hydrophobicity, and enhance moldability and flexibility [398, 399]. This is an important property of PHA where it can be tailored to specific applications toward the body. Biodegradation is one of its salient property, and its degradation products are considered not harmful to the body [400]. Its degradation rate is approximately 3 months, but that rate is also highly dependent on the molecular weight, crystallinity, and surface area of the material. Higher molecular weight with high crystallinity tends to make it more resilient to hydrolysis while increasing its surface area can significantly increase it. PHA once hydrolyzed turns into water-soluble oligomers and monomers inside the body and ultimately converted under aerobic conditions to carbon dioxide and water, these and then excreted by the body through exhalation and urination, respectively.

2.3.2.2.1 Polyhydroxybutyrate (PHB) and poly(3-hydroxybutyrate-co-valerate) (PHBV)

PHB can be derived from multiple bacterial sources: *Pseudomonas pseudomallei*, *Azotobacter chroococcum*, *Pseudomonas putida*, *P. oleovorans*, *Alcaligenes eutrophus*, *Zoogloea ramigera*, *Alcaligenes* sp., and *Ralstonia eutropha*. Its biodegradability together with its robust mechanical properties compared to other naturally derived polymers (polysaccharides and polypeptides) makes it an ideal material for long-term implant and tissue repair applications. PHB is known to be able to form fibers that can be woven to create flexible 2D and 3D materials useful for various tissue repair applications [401]. An iteration of PHB called PH4B has seen recent commercialization success as a degradable suture material (ThephaFLEX) [402] and got US FDA approval in 2007. Another commercially available PH4B-based material is called BioFiber™ [403]; an orthopedic soft tissue scaffold reinforces the suture–tendon interface and mechanically reinforces tendon repairs. And GalaFLEX [404], a P4HB 2D mesh scaffold, is used for reconstructive surgery that supports, elevates, and reinforces soft tissue.

PHBV is a copolymer of PHB and PHV [405, 406]. It came about due to the need to increase PHB's ductility and processability as a bulk material. PHBV still possesses the same biodegradation, biocompatibility, and bioactivity but with superior flexibility allowing it to be used in load-bearing applications such as in bone tissue engineering [407]. PHBV as a bone tissue repair material has been reported. Here, PHBV has been incorporated with HA to induce the proliferation of osteocytes into the scaffold with successful results. This also shows that PHBV can be made into composites without sacrificing its mechanical properties.

Lastly, PH4B has also been reported to be able to cross the blood–brain barrier, leading it to be used for brain-targeted delivery of large molecules [408]. Both PHB and PHBV have been used in nanomedicine as nanoparticles for cancer therapy to deliver different kinds of chemotherapeutics [409, 410]. Its functional chemical structure allows it to entrap the therapeutics and its nanosized formulation can increase the pharmacokinetic and pharmacodynamic properties of these drugs. Once it enacts its role as a drug delivery vehicle, its degradation product forms 3-hydroxybutyrate, which is a natural component of blood which makes it hemocompatible and easily eliminated by the body.

2.3.2.3 Polyanhydrides (PAH)

Polyanhydrides (PAH) are biodegradable polymers characterized by anhydride bonds that are present in its polymeric chain [411, 412]. Generally, there are three classes of PAHs: aliphatic, unsaturated, and aromatic. These classes are determined by looking at their R groups. Aliphatic PAH is characterized by having linear R groups containing carbon atoms bonded in a straight or branched chain. Unsaturated PAH has R groups with one of more double bonds, while aromatic PAH contains R groups with benzene (aromatic) rings. The characteristic anhydride bonds in polyanhydrides are highly susceptible for hydrolysis, which means that the degradation of PAH happens at these anhydride linkages on the polymeric backbone chain. The degradation products of PAH are carboxylic acids which are nontoxic and easily metabolized in the body.

Since PAH is known to degrade at a fast rate, it is usually used for short-term release of therapeutics (1 week) [413–415]. It undergoes surface erosion and passive diffusion to allow the release of whatever therapeutic is entrapped in its polymeric network. The drug release rate of PAH-based platforms can be modified by adding hydrophilic groups to its polymeric backbone to make a copolymer that is less susceptible for hydrolysis. This in turn allows water to penetrate the polymeric network of PAH to make diffusion of the governing release mechanism for the drug payload. PAH has seen commercial success as a drug delivery device for treating brain cancer in a commercial product called Gliadel® [416]. Here, PAH made up of 1,3-bis(4-carboxyphenoxy) propane and sebacic acid (PA-CPP-SA) is formed into erodable wafers that are implanted directly to the brain [412], where a segmented tumor was previously located. Multiple wafers are needed to completely cover the area of the cavity, and the wafers can be easily broken down into pieces to conform to the configuration of the surface of the cavity. The wafers loaded with carmustine, a chemotherapeutic drug known to treat glioma and glioblastoma, are released over time of around 5 days. In addition, multiple oxidized cellulose strips and the use of bioadhesives and sealant are needed to ensure that the wafers are secured in place and not obstruct the circulation of the cerebrospinal fluid in the brain. The elimination of PAH wafer in the brain was also proven, and the degradation products do not incur any severe inflammatory and immune response. Complete elimination of the PAH wafers occurs at around 4 weeks' time.

One significant limitation of PAH-based materials is its storage stability as it requires refrigeration due to its water-labile anhydride backbone. Another limitation is the weak mechanical properties of PAH, and strategies of introducing other monomers in its polymeric backbone (e.g., imide) can increase its mechanical strength to apply it to other load-bearing implantable drug delivery platforms.

2.3.2.4 Poly(orthoesters) (POE)

Poly(orthoesters) are a family of synthetic degradable polymers that are characterized by having a general structure of $-[-R-O-C(R_1, OR_2)-O-R_3-]_n-$. Polyorthoesters are synthesized by the transesterification of orthoesters with diols or by polyaddition of a diketene acetate and diol [417]. POEs are polymers mainly used for hydrophobic implants and have been shown to be useful for the controlled delivery of drugs and other therapeutics. There are four major types of poly(orthoesters). First, poly(orthoesters) that were prepared by the condensation of 2,2-diethoxytetrahydrofuran and a dialcohol will yield polymers that upon hydrolysis, degrade and form acidic by-products that accelerate the degradation process by being an autocatalyst [418]. While the second type of POE (aka DETOSU) is based on the reaction of 3,9-bis(ethylidene-2,4,8,10-tetraoxaspiroundecane) with various dialcohols in contrast to the first type, does not release acidic by-products on hydrolysis and thus does not exhibit autocatalytically increasing the degradation rates. The third type of POE is prepared similar to type I by transesterification, in this case a triol (preferably 1,2,6-hexanetriol) with an orthoester (e.g., triethylorthoacetate). Polyorthoesters type III degradation occurs at the surface by cleavage of the hydrolytically labile bonds in the polymeric backbone and has semisolid to ointment-like flexible properties that are suitable for topical and ophthalmic applications where sudden release and absorption of active ingredients are avoided. Lastly, type IV POEs is a further development of the type of POE type II, which is formed of the diketene acetal DETOSU with a diol which is modified by short sequences of polyglycolide or polylactide. This type of POE can be made into gel or solid preparations depending on the type of diol used in the polymerization reaction. However, unlike type II POE, type IV POE does not require the addition of acidic excipients to facilitate the hydrolytic degradation of the polymer that may create an erratic erosion process. They in turn create glycolic acid or lactic acid as by-products that can further facilitate the eventual complete degradation of the polymer. The degradation rate can be controlled by varying the proportion of glycolic and lactic acid in the sequence. Implants made of polyorthoester type IV show surface erosion while being highly biocompatible with degradation times from days to months and can thus also be used as a long-term drug depot.

Salient point of poly(orthoesters) as bioerodible drug delivery devices [419] is in their acid sensitivity which allows tunable devices with degradation rates lasting from hours to many months by adding acidic or basic excipients physically incorporated into the polymeric matrix. Hydrophobic nature of POE allows for surface degradation which is beneficial for a more controlled release of any therapeutics as it does not allow water to diffuse to the center of the polymer. This in turn makes it possible to achieve zero-order release mechanism for the drug entrapped inside the matrix.

2.3.2.5 Polyketals

The polyketals are synthetic polymers with ketal linkages in their backbone. They are one of the most recent polymeric materials to be synthesized and utilized for their potential for drug delivery as they possess acid sensitivity and biocompatible degradation products. These properties make them a pH-sensitive material that can deliver therapeutics in the field of cancer and phagocytotic-mediated cell therapy [420]. Polyketal synthesis is done via acetal exchange reaction to create ketal linkages on the polymeric chain backbone. These ketal linkages are degraded hydrolytically and catalyzed by the acidic pH environment. The very first report of polyketal synthesis using 1,4-phenyleneacetone dimethylene ketal was done and used to encapsulate dexamethasone to form nanoparticles that range from 200 to 600 nm in size [421]. Further advancements on polyketal micro- and nanoparticle preparations had proven to be able to encapsulate and deliver in a pH-specific manner more sensitive types of therapeutics (protein catalase, superoxide dismutase, and p38 inhibitor) due to the use of mild synthesis and encapsulation process.

A salient property of polyketal particles is that they do not generate acidic degradation products after hydrolysis, compared to polyester-based biomaterials [420, 422]. After protonated hydrolysis, the degradation products that polyketals produce are mainly 1,4-benzenedimethanol and acetone, which are easily metabolized in the liver and eliminated in the body. These degradation products do not create acidification on the tissues that could lead to acute inflammatory and immune responses. This also makes the polyketal more biocompatible than polyester-based biomaterials. Furthermore, polyketal decomposes on a timescale that is faster than PLGA, but slower than poly(orthoesters), and therefore provides greater control of the release of the entrapped therapeutics.

2.3.2.6 Polycyanoacrylates (PCA)

Cyanoacrylates are a family of strong adhesives that polymerize in contact with water or other moisture-containing surfaces [423]. There are different types of cyanoacrylates mainly methyl 2-cyanoacrylate, ethyl 2-cyanoacrylate (ECA aka super glue), and other biomedical utility variants called *n*-butyl cyanoacrylate (*n*-BCA), octyl cyanoacrylate, and 2-octyl cyanoacrylate. The acryl groups of cyanoacrylates rapidly undergo chain-growth polymerization in the presence of water (specifically hydroxide ions), forming long, strong chains, and bonding surfaces together. This reaction is very sensitive and even a small amount of ambient moisture can initiate the polymerization process of cyanoacrylates. Seeing this property of cyanoacrylate, it has been explored to be used as an emergency wound closing material in the battle fields of Vietnam. Cyanoacrylates form strong adhesion between the damaged tissue and create a film that can stop the bleeding caused by a trauma [424, 425]. Historically, *n*-BCA was used for this application, but skin irritation from *n*-BCA has caused the development of 2-octyl cyanoacrylate (Dermabond) as a solution for a more skin-friendly adhesive.

Since then, the application of cyanoacrylates had diverged from being an adhesive to be a drug carrier as it shows biocompatibility and hemocompatibility. Here, nanoparticles using poly-ethyl-2-cyanoacrylate (PECA) and poly-alkyl-cyanoacrylates (PACA) are prepared via emulsion polymerization method (aka nanoprecipitation) [426, 427]. These PECA and PACA nanoparticles carry therapeutics such as cisplatin, gemcitabine, and dexamethasone [428–430] inside their polymeric network and subsequently release the drug payload when the particles are degraded inside the cells via endosomal degradation. PECA and PACA can also be copolymerized with other synthetic polymers (e.g., PEG) to improve their circulation time [431, 432]. This iteration of cyanoacrylate–ethylene glycol copolymer is amphiphilic meaning it can self-assemble and entrap therapeutics inside its polymeric network. The synthesis process is quite facile and could entrap with high-efficiency hydrophobic drugs.

Recently, PECA has also been used as a macro-zwitterionic surface coating for reducing the adhesion of biological elements onto the surface of biomedical devices, e.g., meshes and vascular stents [433–435]. This is an interesting property of ECA as it can undergo neutral surface charge formation when reacted with dimethyl sulfoxide instead of water. In detail, the zwitterions of the ECA monomer create new bonds forming macro-zwitterions in a propagation loop that continues indefinitely due to the absence of hydroxide ions, and this propagation step is continued yielding a high-molecular-weight PECA that can be electrospun together with PU to achieve a flexible 2D mesh. The ambient moisture in the electrospinning chamber can terminate this reaction and yield a well-defined core–shell fiber morphology [434]. This electrospun mat possesses properties that have low friction coefficient and low surface free energy which are suitable in attenuating cell and platelet adhesion. This material could then be used in intraperitoneal mesh implants to avoid tissue adhesion during recovery and it could be used in vascular stenting to avoid cases of restenosis.

2.3.2.7 Polyethylenimine (PEI)

Polyethylenimine (PEI) is an organic cationic polymer that contains repeating units composed of an amine group and a CH_2CH_2 spacer and has the highest positive charge density potential as every third atom of this polymer is a protonable amino nitrogen [436]. There are two types of PEI (linear and branched) and have been reported to deliver gene and oligonucleotide. This is called "nonviral" vector for gene therapy where nucleotides are encapsulated in nanoparticles made from PEI complexes [437–439] and delivered to targeted cells to instruct the cells to create proteins or induce apoptosis. PEIs have proven to be highly efficient and versatile agents for nucleic acid delivery in vitro and in vivo due to their capability to form positive-charged complex with nucleic acids with high efficiency and provide efficient transfection and protection against nuclease-mediated degradation inside the target cells. Once inside the cells, PEIs possess a stock of amines that can undergo protonation during the acidification inside the endosomes and create a buffering effect [440, 441] that could enhance the endosomal escape of the nucleotide payload. This is also known as the proton sponge effect [440, 442], where the acidified internal environment of endosomes is neutralized, and water rushed inside the endosome vesicle will cause it to rupture and release its contents. Most nanoparticle systems use this route to achieve and enhance the therapeutic effect and to liberate whatever active ingredient it carries. In gene therapy, this effect must be done to liberate the DNA, siRNA, mRNA, or nucleotide that complexed with its vector [437–439]. Viruses do this naturally, but PEIs unique chemical structure can mimic this effect in an artificial manner, thus removing the need for manipulating viruses to be used in gene therapy. The presence of numerous amine groups makes it easy for PEI to be modified or conjugate different targeting moieties, ligands, and hydrophilic polymers that allow for increased circulation time and specific targeting of tissues and cells [443, 444], just like what a virus might have on its surface.

A wide range of molecular weight PEI has been studied and transfection efficiency increased with molecular weight, from 0.6 to 70 kDa. However, high-molecular-weight PEI causes higher toxicity due to aggregation and adherence on cell surface [445, 446], which results in necrosis. Comparing the two types of PEI (linear and branched), linear PEIs are found to be more effective as a gene delivery agent compared to branched PEI, but linear PEIs are less effective in forming complexes with longer DNA sequences compared to branched PEI at the same molecular weight. Knowing this, branched PEIs are normally used in the delivery of larger oligonucleotides, plasmid DNA, Epstein–Barr virus-based plasmid vectors, RNA, and intact ribozymes, whereas linear PEIs have seen commercial success as ExGen500® and jet-PEI® [436, 447].

2.3.3 Bioceramics

For centuries, humans have been using ceramics to better their lives, from creating bowls to crucibles that can handle intense heat, and ceramics had always been a reliable material. Now, as technology advances so does the field of ceramics, we have found that ceramics may also be used inside the body to augment and help repair body parts that share the same mechanical properties as native ceramic materials. This new breed of ceramics is called "bioceramics," which are inorganic materials intended to be used inside the body for repairing and augmenting weakened structures especially focused on the skeletal system. It is a subset in the field of biomaterials that are used for repairing or replacing damaged bone brought by diseases or injuries. Bioceramic properties are being stable in contact with tissues and biological elements, biocompatible as they do not create toxic by-products, and aesthetics make them suitable for various medical applications.

There are two types of bioceramics: (1) bioinert ceramics which are made up of nonreactive materials that do not interact with the surrounding tissues. These kinds of ceramics only facilitate structural support to the weakened structure, mostly by acting as fillers, covers, and as a replacement material. (2) Bioactive ceramics are intended to facilitate enhanced growth and proliferation of nascent bone tissue and eventually get reabsorbed by the body once enough bone tissue has been regenerated. However, like all biomaterials, morphology and composition affect how the body may react to a ceramic material. Hence, when designing a bioceramic material for a certain specific application, other properties may be modified (e.g., surface morphology, surface roughness, porosity, hardness, crystallinity, and microstructure) to best suit the requirements of whatever application it may be. These modifications are achieved primarily by doing heat treatments to change the crystal lattice of the ceramic, subsequently changing their other properties as well. Varied heat treatment (sintering) parameters may yield different phases of bioceramics, e.g., oxides are formed when heat treatment is done under the presence of oxygen, while sometimes inert gasses are used to maintain the purity of the ceramic material.

Utilization of ceramics in the medical field also encompasses extracorporeal applications (e.g., blood dialysis, gas separator, and biosensors); however, in this chapter we focus only on their application inside the body. One of its desirable properties is its mechanical properties. As it can mimic the skeletal system's native physical properties, in fact the nonliving component of our skeletal tissue is considered as a ceramic material (calcium phosphate); hence; it is only logical to replace and reinforce weakened skeletal tissues with materials that have similar physicochemical properties. Furthermore, bioceramics exhibit better hard tissue response when compared to polymers or metals used in the same application. Newer applications of bioceramics are not just limited for implantable materials but also for drug delivery applications as nano- and microparticles and ceramic drug composites, and these new iterations of bioceramics can be functionalized with targeting ligands and drugs to achieve multifunctionality. In this part of the chapter, we discuss the different ceramics and elucidate examples on how they are utilized inside the body.

2.3.3.1 Bioinert ceramics

Bioinertness is a property of a material of not soliciting any response when interfaced with biological elements. In ceramics point of view, it means that the material does not induce osteoinduction, integration, and subsequent absorption all throughout the material's life cycle. Bioinert ceramics also possess biocompatibility in a sense that they are both cytocompatible and hemocompatible, and the body can tolerate their presence without eliciting immune or inflammatory responses (a property that is also useful for drug delivery applications) [448]. They also have long-term stability in physical (e.g., load bearing and shear stress) and biochemical conditions (e.g., exposure to hydrolysis and enzymes) to avoid creating complications brought about by producing toxic leachable (e.g., ions and degradation by-products) in the body.

2.3.3.1.1 Alumina

Alumina-based biomaterials possess excellent biomechanical and biocompatibility properties and play a key role in various biomedical applications [449, 450]. In the 1980s, alumina ceramics were used for osteosynthetic devices (alumina monocrystal) or to fabricate bone and joint prostheses (alumina monocrystal and polycrystal). However, due to their brittleness as compared to those of human bone, they are seldom used for these purposes in recent times. Their salient point is that they can be made into a ball that exhibits wear resistance which is highly suited for articular joint replacement. This is done by polishing it to achieve a smooth surface with low tolerances (0.02 μm), thus creating smooth articulation and reviving the range of motion of the affected joint. Since their brittleness makes them unsuitable for high-compliance device applications, they are mostly used in bearing surfaces of joint prosthesis.

Alumina has also been used for drug delivery applications as a nanoparticle preparation [451], and the understanding on how nanoparticles made up of alumina are absorbed by the cells is limited but some reports had shown that, when orally delivered, these nanoparticles translocate almost immediately to the systemic circulation and get accumulated in the liver, spleen, and kidneys [452–454]. Al_2O_3 and Al_0 have shown some differences in the rate of translocation but nevertheless they are both effectively absorbed by the body. However, a tolerable weekly intake of 1 mg/kg body weight was implemented by the European Food Safety Authority, while a provisional tolerable weekly intake of 2 mg/kg body weight has been set by the Joint FAO/WHO Expert Committee on Food Additives. This is since alumina can impact the human health due to its physicochemical properties, e.g., size, high specific surface area, and high reactivity. Some researchers had used this ability of alumina nanoparticles (AlNP) to get absorbed by the body to deliver therapeutics via oral delivery. AlNPs have been used in the form of ordered mesoporous aluminum oxide for improved oral delivery of Telmisartan (a poor-water-soluble compound) [455] to increase their bioavailability and to better control high blood pressure. In other cases, it is used as an anticancer agent via systemic delivery by inducing reactive oxygen species toward PC-3 prostate cancer cells, and other examples used AlNP to stimulate T-cell activation by delivering antigens to dendritic cells which are beneficial for immune therapy [456]. AlNP also possesses antibacterial properties by damaging the cell wall of bacteria due to electrostatic interaction [457]. AlNPs' utility in biochemistry has been proven as a nanoplatform for correcting misfolded protein molecules [458]. It is used as a renaturing material, by interacting electrostatically with denatured negatively charged proteins, and prevents its aggregation and misfolding. The reactivity of AlNP can control the refolding and misfolding of proteins and has a significant value for designing highly stable peptide drugs or correcting malformed proteins inside diseased cells. This is just newer application of alumina and has been expanding ever since.

2.3.3.1.2 Zirconia

Zirconia-based ceramics are used traditionally for femoral head in total hip arthroplasty as part of the articulating component of the prosthesis [459]. This is because zirconia ceramics are superior to alumina ceramics in terms of mechanical strength and fracture toughness. Zirconia ceramics has a low Young's modulus which is desirable for load-bearing applications, meaning that it can withstand deformation without breaking. But this is only in relation to other alternatives such as alumina, and nonmodified zirconia ceramics are still generally considered as a brittle material, and the varied phase transitions on bulk zirconia make them unreliable in long-term load-bearing applications. In order to make them viable for bone implant applications, sintering process should be done to modify the phases (crystal lattices). Zirconia has three crystalline phases [460] (monoclinic, tetragonal, and cubic crystal structures), whereas the cubic crystal lattice is the most desirable for creating a strong material. But the gradual cooling of zirconia allows it to revert to monoclinic phase. So, in order to stabilize it, dopant of oxides such as yttria (Y_2O_3) and calcia (CaO) are added to stabilize the crystal lattice at the cubic structure even at room temperature. This is called "stabilized zirconia" [461, 462]. The low Young's modulus of stabilized zirconia is desirable because the bone tissue around a stiff material tends to get reabsorbed and weakened due to stress shielding effect, in order to maintain a healthy bone density around an implant it should allow some stress transfer to the native bone.

The ability of zirconia to have modifiable surface properties (porosity and surface charge) [463, 464] makes it an ideal nanoparticle-formulated drug delivery system [465, 466]. Porosity plays a role in bioceramic-based nanoparticle systems as they can carry more drugs per volume as compared to a surface-carrying nanoparticle. Hence, creating mesoporous zirconia allows for an efficient loading and subsequent release of drug compounds. Hollow mesoporous zirconia nanoparticles have also been synthesized and loaded with doxorubicin (chemotherapeutic) [467–469]. This iteration of the zirconia nanoparticle has a highly efficient drug-loading capability due to its increased surface-to-volume ratio. The surface charge of zirconia nanoparticles also changes in relation to the pH of the environment allowing for a pH-responsive drug-release mechanism. These are highly utilized in cancer therapy as cancer cells are known to have a more acidic microenvironment compared to normal tissue. However, even though zirconia bioceramic is known to be bioinert, reports have shown that it can induce toxicity in zebra fish embryos as manifested by malformation and mortality when exposed to 0.5–5 µg/mL concentration [470]. Zebrafish in vivo model is a reliable and suitable way for evaluating potential nanotoxicity induced by ZrO_2NPs exposure. Thus, like AlNPs, it should be used sparingly until further investigation of safety and toxicity is validated.

2.3.3.1.3 Carbon

One of the earliest used pure carbon-based materials is pyrolytic carbon. It is a bioinert ceramic that was first introduced in the 1960s. Generally, pyrolytic carbon is produced through thermal deposition (pyrolysis) of gaseous hydrocarbon compounds such as methane or acetylene at its decomposition temperature without the presence of oxygen. This creates pure carbon residues that can be made into various forms that have different microstructures (isotropic or lamellar) depending on the parameters of pyrolysis (temperature, concentration of carbon precursor, gas flow rate, etc.). Other parameter alterations in pyrolytic carbon-processing may be used to modify density, porosity, as well as mechanical properties. There are varied types of pyrolytic carbon used for biomedical applications [471, 472]. These are low-temperature isotropic (LTI) pyrolytic carbon and ultra-LTI (ULTI) pyrolytic carbon. Dense and high strength LTI pyrolytic carbon is generally deposited onto implants via chemical vapor deposition from a hydrocarbon gas at a high temperature in a fluidized bed [471]. The ULTI pyrolytic carbon is usually deposited from a carbon precursor via a hybrid vacuum process in the presence of a catalyst [473].

Pyrolytic carbon is used as a versatile coating due to its good blood and tissue biocompatibility. It can be deposited onto synthetic blood vessels [474] to form a thin layer that has excellent blood compatibility without compromising the flexibility of the grafts. These superb mechanical properties also make it an ideal component for arthroplasty implants [475, 476]. Most interestingly, it is also used in mechanical heart valves as it possesses good fatigue resistance, wettability, antithrombogenicity, and wear properties that are all beneficial for materials that are in constant contact with the blood. Heart valve disease is caused by the weakening of the ligaments that support the main bicuspid or tricuspid valves (valve prolapse) or by calcification of the valves itself (stenotic valve). This in turn makes the blood flow backward limiting the cardiac blood output. Mechanical heart valves are implanted to replace the damaged heart valve, and pyrolitic carbon is one of its vital components as it can serve as a bioinert interfacing material for the blood and provide mechanical support to the flaps of the valve.

Carbon in its pure form has exciting physicochemical properties and has also been explored for drug delivery applications in cancer treatment [477, 478]. Carbon nanotubes and sheets are different forms of graphene that can be functionalized with anticancer drugs, targeting ligands, and imaging molecules to specifically target malignancies in the body. However, there is still a debate in the safety of using carbon-based nanomaterials [479, 480] as contradicting reports had shown different cytotoxic and phenotypic alterations to the cells.

2.3.3.2 Bioactive/bioresorbable ceramics

Bioactive materials allow for biological activity to thrive and induce various responses at their interface, which means that they can activate the proliferation of nascent cells, or without causing any toxic events to the body. In bioceramic bone tissue applications, this material can conduct new bone tissue to gradually repair the damaged or defective bone. There are two types of bioactive ceramics: osteoconductive and osteoproductive [481]. Osteoconductive materials interact with hard tissue (bone) and stimulate bone growth along the surface of the bioactive material, whereas osteoproductive materials stimulate the growth of new bone on the material away from the bone/implant interface, which can also bond to soft tissue such as gingival (gum) and cartilage. Bioactive ceramics integrate well into the bone tissue where they are implanted and get gradually replaced by newly formed bones. This is also known as bioresorbability. This property is highly valuable in creating an implant particularly adapted to the transient transfection of bone cells (osteoblasts and/or osteoclasts) which are functionally deficient in some genetic diseases that affect bone growth and remodeling.

2.3.3.2.1 Silica

Silica is an oxide form of silicon found mostly in nature as quartz and is part of some living organisms [482]. As a biomaterial silica possesses biocompatibility and is considered as bioactive as it induces certain biological reactions to the body. It has intrinsic biologic activity toward bone tissue by promoting the differentiation of bone-forming osteoblasts while inhibiting the differentiation of bone-resorbing osteoclasts [483, 484]. Most contemporary research about the effects of silica ceramic revolves around it being a nanoparticle. Silica nanomaterials represent a promising biomaterial for beneficial effects on bone mass and quality as well as regenerative tissue engineering [485, 486] and are currently being investigated for intrinsic bioactivity toward the primary cells responsible for skeletal homeostasis. Properties such as size, surface charge, and composition affect how biological tissues react when interfaced with silica. Silica's surface properties make them easy to conjugate other bioactive molecules that can further enhance bone tissue proliferation. This modifiability makes it also an attractive material for applications beyond bone tissue engineering. Silica can be made into various geometric shapes, e.g., spheres, rods, and spikes, all of which have varied biological reactions toward the body. But the most interesting geometric nanoshape is it being mesoporous [486, 487], as like other ceramic materials, and it means that it can load active molecules at higher efficiency due to the increase of surface-to-volume ratio.

In detail, silica can increase osteoblast differentiation and is highly related to composition and size, with 50 nm silica nanoparticles being optimal. If made to have negative surface charge, it can then inhibit osteoclast activity by downregulating key osteoclastogenic transcriptional regulator Nfatc1 and its target gene RANK [488], thereby limiting the reabsorption of nascent bone tissue at the early stages of bone formation. This effect can accelerate the bone tissue growth as it favors mostly the induction of new bone material in contrast to reabsorbing it.

In drug delivery applications, mesoporous silica nanoparticles (MSN) facilitate effective loading of drugs and their subsequent controlled release of the target site [489, 490]. Like other mesoporous ceramic particle systems (e.g., zirconia), MSN has high drug loading and porosity, as well as the surface properties that can be altered depending on additives used to prepare the nanoparticles.

2.3.3.2.2 Hydroxyapatite

HA is a naturally occurring mineral form of calcium apatite that exhibits biocompatibility and is the preferred material for bone repair and substitution due to its similarity to the native calcium mineral in the body [491]. HA exists as a specific ratio of calcium and phosphate at Ca/P set at 1.67 with a chemical formula $Ca_{10}(PO_4)_6\,(OH)_2$. The human bone consists of 40% inorganic component (HA), while the rest is made up of 25% water and 35% organic component (proteins). So, logic follows that HA is one of the best suitable materials for repairing damaged bone tissue, as it can be synthesized outside the body and be used as a functional filler, scaffold, or implant coating [492]. HA can be synthesized using different techniques such as precipitation, hydrolysis, and hydrothermal synthesis using easily sourced material from animal sources (e.g., fish bones, seashells, animal bones, and eggshells)

HA has been one of the most used materials for increasing bone regeneration rate in bone defects, and fillers usually made up of polymeric materials due to their injectability have been proven to increase bioactivity if HA has been incorporated in its matrix [493–495]. The porous structure of this polymeric filler acts as a transient ECM for bone cells to attach to, and with the presence of HA, it can stimulate adhesion, proliferation, and migration of osteocytes onto this scaffold. Macropores facilitate osteoid formation and mineralization by increasing the osteoblast and osteoprogenitor migration into the scaffold, whereas interconnected micropores can enhance vascularization and nutrient diffusion during bone reconstruction. Polymer–ceramic composites have the combination of compressive strength of the inorganic ceramic phase as well as toughness and flexibility of the polymer, making them suitable for various load bearing filler applications. Crystalline HA has the lowest rate of degradation in the body and thus could be used for prolonged period if the defect is too large, or longer healing times are required. Filler with up to 70% HA content can facilitate more differentiation and mineralization. As discussed earlier, there should be some tension transfer from the scaffold to the native tissue in order to avoid bone resorption [496, 497]. Here, polymer–ceramic fillers are highly effective in giving this required tension to the native bone due to their balanced elasticity and compressive strength properties. Adding other drugs and bioactive components (growth factors) in these polymer–ceramic composites are also widely used in commercially available composite fillers and scaffolds.

Arthroplasty implants are also coated with HA on their stem components [498]. In some cases, these prostheses suffer from bone mass loss around the stem area due to stress shielding, a phenomenon known that increases bone resorption due to the reduction of experienced stress by the native bone. Bone remodeling is signaled by physical cue where stress is needed to stimulate the osteoblasts to create and calcify into dense bone mass [499, 500]; hence, exercise and physical activities help in increasing bone density, while being sedentary or bedridden makes a person lose bone mass. This effect is notably seen in humans who are stationed in the International Space Station, where astronauts have prolonged exposure to microgravity, thus making their bone decrease in mass due to the significant load being loss by not experiencing gravity [501, 502]. Arthroplasty prosthesis, e.g., hip and knee replacement, creates this stress shielding environment where it is implanted (long bones), and over time bones surrounding the prosthesis disintegrate, and the addition of HA on the surfaces of this prosthesis can reduce this effect by stimulating the bone growth around the stem of the prosthesis and creating a reinforced dense hard tissue that provides enough mass for stable anchorage (osteointegration). HA coatings on the implant surface can be prepared by many different techniques such as plasma spraying, electrophoretic deposition, sol–gel deposition, sputtering techniques, and ion beam deposition [503, 504]. These techniques have varied effects on the effectivity of coatings. Depending on which coating technique is used, crystallinity, coating thickness, adhesion and bonding strength, and cost-efficiency may vary. It is also known that the arthroplasty procedure is prone to implant-related infection, thus creating an antibacterial coating that is beneficial to avoid such complications, and HA can be doped with antibacterial agents such as silver, zinc, and magnetite nanoparticles [505, 506] so that it can decrease the adhesion of bacterial species onto the surface of the prosthesis. HA by itself has no inherent antibacterial effects; however, it can be easily modified to contain these nanoparticles and still preserve its osteoconductive properties.

As a drug delivery system, porous and nanocrystalline HA is valued due to its nontoxicity and biocompatibility [492, 507]. Like other ceramic-based nanoparticle systems, HA can be made into porous or hollow nanosized particle with high surface-to-volume ratio that can be loaded with various therapeutics such as antibiotics, anticancer, anti-inflammatory, and other active substances with high efficiency [508–510]. HA is easily solubilized in acidic environments; thus, it can release its drug payload if absorbed by the cells via phagocytosis. This is the usual cell uptake mechanism for nanoparticles, and it allows for HA to be solubilized inside the endosomal environment ($pH < 5.0$). Once solubilized, Ca and phosphate ions are formed, thus acting as a buffer (weak basic molecule) to create the proton sponge effect that allows water to rush into the endosome causing it to burst and subsequently release the drug payload into the cell.

In localized drug delivery, HA implants are loaded with antibacterial and growth factors that when combined with the osteoconductive properties of HA greatly enhances the bone repair process. The local and sustained release of active substances allows for shortening of the prolonged therapies and can accelerate the bone healing process, as well as minimize the extent of surgical removal of the affected bone.

2.3.3.2.3 Beta-Tricalcium phosphate (β-TCP)

Another calcium phosphate bioceramic is called beta-tricalcium phosphate (β-TCP). Like HA, β-TCP exists as a ratio of calcium and phosphoric acid at 1.5 Ca/P and with a chemical formula $Ca_3(PO_4)_2$ [511]. In comparison to HA that may exhibit poor absorption due to its crystalline phases and stability in neutral pH, β-TCP is relatively balanced between absorption and new bone formation [512], and can also release a large quantity of calcium ion (Ca^{2+}) and sulfate ion (SO_4^{2-}) that are important salts that increase new bone formation. However, β-TCP has some disadvantages, first its absorption does not completely agree with the new bone absorption leading to poor osteoconduction and osteogenic in vivo [513]. Second, its mechanical properties are poor and exhibit slight brittleness, which makes it unable to resist against fatigue and insufficient holding power, thereby making it susceptible to scaffold collapse and severely limiting it to nonload-bearing applications [511]. Hence, to make β-TCP clinically relevant it must be used in combination with other bioceramics (HA) or with a polymer–ceramic composite. The β-TCP/HA composite material is also named as biphasic calcium phosphate (BCP) [514]. As a composite, contents or ratio of β-TCP and HA could affect new bone mass and material degradation rate and the ratio of 85:15 β-TCP:HA was found optimal for new bone formation. Combining the two most common nonorganic components of the bone as a bioceramic material to help it to regenerate makes a lot of sense. BCP could produce more new bone mass for shorter duration of time as it creates synergies that make up for the deficiencies of too rapid absorption of single β-TCP and hard absorption of HA.

Other combinations of β-TCP are also used to enhance its effects in vivo, for example, adding it to the polymeric network of PCL [515, 516] makes flexible scaffolding with osteoconductive properties. Polymer–ceramic composite is a recognized material that utilizes both inherent bioactivities of the ceramic and mechanical properties of the polymer. This in turn yields a material that can be used in load-bearing applications in which pure β-TCP is not suited for. β-TCP/PCL composite fulfills the requirements for new bone formation and simultaneously enhances the mechanical strength. Incorporation of metal ions such as Zn^{2+} both promotes osteoclast apoptosis and attenuates osteoclast activity, thereby accelerating new bone formation and reducing bone resorption activity. It can also delay the absorption of the β-TCP, which increases the life span of the scaffold, prolonging the bone regenerative properties of the material. β-TCP can also be used together with protein-based materials, e.g., platelet-rich plasma (PRP) and plasmatransglutaminase (F XIII) [517–519]. PRP is rich with growth factors such as platelet-derived growth factor, transforming growth factor beta, VEGF, and epithelial growth factor that can enhance the growth of new vasculature that is crucial for transporting nutrients and excretion of waste and also increases the proliferation of nascent bone cells on the scaffold material.

2.3.3.2.4 Calcium sodium phosphosilicate (Bioglass 45S5)

Calcium sodium phosphosilicate, commonly referred to as bioglass, is specifically composed of 45 wt% SiO_2, 24.5 wt% CaO, 24.5 wt% Na_2O, and 6.0 wt% P_2O_5. In comparison to normal glass (lime glass), bioglass is composed of less silica and higher amounts of phosphorus and calcium. Like other glass materials, bioglass is a noncrystalline amorphous solid that has the capability to be not just biocompatible but also bioactive. The 45S5 name signifies glass with 45 wt% of SiO_2 and 5:1 molar ratio of calcium to phosphorus. The bioactivity of bioglass happens when it reacts with the fluids in the body and dissolves into ions (e.g., calcium and silica ions) [520]. An amorphous Ca^{2+} and PO_4^{3-} gather at the silica-rich layer from both the surrounding bodily fluid and the bulk of the bioglass. This creates a layer composed primarily of $CaO–P_2O_5$ on top of the silica layer, and this layer of calcium and phosphate is called a mixed carbonated hydroxyl apatite (HCA). This allows for new calcification sites (tissue hardening) that provide structural support to weakened bone structures in the body. Furthermore, since this layer has structural and chemical similarities to native HA, it can absorb growth factors that can promote healing and initiate the migration of osteoprogenitor cells, and MSCs to implantation site. In addition, bioglass can activate M2 macrophages [521] which are essential in the healing process as it activates MSCs and osteoprogenitor cell migration to the bioglass surface and attach to the HCA layer, whereas a nonbiocompatible material activates M1 macrophages which can trigger inflammation or implant rejection. The recruited MSCs and osteoprogenitor cells differentiate into osteogenic cells and adds to the population of nascent osteoblast that will then generate and deposit ECM components, primarily type I collagen, the main protein component of bone. These proteins will then further be mineralized with the help of HA nanocrystals and reinforce the surface of the implant. Overtime, as the implant degrades it is then slowly replaced by newly mineralized bone tissue and repair the damage or defect.

The glassy structure of this ceramic material allows for different kinds of geometric shape to be formed as it can be machined or even 3D printed to different shapes and sizes it can also be ground up into ultra-fine powder that can be used for coatings of metal implants (screws or plates) or as an additive to toothpaste to decrease dentin oversensitivity. Its versatility is well known in the orthopedic field as it serves as one of the most effective active components on regenerating bone damage and defects [522, 523]. Its degradation products which are ions that can increase the pH of the substrate can inhibit the formation of biofilms from bacterial sources. Furthermore, bioglass debris can adhere on the bacterial cell wall causing it to rupture and die. This is a very crucial property of bioglass as it can avoid nosocomial infection as a result of aseptic failure during implantation. A setback to using Bioglass 45S5 is that it is difficult to process into porous 3D scaffolds due to its inability to fully sinter. Sintering is one of the most common ways to create porosity in a ceramic material. Porosity is an important factor in creating and affecting tissue scaffold as it allows for an efficient cell migration and exchange of nutrients. Also, Bioglass 45S5 glass has a slow degradation and rate of conversion to an HA-like material which makes it more difficult for the degradation rate of the scaffold to coincide with the rate of tissue formation. Increasing porosity may answer these setbacks and with the advent of 3D bioprinting, creating polymer–bioglass composites are possible [524, 525]. This combines the bioactivity of bioglass with the relative flexibility and printability of different polymers. The printability of these polymer–bioglass composites allows for the porosity to be designed and printed without the use of harsh sintering processes.

2.3.4 Metallic biomaterials

Metal as a material has been used in various fields and had catapulted us humans to great heights. Its strength, luster, and resilience make it an ideal material for various industrial applications. Metals as biomaterials have been used for a long time. An example is the addition of molten gold or mercury to replace damage teeth in ancient Egypt. Metallic biomaterials, in general, are strong, malleable, and resistant to wear and corrosion. Most implants for bone reconstruction, e.g., screws, plates, prosthesis, and electronic and nonelectronic biomedical devices (e.g., vascular stents, heart valves, cochlear implants, and pacemakers) are mainly composed of metallic materials [526, 527]. Metals are easy to fabricate into various shapes and forms, and can be machined, stretched, grind, and polished into whatever geometric shape and size meant for a specific application. Besides its mechanical properties, its chemical properties are also useful in biomedical applications as it can be easily functionalized with bioactive coatings that can promote various biological activities [528, 529]. However, not all metals are applicable to be used inside the body. Only a few metallic compounds and alloys are known to be stable enough to be used as a biomaterial. In this part of the chapter, we identify different types and classes of metallic biomaterials and their applications on the human body.

2.3.4.1 Metallic alloys

Metallic alloys are combination of different metals, or a metal combined with one or more other elements that are forged or smelted into one metallic grain (crystals). The resulting alloy typically differs from those of the pure metals, such as increased strength or hardness. In biomaterials, alloys of stainless steel, titanium, nickel–titanium, chromium cobalt, tantalum, magnesium, and zinc are used for various applications [526, 527]. We can further divide this classification of metallic biomaterials into permanent and bioresorbable alloys, whereas permanent metallic alloy materials are used for long-term (or lifetime) implant applications, and bioresorbable metallic alloys are used for transient implant applications [530] (e.g., tissue engineering and local drug delivery). Most metal implants both permanent and bioresorbable are coated, or surface modified with biocompatible materials using techniques such as dip-, spray-, or spin-coating or solvent casting and newly developed methods such as electrodeposition, UV-light-assisted coating, and self-assembly monolayers [528, 529]. For permanent metal implants, the goal is to integrate active compounds onto their surface to add functionality for permanent alloys (bioactivity, antibacterial, and anti-inflammatory), whereas for bioresorbable metals, it is to control the degradation rate and release of ions to the systemic circulation.

2.3.4.1.1 Permanent alloys

For permanent metallic implants, important qualities besides biocompatibility are corrosion resistance, mechanically stability, and bioinertness. All these properties are intertwined as to provide an implant material that could last for a lifetime. Most implants made from these alloys must survive the highly aggressive microenvironment that may lead to material degradation and associated release of unwanted metallic ions and should not experience any kind of mechanical failure of the device (e.g., wear, fractures, and cracking). Therefore, only highly stable alloys with long-term stability under highly reactive in vivo conditions and excellent mechanical properties such as biomedical-grade stainless steel 316L, titanium, nickel–titanium, tantalum, and chromium–cobalt alloys are used in this kind of biomaterial applications (fracture fixation, angioplasty, and bone remodeling).

Stainless steel (biomedical-grade 316L)

316L Stainless steel is composed of chromium, nickel, and molybdenum alloy that exhibits high corrosion resistance, adequate mechanical properties, and ductility. The denotation L means that the carbon content in the alloy is less than 0.03%, comparatively lower than other metallic biomaterials but it still possesses enough mechanical properties. The "biomedical-grade" stainless steel does not contain any nickel in its alloy as it has been proven that nickel may cause allergic and inflammatory reactions to some individuals [531]. In lieu of nickel, nitrogen is used with the alloying of Ni-free stainless steel to maintain low Ni content. The high corrosion resistance of stainless-steel comes from its possession of a high chromium content (>12 wt%) in its alloy with minimal carbon to limit the formation of chromium carbides, and formation of these carbides decreases corrosion resistance. Hence, biomedical-grade stainless steel has low carbon content. The low carbon content also reduces the hardness of the material and makes it more ductile, which is beneficial for bone regeneration application as it avoids stress shielding [532, 533]. However, stainless steel has low osteointegration activity compared to other alloys such as titanium. In this sense, surface functionalization of stainless steel is required to enhance bone integration toward the implant [534, 535]. Common strategies employ the addition of ceramic materials such as HA, β-TCP, and bioglass to stimulate to proliferation of bone cells on the surface of the stainless steel. Other surface functionalization by adding antibacterial or bioactive molecules are also employed in stainless steel [204, 536].

Early generation of vascular stents are made up of 316L stainless steel struts and has been commercialized with relative success [537], but its surface properties can induce thrombosis via platelet and fibrinogen deposition [538]. Also, since it has noncompliant mechanical properties, the stent can cause trauma to the vessels that leads to scar formation. The prolonged used of stainless steel stents can activate local inflammatory response and cytokine production. This in turn will recruit immune cells that will eventually cause in-stent restenosis. This is considered as a device failure for the stent, as the blood flow is again obstructed. Further surgical intervention is needed to alleviate it and additional angioplasty is required to repair the vessel giving more burden to the patient. Thus, in order to avoid this type of scenarios, surface modification and addition of polymeric drug agents are used to attenuate the adhesion of platelets and fibrinogen and reduce neointimal growth of vascular endothelial cells [204, 539, 540].

Titanium

Titanium alloys are lightweight and have far superior toughness and durability. It is also highly corrosion resistant and could last all throughout the patient's lifetime [541]. Typical titanium alloy used for medical applications is called Ti–6Al–4V, denoting 5.5–6.5 wt% and 3.5–4.5 wt% of aluminum and vanadium, respectively. Increasing the aluminum content on the alloy makes it harder by almost 32% without sacrificing its other mechanical properties. Other alloys of titanium are also made to increase wear resistance by adding niobium, as in $Ti_{15}Nb_4Ta_4Zr$ alloy. It can also reduce the friction coefficient of the alloy due to the formation of Nb_2O_5 layer which is beneficial in articulating and moving prosthesis. However, alloying pure Ti with other elements may alter mechanical properties such as ductility [542, 543], which may be undesirable for applications where bending and compliance are parts of the implant function. It could also create stress shielding if it is intended for bone prosthesis and for which unalloyed Ti may present a more suitable alternative. The weight of Ti alloys makes it an attractive material for bone prosthesis [544, 545] (hip and knee arthroplasty) and other osteo-related implants [546] (e.g., cranial and dental plates and screws) because it is lightweight and highly durable and may cause less trauma or distress to the implantation site and facilitate healing due to the high dielectric constant of titanium dioxide layer that can form on the surface of the implant that promotes cell integration, enabling slight osteointegration without further modifications.

However, that does not stop further material research to improve Ti alloy's utility in biomedicine. Sintering can create porosity toward 3D-fabricated Ti alloys (Ti–6Al–4V) [547], and porous structure is highly valuable to conserve weight and increase volume-to-surface area ratio that is highly beneficial to drive the growth of various cells. The advent of metal 3D printing also enables biomedical engineers to design prosthesis that would allow optimal structural properties, thus further saving on weight and material needed for a specific prosthesis or implant. In dentistry, low-density high-strength materials are valuable in guiding the growth of soft and hard tissues. A specific technology called direct laser metal sintering (DLMS) as a 3D printing technology is utilized as a novel fabrication technique for various titanium alloys [548, 549]. The salient points of DLMS 3D-printed titanium (alloy)-based medical devices are the excellent mechanical properties due to the low density and versatility that mimics the anatomical microenvironment of hard tissues. It alloys for producing porous microarchitecture on the implant that serves as anchorage point for bone cells in the jaw and is highly beneficial for reconstructing and supporting weakened areas of the jawbone. The surface of Ti alloys can also be further functionalized with other active compounds to avoid complications associated with metallic implants [550]. Adding drugs to fight infection [551] is a common strategy to avoid bacterial colonization, and immobilization of other nondrug components such as HA [552] and AgNPs [553] can both increase bone cell induction and bacterial colonization, respectively.

Cobalt

Cobalt–chromium (Co–Cr) alloys such as Co–28Cr–6Mo and Co–20Cr–15W–10Ni alloys are highly utilized in biomedicine due to their high corrosion and wear resistance, elasticity, besides from being bioinert and biocompatible [554, 555]. Cobalt (Co)-based implants have higher wear resistance compared to Ti alloys, which makes them ideal in artificial hip joints prosthesis, where the direct contact between femoral head and the bone or plate over time may lead to wear. In biomedical applications, alloy of Co–Cr is one of the most used alloys due to a favorable combination of high strength and high ductility. However, Co–Cr has higher stiffness and density compared to other metallic alloys which makes it less favorable for using it as the stem of prosthesis. It could lead to greater stress shielding that could render bones to disintegrate over time. Also, biocompatibility of and osseointegration capacity of Co–Cr is also lower than that of Ti, thus limiting its application to only parts of the prosthesis that is in constant friction and articulation. Co–Cr is the choice material for those elements that are not interfacing with the bone tissue, while Ti alloys are used for tissue interfacing components of prosthesis. Yet, such designs give rise to metal corrosion due to mechanically assisted crevice corrosion, and shredding at the site of the contact between Co–Cr and Ti. Furthermore, recent findings make some clinicians to believe that ions of Co–Cr can leak over time and lead to various toxicities [556].

Co–Cr is also used for vascular stent applications [557, 558] as it has high elastic modulus and hemocompatibility compared to stainless steel, and Co–Cr vascular stents can be deployed with less force using balloon-inflated catheter. It is considered a safer option because it can comply to the forces that the vasculature renders to it, thus limiting trauma to the vessels. It is also used for dental prosthesis as a metal framework for dental partials. Co–Cr alloy's high elasticity and cyclic fatigue resistance make it useful for such applications.

Nickel–titanium (nitinol)

Nitinol is an alloy of nickel and titanium at 1:1 ratio that has super elastic shape memory properties. This super elasticity allows it to be used in various applications such as stents, heart valve tools, bone anchors, staples, septal defect devices [559–561]. Even though nickel is considered to cause allergic reactions to some patients, this type of alloy makes it viable as the titanium component creates a stable passivation oxide layer that makes it inert. In alloys, such as nitinol, the formation of an oxide layer not only protects against corrosion but also removes Ni atoms from the surface of the material. This passivation property of the alloy is very important to remove toxic implant elements from the implant–body interface [562, 563]. Other active passivation techniques are used to control the surface properties of nitinol (e.g., electropolishing, and polymeric coatings) and these techniques add another layer of inert surfaces in order to limit the leaching of toxic ions and direct contact of biological tissues toward the implant. These techniques will be further discussed in another chapter in detail.

Nitinol as a stent material is highly useful as it does not require a balloon expansion catheter in order to be deployed. Since it has shape memory and elastic properties, it can be crimped into smaller diameter and expand by unsheathing the catheter. This mechanism of deployment is easier for clinicians as it avoids further inflation of a catheter with an elastomeric balloon. This in turn avoids unnecessary trauma that may be caused by over-inflation if the stent is designed for vascular applications. These shape memory properties of nitinol come from its ability to undergo deformation at one temperature, stay in its deformed shape when the external force is removed, and then recover its original, undeformed shape upon heating above its "transformation temperature" [564], whereas this transformation temperature can be tailored to be equal to the human body. It must be noted that below the transformation temperature it exhibits the shape memory effect, and above that temperature it behaves super elastically. Nitinol's unusual properties are derived from a reversible solid-state phase transformation known as a martensitic transformation between two different martensite crystal phases. At high temperatures, nitinol assumes an interpenetrating simple cubic structure referred to as austenite (also known as the parent phase). At low temperatures, nitinol spontaneously transforms to a more complicated monoclinic crystal structure known as martensite (daughter phase). Under the martensite phase, if nitinol is exposed to mechanical deformation it will retain that structure until it is heated up again and it will revert to austenite phase. Above this temperature point, it is superelastic and can withstand deformation without transitioning to martensite phase again. This property of nitinol makes it highly suited for load-bearing applications inside the body. However, there are some concerns about the durability of nitinol-based implants as they are heavily applied in applications that require enormous flexibility and motion (e.g., vascular and nonvascular stents, heart valves, and septal defect implants). While the fatigue performance of nitinol is better compared to stainless steel and cobalt-based alloys, fatigue failures are still observed in the most demanding applications. Hence, there is an enormous research effort underway trying to better understand and define the durability limits of nitinol.

Tantalum

Tantalum alloy is used for biomedical applications due to its biocompatibility and outstanding corrosion resistance even in acidic media [565, 566]. This stability is attributed to the formation of passive oxide layer of Ta_2O_5 that forms on its surface allowing for low reactivity when interfaced with the harsh conditions of the body. However, the metal alloy has high elastic modulus that is not compatible to the modulus of the native cortical bone that may be detrimental for bulk implant applications. Thus, tantalum metal alloys are usually used in conjunction only as a surface coating material for stainless steel implants [567, 568] that have higher stiffness to improve corrosion resistance and osteointegration. Tantalum in a porous state has excellent bone induction properties [569].

Several limitations for widespread use of tantalum also arise such as the difficulty in manufacturing due to the very high melting temperature needed to construct this alloy and its high temperature conductivity that when used for cranial plates leads patients to experience temperature-dependent headaches [570]. But with the advent of selective laser melting (SLM), porous alloys of titanium and tantalum can be manufactured, and this can potentially solve the said limitations and when compared to commercial pure Ti and Ti–6Al–4V alloy, SLM Ti–Ta alloy demonstrated higher strength-to-modulus ratio.

2.3.4.1.2 Bioresorbable alloys

Bioresorbable metal alloys are known to degrade overtime, and their degradation rate and degradation products are controlled and nontoxic. These metals are made up of magnesium, zinc, and iron alloys which in controlled doses are not toxic to the body. Therefore, it must be considered that the degradation rate of these alloys must be gradual not only for toxicity purposes but also to be able to maintain structural integrity of the implant. In most applications, bioabsorbable metals are used where load is needed. If the degradation rate of the alloy is fast, then the structural integrity of the implant or device is compromised and it will lead to a device failure. In comparison to biodegradable polymers, metal alloys used in contrast toward the same applications have significant advantages in aspects of mechanical stability, and degradation due to the main by-products of bioresorbable alloys are used by the body to maintain some bodily functions. Magnesium helps in maintaining normal nerve and muscle function, supports a healthy immune system, keeps the heartbeat steady, and helps bones remain strong. It also helps adjust blood glucose levels. Zinc is needed for the body's defensive (immune) system to properly work. It plays a role in cell division, cell growth, wound healing, and the breakdown of carbohydrates. Zinc is also needed for the senses of smell and taste, while iron is essential in hemoglobin and myoglobin production. However, too much of these elements can cause toxicities; hence, it must be emphasized that a well-controlled degradation of these metallic-based biomaterials must be ensured.

Magnesium

Magnesium alloy has superb mechanical properties with low elasticity and low density. Various forms of screws, rods, metal plates, stents, and staples can be produced with magnesium alloy, and these constructs are meant to degrade over a period giving space for biological tissues to grow and removing the need to remove the implant once its function has been served [571–573]. Magnesium alloys have almost similar mechanical properties to native cortical bone; hence, it is used for osteo-related implants to repair bone defects and damages. And since magnesium is essential in bone regeneration, its gradual degradation enhances the proliferation of osteocytes. However, magnesium alloys have an almost instantaneous degradation process in physiological environments, leading to excessive leakage of Mg^{2+} ions to the implantation site and to the systemic circulation [574, 575]. This may cause magnesium toxicity (hypermagnesemia) which usually develop after serum concentrations exceed 1.74–2.61 mmol/L. Patients may experience various symptoms that if left untreated may cause detrimental effects. Furthermore, the degradation reaction also leads to formation of hydrogen gas bubbles that may form around the implant. This can acidify the implantation site and cause inflammation, and in some cases cause an embolus to form which can be deadly. Lastly, the fast degradation of the material may lead to reduced mechanical strength due to various corrosion mechanisms that may alter the implant. All of this may hinder the bone regeneration, thus defeating the purpose of the implant. Therefore, surface modifications are needed to ensure that the magnesium alloy is being degraded in a controlled manner. Various surface modifications have been explored with various strategies to attenuate the corrosion rate of magnesium alloys [576, 577]. But due to the high reactivity of magnesium, normal passivation techniques are not enough to inhibit the corrosion rate of the alloy. First, and most simple may be the addition of oxide layers on the surface of magnesium by chemical conversion treatment [578, 579]. This is done by adding passivation inhibitors such as chromate, permanganates, nitrite, and H_2O_2, to form oxide layers on the surface of a metal. However, these passivating-type inhibitors only work on conventional metals, if they are used on Mg alloys, and the addition amounts should be large enough to effectively mitigate Mg alloy corrosion. Up to this date, no effective chemical conversion treatment has been reported that can passivize magnesium at a small amount of addition. Other types of corrosion inhibition are addition of a polymeric layer by either dip coating [580], sol–gel method [581], electrospinning [582], electrospraying, and electropolymerization [583]. These techniques add whatever type of polymer it may be to create a layer that can prevent the contact of biological fluids on the alloy while still maintaining the mechanical properties of it. This polymeric layer should persist all throughout the life span of the implant because if it degrades faster than the alloy then the uncontrolled degradation of magnesium may occur, thus defeating the implants' regenerative purpose. One effective technique may be adding highly more stable elements on the alloy itself. Some researchers have proven that the addition of calcium to the crystal lattice of the magnesium alloy may slow down the degradation of the material without the formation of an oxide layer [584].

Zinc

Zinc alloys like magnesium alloys are being used for applications due to their bioabsorbability and mechanical properties. Since they are bioabsorbable and are used by the body for maintaining various homeostatic functions, they can be made to be a transient load-bearing device (stent) or to assist in regenerating bone defects (bone scaffold, screws, and plates) [585, 586]. The degradation mechanism of zinc alloys is associated with the reaction to dissolved singlet oxygen in the body. The corrosion is also mediated by the environmental pH surrounding the alloy, as at neutral pH (7.3), zinc chloride and zinc oxide begin to form and accelerate the corrosion process. However, in comparison to magnesium alloys, zinc alloys have a lower corrosion rate and do not involve the evolution of hydrogen gas, making it well suited for vascular stent applications. Zinc alloys are predominantly being explored for use for third-generation vascular stenting [587]. These are stents that are absorbed by the body once their purpose had been served. Zinc is a safer option for this application as most tissues in the body can tolerate higher concentration of zinc ions compared to magnesium ions once the degradation process of the stent starts. In vivo animal studies have shown that zinc alloys implanted in arteries do not cause local inflammatory response and thrombosis. They can also integrate partially to the arterial lumen without any adverse effects to the animal [588, 589]. Zinc alloys had also been mixed with elements such as Mg, Ca, Sr, Li, Mn, Fe, Cu, and Ag for bone tissue scaffolds [590]. And it is found that zinc with manganese and lithium possesses superb mechanical properties and elastic modulus that is comparable to native cortical bone. The use of these elements compared to magnesium alloys delays the corrosion rate and inhibits the production of hydrogen gasses on the implantation site that may cause acute inflammation. Furthermore, zinc's presence in the bone stimulates osteoblast formation, meanwhile, inhibiting osteoclast differentiation and results in accelerated bone regeneration, making it highly attractive for osteo-related implants and devices.

Iron

Iron alloys are mechanically superior to both magnesium and zinc alloys [566]. Like zinc, its degradation is also influenced by the presence of dissolved oxygen on the implant environment. Among the discussed bioabsorbable metals, iron has the lowest dissolution properties and higher corrosion resistance [591]. This is due to the ability of iron to form a relatively stable oxide layer on its surface preventing the rest of the metal to be exposed in biological media. Iron-based medical implants have higher mechanical properties making them possible to use less material in applications such as stenting [592]. Thin stent struts allow for less chance for thrombosis to happen due to the reduced surface interaction between the blood and the metal. It is also more ductile and compliant to the changing radial forces that the vessel may apply to it. In vivo animal studies had shown that the use of bioabsorbable iron-based stents does not create inflammatory response, toxicity related to ion leakage, and thrombosis. Even though the reduced degradation rate of iron-based implants makes it more stable than zinc, it can cause incomplete corrosion within the vessel. This can create problems as it could block the blood flow in the vessel or cause perforation of sharp edges, or pointy ends are created [593]. Unlike, zinc-based alloy that can reintegrate to the vessel wall, iron-based stents do not show this effect. To solve this, the corrosion rate of iron must be increased by alloying it with zinc. To increase the rate of degradation of the alloy means that there is a risk for over-leaking of ferric ions to the blood stream. The normal daily intake of iron is set at 6–20 mg but chronic exposure to high levels of iron in the blood stream may the condition called hemochromatosis and may lead to life-threatening conditions, such as liver failure, heart problems, and diabetes. This is because iron in the blood tends to concentrate more in the liver, heart, and pancreas. The application of iron-based alloys in orthopedic implants is quite limited as its mechanical properties are too rigid and could cause stress shielding that may increase bone resorption around the implant. However, like other alloys that are too rigid, iron-based alloys can be 3D printed to create porosity to reduce its density and allow for some flexibility and transfer stress around the implantation site [594]. The porosity also allows for more surface area for cells to adhere and migrate.

Iron oxide nanoparticles made up of, e.g., Fe_3O_4 and Fe_2O_3 have also been explored for drug delivery applications, and their reactive surface makes them an ideal platform for attaching bioactive compounds (e.g., chemotherapeutic drugs, growth factors, and peptides) [595, 596]. This has been the nanoparticle of choice for an emerging cancer therapy called hyperthermia. A short description of hyperthermia therapy is the increase of temperature from normal body temperature to 43–45 °C on malignant tumor [597, 598], and this increase in temperature causes apoptosis and necrosis to cancer cells as it has been found that the thermal regulation of malignant cells is highly inefficient [599, 600]. Iron oxide nanoparticles are ferromagnetic, which mean that they respond to the influence of magnetic field. If the magnetic field is oscillating, the magnetic moment of the ferromagnetic domains in the nanoparticles flips and subsequent oscillation causes hysteresis energy loss that is converted into heat. If the iron oxide nanoparticles are designed to target cancer cells, the combined effect of the drug release and hyperthermia therapy leads to higher cell death.

2.3.4.2 Noble metals

Noble metals such as gold and silver have also been used as biomaterials. Notably, with the advent of nanotechnology, synthesizing nanoparticles out of these noble metals had been an interest both in tissue engineering and drug delivery. Gold exhibits superb chemisorption properties in nanoscale, allowing a wide array of molecules to attach to its surface. Thus, this gives a wide potential for delivering therapeutics to target tissues. AgNPs, on the other hand, is an important component for giving materials' antibacterial properties. Silver has a strong electron transfer capability that damages the cell wall of pathogens, and due to its nanosize, it allows for increased contact points toward the bacteria enhancing its pathogen killing effects that may be used in preventing infection.

2.3.4.2.1 Gold

Gold traditionally has been used in dental applications as a tooth cavity filler or shaped into a tooth and acts as a replacement for extracted or missing teeth [601, 602]. Contemporary research and the emergence of nanotechnology extended the application of gold to drug delivery and biosensing. AuNP can be easily synthesized using various chemistry reactions that allow for different geometrical variations on the morphology of AuNPs [603]. As a bulk material, gold is very inert but at the nanoscale, it possesses superb substrate capabilities. The substrate activity of AuNPs depends on several factors and is primarily mediated by particle size, that is, the surface area of nanoparticles which is inversely proportional to the particle size. In both drug delivery and biosensing applications, surface area of the particle must be optimized to allow the maximum drug loading and sensing efficiency, respectively [604].

As a biosensor, AuNP is considered as a material due to its conductivity, and optical properties which are both integral in creating a highly sensitive detection method of biological molecules [605, 606]. Immobilization of selective receptors for detecting proteins, antibodies, and biomolecules is easily done due to its unique chemisorptive properties. Chemisorption is a kind of adsorption which involves a chemical reaction between the surface and the adsorbate. AuNP on the nanoscale has a highly polar surface charge, and molecules (anionic and cationic) can easily bind to its surface. As a drug delivery vehicle, AuNP is chosen because of its high biocompatibility and chemisorptive properties, and due to its high density, it can also be used as an imaging agent using conventional techniques (e.g., computed tomography, magnetic resonance imaging, and X-ray). The ability of a drug delivery platform to be used in both therapeutics and diagnosis is called theranostics [607].

Theranostics is used in cancer treatment as it allows for simultaneous imaging and tracing of affected tissues in the body while also delivering chemotherapeutics.

2.3.4.2.2 Silver

Silver as a biomaterial is mostly used for its antibacterial property. AgNP is readily synthesized using various methods [608, 609]. As a nanoparticle, AgNP possesses almost similar properties with AuNP but its salient point is its ability to disrupt various biological mechanisms of bacteria. With the emergence of drug-resistant pathogenic strains due to the overuse of pharmaceutical antibiotics, researchers are finding non-pharmaceutical-derived materials to avoid infection in the body. Wound dressing materials [194, 610] and bone implants [611, 612] can benefit from this as AgNP can be deposited onto the surface of these biomedical devices to increase the wound healing process and reduce the risk of implant-related infections, respectively. There are various mechanisms on how the antibacterial property of AgNP arises. The most common mechanism is that it can disrupt the ability of the bacteria to form its protective cell wall. As the AgNP encounter the bacteria, they adhere to its cell wall and cell membrane, and the high electron transfer capability of AgNP can interact with the phosphate-containing compounds inside the bacteria and sulfur-containing proteins on its membrane. First, silver ions from the particle can interact with the DNA and RNA of the bacteria causing them to condense and disrupt their sequences. Second, the silver–sulfur interactions at the membrane cause the cell wall to undergo structural changes, like the formation of pits and pores making the bacteria unable to repel extracellular fluids causing it to burst and leak intracellular fluids simultaneously. This effect is greatly enhanced if the size of AgNP is further reduced to around 10 nm as there are more contact points for the bacterium and AgNP. There had been reports that silver can enhance the effectivity of pharmaceutical-based antibiotics due to these mechanisms. It has been reported that AgNP can have synergistic effects together with penicillin G, ampicillin, erythromycin, clindamycin, and vancomycin against gram-positive and gram-negative bacteria (*E. coli* and *S. aureus*).

2.4 Composite biomaterials

Combining the different distinct classes of materials can create a composite. Composite materials solve shortcomings of a pure or pristine material usually experiences. This may be shortcomings in mechanical properties, chemical stability, degradation rate, biocompatibility, and bioactivity. Applications of biomaterials inside the body rely on the perfect balance of the above properties and no single material possesses all of this. Thus, researchers are combining distinct materials that possess individual salient properties to create medical devices, drug delivery platforms, and tissue scaffolds that can perform based on the application requirements. Most materials used for advanced biomedical applications rely heavily on the development of new composites as it can give rise to multifunctionality and solve problems that plague the field.

2.4.1 Polymer–ceramic composites

Combination of polymers and ceramics is applied mostly in hard tissue engineering. This is because the use of polymers can allow the transmission of stress toward the surrounding bone tissue preventing increased bone resorption where other ceramic materials cause because of their high rigidity while allowing calcification sites provided by the bioceramics [613, 614]. Polymers can also increase the degradation rate of an implant by allowing both hydrolytic and enzyme-mediated degradation to happen. Furthermore, polymers make them easier to produce porosity without the need for sintering, or other thermal treatment that pure ceramics require to produce a porous structure. The polymers can also serve as a platform for encapsulating and subsequently release growth factors that may be otherwise impossible if only using pure ceramic materials. In contrast, pure polymeric materials are too elastic and ductile compared to the native cortical bone, thus adding ceramic components that can enhance their rigidity allowing for a more biosimilar mechanical property.

Different types of polymers can be mixed with bioceramics, which may be biostable polymers, bioabsorbable or biodegradable, depending on the goal of the implant. The most common bioceramics used for this application are HA, β-TCP, and bioglass granules and powders [524, 615–617]. One example is the combination of PE and HA as a replacement for load-bearing joint prosthesis [618]. The addition of HDPE reduced protein adhesion on the surface of the implant, making it less likely to form clots and increase surface hydrophilicity that is beneficial for reducing the risk of formation of biofilms. While HA gives HDPE increased creep resistance and bone induction capabilities, bioabsorbable polymers together with bioceramics allow for the development of resorbable cements that facilitate bone induction [619]. Here, polyesters are used due to the ability of these types of polymer to degrade faster than other polymer due to the ester linkages that are highly susceptible to hydrolysis. But in order to enhance osteoinduction, adding HA is needed as polyesters do not possess this quality. The polyesters can act as a binder and improve rheological properties of the bone cement allowing to be injectable, thus making implantation less invasive. Using this strategy, it creates a slow degradation profile for the cement and it could mean that adding and releasing other active molecules such as growth factors are possible due to the entrapment efficacy of a polymeric material compared to using pure bioceramic material that will release the growth factors in an uncontrolled manner. Biopolymers such as polysaccharides and polypeptides can also be used for a polymer–ceramic composite as it increases biocompatibility as its degradation products are less toxic compared to synthetic polymers [620]. It can also act as attachment points for cells as it has functional moieties that mimic the biomolecular structure of the ECM. It has been proven that the combination of polymers and bioceramics creates a wider range of application, and its properties are more viable for clinical applications compared to just using each material alone.

2.4.2 Ceramic–metal composites

Combining bioceramics on metallic implants is exclusively applied to bone tissue implants and devices [621, 622]. The standard permanent metallic alloys such as stainless steel, titanium, tantalum, chromium–cobalt, and nickel–titanium have no intrinsic osteoconduction properties. They are chosen due to their mechanical and corrosion stabilities that are essential properties for load-bearing implants and devices. But problems arise when there is very little bone growth that could anchor these implants and devices because of stress shielding and the lack of affinity of cells toward these alloys. It is only logical to create a material that could induce cell activity so that these alloys are well integrated in the implantation site. Bioceramics such as HA, beta-TCP, and bioglass are used to treat the surfaces of these metallic alloys to induce cell growth on its surface [552, 623, 624]. These ceramics act as an anchor point for the cells and proliferate on the surface of implant. It must be considered that factors such as surface area and hydrophilicity play crucial roles in effect of this surface treatment. The mechanical properties must also be equal to the native cortical bone tissues to avoid stress shielding. This is done by modifying the density of metallic alloys and alloying ratios. But, first, the metal must be pretreated with chemical etching or plasma to create hydroxyl groups on its surface so that the bioceramics can have strong adherence on its surface. Porosity can also play a factor in how well the implant can integrate to the surrounding tissues. Bioceramics can only constitute the surfaces of this implant as their mechanical properties are limited and uncapable of load-bearing stresses that a metallic implant can endure. Osteoplasty prostheses such as hip and knee implants require high load-bearing endurance, and the most suitable material for this is a permanent alloy such as titanium and stainless steel. Thus, combining the osteoconductive properties of ceramics and mechanical and corrosion stabilities of metals is considered as one of the best solutions to solve implant integration.

In bioresorbable metallic alloys (e.g., zinc, magnesium, and iron), adding of ceramics is done to control the corrosion rate of the metal. Since there is invariance to the corrosion stability for both materials [625–627], varying the ratios of the metal to the ceramic during fabrication (alloying) can significantly alter its corrosion rate. In consequence, it also alters other properties such as mechanical, biological, and chemical reactivities. An example is that magnesium alloys are notoriously known for their fast degradation properties and combining them with ceramics such as HA can slow down this degradation process and by doing so, they also increase the osteoconductivity of the material. The opposite is the case for iron-based alloys where the degradation rate is very slow and adding bioceramics can speed up this process.

2.4.3 Metal–polymer composites

Metal and polymers are heavily used in medical devices when the goal is to increase the biocompatibility, control corrosion rate, and add functionality (e.g., drug delivery) [628, 629]. Polymers are the most versatile among the classes of materials, and metals are the least versatile but combining both gives rise to materials that can be used in devices, and drug delivery platforms are very utilitarian [630]. Biomedical devices that are alloy based are coated with polymers to add functionality. This may be to control the cell activity on the surface of the device, create a drug delivery platform, create an antimicrobial surface, or attenuate the corrosion rate of the alloy. Polymers can add this functionality due to their interesting chemical properties. The metal component on the other hand gives structural stability to a device which a pure polymeric material cannot be used. Polymers which may be otherwise overlooked for a specific biomedical application can be utilized if combined with metallic materials.

Medical devices used in osteoplasty require attenuation of microbial infection on its surface. This is because implant-related infection is a high burden for both the patient and the healthcare system. Here, polymers with the properties of encapsulating antimicrobials can be used to ensure a controlled and sustained release of antibiotics to kill pathogens that might adhere to the implant surface. In more advanced strategies, polymers with reducing properties of silver ions are used to create AgNPs onto the surface of the implant to form a non-pharmaceutical-based antimicrobial surface [631, 632]. And with this strategy we can further avoid the emergence of drug-resistant strains of bacteria that may be more detrimental. Other polymeric surface coating strategies where they can induce cell proliferation have also been used. This is done with the use of biopolymers made up of polysaccharides and polypeptides. Due to the similarities to native ECM physicochemical properties, the attachment of cells can be created to integrate the implants in the surrounding tissue.

In other metallic devices such as vascular and non-vascular stents, polymeric coatings and covers are employed to control the proliferation and adhesion of cells to its surfaces [266, 267, 633]. Restenosis is still a recurring issue on this biomedical device, and to maintain the growth of the surrounding tissue is one of the main objectives in keeping the device functioning smoothly. In vascular stents, the effect of thrombosis arises when the surface of the metal induces a myriad of biological cascades, and platelets and fibrin can adhere on its surfaces and deposit over time causing the recruitment of more proinflammatory factors such as cytokines and immune cells [537, 538]. In order to mitigate this, polymeric coatings are used to encapsulate either antiproliferative drugs (e.g., paclitaxel) and antithrombotic drugs (e.g., heparin and warfarin) [260, 261]. The diffusion of this drug can slow down and even inhibit the proliferation of endothelial cells and deposition of platelets and fibrin, whereas in non-vascular stents, the use of polymeric covers can physically block the ingrowth of the tumor into the stent struts, allowing for a localized release of antiproliferative drugs (e.g., chemotherapeutics) [634]. Different release mechanisms of drugs can be achieved depending on the class of polymer used, as the drug release rate is crucial in maintaining an effective response toward cell proliferation. The release rate of drugs encapsulated in this polymeric cover must be always in its therapeutic levels. Thus, active release mechanisms are used in which different stimuli (pH gradient, temperature change, presence of enzymes, etc.) [241, 635] can trigger the release of these drugs giving an on-demand release. Here, stimuli-responsive polymers are utilized as they can degrade, swell, or undergo phase transition depending on the external stimuli present in the body. First generation of these stent devices are made up of pure metallic alloys and suffer greatly from these shortcomings, and with the help of polymers, problems such as restenosis can be mitigated.

In systemic drug delivery where metal nanoparticles are used, it is also common to use polymers to increase their biocompatibility and add drug encapsulating capabilities. Pure metallic particles such as gold or iron nanoparticles had seen recent advances in their capabilities to be used for drug delivery applications, especially in cancer treatment. AuNP and iron nanoparticles have a decent biocompatibility but in order to be able to target specific tumor cells and circulate inside the body for prolonged periods, they must be coated with polymers. PEGylation or the use of PEG is a common strategy to increase the circulation time and biocompatibility of the nanoparticle [636, 637]. It does so by avoiding protein adsorption and reduce opsonization (particle removal mechanism of the body). Polymers can coat the metallic nanoparticles and act as a conjugation platform. Therefore, functionalization by adding targeting ligands can also be achieved by having moieties that can conjugate these active molecules. The same functionalization can also increase the drug-carrying capabilities of the nanoparticles. The use of these metallic nanoparticles can also complement the use of polymeric materials as gold can be used for theranostics (imaging agent for computed tomography, MRI, and X-ray) and iron can create hyperthermia conditions in the body once under the influence of alternating magnetic field. A special property of polymers, where it is used as an in situ reducing agent for the formation of nanoparticles of gold and silver, has also been used in biomedical science. This reduction mechanism arises from polymers that contain moieties acting as a nucleation point for silver and gold ions under aqueous media. In biosensors, AgNP can be deposited in situ onto a polymeric substrate and for wound dressing and implant applications. AgNPs can be added via one-step synthesis, removing the need to synthesize nanoparticles separately and adding it to the polymeric network.

2.5 Conclusion

Understanding different classes of materials that can be used inside the body is crucial to design and create devices, drug delivery platforms, and bioscaffolds. The materials discussed in this chapter only scratch the surface of what may be used for biomaterials. Progress in polymer chemistry, biomolecule engineering, ceramics manufacturing, and metallurgy can still further broaden the catalog of materials that we can use for biomedical applications. The job of a biomaterialist is to recognize different material properties that may be used to further improve and solve shortcomings in their field and how those properties can affect the body. An effective biomaterialist knows how to combine different properties in order to synthesize a material that is useful and does not cause harm toward the end user. The complexity of the human body adds to the challenge on designing a material that can solve most of its problems. But with enough rigor and better understanding, we can advance this field toward greater heights and slowly but surely create biomaterials that are safe and effective. In the next chapter, we will discuss the different biomaterial synthesis techniques and technologies. It is an important tool for a biomaterialist to understand these various techniques for designing and creating materials that use less energy and less toxic substances. Understanding different fabrication techniques can also give useful insights on how to translate useful biomaterials to the clinic.

References

[1] Aspinall GO. The Polysaccharides: Academic Press; 2014.

[2] Boddohi S, Kipper MJ. Engineering nanoassemblies of polysaccharides. Advanced Materials. 2010;22(28):2998–3016.

[3] Rinaudo M. Main properties and current applications of some polysaccharides as biomaterials. Polymer International. 2008;57(3):397–430.

[4] Mizrahy S, Peer D. Polysaccharides as building blocks for nanotherapeutics. Chemical Society Reviews. 2012;41(7):2623–40.

[5] Warrand J. Healthy Polysaccharides. Food Technology & Biotechnology. 2006;44(3).

[6] Draget KI. Alginates. Handbook of Hydrocolloids: Elsevier; 2009. p. 807–28.

[7] Sutherland IW. Alginates. Biomaterials: Springer; 1991. p. 307–31.

[8] Becker TA, Kipke DR, Brandon T. Calcium alginate gel: A biocompatible and mechanically stable polymer for endovascular embolization. Journal of Biomedical Materials Research: An Official Journal of the Society for Biomaterials and the Japanese Society for Biomaterials. 2001;54(1):76–86.

[9] Helgerud T, Gaserod O, Fjæreide T, Andersen P, Larsen C. 4. Alginates. Food Stabilisers, Thickeners and Gelling Agents Wiley-Blackwell. Oxford; 2010. p. 50–72.

[10] Heggset EB, Strand BL, Sundby KW, Simon S, Chinga-Carrasco G, Syverud K. Viscoelastic properties of nanocellulose based inks for 3D printing and mechanical properties of CNF/alginate biocomposite gels. Cellulose. 2019;26(1):581–95.

[11] Lone IH, Kowsalya E, Rebecca LJ. Alginate fiber from brown algae. Der Pharmacia Lettre. 2016;8(8):68–71.

[12] Anis M, Ahmed S, Hasan M. Algae as nutrition, medicine and cosmetic: The forgotten history, present status and future trends. World Journal of Pharmacy and Pharmaceutical Sciences. 2017;6(6):1934–59.

[13] Mir SA, Gull A, Akhtar G, Rather SA, Masoodi FA, Ganaie TA, et al. Fabrication and characterization of alginate-based green materials. Advanced Green Materials: Elsevier; 2021. p. 85–108.

[14] Peppas NA, Hoffman AS. Hydrogels. Biomaterials Science: Elsevier; 2020. p. 153–66.

[15] Augst AD, Kong HJ, Mooney DJ. Alginate hydrogels as biomaterials. Macromolecular Bioscience. 2006;6(8):623–33.

[16] Lee KY, Mooney DJ. Alginate: Properties and biomedical applications. Progress in Polymer Science. 2012;37(1):106–26.

[17] Munarin F, Kabelac C, Coulombe KL. Heparin-modified alginate microspheres enhance neovessel formation in hiPSC-derived endothelial cells and heterocellular in vitro models by controlled release of vascular endothelial growth factor. Journal of Biomedical Materials Research Part A. 2021.

[18] Zisch AH, Lutolf MP, Hubbell JA. Biopolymeric delivery matrices for angiogenic growth factors. Cardiovascular Pathology. 2003;12(6):295–310.

[19] Kim P-H, Yim H-G, Choi Y-J, Kang B-J, Kim J, Kwon S-M, et al. Injectable multifunctional microgel encapsulating outgrowth endothelial cells and growth factors for enhanced neovascularization. Journal of Controlled Release. 2014;187:1–13.

[20] Draget KI, Taylor C. Chemical, physical and biological properties of alginates and their biomedical implications. Food Hydrocolloids. 2011;25(2):251–56.

[21] Espevik T, Rokstad AM, Kulseng B, Strand B, Skjåk-Bræk G. Mechanisms of the host immune response to alginate microcapsules. In: Halle JP, de Vos P, Rosenberg L, editors. The Bioartificial Pancreas and Other Biohybrid Therapies Transworld Research; 2009. p. 279–90.

[22] Torres ML, Fernandez J, Dellatorre FG, Cortizo AM, Oberti TG. Purification of alginate improves its biocompatibility and eliminates cytotoxicity in matrix for bone tissue engineering. Algal Research. 2019;40:101499.

[23] Tam SK, Dusseault J, Polizu S, Ménard M, Hallé J-P. Impact of residual contamination on the biofunctional properties of purified alginates used for cell encapsulation. Biomaterials. 2006;27(8):1296–305.

[24] Szekalska M, Puciłowska A, Szymańska E, Ciosek P, Winnicka K. Alginate: Current use and future perspectives in pharmaceutical and biomedical applications. International Journal of Polymer Science. 2016;2016.

[25] Kapishon V, Whitney RA, Champagne P, Cunningham MF, Neufeld RJ. Polymerization induced self-assembly of alginate based amphiphilic graft copolymers synthesized by single electron transfer living radical polymerization. Biomacromolecules. 2015;16(7):2040–48.

[26] Yang J-S, Xie Y-J, He W. Research progress on chemical modification of alginate: A review. Carbohydrate Polymers. 2011;84(1):33–39.

[27] Pawar SN, Edgar KJ. Alginate derivatization: A review of chemistry, properties and applications. Biomaterials. 2012;33(11):3279–305.

[28] De Boisseson MR, Leonard M, Hubert P, Marchal P, Stequert A, Castel C, et al. Physical alginate hydrogels based on hydrophobic or dual hydrophobic/ionic interactions: Bead formation, structure, and stability. Journal of Colloid and Interface Science. 2004;273(1):131–39.

[29] Thakar H, Sebastian SM, Mandal S, Pople A, Agarwal G, Srivastava A. Biomolecule-conjugated macroporous hydrogels for biomedical applications. ACS Biomaterials Science & Engineering. 2019;5(12):6320–41.

[30] Liao J, Jia Y, Wang B, Shi K, Qian Z. Injectable hybrid poly (ε-caprolactone)-b-poly (ethylene glycol)-b-poly (ε-caprolactone) porous microspheres/alginate hydrogel cross-linked by calcium gluconate crystals deposited in the pores of microspheres improved skin wound healing. ACS Biomaterials Science & Engineering. 2018;4(3):1029–36.

[31] Alzahrani E, Abo-Dief HM, Algethami F. Electrochemical investigations of hydrochloric acid corrosion for carbon steel and coating effect by Poly (butyl Methacrylate)-grafted alginate/ Fe3O4. Arabian Journal of Chemistry. 2021;14(5):103100.

[32] Jeon O, Alsberg E. Photofunctionalization of alginate hydrogels to promote adhesion and proliferation of human mesenchymal stem cells. Tissue Engineering Part A. 2013;19 (11–12):1424–32.

[33] Rowley JA, Madlambayan G, Mooney DJ. Alginate hydrogels as synthetic extracellular matrix materials. Biomaterials. 1999;20(1):45–53.

[34] Fan L, Cao M, Gao S, Wang T, Wu H, Peng M, et al. Preparation and characterization of sodium alginate modified with collagen peptides. Carbohydrate Polymers. 2013;93 (2):380–85.

[35] Passemard S, Szabó L, Noverraz F, Montanari E, Gonelle-Gispert C, Bühler L, et al. Synthesis strategies to extend the variety of alginate-based hybrid hydrogels for cell microencapsulation. Biomacromolecules. 2017;18(9):2747–55.

[36] Sakai S, Kawakami K. Synthesis and characterization of both ionically and enzymatically cross-linkable alginate. Acta Biomaterialia. 2007;3(4):495–501.

[37] Fraczyk J, Wasko J, Walczak M, Kaminski ZJ, Puchowicz D, Kaminska I, et al. Conjugates of copper alginate with arginine-glycine-aspartic acid (RGD) for potential use in regenerative medicine. Materials. 2020;13(2):337.

[38] Linh NT, Paul K, Kim B, Lee B-T. Augmenting in vitro osteogenesis of a glycine–arginine–glycine–aspartic-conjugated oxidized alginate–gelatin–biphasic calcium phosphate hydrogel composite and in vivo bone biogenesis through stem cell delivery. Journal of Biomaterials Applications. 2016;31(5):661–73.

[39] De Guzman R, Van Dyke M. Incorporation of active factors (pharmacological substances) in biomaterials for tissue engineering. Regenerative Pharmacology. 2013:167.

[40] Bonino CA, Krebs MD, Saquing CD, Jeong SI, Shearer KL, Alsberg E, et al. Electrospinning alginate-based nanofibers: From blends to crosslinked low molecular weight alginate-only systems. Carbohydrate Polymers. 2011;85(1):111–19.

[41] Cardoso DA, Ulset AS, Bender J, Jansen JA, Christensen BE, Leeuwenburgh SC. Effects of physical and chemical treatments on the molecular weight and degradation of alginate–hydroxyapatite composites. Macromolecular Bioscience. 2014;14(6):872–80.

[42] Eiselt P, Yeh J, Latvala RK, Shea LD, Mooney DJ. Porous carriers for biomedical applications based on alginate hydrogels. Biomaterials. 2000;21(19):1921–27.

[43] Chirapart A, Ohno M, Ukeda H, Sawamura M, Kusunose H. Chemical composition of agars from a newly reported Japanese agarophyte, Gracilariopsis lemaneiformis. Journal of Applied Phycology. 1995;7(4):359–65.

[44] Normand V, Lootens DL, Amici E, Plucknett KP, Aymard P. New insight into agarose gel mechanical properties. Biomacromolecules. 2000;1(4):730–38.

[45] Pollot BE, Rathbone CR, Wenke JC, Guda T. Natural polymeric hydrogel evaluation for skeletal muscle tissue engineering. Journal of Biomedical Materials Research Part B: Applied Biomaterials. 2018;106(2):672–79.

[46] Tan YJ, Tan X, Yeong WY, Tor SB. Hybrid microscaffold-based 3D bioprinting of multi-cellular constructs with high compressive strength: A new biofabrication strategy. Scientific Reports. 2016;6(1):1–13.

[47] Prajapati VD, Maheriya PM, Jani GK, Solanki HK. RETRACTED: Carrageenan: A natural seaweed polysaccharide and its applications. Carbohydrate Polymers. 2014;105:97–112.

[48] Kim D, Kang SM. Red algae-derived carrageenan coatings for marine antifouling applications. Biomacromolecules. 2020;21(12):5086–92.

[49] Shumilina E, Shchipunov YA. Chitosan–carrageenan gels. Colloid Journal. 2002;64 (3):372–78.

[50] Fauzi MARD, Pudjiastuti P, Wibowo AC, Hendradi E. Preparation, properties and potential of carrageenan-based hard capsules for replacing gelatine: A review. Polymers. 2021;13 (16):2666.

[51] Medina-Torres L, Brito-De La Fuente E, Torrestiana-Sanchez B, Alonso S. Mechanical properties of gels formed by mixtures of mucilage gum (Opuntia ficus indica) and carrageenans. Carbohydrate Polymers. 2003;52(2):143–50.

[52] Padhi JR, Nayak D, Nanda A, Rauta PR, Ashe S, Nayak B. Development of highly biocompatible Gelatin & i-Carrageenan based composite hydrogels: In depth physiochemical analysis for biomedical applications. Carbohydrate Polymers. 2016;153:292–301.

[53] Yousefi M, Jafari SM. Recent advances in application of different hydrocolloids in dairy products to improve their techno-functional properties. Trends in Food Science & Technology. 2019;88:468–83.

[54] Derkach SR, Ilyin SO, Maklakova AA, Kulichikhin VG, Malkin AY. The rheology of gelatin hydrogels modified by κ-carrageenan. LWT-Food Science and Technology. 2015;63(1):612–19.

[55] Li B, Lu F, Wei X, Zhao R. Fucoidan: Structure and bioactivity. Molecules. 2008;13(8):1671–95.

[56] Van Weelden G, Bobiński M, Okła K, Van Weelden WJ, Romano A, Pijnenborg J. Fucoidan structure and activity in relation to anti-cancer mechanisms. Marine Drugs. 2019;17(1):32.

[57] Amin ML, Mawad D, Dokos S, Koshy P, Martens PJ, Sorrell CC. Immunomodulatory properties of photopolymerizable fucoidan and carrageenans. Carbohydrate Polymers. 2020;230:115691.

[58] Tran PH, Duan W, Tran TT. Fucoidan-based nanostructures: A focus on its combination with chitosan and the surface functionalization of metallic nanoparticles for drug delivery. International Journal of Pharmaceutics. 2020;575:118956.

[59] Kim H, Van Phuc Nguyen PM, Jung MJ, Kim SW, Oh J, Kang HW. Doxorubicin-fucoidan-gold nanoparticles composite for dual-chemo-photothermal treatment on eye tumors. Oncotarget. 2017;8(69):113719.

[60] Oliveira C, Neves NM, Reis RL, Martins A, Silva TH. A review on fucoidan antitumor strategies: From a biological active agent to a structural component of fucoidan-based systems. Carbohydrate Polymers. 2020;239:116131.

[61] Jang B, Moorthy MS, Manivasagan P, Xu L, Song K, Lee KD, et al. Fucoidan-coated CuS nanoparticles for chemo-and photothermal therapy against cancer. Oncotarget. 2018;9 (16):12649.

[62] Heinze T Cellulose: Structure and properties. Cellulose chemistry and properties: fibers, nanocelluloses and advanced materials. 2015:1–52.

[63] Zainal SH, Mohd NH, Suhaili N, Anuar FH, Lazim AM, Othaman R. Preparation of cellulose-based hydrogel: A review. Journal of Materials Research and Technology. 2021;10:935–52.

[64] Radotić K, Mićić M. Methods for Extraction and Purification of Lignin and Cellulose from Plant Tissues. Sample Preparation Techniques for Soil, Plant, and Animal Samples: Springer; 2016. p. 365–76.

[65] Krystynowicz A, Czaja W, Wiktorowska-Jezierska A, Gonçalves-Miśkiewicz M, Turkiewicz M, Bielecki S. Factors affecting the yield and properties of bacterial cellulose. Journal of Industrial Microbiology and Biotechnology. 2002;29(4):189–95.
[66] Wang J, Tavakoli J, Tang Y. Bacterial cellulose production, properties and applications with different culture methods–A review. Carbohydrate Polymers. 2019;219:63–76.
[67] Gopi S, Balakrishnan P, Chandradhara D, Poovathankandy D, Thomas S. General scenarios of cellulose and its use in the biomedical field. Materials Today Chemistry. 2019;13:59–78.
[68] Moohan J, Stewart SA, Espinosa E, Rosal A, Rodríguez A, Larrañeta E, et al. Cellulose nanofibers and other biopolymers for biomedical applications. A Review. Applied Sciences. 2020;10(1):65.
[69] Wan W, Hutter J, Milton L, Guhados G. Bacterial Cellulose and Its Nanocomposites for Biomedical Applications: ACS Publications; 2006.
[70] Bashari A, Rouhani Shirvan A, Shakeri M. Cellulose-based hydrogels for personal care products. Polymers for Advanced Technologies. 2018;29(12):2853–67.
[71] Rigby A, Anand S, Horrocks A. Textile materials for medical and healthcare applications. Journal of the Textile Institute. 1997;88(3):83–93.
[72] Hoenich NA. Cellulose for medical applications: Past, present, and future. BioResources. 2006;1(2):270–80.
[73] Adolfsson KH, Melilli G, Hakkarainen M. Oxidized carbonized cellulose-coated filters for environmental contaminant adsorption and detection. Industrial & Engineering Chemistry Research. 2020;59(30):13578–87.
[74] Gwon J-G, Cho H-J, Chun S-J, Lee S, Wu Q, Li M-C, et al. Mechanical and thermal properties of toluene diisocyanate-modified cellulose nanocrystal nanocomposites using semi-crystalline poly (lactic acid) as a base matrix. RSC Advances. 2016;6(77):73879–86.
[75] Lin Y, Bilotti E, Bastiaansen CW, Peijs T. Transparent semi-crystalline polymeric materials and their nanocomposites: A review. Polymer Engineering & Science. 2020;60(10):2351–76.
[76] Usov I, Nyström G, Adamcik J, Handschin S, Schütz C, Fall A, et al. Understanding nanocellulose chirality and structure–properties relationship at the single fibril level. Nature Communications. 2015;6(1):1–11.
[77] Mariano M, El Kissi N, Dufresne A. Cellulose nanocrystals and related nanocomposites: Review of some properties and challenges. Journal of Polymer Science Part B: Polymer Physics. 2014;52(12):791–806.
[78] Luo H, Cha R, Li J, Hao W, Zhang Y, Zhou F. Advances in tissue engineering of nanocellulose-based scaffolds: A review. Carbohydrate Polymers. 2019;224:115144.
[79] Nguyen LH, Naficy S, Chandrawati R, Dehghani F. Nanocellulose for sensing applications. Advanced Materials Interfaces. 2019;6(18):1900424.
[80] Shavandi A, Hosseini S, Okoro OV, Nie L, Eghbali Babadi F, Melchels F. 3D bioprinting of lignocellulosic biomaterials. Advanced Healthcare Materials. 2020;9(24):2001472.
[81] Round AN, Rigby NM, MacDougall AJ, Morris VJ. A new view of pectin structure revealed by acid hydrolysis and atomic force microscopy. Carbohydrate Research. 2010;345(4):487–97.
[82] Ropartz D, Ralet M-C. Pectin Structure. Pectin: Technological and Physiological Properties: Springer; 2020. p. 17–36.
[83] Noreen A, Akram J, Rasul I, Mansha A, Yaqoob N, Iqbal R, et al. Pectins functionalized biomaterials; a new viable approach for biomedical applications: A review. International Journal of Biological Macromolecules. 2017;101:254–72.
[84] Lapomarda A, De Acutis A, Chiesa I, Fortunato GM, Montemurro F, De Maria C, et al. Pectin-GPTMS-based biomaterial: Toward a sustainable bioprinting of 3D scaffolds for tissue engineering application. Biomacromolecules. 2019;21(2):319–27.

[85] Andriotis EG, Eleftheriadis GK, Karavasili C, Fatouros DG. Development of bio-active patches based on pectin for the treatment of ulcers and wounds using 3D-bioprinting technology. Pharmaceutics. 2020;12(1):56.
[86] Xu J, Zheng S, Hu X, Li L, Li W, Parungao R, et al. Advances in the research of bioinks based on natural collagen, polysaccharide and their derivatives for skin 3D bioprinting. Polymers. 2020;12(6):1237.
[87] Bertoft E. Understanding starch structure: Recent progress. Agronomy. 2017;7(3):56.
[88] Zheng L, Yu Y, Tong Z, Zou Q, Han S, Jiang H. The characteristics of starch gels molded by 3D printing. Journal of Food Processing and Preservation. 2019;43(7):e13993.
[89] Maniglia BC, Lima DC, da Matta Junior M, Oge A, Le-Bail P, Augusto PE, et al. Dry heating treatment: A potential tool to improve the wheat starch properties for 3D food printing application. Food Research International. 2020;137:109731.
[90] Lille M, Nurmela A, Nordlund E, Metsä-Kortelainen S, Sozer N. Applicability of protein and fiber-rich food materials in extrusion-based 3D printing. Journal of Food Engineering. 2018;220:20–27.
[91] Sarder R, Piner E, Rios DC, Chacon L, Artner MA, Barrios N, et al. Copolymers of starch, a sustainable template for biomedical applications: A review. Carbohydrate Polymers. 2021:118973.
[92] Neely WB. Dextran: Structure and synthesis. Advances in Carbohydrate Chemistry. 1961;15:341–69.
[93] Purama RK, Goswami P, Khan AT, Goyal A. Structural analysis and properties of dextran produced by Leuconostoc mesenteroides NRRL B-640. Carbohydrate Polymers. 2009;76(1):30–35.
[94] Gibbons R, Fitzgerald R. Dextran-induced agglutination of Streptococcus mutans, and its potential role in the formation of microbial dental plaques. Journal of Bacteriology. 1969;98(2):341–46.
[95] Teixeira JP, Van Sant LM, Nielsen ND. Pharmacology and clinical use of plasma expanders. Anaesthesia & Intensive Care Medicine. 2020.
[96] Bhavani AL, Nisha J. Dextran – the polysaccharide with versatile uses. International Journal of Pharmacy and Biological Science. 2010;1(4):569–73.
[97] Alibolandi M, Mohammadi M, Taghdisi SM, Ramezani M, Abnous K. Fabrication of aptamer decorated dextran coated nano-graphene oxide for targeted drug delivery. Carbohydrate Polymers. 2017;155:218–29.
[98] Nguyen HV, Campbell K, Painter GF, Young SL, Walker GF. Nanoparticle system based on amino-dextran as a drug delivery vehicle: Immune-stimulatory CpG-oligonucleotide loading and delivery. Pharmaceutics. 2020;12(12):1150.
[99] Prusty K, Swain SK. Nano silver decorated polyacrylamide/dextran nanohydrogels hybrid composites for drug delivery applications. Materials Science and Engineering: C. 2018;85:130–41.
[100] Rai R, Alwani S, Badea I. Polymeric nanoparticles in gene therapy: New avenues of design and optimization for delivery applications. Polymers. 2019;11(4):745.
[101] Erdogan S, Silindir M. Nanoparticulate Contrast Agents for CT, SPECT and PET Imaging. Handbook of Nanobiomedical Research: Fundamentals, Applications and Recent Developments: Volume 3 Applications in Diagnostics: World Scientific; 2014. p. 47–76.
[102] Singh RS, Saini GK, Kennedy JF. Pullulan: Microbial sources, production and applications. Carbohydrate Polymers. 2008;73(4):515–31.
[103] More VS, Ebinesar A, Prakruthi A, Praveen P, Fasim A, Rao A, et al. Isolation and Purification of Microbial Exopolysaccharides and Their Industrial Application. Microbial Polymers: Springer; 2021. p. 69–86.

[104] Prajapati VD, Chaudhari AM, Gandhi AK, Maheriya P. Pullulan based oral thin film formulation of zolmitriptan: Development and optimization using factorial design. International Journal of Biological Macromolecules. 2018;107:2075–85.
[105] Gherman S, Zavastin D, Ochiuz L, Biliuta G, Coseri S. Enalapril maleate loaded pullulan film for mucoadhesive buccal drug delivery applications. Cellular Chemical and Technology. 2016;50:593–600.
[106] Dragojevic S, Ryu JS, Raucher D. Polymer-based prodrugs: Improving tumor targeting and the solubility of small molecule drugs in cancer therapy. Molecules. 2015;20(12):21750–69.
[107] Ganie SA, Rather LJ, Li Q. A review on anticancer applications of pullulan and pullulan derivative nanoparticles. Carbohydrate Polymer Technologies and Applications. 2021;2:100115.
[108] Singh RS, Kaur N, Rana V, Kennedy JF. Recent insights on applications of pullulan in tissue engineering. Carbohydrate Polymers. 2016;153:455–62.
[109] Atila D, Keskin D, Tezcaner A. Crosslinked pullulan/cellulose acetate fibrous scaffolds for bone tissue engineering. Materials Science and Engineering: C. 2016;69:1103–15.
[110] Fujioka-Kobayashi M, Ota MS, Shimoda A, Nakahama K-I, Akiyoshi K, Miyamoto Y, et al. Cholesteryl group-and acryloyl group-bearing pullulan nanogel to deliver BMP2 and FGF18 for bone tissue engineering. Biomaterials. 2012;33(30):7613–20.
[111] Younes I, Rinaudo M. Chitin and chitosan preparation from marine sources. Structure, Properties and Applications. Marine Drugs. 2015;13(3):1133–74.
[112] Jang MK, Kong BG, Jeong YI, Lee CH, Nah JW. Physicochemical characterization of α-chitin, β-chitin, and γ-chitin separated from natural resources. Journal of Polymer Science Part A: Polymer Chemistry. 2004;42(14):3423–32.
[113] Moussian B. Chitin: Structure, chemistry and biology. Targeting Chitin-containing Organisms. 2019:5–18.
[114] Yang T-L. Chitin-based materials in tissue engineering: Applications in soft tissue and epithelial organ. International Journal of Molecular Sciences. 2011;12(3):1936–63.
[115] Nagahama H, Nwe N, Jayakumar R, Koiwa S, Furuike T, Tamura H. Novel biodegradable chitin membranes for tissue engineering applications. Carbohydrate Polymers. 2008;73 (2):295–302.
[116] Peter M, Kumar PTS, Binulal NS, Nair SV, Tamura H, Jayakumar R. Development of novel α-chitin/nanobioactive glass ceramic composite scaffolds for tissue engineering applications. Carbohydrate Polymers. 2009;78(4):926–31.
[117] Madhumathi K, Kumar PS, Kavya K, Furuike T, Tamura H, Nair S, et al. Novel chitin/nanosilica composite scaffolds for bone tissue engineering applications. International Journal of Biological Macromolecules. 2009;45(3):289–92.
[118] Doulabi AH, Mirzadeh H, Imani M, Samadi N. Chitosan/polyethylene glycol fumarate blend film: Physical and antibacterial properties. Carbohydrate Polymers. 2013;92(1):48–56.
[119] Dev A, Binulal N, Anitha A, Nair S, Furuike T, Tamura H, et al. Preparation of poly (lactic acid)/chitosan nanoparticles for anti-HIV drug delivery applications. Carbohydrate Polymers. 2010;80(3):833–38.
[120] Hasnain MS, Nayak AK. Chitosan as Responsive Polymer for Drug Delivery Applications. Stimuli Responsive Polymeric Nanocarriers for Drug Delivery Applications, Vol. ume 1: Elsevier; 2018. p. 581–605.
[121] Ailincai D, Mititelu-Tartau L, Marin L. Citryl-imine-PEG-ylated chitosan hydrogels–promising materials for drug delivery applications. International Journal of Biological Macromolecules. 2020;162:1323–37.

[122] Gooneh-Farahani S, Naimi-Jamal MR, Naghib SM. Stimuli-responsive graphene-incorporated multifunctional chitosan for drug delivery applications: A review. Expert Opinion on Drug Delivery. 2019;16(1):79–99.
[123] Homayoni H, Ravandi SAH, Valizadeh M. Electrospinning of chitosan nanofibers: Processing optimization. Carbohydrate Polymers. 2009;77(3):656–61.
[124] Islam S, Bhuiyan MR, Islam M. Chitin and chitosan: Structure, properties and applications in biomedical engineering. Journal of Polymers and the Environment. 2017;25(3):854–66.
[125] Domard A, Domard M. Chitosan: Structure-properties relationship and biomedical applications. Polymeric Biomaterials. 2002;9:187–212.
[126] Ibrahim H, El-Zairy E. Chitosan as a biomaterial – structure, properties, and electrospun nanofibers. Concepts, Compounds and the Alternatives of Antibacterials. 2015:81–101.
[127] Ahmed S, Ahmad M, Ikram S. Chitosan: A natural antimicrobial agent-a review. Journal of Applicable Chemistry. 2014;3(2):493–503.
[128] Custódio CA, Alves C, Reis R, Mano J. Immobilization of fibronectin in chitosan substrates improves cell adhesion and proliferation. Journal of Tissue Engineering and Regenerative Medicine. 2010;4(4):316–23.
[129] Cuy JL, Beckstead BL, Brown CD, Hoffman AS, Giachelli CM. Adhesive protein interactions with chitosan: Consequences for valve endothelial cell growth on tissue-engineering materials. Journal of Biomedical Materials Research Part A: An Official Journal of the Society for Biomaterials, the Japanese Society for Biomaterials, and the Australian Society for Biomaterials and the Korean Society for Biomaterials. 2003;67(2):538–47.
[130] Laurencin CT, Jiang T, Kumbar SG, Nair LS. Biologically active chitosan systems for tissue engineering and regenerative medicine. Current Topics in Medicinal Chemistry. 2008;8(4):354–64.
[131] Prego C, Torres D, Alonso MJ. The potential of chitosan for the oral administration of peptides. Expert Opinion on Drug Delivery. 2005;2(5):843–54.
[132] Şenel S, Kremer M, Kaş S, Wertz P, Hıncal A, Squier C. Enhancing effect of chitosan on peptide drug delivery across buccal mucosa. Biomaterials. 2000;21(20):2067–71.
[133] Wong CY, Al-Salami H, Dass CR. The role of chitosan on oral delivery of peptide-loaded nanoparticle formulation. Journal of Drug Targeting. 2018;26(7):551–62.
[134] Hwang SR, Byun Y. Advances in oral macromolecular drug delivery. Expert Opinion on Drug Delivery. 2014;11(12):1955–67.
[135] Aguirre TA, Teijeiro-Osorio D, Rosa M, Coulter I, Alonso M, Brayden D. Current status of selected oral peptide technologies in advanced preclinical development and in clinical trials. Advanced Drug Delivery Reviews. 2016;106:223–41.
[136] Richard J. Challenges in oral peptide delivery: Lessons learnt from the clinic and future prospects. Therapeutic Delivery. 2017;8(8):663–84.
[137] Durán-Lobato M, Niu Z, Alonso MJ. Oral delivery of biologics for precision medicine. Advanced Materials. 2020;32(13):1901935.
[138] Selyanin MA, Boykov PY, Khabarov VN, Polyak F. Hyaluronic Acid: John Wiley & Sons, Ltd; 2015.
[139] Atkins E, Sheehan J. The molecular structure of hyaluronic acid. Biochemical Journal. 1971;125(4):92P.
[140] Sze JH, Brownlie JC, Love CA. Biotechnological production of hyaluronic acid: A mini review. 3 Biotech. 2016;6(1):67.
[141] Beasley KL, Weiss MA, Weiss RA. Hyaluronic acid fillers: A comprehensive review. Facial Plastic Surgery. 2009;25(02):086–94.
[142] Bukhari SNA, Roswandi NL, Waqas M, Habib H, Hussain F, Khan S, et al. Hyaluronic acid, a promising skin rejuvenating biomedicine: A review of recent updates and pre-clinical and

clinical investigations on cosmetic and nutricosmetic effects. International Journal of Biological Macromolecules. 2018;120:1682–95.
[143] Sciabica S, Tafuro G, Semenzato A, Traini D, Silva DM, Reis LGD, et al. Design, synthesis, characterization, and in vitro evaluation of a new cross-linked hyaluronic acid for pharmaceutical and cosmetic applications. Pharmaceutics. 2021;13(10):1672.
[144] Graça MF, Miguel SP, Cabral CS, Correia IJ. Hyaluronic acid – Based wound dressings: A review. Carbohydrate Polymers. 2020;241:116364.
[145] Casale M, Moffa A, Vella P, Sabatino L, Capuano F, Salvinelli B, et al. Hyaluronic acid: Perspectives in dentistry. A systematic review. International Journal of Immunopathology and Pharmacology. 2016;29(4):572–82.
[146] Collins MN, Birkinshaw C. Hyaluronic acid based scaffolds for tissue engineering – A review. Carbohydrate Polymers. 2013;92(2):1262–79.
[147] Pan NC, Pereira HCB, da Silva M, Vasconcelos AFD, Celligoi MAPC. Improvement production of hyaluronic acid by Streptococcus zooepidemicus in sugarcane molasses. Applied Biochemistry and Biotechnology. 2017;182(1):276–93.
[148] Hemshekhar M, Thushara RM, Chandranayaka S, Sherman LS, Kemparaju K, Girish KS. Emerging roles of hyaluronic acid bioscaffolds in tissue engineering and regenerative medicine. International Journal of Biological Macromolecules. 2016;86:917–28.
[149] Burdick JA, Prestwich GD. Hyaluronic acid hydrogels for biomedical applications. Advanced Materials. 2011;23(12):H41–H56.
[150] Inuyama Y, Kitamura C, Nishihara T, Morotomi T, Nagayoshi M, Tabata Y, et al. Effects of hyaluronic acid sponge as a scaffold on odontoblastic cell line and amputated dental pulp. Journal of Biomedical Materials Research Part B: Applied Biomaterials: An Official Journal of the Society for Biomaterials, the Japanese Society for Biomaterials, and the Australian Society for Biomaterials and the Korean Society for Biomaterials. 2010;92(1):120–28.
[151] Kim IL, Khetan S, Baker BM, Chen CS, Burdick JA. Fibrous hyaluronic acid hydrogels that direct MSC chondrogenesis through mechanical and adhesive cues. Biomaterials. 2013;34 (22):5571–80.
[152] Xu S, Li J, He A, Liu W, Jiang X, Zheng J, et al. Chemical crosslinking and biophysical properties of electrospun hyaluronic acid based ultra-thin fibrous membranes. Polymer. 2009;50(15):3762–69.
[153] Schanté CE, Zuber G, Herlin C, Vandamme TF. Chemical modifications of hyaluronic acid for the synthesis of derivatives for a broad range of biomedical applications. Carbohydrate Polymers. 2011;85(3):469–89.
[154] Tiwari S, Bahadur P. Modified hyaluronic acid based materials for biomedical applications. International Journal of Biological Macromolecules. 2019;121:556–71.
[155] Brenner EK, Schiffman JD, Thompson EA, Toth LJ, Schauer CL. Electrospinning of hyaluronic acid nanofibers from aqueous ammonium solutions. Carbohydrate Polymers. 2012;87 (1):926–29.
[156] Noh I, Kim N, Tran HN, Lee J, Lee C. 3D printable hyaluronic acid-based hydrogel for its potential application as a bioink in tissue engineering. Biomaterials Research. 2019;23 (1):1–9.
[157] Miranda DG, Malmonge SM, Campos DM, Attik NG, Grosgogeat B, Gritsch K. A chitosan-hyaluronic acid hydrogel scaffold for periodontal tissue engineering. Journal of Biomedical Materials Research Part B: Applied Biomaterials. 2016;104(8):1691–702.
[158] Nair S, Remya N, Remya S, Nair PD. A biodegradable in situ injectable hydrogel based on chitosan and oxidized hyaluronic acid for tissue engineering applications. Carbohydrate Polymers. 2011;85(4):838–44.

[159] Yamane S, Iwasaki N, Majima T, Funakoshi T, Masuko T, Harada K, et al. Feasibility of chitosan-based hyaluronic acid hybrid biomaterial for a novel scaffold in cartilage tissue engineering. Biomaterials. 2005;26(6):611–19.
[160] Bauer C, Berger M, Baumgartner RR, Höller S, Zwickl H, Niculescu-Morzsa E, et al. A novel cross-linked hyaluronic acid porous scaffold for cartilage repair: An in vitro study with osteoarthritic chondrocytes. Cartilage. 2016;7(3):265–73.
[161] Shoulders MD, Raines RT. Collagen structure and stability. Annual Review of Biochemistry. 2009;78:929–58.
[162] Bama P, Vijayalakshimi M, Jayasimman R, Kalaichelvan P, Deccaraman M, Sankaranarayanan S. Extraction of collagen from cat fish (Tachysurus maculatus) by pepsin digestion and preparation and characterization of collagen chitosan sheet. International Journal of Pharmacy and Pharmaceutical Sciences. 2010;2(4):133–37.
[163] Parekh A, Weaver AM. Regulation of cancer invasiveness by the physical extracellular matrix environment. Cell Adhesion & Migration. 2009;3(3):288–92.
[164] Hernandez-Gordillo V, Chmielewski J. Mimicking the extracellular matrix with functionalized, metal-assembled collagen peptide scaffolds. Biomaterials. 2014;35(26):7363–73.
[165] Lin K, Zhang D, Macedo MH, Cui W, Sarmento B, Shen G. Advanced collagen-based biomaterials for regenerative biomedicine. Advanced Functional Materials. 2019;29(3):1804943.
[166] Rodrigues C, Serricella P, Linhares A, Guerdes R, Borojevic R, Rossi M, et al. Characterization of a bovine collagen–hydroxyapatite composite scaffold for bone tissue engineering. Biomaterials. 2003;24(27):4987–97.
[167] Semyari H, Salehi M, Taleghani F, Ehterami A, Bastami F, Jalayer T, et al. Fabrication and characterization of collagen–hydroxyapatite-based composite scaffolds containing doxycycline via freeze-casting method for bone tissue engineering. Journal of Biomaterials Applications. 2018;33(4):501–13.
[168] Muthukumar T, Sreekumar G, Sastry T, Chamundeeswari M. Collagen as a potential biomaterial in biomedical applications. Reviews on Advanced Materials Science. 2018;53(1):29–39.
[169] Wong FSY, Tsang KK, Chu AMW, Chan BP, Yao KM, Lo ACY. Injectable cell-encapsulating composite alginate-collagen platform with inducible termination switch for safer ocular drug delivery. Biomaterials. 2019;201:53–67.
[170] An B, Lin Y-S, Brodsky B. Collagen interactions: Drug design and delivery. Advanced Drug Delivery Reviews. 2016;97:69–84.
[171] Jayaraman A, Price C, Sullivan MO, Kiick KL. Collagen-Peptide-Based Drug Delivery Strategies. Technology & Innovation. 2020;21(4):1–20.
[172] Johnston-Banks F. Gelatine. Food Gels: Springer; 1990. p. 233–89.
[173] Schrieber R, Gareis H. Gelatine Handbook: Theory and Industrial Practice: John Wiley & Sons; 2007.
[174] Schwartz R, Malpica M, Thompson GL, Miri AK. Cell encapsulation in gelatin bioink impairs 3D bioprinting resolution. Journal of the Mechanical Behavior of Biomedical Materials. 2020;103:103524.
[175] Yin J, Yan M, Wang Y, Fu J, Suo H. 3D bioprinting of low-concentration cell-laden gelatin methacrylate (GelMA) bioinks with a two-step cross-linking strategy. ACS Applied Materials & Interfaces. 2018;10(8):6849–57.
[176] Singh YP, Bandyopadhyay A, Mandal BB. 3D bioprinting using cross-linker-free silk–gelatin bioink for cartilage tissue engineering. ACS Applied Materials & Interfaces. 2019;11(37):33684–96.

[177] Lai P, Daear W, Löbenberg R, Prenner EJ. Overview of the preparation of organic polymeric nanoparticles for drug delivery based on gelatine, chitosan, poly (d, l-lactide-co-glycolic acid) and polyalkylcyanoacrylate. Colloids and Surfaces B: Biointerfaces. 2014;118:154–63.
[178] Coradin T, Bah S, Livage J. Gelatine/silicate interactions: From nanoparticles to composite gels. Colloids and Surfaces B: Biointerfaces. 2004;35(1):53–58.
[179] Zhang J, Chen L, Tse WH, Bi R, Chen L. Inorganic nanoparticles: Engineering for biomedical applications. IEEE Nanotechnology Magazine. 2014;8(4):21–28.
[180] Gérard VA, Maguire CM, Bazou D, Gun'ko YK. Folic acid modified gelatine coated quantum dots as potential reagents for in vitro cancer diagnostics. Journal of Nanobiotechnology. 2011;9(1):1–7.
[181] Xue J, Liu Y, Darabi MA, Tu G, Huang L, Ying L, et al. An injectable conductive gelatin-PANI hydrogel system serves as a promising carrier to deliver BMSCs for Parkinson's disease treatment. Materials Science and Engineering: C. 2019;100:584–97.
[182] Ren Y, Zhao X, Liang X, Ma PX, Guo B. Injectable hydrogel based on quaternized chitosan, gelatin and dopamine as localized drug delivery system to treat Parkinson's disease. International Journal of Biological Macromolecules. 2017;105:1079–87.
[183] Wang B, Yang W, McKittrick J, Meyers MA. Keratin: Structure, mechanical properties, occurrence in biological organisms, and efforts at bioinspiration. Progress in Materials Science. 2016;76:229–318.
[184] McKittrick J, Chen P-Y, Bodde S, Yang W, Novitskaya E, Meyers M. The structure, functions, and mechanical properties of keratin. Jom. 2012;64(4):449–68.
[185] Ramadoss P, Arul KT, Ramya JR, Begam MR, Chandra VS, Manikandan E. Enhanced mechanical strength and sustained drug release of gelatin/keratin scaffolds. Materials Letters. 2017;186:109–12.
[186] Navarro J, Swayambunathan J, Santoro M, Fisher J Assessment of the effects of energy density in crosslinking of keratin-based photo-sensitive resin. 2018 IX International Seminar of Biomedical Engineering (SIB): IEEE; 2018. p. 1–6.
[187] Feroz S, Muhammad N, Ratnayake J, Dias G. Keratin-based materials for biomedical applications. Bioactive Materials. 2020;5(3):496–509.
[188] Koh L-D, Cheng Y, Teng C-P, Khin Y-W, Loh X-J, Tee S-Y, et al. Structures, mechanical properties and applications of silk fibroin materials. Progress in Polymer Science. 2015;46:86–110.
[189] Melke J, Midha S, Ghosh S, Ito K, Hofmann S. Silk fibroin as biomaterial for bone tissue engineering. Acta Biomaterialia. 2016;31:1–16.
[190] Hajiabbas M, Alemzadeh I, Vossoughi M. A porous hydrogel-electrospun composite scaffold made of oxidized alginate/gelatin/silk fibroin for tissue engineering application. Carbohydrate Polymers. 2020;245:116465.
[191] Lammel AS, Hu X, Park S-H, Kaplan DL, Scheibel TR. Controlling silk fibroin particle features for drug delivery. Biomaterials. 2010;31(16):4583–91.
[192] Zhu B, Wang H, Leow WR, Cai Y, Loh XJ, Han MY, et al. Silk fibroin for flexible electronic devices. Advanced Materials. 2016;28(22):4250–65.
[193] Yavuz B, Chambre L, Harrington K, Kluge J, Valenti L, Kaplan DL. Silk fibroin microneedle patches for the sustained release of levonorgestrel. ACS Applied Bio Materials. 2020;3(8):5375–82.
[194] Brako F, Luo C, Craig DQ, Edirisinghe M. An inexpensive, portable device for point-of-need generation of silver-nanoparticle doped cellulose acetate nanofibers for advanced wound dressing. Macromolecular Materials and Engineering. 2018;303(5):1700586.
[195] Song Y, Wang H, Yue F, Lv Q, Cai B, Dong N, et al. Silk-based biomaterials for cardiac tissue engineering. Advanced Healthcare Materials. 2020;9(23):2000735.

[196] Hernández-Rangel A, Martin-Martinez ES. Collagen based electrospun materials for skin wounds treatment. Journal of Biomedical Materials Research Part A. 2021.
[197] Lee A, Hudson A, Shiwarski D, Tashman J, Hinton T, Yerneni S, et al. 3D bioprinting of collagen to rebuild components of the human heart. Science. 2019;365(6452):482–87.
[198] Harborne JB. Phenolic Compounds. Phytochemical Methods: Springer; 1973. p. 33–88.
[199] Laura A, Moreno-Escamilla JO, Rodrigo-García J, Alvarez-Parrilla E. Phenolic Compounds. Postharvest Physiology and Biochemistry of Fruits and Vegetables: Elsevier; 2019. p. 253–71.
[200] Magnani C, Isaac VLB, Correa MA, Salgado HRN. Caffeic acid: A review of its potential use in medications and cosmetics. Analytical Methods. 2014;6(10):3203–10.
[201] Aguilar LE, Jang SR, Park CH, Lee KM. Supramolecular caffeic acid and bortezomib nanomedicine: Prodrug inducing reactive oxygen species and inhibiting cancer cell survival. Pharmaceutics. 2020;12(11):1082.
[202] Aguilar LE, Chalony C, Kumar D, Park CH, Kim CS. Phenol-Boronic surface functionalization of gold nanoparticles; to induce ROS damage while inhibiting the survival mechanisms of cancer cells. International Journal of Pharmaceutics. 2021;596:120267.
[203] Lee JY, Aguilar LE, Park CH, Kim CS. UV light assisted coating method of polyphenol caffeic acid and mediated immobilization of metallic silver particles for antibacterial implant surface modification. Polymers. 2019;11(7):1200.
[204] Aguilar LE, Lee JY, Park CH, Kim CS. Biomedical grade stainless steel coating of polycaffeic acid via combined oxidative and ultraviolet light-assisted polymerization process for bioactive implant application. Polymers. 2019;11(4):584.
[205] Sanz VC, Mena ML, González-Cortés A, Yanez-Sedeno P, Pingarrón J. Development of a tyrosinase biosensor based on gold nanoparticles-modified glassy carbon electrodes: Application to the measurement of a bioelectrochemical polyphenols index in wines. Analytica Chimica Acta. 2005;528(1):1–8.
[206] Lanzellotto C, Favero G, Antonelli M, Tortolini C, Cannistraro S, Coppari E, et al. Nanostructured enzymatic biosensor based on fullerene and gold nanoparticles: Preparation, characterization and analytical applications. Biosensors and Bioelectronics. 2014;55:430–37.
[207] Lestari ML, Indrayanto G. Curcumin. Profiles of Drug Substances, Excipients and Related Methodology. 2014;39:113–204.
[208] Anjum S, Gupta A, Sharma D, Gautam D, Bhan S, Sharma A, et al. Development of novel wound care systems based on nanosilver nanohydrogels of polymethacrylic acid with Aloe vera and curcumin. Materials Science and Engineering: C. 2016;64:157–66.
[209] Guadarrama-Acevedo MC, Mendoza-Flores RA, Prado-Audelo D, María L, Urbán-Morlán Z, Giraldo-Gomez DM, et al. Development and evaluation of alginate membranes with curcumin-loaded nanoparticles for potential wound-healing applications. Pharmaceutics. 2019;11 (8):389.
[210] Gupta N, Verma K, Nalla S, Kulshreshtha A, Lall R, Prasad S. Free radicals as a double-edged sword: The cancer preventive and therapeutic roles of curcumin. Molecules. 2020;25 (22):5390.
[211] Abdallah AA, Nasr El-Deen NA, Abd El-Aziz HI, Neamat-Allah AN. Effect of the aqueous root extract of Curcuma longa L. (turmeric) against thermally oxidized oil-induced hematological, biochemical and histopathological alterations. Comparative Clinical Pathology. 2020;29:837–45.
[212] El-Naggar ME, Al-Joufi F, Anwar M, Attia MF, El-Bana MA. Curcumin-loaded PLA-PEG copolymer nanoparticles for treatment of liver inflammation in streptozotocin-induced diabetic rats. Colloids and Surfaces B: Biointerfaces. 2019;177:389–98.

[213] Dai Q, Di Zhou LX, Song X. Curcumin alleviates rheumatoid arthritis-induced inflammation and synovial hyperplasia by targeting mTOR pathway in rats. Drug Design, Development and Therapy. 2018;12:4095.
[214] Zhang Z, Leong DJ, Xu L, He Z, Wang A, Navati M, et al. Curcumin slows osteoarthritis progression and relieves osteoarthritis-associated pain symptoms in a post-traumatic osteoarthritis mouse model. Arthritis Research & Therapy. 2016;18(1):1–12.
[215] Manca ML, Castangia I, Zaru M, Nácher A, Valenti D, Fernàndez-Busquets X, et al. Development of curcumin loaded sodium hyaluronate immobilized vesicles (hyalurosomes) and their potential on skin inflammation and wound restoring. Biomaterials. 2015;71:100–09.
[216] Nagle DG, Ferreira D, Zhou Y-D. Epigallocatechin-3-gallate (EGCG): Chemical and biomedical perspectives. Phytochemistry. 2006;67(17):1849–55.
[217] Das A, Haque I, Ray P, Ghosh A, Dutta D, Quadir M, et al. CCN5 activation by free or encapsulated EGCG is required to render triple-negative breast cancer cell viability and tumor progression. Pharmacology Research & Perspectives. 2021;9(2):e00753.
[218] Yan X, Zhang X, McClements DJ, Zou L, Liu X, Liu F. Co-encapsulation of epigallocatechin gallate (EGCG) and curcumin by two proteins-based nanoparticles: Role of EGCG. Journal of Agricultural and Food Chemistry. 2019;67(48):13228–36.
[219] Chu C, Wang Y, Wang Y, Yang R, Liu L, Rung S, et al. Evaluation of epigallocatechin-3-gallate (EGCG) modified collagen in guided bone regeneration (GBR) surgery and modulation of macrophage phenotype. Materials Science and Engineering: C. 2019;99:73–82.
[220] Chu C, Liu L, Wang Y, Wei S, Wang Y, Man Y, et al. Macrophage phenotype in the epigallocatechin-3-gallate (EGCG)-modified collagen determines foreign body reaction. Journal of Tissue Engineering and Regenerative Medicine. 2018;12(6):1499–507.
[221] Song C, Xu S, Chang L, Zhao X, Mei X, Ren X, et al. Preparation of EGCG decorated, injectable extracellular vesicles for cartilage repair in rat arthritis. Regenerative Biomaterials. 2021;8(6):rbab067.
[222] Zhang J, Li L, Kim S-H, Hagerman AE, Lü J. Anti-cancer, anti-diabetic and other pharmacologic and biological activities of penta-galloyl-glucose. Pharmaceutical Research. 2009;26(9):2066–80.
[223] Cao Y, Himmeldirk KB, Qian Y, Ren Y, Malki A, Chen X. Biological and biomedical functions of Penta-O-galloyl-D-glucose and its derivatives. Journal of Natural Medicines. 2014;68(3):465–72.
[224] Sierad LN, Simionescu A, Albers C, Chen J, Maivelett J, Tedder ME, et al. Design and testing of a pulsatile conditioning system for dynamic endothelialization of polyphenol-stabilized tissue engineered heart valves. Cardiovascular Engineering and Technology. 2010;1(2):138–53.
[225] Chuang T-H, Stabler C, Simionescu A, Simionescu DT. Polyphenol-stabilized tubular elastin scaffolds for tissue engineered vascular grafts. Tissue Engineering Part A. 2009;15(10):2837–51.
[226] Pennel T, Fercana G, Bezuidenhout D, Simionescu A, Chuang T-H, Zilla P, et al. The performance of cross-linked acellular arterial scaffolds as vascular grafts; pre-clinical testing in direct and isolation loop circulatory models. Biomaterials. 2014;35(24):6311–22.
[227] Deborde C, Simionescu DT, Wright C, Liao J, Sierad LN, Simionescu A. Stabilized collagen and elastin-based scaffolds for mitral valve tissue engineering. Tissue Engineering Part A. 2016;22(21–22):1241–51.
[228] Shavandi A, Bekhit AE-DA, Saeedi P, Izadifar Z, Bekhit AA, Khademhosseini A. Polyphenol uses in biomaterials engineering. Biomaterials. 2018;167:91–106.

[229] Bladé C, Aragonès G, Arola-Arnal A, Muguerza B, Bravo FI, Salvadó MJ, et al. Proanthocyanidins in health and disease. Biofactors. 2016;42(1):5–12.

[230] Han B, Jaurequi J, Tang BW, Nimni ME. Proanthocyanidin: A natural crosslinking reagent for stabilizing collagen matrices. Journal of Biomedical Materials Research Part A: An Official Journal of the Society for Biomaterials, the Japanese Society for Biomaterials, and the Australian Society for Biomaterials and the Korean Society for Biomaterials. 2003;65 (1):118–24.

[231] Balalaie A, Rezvani MB, Basir MM. Dual function of proanthocyanidins as both MMP inhibitor and crosslinker in dentin biomodification: A literature review. Dental Materials Journal. 2018;37(2):173–82.

[232] Fawzy A, Priyadarshini B, Selvan S, Lu TB, Neo J. Proanthocyanidins-loaded nanoparticles enhance dentin degradation resistance. Journal of Dental Research. 2017;96(7):780–89.

[233] Ryu JH, Messersmith PB, Lee H. Polydopamine surface chemistry: A decade of discovery. ACS Applied Materials & Interfaces. 2018;10(9):7523–40.

[234] Liebscher J, Mrówczyński R, Scheidt HA, Filip C, Hădade ND, Turcu R, et al. Structure of polydopamine: A never-ending story?. Langmuir. 2013;29(33):10539–48.

[235] GhavamiNejad A, Aguilar LE, Ambade RB, Lee S-H, Park CH, Kim CS. Immobilization of silver nanoparticles on electropolymerized polydopamine films for metal implant applications. Colloids and Interface Science Communications. 2015;6:5–8.

[236] Yi Z, Zhao B, Liao M, Qin Z. Fabrication of superhydrophobic wood surface by etching polydopamine coating with sodium hydroxide. Coatings. 2020;10(9):847.

[237] Kwon IS, Bettinger CJ. Polydopamine nanostructures as biomaterials for medical applications. Journal of Materials Chemistry B. 2018;6(43):6895–903.

[238] Aguilar LE, Thomas RG, Moon MJ, Jeong YY, Park CH, Kim CS. Implantable chemothermal brachytherapy seeds: A synergistic approach to brachytherapy using polymeric dual drug delivery and hyperthermia for malignant solid tumor ablation. European Journal of Pharmaceutics and Biopharmaceutics. 2018;129:191–203.

[239] Aguilar LE, Tumurbaatar B, Ghavaminejad A, Park CH, Kim CS. Functionalized non-vascular nitinol stent via electropolymerized polydopamine thin film coating loaded with bortezomib adjunct to hyperthermia therapy. Scientific Reports. 2017;7(1):1–11.

[240] Wu Q, Niu M, Chen X, Tan L, Fu C, Ren X, et al. Biocompatible and biodegradable zeolitic imidazolate framework/polydopamine nanocarriers for dual stimulus triggered tumor thermo-chemotherapy. Biomaterials. 2018;162:132–43.

[241] Aguilar LE, GhavamiNejad A, Park CH, Kim CS. On-demand drug release and hyperthermia therapy applications of thermoresponsive poly-(NIPAAm-co-HMAAm)/polyurethane core-shell nanofiber mat on non-vascular nitinol stents. Nanomedicine: Nanotechnology, Biology and Medicine. 2017;13(2):527–38.

[242] Yang K, Lee JS, Kim J, Lee YB, Shin H, Um SH, et al. Polydopamine-mediated surface modification of scaffold materials for human neural stem cell engineering. Biomaterials. 2012;33(29):6952–64.

[243] Salazar P, Martín M, González-Mora J Polydopamine-modified surfaces in biosensor applications. Polymer science: research advances, practical applications and educational aspects Formatex Research Center SL (Spain). 2016:385–96.

[244] Palladino P, Bettazzi F, Scarano S. Polydopamine: Surface coating, molecular imprinting, and electrochemistry – successful applications and future perspectives in (bio) analysis. Analytical and Bioanalytical Chemistry. 2019;411(19):4327–38.

[245] Shafiei-Irannejad V, Soleymani J, Azizi S, KhoubnasabJafari M, Jouyban A, Hasanzadeh M. Advanced nanomaterials towards biosensing of insulin: Analytical approaches. TrAC Trends in Analytical Chemistry. 2019;116:1–12.

[246] Xia J, Liu Y, Ran M, Lu W, Bi L, Wang Q, et al. The simultaneous detection of the squamous cell carcinoma antigen and cancer antigen 125 in the cervical cancer serum using nano-Ag polydopamine nanospheres in an SERS-based lateral flow immunoassay. RSC Advances. 2020;10(49):29156–70.
[247] Li J, Baird MA, Davis MA, Tai W, Zweifel LS, Waldorf KMA, et al. Dramatic enhancement of the detection limits of bioassays via ultrafast deposition of polydopamine. Nature Biomedical Engineering. 2017;1(6):1–12.
[248] Jirofti N, Golandi M, Movaffagh J, Ahmadi FS, Kalalinia F. Improvement of the Wound-Healing Process by Curcumin-Loaded Chitosan/Collagen Blend Electrospun Nanofibers: In Vitro and In Vivo Studies. ACS Biomaterials Science & Engineering. 2021;7(8):3886–97.
[249] Rezk AI, Obiweluozor FO, Choukrani G, Park CH, Kim CS. Drug release and kinetic models of anticancer drug (BTZ) from a pH-responsive alginate polydopamine hydrogel: Towards cancer chemotherapy. International Journal of Biological Macromolecules. 2019;141:388–400.
[250] Jenkins M. Biomedical Polymers: Elsevier; 2007.
[251] Yalcin Enis I, Gok Sadikoglu T. Design parameters for electrospun biodegradable vascular grafts. Journal of Industrial Textiles. 2018;47(8):2205–27.
[252] Wang Z, Liu L, Mithieux SM, Weiss AS. Fabricating organized elastin in vascular grafts. Trends in Biotechnology. 2021;39(5):505–18.
[253] Imazato S, Kohno T, Tsuboi R, Thongthai P, Xu HH, Kitagawa H. Cutting-edge filler technologies to release bio-active components for restorative and preventive dentistry. Dental Materials Journal. 2020:2019–350.
[254] Nazeer N, Ahmed M. Polymers in Medicine. Polymer Science and Nanotechnology: Elsevier; 2020. p. 281–323.
[255] Zhang J-Z, Zhou Q-H, Yang L-Q, Zhang W, Li J-X, Jin Y, et al. Non-biodegradable and biodegradable long-term subdermal contraceptive implants: Advances in theoretical research and application. Chinese Journal of Tissue Engineering Research. 2017;21(22):3595.
[256] Cobb W, Peindl R, Zerey M, Carbonell A, Heniford B. Mesh terminology 101. Hernia. 2009;13 (1):1–6.
[257] Hepburn C. Polyurethane Elastomers: Springer Science & Business Media; 2012.
[258] Uttayarat P, Perets A, Li M, Pimton P, Stachelek SJ, Alferiev I, et al. Micropatterning of three-dimensional electrospun polyurethane vascular grafts. Acta Biomaterialia. 2010;6 (11):4229–37.
[259] Gostev AA, Chernonosova VS, Murashov IS, Sergeevichev DS, Korobeinikov AA, Karaskov AM, et al. Electrospun polyurethane-based vascular grafts: Physicochemical properties and functioning in vivo. Biomedical Materials. 2019;15(1):015010.
[260] Tepe G, Schmehl J, Wendel HP, Schaffner S, Heller S, Gianotti M, et al. Reduced thrombogenicity of nitinol stents – in vitro evaluation of different surface modifications and coatings. Biomaterials. 2006;27(4):643–50.
[261] Nikam SP, Chen P, Nettleton K, Hsu Y-H, Becker ML. Zwitterion surface-functionalized thermoplastic polyurethane for antifouling catheter applications. Biomacromolecules. 2020;21(7):2714–25.
[262] Gu SY, Chang K, Jin SP. A dual-induced self-expandable stent based on biodegradable shape memory polyurethane nanocomposites (PCLAU/Fe3O4) triggered around body temperature. Journal of Applied Polymer Science. 2018;135(3):45686.
[263] Wheatley D, Raco L, Bernacca G, Sim I, Belcher P, Boyd J. Polyurethane: Material for the next generation of heart valve prostheses?. European Journal of Cardio-thoracic Surgery. 2000;17 (4):440–48.
[264] Kütting M, Roggenkamp J, Urban U, Schmitz-Rode T, Steinseifer U. Polyurethane heart valves: Past, present and future. Expert Review of Medical Devices. 2011;8(2):227–33.

[265] Obiweluozor FO, Maharjan B, Emechebe AG, Park CH, Kim CS. Mussel-inspired elastic interpenetrated network hydrogel as an alternative for anti-thrombotic stent coating membrane. Chemical Engineering Journal. 2018;347:932–43.
[266] Aguilar LE, Unnithan AR, Amarjargal A, Tiwari AP, Hong ST, Park CH, et al. Electrospun polyurethane/Eudragit® L100-55 composite mats for the pH dependent release of paclitaxel on duodenal stent cover application. International Journal of Pharmaceutics. 2015;478(1):1–8.
[267] Zia F, Zia KM, Zuber M, Tabasum S, Rehman S. Heparin based polyurethanes: A state-of-the-art review. International Journal of Biological Macromolecules. 2016;84:101–11.
[268] Joshi M, Adak B, Butola B. Polyurethane nanocomposite based gas barrier films, membranes and coatings: A review on synthesis, characterization and potential applications. Progress in Materials Science. 2018;97:230–82.
[269] Nakazawa CT, Higuchi A, Asano A, Kameda T, Aytemiz D, Nakazawa Y. Solid-state NMR studies for the development of non-woven biomaterials based on silk fibroin and polyurethane. Polymer Journal. 2017;49(7):583–86.
[270] Scholten E, Bromberg L, Rutledge GC, Hatton TA. Electrospun polyurethane fibers for absorption of volatile organic compounds from air. ACS Applied Materials & Interfaces. 2011;3(10):3902–09.
[271] Gaaz TS, Sulong AB, Kadhum AAH. Effect of HNTs addition in the injection moulded thermoplastic polyurethane matrix on the mechanical and thermal properties. Sains Malays. 2016;45:1235–42.
[272] Adipurnama I, Yang M-C, Ciach T, Butruk-Raszeja B. Surface modification and endothelialization of polyurethane for vascular tissue engineering applications: A review. Biomaterials Science. 2017;5(1):22–37.
[273] Mata A, Fleischman AJ, Roy S. Characterization of polydimethylsiloxane (PDMS) properties for biomedical micro/nanosystems. Biomedical Microdevices. 2005;7(4):281–93.
[274] Yoshida SH, Chang CC, Teuber SS, Gershwin ME. Silicon and silicone: Theoretical and clinical implications of breast implants. Regulatory Toxicology and Pharmacology. 1993;17(1):3–18.
[275] McDonald JC, Whitesides GM. Poly (dimethylsiloxane) as a material for fabricating microfluidic devices. Accounts of Chemical Research. 2002;35(7):491–99.
[276] Zhang M, Wu J, Wang L, Xiao K, Wen W. A simple method for fabricating multi-layer PDMS structures for 3D microfluidic chips. Lab on a Chip. 2010;10(9):1199–203.
[277] Park J, Kim B. 3D micro patterning on a concave substrate for creating the replica of a cylindrical PDMS stamp. Microelectronic Engineering. 2012;98:540–43.
[278] Chatterjee T, Wiessner S, Naskar K, Heinrich G. Exploring a novel cyclic monofunctional peroxide for curing of silicone rubber at elevated temperature. Polymer Engineering & Science. 2017;57(10):1073–82.
[279] Car A, Baumann P, Duskey JT, Chami M, Bruns N, Meier W. pH-responsive PDMS-b-PDMAEMA micelles for intracellular anticancer drug delivery. Biomacromolecules. 2014;15(9):3235–45.
[280] Aiwazian J, Baker C, Nissanka N, Sarker U Long Term Flux Profile of Implanon Birth Control Implant. 2010.
[281] Lloyd AW, Faragher RG, Denyer SP. Ocular biomaterials and implants. Biomaterials. 2001;22 (8):769–85.
[282] Ali U, Karim KJBA, Buang NA. A review of the properties and applications of poly (methyl methacrylate)(PMMA). Polymer Reviews. 2015;55(4):678–705.
[283] Bhat SV. Biomaterials in Ophthalmology. Biomaterials: Springer; 2002. p. 163–70.
[284] Hollick EJ, Spalton DJ, Ursell PG, Meacock WR, Barman SA, Boyce JF. Posterior capsular opacification with hydrogel, polymethylmethacrylate, and silicone intraocular lenses: Two-year results of a randomized prospective trial. American Journal of Ophthalmology. 2000;129 (5):577–84.

[285] Zafar MS. Prosthodontic applications of polymethyl methacrylate (PMMA): An update. Polymers. 2020;12(10):2299.
[286] Luo K, Jiang G, Zhu J, Lu B, Lu J, Zhang K, et al. Poly (methyl methacrylate) bone cement composited with mineralized collagen for osteoporotic vertebral compression fractures in extremely old patients. Regenerative Biomaterials. 2020;7(1):29–34.
[287] Zhu J, Yang S, Cai K, Wang S, Qiu Z, Huang J, et al. Bioactive poly (methyl methacrylate) bone cement for the treatment of osteoporotic vertebral compression fractures. Theranostics. 2020;10(14):6544.
[288] Leggat PA, Smith DR, Kedjarune U. Surgical applications of methyl methacrylate: A review of toxicity. Archives of Environmental & Occupational Health. 2009;64(3):207–12.
[289] Fini M, Giavaresi G, Aldini NN, Torricelli P, Botter R, Beruto D, et al. A bone substitute composed of polymethylmethacrylate and α-tricalcium phosphate: Results in terms of osteoblast function and bone tissue formation. Biomaterials. 2002;23(23):4523–31.
[290] Jang JS, Lee SH, Rhee CH, Lee SH. Polymethylmethacrylate-augmented screw fixation for stabilization in metastatic spinal tumors. Journal of Neurosurgery: Spine. 2002;96(1):131–34.
[291] Moon BJ, Cho BY, Choi EY, Zhang HY. Polymethylmethacrylate-augmented screw fixation for stabilization of the osteoporotic spine: A three-year follow-up of 37 patients. Journal of Korean Neurosurgical Society. 2009;46(4):305.
[292] Maddah HA. Polypropylene as a promising plastic: A review. American Journal of Polymer Science. 2016;6(1):1–11.
[293] Halm J, De Wall L, Steyerberg E, Jeekel J, Lange J. Intraperitoneal polypropylene mesh hernia repair complicates subsequent abdominal surgery. World Journal of Surgery. 2007;31 (2):423–29.
[294] Larouche M, Geoffrion R, Walter J-E. No. 351-Transvaginal mesh procedures for pelvic organ prolapse. Journal of Obstetrics and Gynaecology Canada. 2017;39(11):1085–97.
[295] Fortelny R, Petter-Puchner A, Glaser K, Offner F, Benesch T, Rohr M. Adverse effects of polyvinylidene fluoride-coated polypropylene mesh used for laparoscopic intraperitoneal onlay repair of incisional hernia. Journal of British Surgery. 2010;97(7):1140–45.
[296] Liang -C-C, Lin Y-H, Chang Y-L, Chang S-D. Urodynamic and clinical effects of transvaginal mesh repair for severe cystocele with and without urinary incontinence. International Journal of Gynecology & Obstetrics. 2011;112(3):182–86.
[297] Kist C, Manna BB, Montes JHM, Bigolin AV, Grossi JVM, Cavazzola LT. Comparative study of intraperitoneal adhesions associated with the use of meshes of polypropylene and polypropylene coated with omega-3 fatty acid. Revista Do Colégio Brasileiro de Cirurgiões. 2012;39:201–06.
[298] Franklin ME, Voeller G, Matthews B, Earle DB. The Benefits of Omega-3 Fatty Acid-Coated Mesh in Ventral Hernia Repair. Specification Representative. 2010;37:1–8.
[299] Ronca S. Polyethylene. Brydson's Plastics Materials: Elsevier; 2017. p. 247–78.
[300] Bistolfi A, Giustra F, Bosco F, Sabatini L, Aprato A, Bracco P, et al. Ultra-high molecular weight polyethylene (UHMWPE) for hip and knee arthroplasty: The present and the future. Journal of Orthopaedics. 2021.
[301] Paxton NC, Allenby MC, Lewis PM, Woodruff MA. Biomedical applications of polyethylene. European Polymer Journal. 2019;118:412–28.
[302] Costantino PD, Friedman CD, Lane A. Synthetic biomaterials in facial plastic and reconstructive surgery. Facial Plastic Surgery. 1993;9(01):1–15.
[303] Paxton NC, Dinoro J, Ren J, Ross MT, Daley R, Zhou R, et al. Additive manufacturing enables personalised porous high-density polyethylene surgical implant manufacturing with improved tissue and vascular ingrowth. Applied Materials Today. 2021;22:100965.

[304] Kanczler JM, Mirmalek-Sani S-H, Hanley NA, Ivanov AL, Barry JJ, Upton C, et al. Biocompatibility and osteogenic potential of human fetal femur-derived cells on surface selective laser sintered scaffolds. Acta Biomaterialia. 2009;5(6):2063–71.
[305] Kumar RM, Gupta P, Sharma SK, Mittal A, Shekhar M, Kumar V, et al. Sustained drug release from surface modified UHMWPE for acetabular cup lining in total hip implant. Materials Science and Engineering: C. 2017;77:649–61.
[306] Richey T, Iwata H, Oowaki H, Uchida E, Matsuda S, Ikada Y. Surface modification of polyethylene balloon catheters for local drug delivery. Biomaterials. 2000;21(10):1057–65.
[307] Bunn C, Cobbold A, Palmer R. The fine structure of polytetrafluoroethylene. Journal of Polymer Science. 1958;28(117):365–76.
[308] Dhanumalayan E, Joshi GM. Performance properties and applications of polytetrafluoroethylene (PTFE) – a review. Advanced Composites and Hybrid Materials. 2018;1(2):247–68.
[309] Lau KK, Caulfield JA, Gleason KK. Structure and morphology of fluorocarbon films grown by hot filament chemical vapor deposition. Chemistry of Materials. 2000;12(10):3032–37.
[310] Casaccia M, Torelli P, Troilo BM, Savelli A, Valente U. Composite mesh repair of a large paraesophageal hiatal hernia. Journal of Laparoendoscopic & Advanced Surgical Techniques. 2006;16(4):381–85.
[311] Memisoglu K, Saribeyoglu K, Pekmezci S, Karahasanoglu T, Sen B, Bayrak İ, et al. Mesh fixation devices and formation of intraperitoneal adhesions. Journal of Laparoendoscopic & Advanced Surgical Techniques. 2006;16(5):439–44.
[312] Retzepi M, Donos N. Guided bone regeneration: Biological principle and therapeutic applications. Clinical Oral Implants Research. 2010;21(6):567–76.
[313] Mardas N, Kostopoulos L, Karring T. Bone and suture regeneration in calvarial defects by e-PTFE-membranes and demineralized bone matrix and the impact on calvarial growth: An experimental study in the rat. Journal of Craniofacial Surgery. 2002;13(3):453–62.
[314] Niamtu J. Advanta ePTFE facial implants in cosmetic facial surgery. Journal of Oral and Maxillofacial Surgery. 2006;64(3):543–49.
[315] Ham J, Miller PJ. Expanded polytetrafluoroethylene implants in rhinoplasty: Literature review, operative techniques, and outcome. Facial Plastic Surgery. 2003;19(04):331–40.
[316] Chertin B, Colhoun E, Velayudham M, Puri P. Endoscopic treatment of vesicoureteral reflux: 11 to 17 years of followup. The Journal of Urology. 2002;167(3):1443–46.
[317] Ku DN, Allen RC Vascular grafts. Tissue Engineering and Artificial Organs. 2013.
[318] Tordoir J, Herman J, Kwan T, Diderich P. Long-term follow-up of the polytetrafluoroethylene (PTFE) prosthesis as an arteriovenous fistula for haemodialysis. European Journal of Vascular Surgery. 1988;2(1):3–7.
[319] Abitha H, Kavitha V, Gomathi B, Ramachandran B. A recent investigation on shape memory alloys and polymers based materials on bio artificial implants-hip and knee joint. Materials Today: Proceedings. 2020;33:4458–66.
[320] Liang M, Ren S, Fan W, Xin X, Shi J, Luo H. Rheological property and stability of polymer modified asphalt: Effect of various vinyl-acetate structures in EVA copolymers. Construction and Building Materials. 2017;137:367–80.
[321] Schneider C, Langer R, Loveday D, Hair D. Applications of ethylene vinyl acetate copolymers (EVA) in drug delivery systems. Journal of Controlled Release. 2017;262:284–95.
[322] Kumar S, Hota S, Pais R, Das K, Santra T, Misra S. Implications of advance biomaterials in development of new contraceptive devices. Journal of Materials NanoScience. 2021;8(1):23–34.
[323] Awwad S, Ahmed AHM, Sharma G, Heng JS, Khaw PT, Brocchini S, et al. Principles of pharmacology in the eye. British Journal of Pharmacology. 2017;174(23):4205–23.

[324] Singh RB, Ichhpujani P, Thakur S, Jindal S. Promising therapeutic drug delivery systems for glaucoma: A comprehensive review. Therapeutic Advances in Ophthalmology. 2020;12:2515841420905740.
[325] Quarterman JC, Geary SM, Salem AK. Evolution of drug-eluting biomedical implants for sustained drug delivery. European Journal of Pharmaceutics and Biopharmaceutics. 2021;159:21–35.
[326] Joo S, Cho IJ, Seo H, Son HF, Sagong H-Y, Shin TJ, et al. Structural insight into molecular mechanism of poly (ethylene terephthalate) degradation. Nature Communications. 2018;9 (1):1–12.
[327] Kawai F, Kawabata T, Oda M. Current state and perspectives related to the polyethylene terephthalate hydrolases available for biorecycling. ACS Sustainable Chemistry & Engineering. 2020;8(24):8894–908.
[328] Amabili M, Balasubramanian P, Ferrari G, Franchini G, Giovanniello F, Tubaldi E. Identification of viscoelastic properties of Dacron aortic grafts subjected to physiological pulsatile flow. Journal of the Mechanical Behavior of Biomedical Materials. 2020;110:103804.
[329] Guo Z, Zhang C, Wang H, Ge S. Peratrial device closure of atrial septal defect under transesophageal echocardiographic guidance without fluoroscopy compared to conventional on-pump surgical closure. Brazilian Journal of Cardiovascular Surgery. 2017;32:38–42.
[330] El-Sisi AM, El-Saiedi SA, Ammar R, Abdelhameed A, Hijazi ZM, Soliman MM. Safety of Occlutech septal occluder ACCELL Flex II for transcatheter closure of secundum atrial septal defects in children: a long-term follow-up. Journal of Interventional Cardiology. 2022;2022.
[331] Zahn EM, Wilson N, Cutright W, Latson LA. Development and testing of the Helex septal occluder, a new expanded polytetrafluoroethylene atrial septal defect occlusion system. Circulation. 2001;104(6):711–16.
[332] Zhu Y-F, Huang X-M, Cao J, Hu J-Q, Bai Y, Jiang H-B, et al. Animal experimental study of the fully biodegradable atrial septal defect (ASD) occluder. Journal of Biomedicine and Biotechnology. 2012;2012.
[333] Ratcliffe A, Butler DL, Dyment NA, Cagle PJ, Proctor CS, Ratcliffe SS, et al. Scaffolds for tendon and ligament repair and regeneration. Annals of Biomedical Engineering. 2015;43 (3):819–31.
[334] Peixoto T, Carneiro S, Fangueiro R, Guedes RM, Paiva MC, Lopes MA. Engineering hybrid textile braids for tendon and ligament repair application. Journal of Applied Polymer Science. 2021:52013.
[335] Guidoin M-F, Marois Y, Bejui J, Poddevin N, King MW, Guidoin R. Analysis of retrieved polymer fiber based replacements for the ACL. Biomaterials. 2000;21(23):2461–74.
[336] Batty LM, Norsworthy CJ, Lash NJ, Wasiak J, Richmond AK, Feller JA. Synthetic devices for reconstructive surgery of the cruciate ligaments: A systematic review. Arthroscopy: The Journal of Arthroscopic & Related Surgery. 2015;31(5):957–68.
[337] Bailey FJ. Poly (ethylene oxide): Elsevier; 2012.
[338] Chen H, Zhang Z, Chen Y, Brook MA, Sheardown H. Protein repellant silicone surfaces by covalent immobilization of poly (ethylene oxide). Biomaterials. 2005;26(15):2391–99.
[339] Mahou R, Wandrey C. Versatile route to synthesize heterobifunctional poly (ethylene glycol) of variable functionality for subsequent pegylation. Polymers. 2012;4(1):561–89.
[340] Harris JM, Martin NE, Modi M. Pegylation. Clinical Pharmacokinetics. 2001;40(7):539–51.
[341] Yadav D, Dewangan HK. PEGYLATION: An important approach for novel drug delivery system. Journal of Biomaterials Science, Polymer Edition. 2021;32(2):266–80.
[342] Huang Z-G, Lv F-M, Wang J, Cao S-J, Liu Z-P, Liu Y, et al. RGD-modified PEGylated paclitaxel nanocrystals with enhanced stability and tumor-targeting capability. International Journal of Pharmaceutics. 2019;556:217–25.

[343] Zhang X, Zhao M, Cao N, Qin W, Zhao M, Wu J, et al. Construction of a tumor microenvironment pH-responsive cleavable PEGylated hyaluronic acid nano-drug delivery system for colorectal cancer treatment. Biomaterials Science. 2020;8(7):1885–96.
[344] Molineux G. Pegylation: Engineering improved biopharmaceuticals for oncology. Pharmacotherapy: The Journal of Human Pharmacology and Drug Therapy. 2003;23 (8P2):3S–8S.
[345] Misbah S, Ahmad A, Butt MH, Khan YH, Alotaibi NH, Mallhi TH. A systematic analysis of studies on corona virus disease 19 (COVID-19) from viral emergence to treatment. Journal of College of Physicians and Surgeons Pakistan. 2020;30(6):9–18.
[346] Ahn D-G, Shin H-J, Kim M-H, Lee S, Kim H-S, Myoung J, et al Current status of epidemiology, diagnosis, therapeutics, and vaccines for novel coronavirus disease 2019 (COVID-19). 2020.
[347] Ozcelik B, Ho KKK, Glattauer V, Willcox M, Kumar N, Thissen H. Poly (ethylene glycol)-based coatings combining low-biofouling and quorum-sensing inhibiting properties to reduce bacterial colonization. ACS Biomaterials Science & Engineering. 2017;3(1):78–87.
[348] Greene GW, Ortiz V, Pozo-Gonzalo C, Moulton SE, Wang X, Martin LL, et al. Lubricin antiadhesive coatings exhibit size-selective transport properties that inhibit biofouling of electrode surfaces with minimal loss in electrochemical activity. Advanced Materials Interfaces. 2018;5(7):1701296.
[349] Ding X, Yang C, Lim TP, Hsu LY, Engler AC, Hedrick JL, et al. Antibacterial and antifouling catheter coatings using surface grafted PEG-b-cationic polycarbonate diblock copolymers. Biomaterials. 2012;33(28):6593–603.
[350] Peng L, Chang L, Liu X, Lin J, Liu H, Han B, et al. Antibacterial property of a polyethylene glycol-grafted dental material. ACS Applied Materials & Interfaces. 2017;9(21):17688–92.
[351] Khan F, Tabassum N, Kim Y-M. A strategy to control colonization of pathogens: Embedding of lactic acid bacteria on the surface of urinary catheter. Applied Microbiology and Biotechnology. 2020:1–14.
[352] Wang D, Ge C, Liang W, Yang Q, Liu Q, Ma W, et al. In vivo enrichment and elimination of circulating tumor cells by using a black phosphorus and antibody functionalized intravenous catheter. Advanced Science. 2020;7(17):2000940.
[353] McBride MC, Malcolm RK, Woolfson AD, Gorman SP. Persistence of antimicrobial activity through sustained release of triclosan from pegylated silicone elastomers. Biomaterials. 2009;30(35):6739–47.
[354] Underhill GH, Chen AA, Albrecht DR, Bhatia SN. Assessment of hepatocellular function within PEG hydrogels. Biomaterials. 2007;28(2):256–70.
[355] Jelinska N, Kalnins M, Tupureina V, Dzene A. Poly(vinyl alcohol)/poly (vinyl acetate). Rigas Tehniskas Universitates Zinatniskie Raksti. 2010;21:55.
[356] Kamoun EA, Kenawy E-RS, Chen X. A review on polymeric hydrogel membranes for wound dressing applications: PVA-based hydrogel dressings. Journal of Advanced Research. 2017;8(3):217–33.
[357] Sciarretta FV. to 8 years follow-up of knee chondral defects treated by PVA-H hydrogel implants. European Review for Medical and Pharmacological Sciences. 2013;17(22):3031–38.
[358] Spiller KL, Maher SA, Lowman AM. Hydrogels for the repair of articular cartilage defects. Tissue Engineering Part B: Reviews. 2011;17(4):281–99.
[359] Aziz SB, Abdullah OG, Hussein SA, Ahmed HM. Effect of PVA blending on structural and ion transport properties of CS: AgNt-based polymer electrolyte membrane. Polymers. 2017;9 (11):622.
[360] Teodorescu M, Bercea M, Morariu S. Biomaterials of PVA and PVP in medical and pharmaceutical applications: Perspectives and challenges. Biotechnology Advances. 2019;37(1):109–31.

[361] Uma Maheshwari S, Govindan K, Raja M, Raja A, Pravin M, Vasanth Kumar S. Preliminary studies of PVA/PVP blends incorporated with HAp and β-TCP bone ceramic as template for hard tissue engineering. Bio-medical Materials and Engineering. 2017;28(4):401–15.
[362] Gökmeşe F, Uslu I, Aytimur A. Preparation and characterization of PVA/PVP nanofibers as promising materials for wound dressing. Polymer-Plastics Technology and Engineering. 2013;52(12):1259–65.
[363] Jem KJ, Tan B. The development and challenges of poly (lactic acid) and poly (glycolic acid). Advanced Industrial and Engineering Polymer Research. 2020;3(2):60–70.
[364] Gilding D, Reed A. Biodegradable polymers for use in surgery – polyglycolic/poly (actic acid) homo-and copolymers: 1. Polymer. 1979;20(12):1459–64.
[365] Elmowafy EM, Tiboni M, Soliman ME. Biocompatibility, biodegradation and biomedical applications of poly (lactic acid)/poly (lactic-co-glycolic acid) micro and nanoparticles. Journal of Pharmaceutical Investigation. 2019;49(4):347–80.
[366] Allen RP, Bolandparvaz A, Ma JA, Manickam VA, Lewis JS. Latent, Immunosuppressive Nature of Poly (lactic-co-glycolic acid) Microparticles. ACS Biomaterials Science & Engineering. 2018;4(3):900–18.
[367] Hedberg EL, Shih CK, Lemoine JJ, Timmer MD, Liebschner MA, Jansen JA, et al. In vitro degradation of porous poly (propylene fumarate)/poly (DL-lactic-co-glycolic acid) composite scaffolds. Biomaterials. 2005;26(16):3215–25.
[368] Gentile P, Chiono V, Carmagnola I, Hatton PV. An overview of poly (lactic-co-glycolic) acid (PLGA)-based biomaterials for bone tissue engineering. International Journal of Molecular Sciences. 2014;15(3):3640–59.
[369] Wang J, Tian L, He L, Chen N, Ramakrishna S, So K-F, et al. Lycium barbarum polysaccharide encapsulated Poly lactic-co-glycolic acid Nanofibers: Cost effective herbal medicine for potential application in peripheral nerve tissue engineering. Scientific Reports. 2018;8 (1):1–14.
[370] Liu X, Nielsen LH, Kłodzińska SN, Nielsen HM, Qu H, Christensen LP, et al. Ciprofloxacin-loaded sodium alginate/poly (lactic-co-glycolic acid) electrospun fibrous mats for wound healing. European Journal of Pharmaceutics and Biopharmaceutics. 2018;123:42–49.
[371] Ertl B, Heigl F, Wirth M, Gabor F. Lectin-mediated bioadhesion: Preparation, stability and caco-2 binding of wheat germ agglutinin-functionalized Poly (D, L-lactic-co-glycolic acid)-microspheres. Journal of Drug Targeting. 2000;8(3):173–84.
[372] Fujimaki H, Matsumine H, Osaki H, Ueta Y, Kamei W, Shimizu M, et al. Dedifferentiated fat cells in polyglycolic acid-collagen nerve conduits promote rat facial nerve regeneration. Regenerative Therapy. 2019;11:240–48.
[373] Inkinen S, Hakkarainen M, Albertsson A-C, Södergård A. From lactic acid to poly (lactic acid) (PLA): Characterization and analysis of PLA and its precursors. Biomacromolecules. 2011;12 (3):523–32.
[374] Abd Alsaheb RA, Aladdin A, Othman NZ, Abd Malek R, Leng OM, Aziz R, et al. Lactic acid applications in pharmaceutical and cosmeceutical industries. Journal of Chemical and Pharmaceutical Research. 2015;7(10):729–35.
[375] Thordarson DB, Samuelson M, Shepherd LE, Merkle PF, Lee J. Bioabsorbable versus stainless steel screw fixation of the syndesmosis in pronation-lateral rotation ankle fractures: A prospective randomized trial. Foot & Ankle International. 2001;22(4):335–38.
[376] Narayanan G, Vernekar VN, Kuyinu EL, Laurencin CT. Poly (lactic acid)-based biomaterials for orthopaedic regenerative engineering. Advanced Drug Delivery Reviews. 2016;107:247–76.
[377] Naahidi S, Jafari M, Logan M, Wang Y, Yuan Y, Bae H, et al. Biocompatibility of hydrogel-based scaffolds for tissue engineering applications. Biotechnology Advances. 2017;35 (5):530–44.

[378] Swider E, Koshkina O, Tel J, Cruz LJ, de Vries IJM, Srinivas M. Customizing poly (lactic-co-glycolic acid) particles for biomedical applications. Acta Biomaterialia. 2018;73:38–51.
[379] Wang S, Zheng F, Huang Y, Fang Y, Shen M, Zhu M, et al. Encapsulation of amoxicillin within laponite-doped poly (lactic-co-glycolic acid) nanofibers: Preparation, characterization, and antibacterial activity. ACS Applied Materials & Interfaces. 2012;4(11):6393–401.
[380] Sequeira JA, Santos AC, Serra J, Veiga F, Ribeiro AJ. Poly (lactic-co-glycolic acid)(PLGA) Matrix Implants. Nanostructures for the Engineering of Cells, Tissues and Organs: Elsevier; 2018. p. 375–402.
[381] Kempen D, Kim CW, Lu L, Dhert W, Currier B, Yaszemski M. Controlled Release from Poly (lactic-co-glycolic acid) Microspheres Embedded in an Injectable, Biodegradable Scaffold for Bone Tissue Engineering. Materials Science Forum. Zurich-Uetikon, Switzerland: Trans Tech Publications Ltd.; 2003. p. 3151–56.
[382] Goldberg D. A review of the biodegradability and utility of poly (caprolactone). Journal of Environmental Polymer Degradation. 1995;3(2):61–67.
[383] Formica FA, Öztürk E, Hess SC, Stark WJ, Maniura-Weber K, Rottmar M, et al. A bioinspired ultraporous nanofiber-hydrogel mimic of the cartilage extracellular matrix. Advanced Healthcare Materials. 2016;5(24):3129–38.
[384] Rameshbabu AP, Datta S, Bankoti K, Subramani E, Chaudhury K, Lalzawmliana V, et al. Polycaprolactone nanofibers functionalized with placental derived extracellular matrix for stimulating wound healing activity. Journal of Materials Chemistry B. 2018;6(42):6767–80.
[385] Gao X, Song J, Zhang Y, Xu X, Zhang S, Ji P, et al. Bioinspired design of polycaprolactone composite nanofibers as artificial bone extracellular matrix for bone regeneration application. ACS Applied Materials & Interfaces. 2016;8(41):27594–610.
[386] López-Esparza J, Espinosa-Cristóbal L, Donohue-Cornejo A, Reyes-López S. Antimicrobial activity of silver nanoparticles in polycaprolactone nanofibers against gram-positive and gram-negative bacteria. Industrial & Engineering Chemistry Research. 2016;55(49):12532–38.
[387] Tran PA, Hocking DM, O'Connor AJ. In situ formation of antimicrobial silver nanoparticles and the impregnation of hydrophobic polycaprolactone matrix for antimicrobial medical device applications. Materials Science and Engineering: C. 2015;47:63–69.
[388] Gorodzha S, Surmeneva M, Selezneva I, Ermakov A, Zaitsev V, Surmenev R. Investigation of the morphology and structure of porous hybrid 3D scaffolds based on polycaprolactone involving silicate-containing hydroxyapatite. Journal of Surface Investigation: X-ray, Synchrotron and Neutron Techniques. 2018;12(4):717–26.
[389] Correa E, Moncada M, Gutiérrez O, Vargas C, Zapata V. Characterization of polycaprolactone/rGO nanocomposite scaffolds obtained by electrospinning. Materials Science and Engineering: C. 2019;103:109773.
[390] Hedayatnasab Z, Dabbagh A, Abnisa F, Daud WMAW. Polycaprolactone-coated superparamagnetic iron oxide nanoparticles for in vitro magnetic hyperthermia therapy of cancer. European Polymer Journal. 2020;133:109789.
[391] Bromberg L. Polymeric micelles in oral chemotherapy. Journal of Controlled Release. 2008;128(2):99–112.
[392] Visscher LE, Dang HP, Knackstedt MA, Hutmacher DW, Tran PA. 3D printed Polycaprolactone scaffolds with dual macro-microporosity for applications in local delivery of antibiotics. Materials Science and Engineering: C. 2018;87:78–89.
[393] Muwaffak Z, Goyanes A, Clark V, Basit AW, Hilton ST, Gaisford S. Patient-specific 3D scanned and 3D printed antimicrobial polycaprolactone wound dressings. International Journal of Pharmaceutics. 2017;527(1–2):161–70.

[394] Teo EY, Ong S-Y, Chong MSK, Zhang Z, Lu J, Moochhala S, et al. Polycaprolactone-based fused deposition modeled mesh for delivery of antibacterial agents to infected wounds. Biomaterials. 2011;32(1):279–87.
[395] Reddy C, Ghai R, Kalia VC. Polyhydroxyalkanoates: An overview. Bioresource Technology. 2003;87(2):137–46.
[396] Brigham CJ, Sinskey AJ. Applications of polyhydroxyalkanoates in the medical industry. International Journal of Biotechnology for Wellness Industries. 2012;1(1):52–60.
[397] Raza ZA, Abid S, Banat IM. Polyhydroxyalkanoates: Characteristics, production, recent developments and applications. International Biodeterioration & Biodegradation. 2018;126:45–56.
[398] Gadgil BST, Killi N, Rathna GV. Polyhydroxyalkanoates as biomaterials. Medicinal Chemical Communication. 2017;8(9):1774–87.
[399] Gumel A, Aris M, Annuar M. Modification of polyhydroxyalkanoates (PHAs). Polyhydroxyalkanoate (PHA) Based Blends, Composites and Nanocomposites. 2014:141–82.
[400] Jendrossek D, Handrick R. Microbial degradation of polyhydroxyalkanoates. Annual Review of Microbiology. 2002;56(1):403–32.
[401] Keirouz A, Chung M, Kwon J, Fortunato G, Radacsi N. 2D and 3D electrospinning technologies for the fabrication of nanofibrous scaffolds for skin tissue engineering: A review. Wiley Interdisciplinary Reviews: Nanomedicine and Nanobiotechnology. 2020;12(4):e1626.
[402] Blanco FG, Hernández N, Rivero-Buceta V, Maestro B, Sanz JM, Mato A, et al. From residues to added-value bacterial biopolymers as nanomaterials for biomedical applications. Nanomaterials. 2021;11(6):1492.
[403] Sahoo S Biologic-and synthetic-based Scaffolds for tendon regeneration. Tendon Regeneration. 2015:243–55.
[404] Goswami M, Rekhi P, Debnath M, Ramakrishna S. Microbial polyhydroxyalkanoates granules: An approach targeting biopolymer for medical applications and developing bone Scaffolds. Molecules. 2021;26(4):860.
[405] Ganzeveld KJ, van Hagen A, van Agteren MH, de Koning W, Uiterkamp AJS. Upgrading of organic waste: Production of the copolymer poly-3-hydroxybutyrate-co-valerate by Ralstonia eutrophus with organic waste as sole carbon source. Journal of Cleaner Production. 1999;7 (6):413–19.
[406] Rivera-Briso AL, Serrano-Aroca Á. Poly (3-Hydroxybutyrate-co-3-Hydroxyvalerate): Enhancement strategies for advanced applications. Polymers. 2018;10(7):732.
[407] Rivera-Briso AL, Aachmann FL, Moreno-Manzano V, Serrano-Aroca Á. Graphene oxide nanosheets versus carbon nanofibers: Enhancement of physical and biological properties of poly (3-hydroxybutyrate-co-3-hydroxyvalerate) films for biomedical applications. International Journal of Biological Macromolecules. 2020;143:1000–08.
[408] de Morais MG, de Morais EG, Silva Vaz B, Goncalves CF, Lisboa C, Vieira Costa JA. Nanoencapsulation of the bioactive compounds of spirulina with a microalgal biopolymer coating. Journal of Nanoscience and Nanotechnology. 2016;16(1):81–91.
[409] Perveen K, Masood F, Hameed A. Preparation, characterization and evaluation of antibacterial properties of epirubicin loaded PHB and PHBV nanoparticles. International Journal of Biological Macromolecules. 2020;144:259–66.
[410] Tebaldi ML, Maia ALC, Poletto F, de Andrade FV, Soares DCF. Poly (–3-hydroxybutyrate-co-3-hydroxyvalerate)(PHBV): Current advances in synthesis methodologies, antitumor applications and biocompatibility. Journal of Drug Delivery Science and Technology. 2019;51:115–26.
[411] Kumar N, Langer RS, Domb AJ. Polyanhydrides: An overview. Advanced Drug Delivery Reviews. 2002;54(7):889–910.

[412] Domb AJ, Elmalak O, Shastri VR, Ta-Shma Z, Masters D, Ringel I, et al. Polyanhydrides. Handbook of Biodegradable Polymers. 1997;7:135–59.
[413] Jain JP, Modi S, Domb A, Kumar N. Role of polyanhydrides as localized drug carriers. Journal of Controlled Release. 2005;103(3):541–63.
[414] Carrillo-Conde BR, Darling RJ, Seiler SJ, Ramer-Tait AE, Wannemuehler MJ, Narasimhan B. Sustained release and stabilization of therapeutic antibodies using amphiphilic polyanhydride nanoparticles. Chemical Engineering Science. 2015;125:98–107.
[415] Kipper MJ, Shen E, Determan A, Narasimhan B. Design of an injectable system based on bioerodible polyanhydride microspheres for sustained drug delivery. Biomaterials. 2002;23 (22):4405–12.
[416] Basu A, Domb AJ. Recent advances in polyanhydride based biomaterials. Advanced Materials. 2018;30(41):1706815.
[417] Heller J, Barr J. Poly (ortho esters) from concept to reality. Biomacromolecules. 2004;5 (5):1625–32.
[418] Heller J. Development of poly (ortho esters): A historical overview. Biomaterials. 1990;11 (9):659–65.
[419] Siepmann J, Goepferich A. Mathematical modeling of bioerodible, polymeric drug delivery systems. Advanced Drug Delivery Reviews. 2001;48(2–3):229–47.
[420] Yang SC, Bhide M, Crispe IN, Pierce RH, Murthy N. Polyketal copolymers: A new acid-sensitive delivery vehicle for treating acute inflammatory diseases. Bioconjugate Chemistry. 2008;19(6):1164–69.
[421] Heffernan MJ, Murthy N. Polyketal nanoparticles: A new pH-sensitive biodegradable drug delivery vehicle. Bioconjugate Chemistry. 2005;16(6):1340–42.
[422] Shenoi RA, Narayanannair JK, Hamilton JL, Lai BF, Horte S, Kainthan RK, et al. Branched multifunctional polyether polyketals: Variation of ketal group structure enables unprecedented control over polymer degradation in solution and within cells. Journal of the American Chemical Society. 2012;134(36):14945–57.
[423] Burns B. Polycyanoacrylates. Encyclopedia of Polymer Science and Technology. 2002:1–27.
[424] Kumar MS, Natta S, Shankar G, Reddy SHK, Visalakshi D, Seshiah G. Comparison between silk sutures and cyanoacrylate adhesive in human mucosa-a clinical and histological study. Journal of International Oral Health: JIOH. 2013;5(5):95.
[425] Gulalp B, Seyhan T, Gursoy S, Altinors MN. Emergency wounds treated with cyanoacrylate and long-term results in pediatrics: A series of cases; what are the advantages and boards?. BMC Research Notes. 2009;2(1):1–5.
[426] Obinu A, Rassu G, Corona P, Maestri M, Riva F, Miele D, et al. Poly (ethyl 2-cyanoacrylate) nanoparticles (PECA-NPs) as possible agents in tumor treatment. Colloids and Surfaces B: Biointerfaces. 2019;177:520–28.
[427] Duffy C, Zetterlund PB, Aldabbagh F. Radical polymerization of alkyl 2-cyanoacrylates. Molecules. 2018;23(2):465.
[428] Lin X, Zhou R, Qiao Y, Jin F, Zhai Y, Xing J, et al. Poly (ethylene glycol)/poly (ethyl cyanoacrylate) amphiphilic triblock copolymer nanoparticles as delivery vehicles for dexamethasone. Journal of Polymer Science Part A: Polymer Chemistry. 2008;46 (23):7809–15.
[429] Vauthier C, Dubernet C, Chauvierre C, Brigger I, Couvreur P. Drug delivery to resistant tumors: The potential of poly (alkyl cyanoacrylate) nanoparticles. Journal of Controlled Release. 2003;93(2):151–60.
[430] Yan X, Gemeinhart RA. Cisplatin delivery from poly (acrylic acid-co-methyl methacrylate) microparticles. Journal of Controlled Release. 2005;106(1–2):198–208.

[431] Calvo P, Gouritin B, Chacun H, Desmaële D, D'Angelo J, Noel J-P, et al. Long-circulating PEGylated polycyanoacrylate nanoparticles as new drug carrier for brain delivery. Pharmaceutical Research. 2001;18(8):1157–66.
[432] Peracchia M, Harnisch S, Pinto-Alphandary H, Gulik A, Dedieu J, Desmaele D, et al. Visualization of in vitro protein-rejecting properties of PEGylated stealth® polycyanoacrylate nanoparticles. Biomaterials. 1999;20(14):1269–75.
[433] Mele E, Heredia-Guerrero JA, Bayer IS, Ciofani G, Genchi GG, Ceseracciu L, et al. Zwitterionic nanofibers of super-glue for transparent and biocompatible multi-purpose coatings. Scientific Reports. 2015;5(1):1–13.
[434] Chalony C, Aguilar LE, Kim JY, Park CH, Kim CS. Development of electrospun core–shell polymeric mat using poly (ethyl-2) cyanoacrylate/polyurethane to attenuate biological adhesion on polymeric mesh implants. Materials Science and Engineering: C. 2021;122:111930.
[435] Chalony C, Aguilar LE, Park CH, Kim CS. Drug free anti-cell proliferative and anti-platelet adhesion coating for vascular stents via polymeric electrospun fibers. Materials Letters. 2021;291:129545.
[436] Demeneix B, Behr JP. Polyethylenimine (PEI). Advances in Genetics. 2005;53:215–30.
[437] Lee H-Y, Mohammed KA, Nasreen N. Nanoparticle-based targeted gene therapy for lung cancer. American Journal of Cancer Research. 2016;6(5):1118.
[438] Xia T, Kovochich M, Liong M, Meng H, Kabehie S, George S, et al. Polyethyleneimine coating enhances the cellular uptake of mesoporous silica nanoparticles and allows safe delivery of siRNA and DNA constructs. ACS Nano. 2009;3(10):3273–86.
[439] Rodriguez M, Lapierre J, Ojha CR, Kaushik A, Batrakova E, Kashanchi F, et al. Intranasal drug delivery of small interfering RNA targeting Beclin1 encapsulated with polyethylenimine (PEI) in mouse brain to achieve HIV attenuation. Scientific Reports. 2017;7(1):1–10.
[440] Richard I, Thibault M, De Crescenzo G, Buschmann MD, Lavertu M. Ionization behavior of chitosan and chitosan–DNA polyplexes indicate that chitosan has a similar capability to induce a proton-sponge effect as PEI. Biomacromolecules. 2013;14(6):1732–40.
[441] Choosakoonkriang S, Lobo BA, Koe GS, Koe JG, Middaugh CR. Biophysical characterization of PEI/DNA complexes. Journal of Pharmaceutical Sciences. 2003;92(8):1710–22.
[442] Benjaminsen RV, Mattebjerg MA, Henriksen JR, Moghimi SM, Andresen TL. The possible "proton sponge" effect of polyethylenimine (PEI) does not include change in lysosomal pH. Molecular Therapy. 2013;21(1):149–57.
[443] Merdan T, Callahan J, Petersen H, Kunath K, Bakowsky U, Kopeckova P, et al. Pegylated polyethylenimine– Fab 'antibody fragment conjugates for targeted gene delivery to human ovarian carcinoma cells. Bioconjugate Chemistry. 2003;14(5):989–96.
[444] Kim SH, Jeong JH, Lee SH, Kim SW, Park TG. LHRH receptor-mediated delivery of siRNA using polyelectrolyte complex micelles self-assembled from siRNA-PEG-LHRH conjugate and PEI. Bioconjugate Chemistry. 2008;19(11):2156–62.
[445] Chiper M, Tounsi N, Kole R, Kichler A, Zuber G. Self-aggregating 1.8 kDa polyethylenimines with dissolution switch at endosomal acidic pH are delivery carriers for plasmid DNA, mRNA, siRNA and exon-skipping oligonucleotides. Journal of Controlled Release. 2017;246:60–70.
[446] Chollet P, Favrot MC, Hurbin A, Coll JL. Side-effects of a systemic injection of linear polyethylenimine–DNA complexes. The Journal of Gene Medicine: A Cross-disciplinary Journal for Research on the Science of Gene Transfer and Its Clinical Applications. 2002;4(1):84–91.
[447] Hao F, Li Y, Zhu J, Sun J, Marshall B, Lee RJ, et al. Polyethylenimine-based formulations for delivery of oligonucleotides. Current Medicinal Chemistry. 2019;26(13):2264–84.

[448] Piconi C. Bioinert Ceramics: State-of-the-art. Key Engineering Materials: Trans Tech Publ; 2017. p. 3–13.
[449] Rahmati M, Mozafari M. Biocompatibility of alumina-based biomaterials–A review. Journal of Cellular Physiology. 2019;234(4):3321–35.
[450] Alasvand N, Banijamali S, Milan PB, Mozafari M. Cellular Response to Alumina. Handbook of Biomaterials Biocompatibility: Elsevier; 2020. p. 335–52.
[451] Kapruwan P, Ferré-Borrull J, Marsal LF. Nanoporous anodic alumina platforms for drug delivery applications: Recent advances and perspective. Advanced Materials Interfaces. 2020;7(22):2001133.
[452] Park E-J, Kim H, Kim Y, Choi K. Repeated-dose toxicity attributed to aluminum nanoparticles following 28-day oral administration, particularly on gene expression in mouse brain. Toxicological & Environmental Chemistry. 2011;93(1):120–33.
[453] Kumar V, Sharma N, Maitra S. In vitro and in vivo toxicity assessment of nanoparticles. International Nano Letters. 2017;7(4):243–56.
[454] Bunderson-Schelvan M, Holian A, Trout KL, Hamilton RF. Translocation, Biodistribution, and Fate of Nanomaterials in the Body. Interaction of Nanomaterials with the Immune System: Springer; 2020. p. 99–125.
[455] Hassanpour P, Panahi Y, Ebrahimi-Kalan A, Akbarzadeh A, Davaran S, Nasibova AN, et al. Biomedical applications of aluminium oxide nanoparticles. Micro & Nano Letters. 2018;13 (9):1227–31.
[456] Roach KA, Anderson SE, Stefaniak AB, Shane HL, Kodali V, Kashon M, et al. Surface area-and mass-based comparison of fine and ultrafine nickel oxide lung toxicity and augmentation of allergic response in an ovalbumin asthma model. Inhalation Toxicology. 2019;31 (8):299–324.
[457] Sutradhar P, Debnath N, Saha M. Microwave-assisted rapid synthesis of alumina nanoparticles using tea, coffee and triphala extracts. Advances in Manufacturing. 2013;1 (4):357–61.
[458] Panzarini E, Inguscio V, Tenuzzo BA, Carata E, Dini L. Nanomaterials and autophagy: New insights in cancer treatment. Cancers. 2013;5(1):296–319.
[459] Ji Y, Zhang X, Wang X, Che Z, Yu X, Yang H. Zirconia bioceramics as all-ceramics crowns material: A review. Reviews on Advanced Material Science. 2013;34(1):72–78.
[460] Kisi EH, Howard C. Crystal structures of zirconia phases and their inter-relation. Key Engineering Materials. Trans Tech Publication. 1998:1–36.
[461] Boulouz M, Martin L, Boulouz A, Boyer A. Effect of the dopant content on the physical properties of Y2O3–ZrO2 and CaO–ZrO2 thin films produced by evaporation and sputtering techniques. Materials Science and Engineering: B. 1999;67(3):122–31.
[462] Kilo M, Taylor MA, Argirusis C, Borchardt G, Lesage B, Weber S, et al. Cation self-diffusion of 44 Ca, 88 Y, and 96 Zr in single-crystalline calcia-and yttria-doped zirconia. Journal of Applied Physics. 2003;94(12):7547–52.
[463] Soon G, Pingguan-Murphy B, Lai KW, Akbar SA. Review of zirconia-based bioceramic: Surface modification and cellular response. Ceramics International. 2016;42(11):12543–55.
[464] Karthigeyan S, Ravindran AJ, Bhat RT, Nageshwarao MN, Murugesan SV, Angamuthu V. Surface modification techniques for zirconia-based bioceramics: A review. Journal of Pharmacy & Bioallied Sciences. 2019;11(Suppl 2):S131.
[465] Arcos D, Vallet-Regi M. Bioceramics for drug delivery. Acta Materialia. 2013;61(3):890–911.
[466] Paul W, Sharma CP. Bioceramics, towards nano-enabled drug delivery: A mini review. Trends in Biomaterials and Artificial Organs. 2005;1:7–11.

[467] Xu C, Chen F, Valdovinos HF, Jiang D, Goel S, Yu B, et al. Bacteria-like mesoporous silica-coated gold nanorods for positron emission tomography and photoacoustic imaging-guided chemo-photothermal combined therapy. Biomaterials. 2018;165:56–65.
[468] Chen L, Zhong H, Qi X, Shao H, Xu K. Modified core–shell magnetic mesoporous zirconia nanoparticles formed through a facile "outside-to-inside" way for CT/MRI dual-modal imaging and magnetic targeting cancer chemotherapy. RSC Advances. 2019;9(23):13220–33.
[469] Feng L, Gai S, He F, Dai Y, Zhong C, Yang P, et al. Multifunctional mesoporous ZrO2 encapsulated upconversion nanoparticles for mild NIR light activated synergistic cancer therapy. Biomaterials. 2017;147:39–52.
[470] Jia H-R, Zhu Y-X, Duan Q-Y, Chen Z, Wu F-G. Nanomaterials meet zebrafish: Toxicity evaluation and drug delivery applications. Journal of Controlled Release. 2019;311:301–18.
[471] Driver M. Coatings for Biomedical Applications: Elsevier; 2012.
[472] More RB, Sines G, Ma L, Bokros JC. Pyrolytic carbon. Encyclopedia of Biomaterials and Biomedical Engineering. 2004:1308–19.
[473] Rezaei B, Hansen TW, Keller SS. Stereolithography-derived three-dimensional pyrolytic carbon/Mn3O4 nanostructures for free-standing hybrid supercapacitor electrodes. ACS Applied Nano Materials. 2021.
[474] Sharp W, Teague P, Scott D. Thromboresistance of pyrolytic carbon grafts. Transactions-American Society for Artificial Internal Organs. 1978;24:223–28.
[475] Bravo CJ, Rizzo M, Hormel KB, Beckenbaugh RD. Pyrolytic carbon proximal interphalangeal joint arthroplasty: Results with minimum two-year follow-up evaluation. The Journal of Hand Surgery. 2007;32(1):1–11.
[476] Chung KC, Ram AN, Shauver MJ. Outcomes of pyrolytic carbon arthroplasty for the proximal interphalangeal joint. Plastic and Reconstructive Surgery. 2009;123(5):1521.
[477] Liu Z, Chen K, Davis C, Sherlock S, Cao Q, Chen X, et al. Drug delivery with carbon nanotubes for in vivo cancer treatment. Cancer Research. 2008;68(16):6652–60.
[478] Liu Z, Robinson JT, Tabakman SM, Yang K, Dai H. Carbon materials for drug delivery & cancer therapy. Materials Today. 2011;14(7–8):316–23.
[479] Mauter MS, Elimelech M. Environmental applications of carbon-based nanomaterials. Environmental Science & Technology. 2008;42(16):5843–59.
[480] Johnson DR, Methner MM, Kennedy AJ, Steevens JA. Potential for occupational exposure to engineered carbon-based nanomaterials in environmental laboratory studies. Environmental Health Perspectives. 2010;118(1):49–54.
[481] Hench L. Bioactive Ceramics: Theory and Clinical Applications. Bioceramics: Elsevier; 1994. p. 3–14.
[482] Flörke OW, Graetsch HA, Brunk F, Benda L, Paschen S, Bergna HE, et al Silica. Ullmann's Encyclopedia of Industrial Chemistry. 2000.
[483] Ha S-W, Viggeswarapu M, Habib MM, Beck GR Jr. Bioactive effects of silica nanoparticles on bone cells are size, surface, and composition dependent. Acta Biomaterialia. 2018;82:184–96.
[484] Beck GR Jr, Ha S-W, Camalier CE, Yamaguchi M, Li Y, Lee J-K, et al. Bioactive silica-based nanoparticles stimulate bone-forming osteoblasts, suppress bone-resorbing osteoclasts, and enhance bone mineral density in vivo. Nanomedicine: Nanotechnology, Biology and Medicine. 2012;8(6):793–803.
[485] Rosenholm JM, Zhang J, Linden M, Sahlgren C. Mesoporous silica nanoparticles in tissue engineering–a perspective. Nanomedicine. 2016;11(4):391–402.
[486] Eivazzadeh-Keihan R, Chenab KK, Taheri-Ledari R, Mosafer J, Hashemi SM, Mokhtarzadeh A, et al. Recent advances in the application of mesoporous silica-based nanomaterials for bone tissue engineering. Materials Science and Engineering: C. 2020;107:110267.

[487] Manzano M, Vallet-Regí M. Mesoporous silica nanoparticles for drug delivery. Advanced Functional Materials. 2020;30(2):1902634.
[488] Ryu J, Kim HJ, Chang EJ, Huang H, Banno Y, Kim HH. Sphingosine 1-phosphate as a regulator of osteoclast differentiation and osteoclast–osteoblast coupling. The EMBO Journal. 2006;25 (24):5840–51.
[489] Möller K, Bein T. Talented mesoporous silica nanoparticles. Chemistry of Materials. 2017;29 (1):371–88.
[490] Kwon S, Singh RK, Perez RA, Abou Neel EA, Kim H-W, Chrzanowski W. Silica-based mesoporous nanoparticles for controlled drug delivery. Journal of Tissue Engineering. 2013;4:2041731413503357.
[491] Lin K, Chang J. Structure and Properties of Hydroxyapatite for Biomedical Applications. Hydroxyapatite (HAp) for Biomedical Applications: Elsevier; 2015. p. 3–19.
[492] Szcześ A, Hołysz L, Chibowski E. Synthesis of hydroxyapatite for biomedical applications. Advances in Colloid and Interface Science. 2017;249:321–30.
[493] Wang Q, Gu Z, Jamal S, Detamore MS, Berkland C. Hybrid hydroxyapatite nanoparticle colloidal gels are injectable fillers for bone tissue engineering. Tissue Engineering Part A. 2013;19(23–24):2586–93.
[494] Bu S, Yan S, Wang R, Xia P, Zhang K, Li G, et al. In situ precipitation of cluster and acicular hydroxyapatite onto porous poly (γ-benzyl-l-glutamate) microcarriers for bone tissue engineering. ACS Applied Materials & Interfaces. 2020;12(11):12468–77.
[495] Ng H-M, Bee S-T, Tin Sin L, Ratnam CT, Rahmat AR. Hydroxyapatite for poly (α-hydroxy esters) biocomposites applications. Polymer Reviews. 2019;59(2):187–239.
[496] Wu S, Liu X, Yeung KW, Liu C, Yang X. Biomimetic porous scaffolds for bone tissue engineering. Materials Science and Engineering: R: Reports. 2014;80:1–36.
[497] Jaecques S, Van Oosterwyck H, Muraru L, Van Cleynenbreugel T, De Smet E, Wevers M, et al. Individualised, micro CT-based finite element modelling as a tool for biomechanical analysis related to tissue engineering of bone. Biomaterials. 2004;25(9):1683–96.
[498] Geesink RG. Osteoconductive coatings for total joint arthroplasty. Clinical Orthopaedics and Related Research®. 2002;395:53–65.
[499] Brindley D, Moorthy K, Lee J-H, Mason C, Kim H-W, Wall I. Bioprocess forces and their impact on cell behavior: Implications for bone regeneration therapy. Journal of Tissue Engineering. 2011;2011.
[500] Klein-Nulend J, Bacabac RG, Mullender MG. Mechanobiology of bone tissue. Pathologie Biologie. 2005;53(10):576–80.
[501] Juhl OJ, Buettmann EG, Friedman MA, DeNapoli RC, Hoppock GA, Donahue HJ. Update on the effects of microgravity on the musculoskeletal system. NPJ Microgravity. 2021;7(1):1–15.
[502] Hargens AR, Bhattacharya R, Schneider SM. Space physiology VI: Exercise, artificial gravity, and countermeasure development for prolonged space flight. European Journal of Applied Physiology. 2013;113(9):2183–92.
[503] Rabiei A, Sandukas S. Processing and evaluation of bioactive coatings on polymeric implants. Journal of Biomedical Materials Research Part A. 2013;101(9):2621–29.
[504] Yang Y, Kim K-H, Ong JL. A review on calcium phosphate coatings produced using a sputtering process – an alternative to plasma spraying. Biomaterials. 2005;26(3):327–37.
[505] Yilmaz B, Alshemary AZ, Evis Z. Co-doped hydroxyapatites as potential materials for biomedical applications. Microchemical Journal. 2019;144:443–53.
[506] Predoi D, Iconaru SL, Predoi MV, Buton N, Motelica-Heino M. Zinc doped hydroxyapatite thin films prepared by sol–gel spin coating procedure. Coatings. 2019;9(3):156.

[507] Liu M, Liu H, Sun S, Li X, Zhou Y, Hou Z, et al. Multifunctional hydroxyapatite/Na (Y/Gd) F4: Yb3+, Er3+ composite fibers for drug delivery and dual modal imaging. Langmuir. 2014;30 (4):1176–82.

[508] Munir MU, Salman S, Javed I, Bukhari SNA, Ahmad N, Shad NA, et al. Nano-hydroxyapatite as a delivery system: Overview and advancements. Artificial Cells, Nanomedicine, and Biotechnology. 2021;49(1):717–27.

[509] Levingstone TJ, Herbaj S, Dunne NJ. Calcium phosphate nanoparticles for therapeutic applications in bone regeneration. Nanomaterials. 2019;9(11):1570.

[510] Parent M, Baradari H, Champion E, Damia C, Viana-Trecant M. Design of calcium phosphate ceramics for drug delivery applications in bone diseases: A review of the parameters affecting the loading and release of the therapeutic substance. Journal of Controlled Release. 2017;252:1–17.

[511] Tanaka T, Komaki H, Chazono M, Kitasato S, Kakuta A, Akiyama S, et al. Basic research and clinical application of beta-tricalcium phosphate (β-TCP). Morphologie. 2017;101 (334):164–72.

[512] Sakai S, Anada T, Tsuchiya K, Yamazaki H, Margolis HC, Suzuki O. Comparative study on the resorbability and dissolution behavior of octacalcium phosphate, β-tricalcium phosphate, and hydroxyapatite under physiological conditions. Dental Materials Journal. 2016;35 (2):216–24.

[513] Westhauser F, Karadjian M, Essers C, Senger A-S, Hagmann S, Schmidmaier G, et al. Osteogenic differentiation of mesenchymal stem cells is enhanced in a 45S5-supplemented β-TCP composite scaffold: An in-vitro comparison of Vitoss and Vitoss BA. PLoS One. 2019;14 (2):e0212799.

[514] Bouler J-M, Pilet P, Gauthier O, Verron E. Biphasic calcium phosphate ceramics for bone reconstruction: A review of biological response. Acta Biomaterialia. 2017;53:1–12.

[515] Lee S, Choi D, Shim J-H, Nam W. Efficacy of three-dimensionally printed polycaprolactone/ beta tricalcium phosphate scaffold on mandibular reconstruction. Scientific Reports. 2020;10 (1):1–9.

[516] Yusof MR, Shamsudin R, Zakaria S, Abdul Hamid MA, Yalcinkaya F, Abdullah Y, et al. Fabrication and characterization of carboxymethyl starch/poly (l-lactide) acid/β-tricalcium phosphate composite nanofibers via electrospinning. Polymers. 2019;11(9):1468.

[517] Haugen HJ, Basu P, Sukul M, Mano JF, Reseland JE. Injectable biomaterials for dental tissue regeneration. International Journal of Molecular Sciences. 2020;21(10):3442.

[518] Lu H, Zhou Y, Ma Y, Xiao L, Ji W, Zhang Y, et al. Current application of beta-tricalcium phosphate in bone repair and its mechanism to regulate osteogenesis. Frontiers in Materials. 2021:277.

[519] Menger MM, Laschke MW, Orth M, Pohlemann T, Menger MD, Histing T. Vascularization strategies in the prevention of nonunion formation. Tissue Engineering Part B: Reviews. 2021;27(2):107–32.

[520] Profeta AC, Mannocci F, Foxton RM, Thompson I, Watson TF, Sauro S. Bioactive effects of a calcium/sodium phosphosilicate on the resin–dentine interface: A microtensile bond strength, scanning electron microscopy, and confocal microscopy study. European Journal of Oral Sciences. 2012;120(4):353–62.

[521] Al-Harbi N, Mohammed H, Al-Hadeethi Y, Bakry AS, Umar A, Hussein MA, et al. Silica-based bioactive glasses and their applications in hard tissue regeneration: A review. Pharmaceuticals. 2021;14(2):75.

[522] Jones JR, Brauer DS, Hupa L, Greenspan DC. Bioglass and bioactive glasses and their impact on healthcare. International Journal of Applied Glass Science. 2016;7(4):423–34.

[523] Sonatkar J, Lolge S. Bioactive glass: A potential component. International Research Journal of Modernization in Engineering Technology and Science. 2020;2(9):1030–34.
[524] Vukajlovic D, Parker J, Bretcanu O, Novakovic K. Chitosan based polymer/bioglass composites for tissue engineering applications. Materials Science and Engineering: C. 2019;96:955–67.
[525] Boccaccini AR, Maquet V. Bioresorbable and bioactive polymer/Bioglass® composites with tailored pore structure for tissue engineering applications. Composites Science and Technology. 2003;63(16):2417–29.
[526] Park JB, Kim YK. Metallic Biomaterials. Biomaterials: CRC press; 2007. p. 1–22.
[527] Pilliar RM. Metallic Biomaterials. Biomedical materials: Springer; 2021. p. 1–47.
[528] Santos-Coquillat A, Martínez-Campos E, Sánchez HM, Moreno L, Arrabal R, Mohedano M, et al. Hybrid functionalized coatings on metallic biomaterials for tissue engineering. Surface and Coatings Technology. 2021;422:127508.
[529] Mahapatro A. Bio-functional nano-coatings on metallic biomaterials. Materials Science and Engineering: C. 2015;55:227–51.
[530] Heiden M, Walker E, Stanciu L. Magnesium, iron and zinc alloys, the trifecta of bioresorbable orthopaedic and vascular implantation-a review. Journal of Biotechnology & Biomaterials. 2015;5(2):1.
[531] Wu Y, Guelorget B, Sun Z, Déturche R, Retraint D. Characterization of gradient properties generated by SMAT for a biomedical grade 316L stainless steel. Materials Characterization. 2019;155:109788.
[532] Amanov A, Lee SW, Pyun YS. Low friction and high strength of 316L stainless steel tubing for biomedical applications. Materials Science and Engineering: C. 2017;71:176–85.
[533] Hermawan H, Ramdan D, Djuansjah JR. Metals for biomedical applications. Biomedical Engineering-from Theory to Applications. 2011;1:411–30.
[534] Latifi A, Imani M, Khorasani MT, Joupari MD. Electrochemical and chemical methods for improving surface characteristics of 316L stainless steel for biomedical applications. Surface and Coatings Technology. 2013;221:1–12.
[535] Bekmurzayeva A, Duncanson WJ, Azevedo HS, Kanayeva D. Surface modification of stainless steel for biomedical applications: Revisiting a century-old material. Materials Science and Engineering: C. 2018;93:1073–89.
[536] Sivaraj D, Vijayalakshmi K, Ganeshkumar A, Rajaram R. Tailoring Cu substituted hydroxyapatite/functionalized multiwalled carbon nanotube composite coating on 316L SS implant for enhanced corrosion resistance, antibacterial and bioactive properties. International Journal of Pharmaceutics. 2020;590:119946.
[537] Holton A, Walsh E, Anayiotos A, Pohost G, Venugopalan R. Comparative mri compatibility of 316l stainless steel alloy and nickel–titanium alloy stents: Original article technical. Journal of Cardiovascular Magnetic Resonance. 2002;4(4):423–30.
[538] Mikhalovska LI, Santin M, Denyer SP, Lloyd AW, Teer DG, Field S, et al. Fibrinogen adsorption and platelet adhesion to metal and carbon coatings. Thrombosis and Haemostasis. 2004;92 (11):1032–39.
[539] Aguilar LE. Nanotechnology in Improving Medical Devices for Smart Drug Delivery. Biomimetic Nanoengineered Materials for Advanced Drug Delivery: Elsevier; 2019. p. 73–90.
[540] Yang Z, Wang J, Luo R, Maitz MF, Jing F, Sun H, et al. The covalent immobilization of heparin to pulsed-plasma polymeric allylamine films on 316L stainless steel and the resulting effects on hemocompatibility. Biomaterials. 2010;31(8):2072–83.
[541] Kaur M, Singh K. Review on titanium and titanium based alloys as biomaterials for orthopaedic applications. Materials Science and Engineering: C. 2019;102:844–62.

[542] Griza S, de Souza Sá DHG, Batista WW, de Blas JCG, Pereira LC. Microstructure and mechanical properties of hot rolled TiNbSn alloys. Materials & Design. 1980–2015;2014(56):200–08.
[543] He G, Hagiwara M. Ti alloy design strategy for biomedical applications. Materials Science and Engineering: C. 2006;26(1):14–19.
[544] Sultan AA, Mahmood B, Samuel LT, Stearns KL, Molloy RM, Moskal JT, et al. Cementless 3D printed highly porous titanium-coated baseplate total knee arthroplasty: Survivorship and outcomes at 2-year minimum follow-up. The Journal of Knee Surgery. 2020;33(03):279–83.
[545] Faizan A, Bhowmik-Stoker M, Alipit V, Kirk AE, Krebs VE, Harwin SF, et al. Development and verification of novel porous titanium metaphyseal cones for revision total knee arthroplasty. The Journal of Arthroplasty. 2017;32(6):1946–53.
[546] Wang M, Chen T, Lu S, Zhao Y, Chen H, Wu Y, et al. Osteogenic comparison on selective laser melting printed and sandblasting-acid-etching Ti substrates for customized implant applications. Science of Advanced Materials. 2017;9(5):705–14.
[547] Barui S, Chatterjee S, Mandal S, Kumar A, Basu B. Microstructure and compression properties of 3D powder printed Ti-6Al-4V scaffolds with designed porosity: Experimental and computational analysis. Materials Science and Engineering: C. 2017;70:812–23.
[548] Kim J-H, Kim M-Y, Knowles JC, Choi S, Kang H, Park S-H, et al. Mechanophysical and biological properties of a 3D-printed titanium alloy for dental applications. Dental Materials. 2020;36(7):945–58.
[549] Shaker N Direct Metal Laser Sintering Titanium Implants: X-ray Photoelectron Spectroscopy Analysis of Reprocessed Powder. 2019.
[550] Schliephake H, Scharnweber D. Chemical and biological functionalization of titanium for dental implants. Journal of Materials Chemistry. 2008;18(21):2404–14.
[551] Qiao S, Wu D, Li Z, Zhu Y, Zhan F, Lai H, et al. The combination of multi-functional ingredients-loaded hydrogels and three-dimensional printed porous titanium alloys for infective bone defect treatment. Journal of Tissue Engineering. 2020;11:2041731420965797.
[552] Pylypchuk IV, Petranovskaya A, Gorbyk P, Korduban A, Markovsky P, Ivasishin O. Biomimetic hydroxyapatite growth on functionalized surfaces of Ti-6Al-4V and Ti-Zr-Nb alloys. Nanoscale Research Letters. 2015;10(1):1–8.
[553] Cheng YF, Zhang JY, Wang YB, Li CM, Lu ZS, Hu XF, et al. Deposition of catechol-functionalized chitosan and silver nanoparticles on biomedical titanium surfaces for antibacterial application. Materials Science and Engineering: C. 2019;98:649–56.
[554] Disegi JA, Kennedy RL, Pilliar R. Cobalt-Base Alloys for Biomedical Applications: ASTM Danvers; 1999.
[555] Al Jabbari YS. Physico-mechanical properties and prosthodontic applications of Co-Cr dental alloys: A review of the literature. The Journal of Advanced Prosthodontics. 2014;6(2):138–45.
[556] Kassapidou M, Hjalmarsson L, Johansson CB, Johansson PH, Morisbak E, Wennerberg A, et al. Cobalt–chromium alloys fabricated with four different techniques: Ion release, toxicity of released elements and surface roughness. Dental Materials. 2020;36(11):e352–e63.
[557] Kapnisis K, Constantinides G, Georgiou H, Cristea D, Gabor C, Munteanu D, et al. Multi-scale mechanical investigation of stainless steel and cobalt–chromium stents. Journal of the Mechanical Behavior of Biomedical Materials. 2014;40:240–51.
[558] Buszman P, Trznadel S, Żurakowski A, Milewski K, Kinasz L, Król M, et al. Original article Prospective registry evaluating safety and efficacy of cobalt-chromium stent implantation in patients with de novo coronary lesions. Kardiologia Polska (Polish Heart Journal). 2007;65 (9):1041–46.
[559] Sharma N, Raj T, Jangra K. Applications of nickel-titanium alloy. Journal of Engineering and Technology. 2015;5(1):1.

[560] Chaudhari R, Vora JJ, Parikh D. A Review on Applications of Nitinol Shape Memory Alloy. Recent Advances in Mechanical Infrastructure: Springer; 2021. p. 123–32.
[561] Biscarini A, Mazzolai G, Tuissi A. Enhanced nitinol properties for biomedical applications. Recent Patents on Biomedical Engineering (Discontinued). 2008;1(3):180–96.
[562] O'Brien B, Carroll W, Kelly M. Passivation of nitinol wire for vascular implants – a demonstration of the benefits. Biomaterials. 2002;23(8):1739–48.
[563] Hassel A. Surface treatment of NiTi for medical applications. Minimally Invasive Therapy & Allied Technologies. 2004;13(4):240–47.
[564] Pelton AR, Russell SM, DiCello J. The physical metallurgy of nitinol for medical applications. Jom. 2003;55(5):33–37.
[565] Maccauro G, Iommetti PR, Muratori F, Raffaelli L, Manicone PF, Fabbriciani C. An overview about biomedical applications of micron and nano size tantalum. Recent Patents on Biotechnology. 2009;3(3):157–65.
[566] Niinomi M, Nakai M, Hieda J. Development of new metallic alloys for biomedical applications. Acta Biomaterialia. 2012;8(11):3888–903.
[567] Macionczyk F, Gerold B, Thull R. Repassivating tantalum/tantalum oxide surface modification on stainless steel implants. Surface and Coatings Technology. 2001;142:1084–87.
[568] Schildhauer TA, Robie B, Muhr G, Köller M. Bacterial adherence to tantalum versus commonly used orthopedic metallic implant materials. Journal of Orthopaedic Trauma. 2006;20(7):476–84.
[569] Han Q, Wang C, Chen H, Zhao X, Wang J. Porous tantalum and titanium in orthopedics: A review. ACS Biomaterials Science & Engineering. 2019;5(11):5798–824.
[570] Prasad K, Bazaka O, Chua M, Rochford M, Fedrick L, Spoor J, et al. Metallic biomaterials: Current challenges and opportunities. Materials. 2017;10(8):884.
[571] Sezer N, Evis Z, Kayhan SM, Tahmasebifar A, Koç M. Review of magnesium-based biomaterials and their applications. Journal of Magnesium and Alloys. 2018;6(1):23–43.
[572] Aljihmani L, Alic L, Boudjemline Y, Hijazi ZM, Mansoor B, Serpedin E, et al. Magnesium-based bioresorbable stent materials: Review of reviews. Journal of Bio-and Tribo-Corrosion. 2019;5(1):26.
[573] Seal C, Vince K, Hodgson M Biodegradable surgical implants based on magnesium alloys–A review of current research. IOP conference series: materials science and engineering: IOP Publishing; 2009. p. 012011.
[574] Zhao N, Zhu D. Endothelial responses of magnesium and other alloying elements in magnesium based stent materials. Metallomics. 2015;7(1):118–28.
[575] Liu H. The effects of surface and biomolecules on magnesium degradation and mesenchymal stem cell adhesion. Journal of Biomedical Materials Research Part A. 2011;99(2):249–60.
[576] Poinern GEJ, Brundavanam S, Fawcett D. Biomedical magnesium alloys: A review of material properties, surface modifications and potential as a biodegradable orthopaedic implant. American Journal of Biomedical Engineering. 2012;2(6):218–40.
[577] Saberi A, Bakhsheshi-Rad HR, Abazari S, Ismail AF, Sharif S, Ramakrishna S, et al. A comprehensive review on surface modifications of biodegradable magnesium-based implant alloy: Polymer coatings opportunities and challenges. Coatings. 2021;11(7):747.
[578] Fan X-L, Huo Y-F, Li C-Y, Kannan MB, Chen X-B, Guan S-K, et al. Corrosion resistance of nanostructured magnesium hydroxide coating on magnesium alloy AZ31: Influence of EDTA. Rare Metals. 2019;38(6):520–31.
[579] Prabhu DB, Gopalakrishnan P, Ravi K. Morphological studies on the development of chemical conversion coating on surface of Mg–4Zn alloy and its corrosion and bio mineralisation behaviour in simulated body fluid. Journal of Alloys and Compounds. 2020;812:152146.

[580] Ostrowski NJ, Lee B, Roy A, Ramanathan M, Kumta PN. Biodegradable poly (lactide-co-glycolide) coatings on magnesium alloys for orthopedic applications. Journal of Materials Science: Materials in Medicine. 2013;24(1):85–96.

[581] Samadianfard R, Seifzadeh D, Habibi-Yangjeh A, Jafari-Tarzanagh Y. Oxidized fullerene/sol-gel nanocomposite for corrosion protection of AM60B magnesium alloy. Surface and Coatings Technology. 2020;385:125400.

[582] Kim J, Mousa HM, Park CH, Kim CS. Enhanced corrosion resistance and biocompatibility of AZ31 Mg alloy using PCL/ZnO NPs via electrospinning. Applied Surface Science. 2017;396:249–58.

[583] Chakraborty Banerjee P, Al-Saadi S, Choudhary L, Harandi SE, Singh R. Magnesium implants: Prospects and challenges. Materials. 2019;12(1):136.

[584] Antoniac I, Blajan A. Effect of alloying elements on the microstructure and degradation properties of magnesium biodegradable alloys for biomedical application. European Cells and Materials. 2015;29(1):1.

[585] Hernández-Escobar D, Champagne S, Yilmazer H, Dikici B, Boehlert CJ, Hermawan H. Current status and perspectives of zinc-based absorbable alloys for biomedical applications. Acta Biomaterialia. 2019;97:1–22.

[586] Kabir H, Munir K, Wen C, Li Y. Recent research and progress of biodegradable zinc alloys and composites for biomedical applications: Biomechanical and biocorrosion perspectives. Bioactive Materials. 2021;6(3):836–79.

[587] Bowen PK, Shearier ER, Zhao S, Guillory RJ, Zhao F, Goldman J, et al. Biodegradable metals for cardiovascular stents: From clinical concerns to recent Zn-alloys. Advanced Healthcare Materials. 2016;5(10):1121–40.

[588] Yang H, Wang C, Liu C, Chen H, Wu Y, Han J, et al. Evolution of the degradation mechanism of pure zinc stent in the one-year study of rabbit abdominal aorta model. Biomaterials. 2017;145:92–105.

[589] Zhou C, Li H-F, Yin Y-X, Shi -Z-Z, Li T, Feng X-Y, et al. Long-term in vivo study of biodegradable Zn-Cu stent: A 2-year implantation evaluation in porcine coronary artery. Acta Biomaterialia. 2019;97:657–70.

[590] Yang H, Jia B, Zhang Z, Qu X, Li G, Lin W, et al. Alloying design of biodegradable zinc as promising bone implants for load-bearing applications. Nature Communications. 2020;11(1):1–16.

[591] Nady H, El-Rabiei M, Bahrawy A, El-Katori EE. Assessment of H2O2/albumin and glucose on the biomedical iron alloys corrosion in simulated body fluid: Experimental, surface, and computational investigations. Journal of Molecular Liquids. 2021;339:116823.

[592] Francis A, Yang Y, Virtanen S, Boccaccini A. Iron and iron-based alloys for temporary cardiovascular applications. Journal of Materials Science: Materials in Medicine. 2015;26(3):1–16.

[593] Tung CK. Application of stent implantation in congenital heart disease. Chinese Medical Journal. 2014;127(21):3814–24.

[594] Putra N, Leeflang M, Minneboo M, Taheri P, Fratila-Apachitei L, Mol J, et al. Extrusion-based 3D printed biodegradable porous iron. Acta Biomaterialia. 2021;121:741–56.

[595] Sangaiya P, Jayaprakash R. A review on iron oxide nanoparticles and their biomedical applications. Journal of Superconductivity and Novel Magnetism. 2018;31(11):3397–413.

[596] Vangijzegem T, Stanicki D, Laurent S. Magnetic iron oxide nanoparticles for drug delivery: Applications and characteristics. Expert Opinion on Drug Delivery. 2019;16(1):69–78.

[597] Mahmoudi K, Bouras A, Bozec D, Ivkov R, Hadjipanayis C. Magnetic hyperthermia therapy for the treatment of glioblastoma: A review of the therapy's history, efficacy and application in humans. International Journal of Hyperthermia. 2018;34(8):1316–28.

[598] Mallory M, Gogineni E, Jones GC, Greer L, Simone CB II. Therapeutic hyperthermia: The old, the new, and the upcoming. Critical Reviews in Oncology/hematology. 2016;97:56–64.
[599] Mancuso MR, Davis R, Norberg SM, O'Brien S, Sennino B, Nakahara T, et al. Rapid vascular regrowth in tumors after reversal of VEGF inhibition. The Journal of Clinical Investigation. 2006;116(10):2610–21.
[600] Shen Y, Liu P, Zhang A, Xu LX. Study on tumor microvasculature damage induced by alternate cooling and heating. Annals of Biomedical Engineering. 2008;36(8):1409.
[601] Zarow M, Ramirez-Sebastia A, Paolone G, de Ribot Porta J, Mora J, Espona J, et al. A new classification system for the restoration of root filled teeth. International Endodontic Journal. 2018;51(3):318–34.
[602] Bayne S, Ferracane J, Marshall G, Marshall S, Van Noort R. The evolution of dental materials over the past century: Silver and gold to tooth color and beyond. Journal of Dental Research. 2019;98(3):257–65.
[603] Elahi N, Kamali M, Baghersad MH. Recent biomedical applications of gold nanoparticles: A review. Talanta. 2018;184:537–56.
[604] Tiwari PM, Vig K, Dennis VA, Singh SR. Functionalized gold nanoparticles and their biomedical applications. Nanomaterials. 2011;1(1):31–63.
[605] Jena BK, Raj CR. Electrochemical biosensor based on integrated assembly of dehydrogenase enzymes and gold nanoparticles. Analytical Chemistry. 2006;78(18):6332–39.
[606] Hwu S, Garzuel M, Forró C, Ihle SJ, Reichmuth AM, Kurdzesau F, et al. An analytical method to control the surface density and stability of DNA-gold nanoparticles for an optimized biosensor. Colloids and Surfaces B: Biointerfaces. 2020;187:110650.
[607] Ovais M, Raza A, Naz S, Islam NU, Khalil AT, Ali S, et al. Current state and prospects of the phytosynthesized colloidal gold nanoparticles and their applications in cancer theranostics. Applied Microbiology and Biotechnology. 2017;101(9):3551–65.
[608] Haider A, Kang I-K. Preparation of silver nanoparticles and their industrial and biomedical applications: A comprehensive review. Advances in Materials Science and Engineering. 2015:2015.
[609] Burduşel A-C, Gherasim O, Grumezescu AM, Mogoantă L, Ficai A, Andronescu E. Biomedical applications of silver nanoparticles: An up-to-date overview. Nanomaterials. 2018;8(9):681.
[610] Chen H, Lan G, Ran L, Xiao Y, Yu K, Lu B, et al. A novel wound dressing based on a Konjac glucomannan/silver nanoparticle composite sponge effectively kills bacteria and accelerates wound healing. Carbohydrate Polymers. 2018;183:70–80
[611] Bakhsheshi-Rad HR, Ismail AF, Aziz M, Akbari M, Hadisi Z, Khoshnava SM, et al. Co-incorporation of graphene oxide/silver nanoparticle into poly-L-lactic acid fibrous: A route toward the development of cytocompatible and antibacterial coating layer on magnesium implants. Materials Science and Engineering: C. 2020;111:110812.
[612] Chen P, Wu Z, Leung A, Chen X, Landao-Bassonga E, Gao J, et al. Fabrication of a silver nanoparticle-coated collagen membrane with anti-bacterial and anti-inflammatory activities for guided bone regeneration. Biomedical Materials. 2018;13(6):065014.
[613] Alizadeh-Osgouei M, Li Y, Wen C. A comprehensive review of biodegradable synthetic polymer-ceramic composites and their manufacture for biomedical applications. Bioactive Materials. 2019;4:22–36.
[614] Sun L, Danoux CB, Wang Q, Pereira D, Barata D, Zhang J, et al. Independent effects of the chemical and microstructural surface properties of polymer/ceramic composites on proliferation and osteogenic differentiation of human MSCs. Acta Biomaterialia. 2016;42:364–77.
[615] Shue L, Yufeng Z, Mony U. Biomaterials for periodontal regeneration: A review of ceramics and polymers. Biomatter. 2012;2(4):271–77.

[616] Sobczak-Kupiec A, Pluta K, Drabczyk A, Włoś M, Tyliszczak B. Synthesis and characterization of ceramic-polymer composites containing bioactive synthetic hydroxyapatite for biomedical applications. Ceramics International. 2018;44(12):13630–38.

[617] Huang B, Caetano G, Vyas C, Blaker JJ, Diver C, Bártolo P. Polymer-ceramic composite scaffolds: The effect of hydroxyapatite and β-tri-calcium phosphate. Materials. 2018;11 (1):129.

[618] Rahbek O, Kold S, Bendix K, Overgaard S, Søballe K. Superior sealing effect of hydroxyapatite in porous-coated implants: Experimental studies on the migration of polyethylene particles around stable and unstable implants in dogs. Acta Orthopaedica. 2005;76(3):375–85.

[619] Burg KJ, Porter S, Kellam JF. Biomaterial developments for bone tissue engineering. Biomaterials. 2000;21(23):2347–59.

[620] Basha RY, TS SK, Doble M. Design of biocomposite materials for bone tissue regeneration. Materials Science and Engineering: C. 2015;57:452–63.

[621] Travitzky N. Processing of ceramic–metal composites. Advances in Applied Ceramics. 2012;111(5–6):286–300.

[622] Bose S, Roy M, Bandyopadhyay A. Recent advances in bone tissue engineering scaffolds. Trends in Biotechnology. 2012;30(10):546–54.

[623] Iannazzo D, Pistone A, Salamò M, Galvagno S. Hybrid Ceramic/Polymer Composites for Bone Tissue Regeneration. Hybrid Polymer Composite Materials: Elsevier; 2017. p. 125–55.

[624] Lam T-N, Trinh M-G, Huang C-C, Kung P-C, Huang W-C, Chang W, et al. Investigation of bone growth in additive-manufactured pedicle screw implant by using Ti-6Al-4V and bioactive glass powder composite. International Journal of Molecular Sciences. 2020;21(20):7438.

[625] Joy-anne NO, Su Y, Lu X, Kuo P-H, Du J, Zhu D. Bioactive glass coatings on metallic implants for biomedical applications. Bioactive Materials. 2019;4:261–70.

[626] Dorozhkin SV. Calcium orthophosphate coatings on magnesium and its biodegradable alloys. Acta Biomaterialia. 2014;10(7):2919–34.

[627] Nilawar S, Uddin M, Chatterjee K. Surface engineering of biodegradable implants: Emerging trends in bioactive ceramic coatings and mechanical treatments. Materials Advances. 2021.

[628] Okulov A, Volegov A, Weissmüller J, Markmann J, Okulov I. Dealloying-based metal-polymer composites for biomedical applications. Scripta Materialia. 2018;146:290–94.

[629] Zare Y, Shabani I. Polymer/metal nanocomposites for biomedical applications. Materials Science and Engineering: C. 2016;60:195–203.

[630] Hassan SS, Kamel AH, Hashem HM, Bary EA. Drug delivery systems between metal, liposome, and polymer-based nanomedicine: A review. European Chemical Bulletin. 2020;9 (3):91–102.

[631] Aisida SO, Akpa PA, Ahmad I, Zhao T-K, Maaza M, Ezema FI. Bio-inspired encapsulation and functionalization of iron oxide nanoparticles for biomedical applications. European Polymer Journal. 2020;122:109371.

[632] Zhang X, Yang C-W, Yu H-Q, Sheng G-P. Light-induced reduction of silver ions to silver nanoparticles in aquatic environments by microbial extracellular polymeric substances (EPS). Water Research. 2016;106:242–48.

[633] Belibel R, Sali S, Marinval N, Garcia-Sanchez A, Barbaud C, Hlawaty H. PDMMLA derivatives as a promising cardiovascular metallic stent coating: Physicochemical and biological evaluation. Materials Science and Engineering: C. 2020;117:111284.

[634] Arafat M, Fouladian P, Blencowe A, Albrecht H, Song Y, Garg S. Drug-eluting non-vascular stents for localised drug targeting in obstructive gastrointestinal cancers. Journal of Controlled Release. 2019;308:209–31.

[635] Yao W, Bao Y, Chen Y. Formation of microcapsules by ultrasound stimulation for use in remote-controlled drug-eluting stents. Medical Engineering & Physics. 2018;56:42–47.
[636] Qian W, Murakami M, Ichikawa Y, Che Y. Highly efficient and controllable PEGylation of gold nanoparticles prepared by femtosecond laser ablation in water. The Journal of Physical Chemistry C. 2011;115(47):23293–98.
[637] Liu D, Wu W, Ling J, Wen S, Gu N, Zhang X. Effective PEGylation of iron oxide nanoparticles for high performance in vivo cancer imaging. Advanced Functional Materials. 2011;21 (8):1498–504.

Chapter 3
Methods in biomaterials

3.1 Biomaterial synthesis and fabrication

Researchers use various techniques to create different kinds of biomaterials and these techniques rely heavily on chemistry, biology, and engineering concepts to achieve material properties that are useful and safe for the human body. These techniques are products of decades of research on material science and they determine how materials are further improved and modified, and how a material can exhibit different properties, depending on the fabrication method. Biomaterials have seen great advances in this aspect. They have highlighted the importance of innovating traditional fabrication to novel methods in creating and designing a safe and effective material, with new properties, which can be used inside the body (Figure 3.1). Biomaterials can be fabricated into different forms using various methods, to make films, fibers, gels, particles, etc., with varied properties, such as density, porosity, surface roughness, and reactivity, to name a few. Understanding the principles of these fabrication techniques is crucial in designing an effective biomaterial. In this chapter, these techniques will be discussed and how they affect the material properties, and, in addition, the subsequent effects on the body.

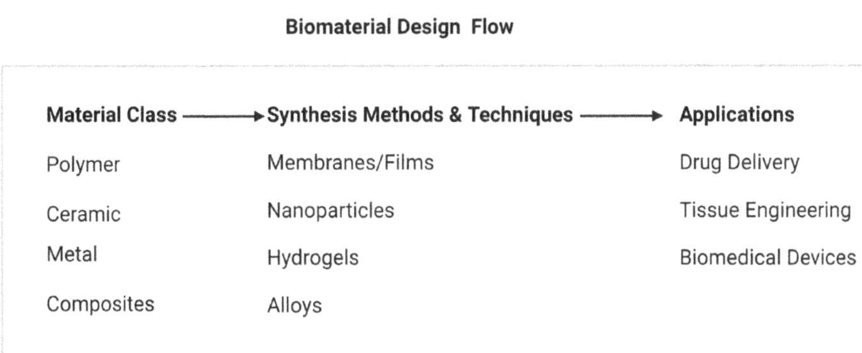

Figure 3.1: Biomaterial design flow. Any material designed to be used by the body is classified as a biomaterial. A material must undergo fabrication and synthesis methods and techniques to be better suited for different bioapplications. Differences in these techniques can vary the form of a material and, consequently, its application and performance.

https://doi.org/10.1515/9783110655377-003

3.2 Electrospinning

Electrospinning is a method of producing non-woven fibers from polymeric materials [1–4]. It uses electric forces to draw out threads from polymeric solutions and melts. The fibers can vary in diameter, depending on the parameters used in the set-up [5–8]. The collected fibers can form a porous membrane that has multiple application potential as a biomaterial [9, 10]. Electrospun fibers can have different properties and can be used as a wound dressing material [11–13], drug delivery platform [13, 14], bioscaffold [15–18], and even in biosensors [19–21]. Since it is nonwoven and has high porosity, the fiber has the capability to be used as a wound dressing material due to its breathability, where gas transport is viable. While the porosity can be tailored to be small enough to prevent the passage of pathogens into the open wound, it can also be used to deliver growth factors via diffusion mechanism, and this could help in accelerating the wound healing process. Polymeric fibers have a high upside on local drug delivery applications as they can encapsulate, at a high efficiency, multiple types of therapeutics and release them via active or passive mechanisms. Lastly, depending on the parameter used, individual fibers can be made at a nanoscale, in which it starts to mimic the extracellular matrix environment [22–24]. Cell proliferation is enhanced by creating adherence points onto the fiber membrane that can be used to engineer different types of tissues.

3.2.1 Basics of the electrospinning setup

The electrospinning process and its setup have these components, Figure 3.2. The polymer solution, which is usually contained in a syringe, is loaded on an injection pump. This pump maintains a constant polymer feed rate to the spinning tip. The spinning tip can be made from a nozzle that is conductive, and the nozzle's inner diameter can be changed, depending on the experiment protocol [5, 8, 25–28]. A direct voltage from a DC power supply is connected to the tip of the nozzle to draw out the polymer by overcoming its viscosity and surface tension. At the tip of the spinning nozzle, the polymer solution is charged and the electrostatic repulsion counteracts the surface tension, and the droplet is stretched. At a critical point, a stream of liquid erupts from the surface. This point of eruption is known as the Taylor cone. If the molecular cohesion of the liquid is sufficiently high or if the molecular weight of the polymer is large enough, this stream can be stretched continuously, first at a slow acceleration under ohmic flow until it reaches a convective flow, where the liquid transitions from a solution to a solid fiber. The jet is then elongated by a whipping process, caused by electrostatic repulsion initiated at small bends in the fiber, until it is finally deposited on the grounded collector. The elongation and thinning of the fiber, resulting from this bending instability, leads to the formation of uniform fibers. In certain cases, where the polymer solution has a lower molecular weight,

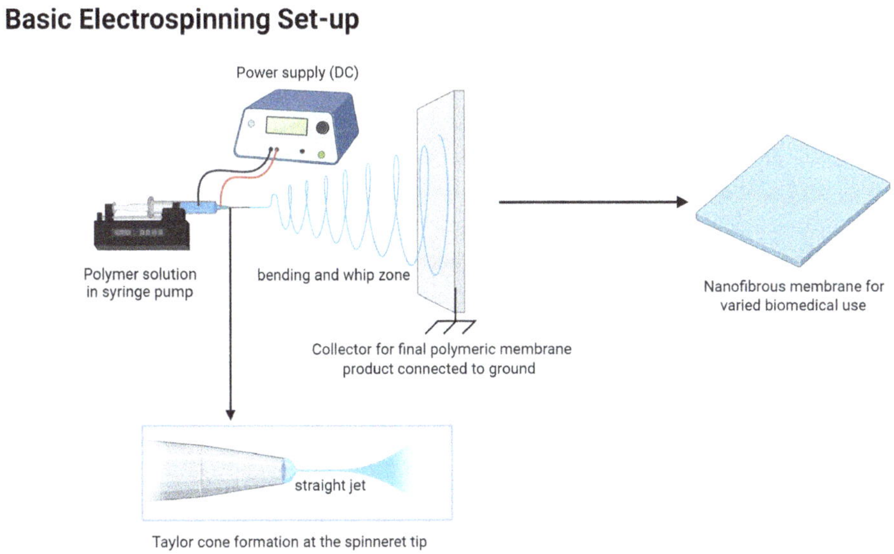

Figure 3.2: Schematics of a basic electrospinning setup showing its basic components needed to produce a nanofibrous membrane material used for various biomedical applications.

lower conductivity, and lower cohesion, droplets may form instead of fibers (this is called electrospraying) that can be used for other applications.

The success of electrospinning starts with choosing the right polymer for the application. Ideally, polymers with high molecular weight are easier to electrospin. This is due to the fact that the intermolecular chain length of the polymer leads to a stable cohesion force when made into a solution, which exhibits just the right viscosity. This viscosity varies, depending on the polymer used, and it has been known that synthetic polymers are easier to use in electrospinning compared to biopolymers because of this reason. But, modifications to the polymer solution and the electrospinning setup have made biopolymers a viable material for this application [27]. Other external factors also play a role in successfully ensuring a uniform fiber distribution; these are ambient temperature and humidity inside the electrospinning chamber, the voltage applied to the polymer solution, the distance from the tip to the collector, the diameter of the nozzle, and the polymer feed rate [25]. Finding the right parameters for a certain polymer is crucial in ensuring that the fibers formed are uniform and without beading. Beading is a phenomenon where instead of fibers, clusters of polymers are formed, and this can significantly affect the physicochemical properties of the resulting membrane [29–31]. This is considered a defect and must always be avoided.

3.2.2 Effect of polymer properties (molecular weight distribution)

Depending on the polymerization process, different polymer properties may arise. Commonly, most polymers are synthesized by the following methods: radical, anionic, cationic, condensation, ring opening, and coordination polymerizations [32, 33]. These polymer syntheses involve a three-stage process: it starts with the initiation process, followed by chain propagation, and lastly, propagation termination. Monomers are exposed to a suitable initiator compound (usually a free radical) to destabilize the stable bonds in its structure. Once these bonds are open, it then attacks other monomers in the solution, creating a propagation loop that continues until it gets saturated or the chain gets exposed to another free radical that stabilizes the polymeric chain (termination) [34, 35]. The resulting polymer molecule in the solution varies in molecular weight. Depending on how long is the propagation step, each molecular chain length may also vary. A well-controlled polymerization process will yield a polymer that has a narrow molecular weight distribution, making a stable polymer that is easy to fabricate into its various forms [36–38]. If the polymerization process is uncontrolled, it may yield a product that contains short and long chained polymers that may cause problems when electrospun. This, in turn, creates an unstable Taylor cone due to the varied intermolecular chain lengths. Beading and nonuniform fiber sizes arise from these polymers and this, in turn, alters the properties of the resulting membrane. Ensuring that the morphology of all fibers is alike is crucial in creating a robust membrane, by avoiding weak and strong points throughout the membrane's dimension [39].

3.2.3 Effect of polymer solution properties (viscosity, conductivity, surface tension) on electrospun fibers

When dissolving a bulk polymer with a suitable solvent, the solution's viscosity plays a role in the electrospinning process and the subsequent fiber morphology [32]. It can be determined whether the feedstock from the syringe will form a coherent flow towards the tip of the nozzle. Polymers with high molecular weight and with long-chain length produce a viscous solution that is suited for this application [40, 41]. If the viscosity is too low, there is less chance for a stable Taylor cone to form due to the decreased intermolecular cohesion in the polymer. However, if the viscosity is too high, it requires a higher voltage to overcome this cohesion and an unstable Taylor cone will also form. The viscosity of the polymer solution can be changed by varying the concentration of the polymer in relation to the solvent. High-molecular-weight polymers tend to use fewer polymers, by weight concentration, compared to lower molecular weight polymers. Surface tension, on the other hand, is determined by the difference of the interactions between the molecules of the material (fluid) and the molecules of the material in contact; a higher viscosity means a greater surface tension and a lower viscosity means a lower surface tension. The conductivity of the polymer must also be ensured in order to apply suitable voltage during electrospinning [6, 8]. Some polymers exhibit low conductivity, which makes it difficult to electrospin, but adding ions to the polymer solution may increase its conductivity. However, by doing so, it can introduce impurities to the yielded polymeric mat that may pose problems in biocompatibility or other functionality. A low conductivity polymer requires a high applied voltage to be able to form fibers, but this severely limits the range of the fiber diameter that it can yield. Also, random fiber sizes may form that can affect the properties of the final membrane.

3.2.4 Effect of applied voltage and feed rate

Application of direct voltage is the main driving force in creating fibers from a polymeric solution [8, 28, 32, 42]. In order to do so, enough voltage is needed to overcome the cohesion forces in the Taylor cone droplet to force it to jet into the collector. Different polymers react differently to the applied voltage, depending on their conductivity and viscosity. Conductive and less viscous polymers form a stable jetting with less applied voltage, while less conductive and high viscosity polymers require higher voltages. A stable jet is achieved once the polymer solution reaches its critical voltage [43, 44]. Higher the voltage, the more erratic the jetting becomes. Once it breaches the supercritical voltage, the polymer starts to spray, and beading and nonuniform fibers may start to form on the collector. Thus, finding the right critical voltage is essential to maintain a uniform fiber distribution. The feed rate plays a factor in successful electrospinning, because in order to maintain a constant and stable Taylor cone, the nozzle must always contain enough polymer solution feed [32, 45, 46]. If the feed rate is too slow compared to the establishment of the jetting, disconnected fibers may form, thus affecting the properties of the final membrane. And, if the feed rate is too fast compared to the jetting, the polymer solution will drip from the nozzle and a stable Taylor cone will be unable to form. These two parameters go hand in hand and must be maintained at a proper level. These parameters are highly dependent on the polymeric properties (viscosity and conductivity); thus, using different polymers may require finding the optimal applied voltage and feed rate ratio.

3.2.5 Effect of distance of nozzle to collector

The distance between the nozzle and the collector can change the properties of the polymeric fibers [28, 32]. The farther the distance, higher is the jetting time for the polymeric solution; thus forming smaller fiber diameter morphology. The opposite happens when the collector is nearer to the nozzle. Depending on the application, researchers may vary this distance to achieve different fiber size morphology. Also, since the porosity of the membrane is dependent on the size of the fiber, researchers can make a highly porous membrane just by increasing the distance of the nozzle to the collector.

3.2.6 Effect of distance of ambient parameters (temperature, humidity in the chamber)

Ambient conditions highly affect the electrospinning setup, as the levels of temperature and humidity may alter the solidification of the polymer jet [47, 48]. At high temperatures, evaporation of the solvent may happen fast; thus disallowing the formation of a proper Taylor cone and polymer jet [49]. At high humidity levels, the evaporation of the solvent slows down, leading to beading or the collection of unevaporated solvent onto the collector; severely affecting the collected fibers. Different polymer solutions might exhibit varied response to temperature and humidity levels. Depending on the solvent selection for different polymers, finding the right temperature and humidity inside the chamber must be considered [50, 51].

3.2.7 Effect of motion of the nozzle and collector rotation

Traditional electrospinning setups use a static nozzle and collector configuration, but recent advances in the field find it useful to move the nozzle and the collector to ensure a well-distributed membrane thickness. In static configurations, the middle area of the membrane collects most of the fibers; creating an uneven thickness. Thus, in order to achieve a uniform thickness, the collector is rotated axially, and the nozzle moves up and down or from left to right. This dynamic configuration is achieved by attaching an electric step motor and programming it to move in a preset direction. The rotation speed of the collector plays a role in achieving aligned fibers [52–54]. At high rotation speeds (upward of 700 rpm), the fibers are collected in an aligned manner and if the rotation speed is slow, a random distribution of fibers is collected.

3.2.8 Effect of nozzle gauge

The diameter of the nozzle can influence the resulting fiber membrane, by allowing for different sizes of Taylor cone to form. The higher the gauge (inverse proportion to diameter), the smaller is the Taylor cone; thus producing fiber of smaller diameter, while the opposite is true at lower nozzle gauges.

3.2.9 Challenges in electrospinning

With all the upside of creating a nonwoven fiber membrane, it still faces certain challenges; thus limiting its widespread translation to the clinical setting [55]. The first and most common challenge for electrospinning is its ability to scale up. Mass production of electrospun fibers is limited because of the inherent slow fabrication time in producing a usable membrane [56]. The time needed to produce an A4-sized mat with enough thickness depends on how fast the stock feed rate is set. This poses a problem as the feed rate for the polymer solution is usually set at around 1–2 mL per hour and the amount of polymer solution needed to form a useful membrane is 8 mL, on an average. This is a big limitation as it needs 4–8 h of electrospinning time, just to create an A4-sized mat.

There are also limitations to what kind of polymers can be used for electrospinning. For example, thermosets have a very high molecular chain length and they require heat to dissolve and to form a solution [57–59]. Whereas biopolymers, which are highly attractive materials to be electrospun, have randomly distributed molecular chain length, and producing nanofibers from them is deemed difficult and almost close to impossible. The use of highly polar solvents may also alter the structure of these biopolymers [60], making them difficult to electrospin. Furthermore, the use of toxic solvents to create the polymeric solution also affects the biocompatibility of the resulting fibers [61, 62]. It must be ensured that all residual solvent is evaporated by putting the fibers inside an oven, usually overnight, just to make sure that there are no solvent impurities left in the fibers. Special modifications are required in the electrospinning setup just to be able to electrospin polymers that have unique requirements. Lastly, the use of direct current from a large DC power supply [63], and a well-controlled ambient humidity and temperature chamber further limits the widespread availability of electrospinning technology. Even though the setup may be simple, the possibility to produce robust fibers from a crude setup may be difficult. Thus, in order to make sure that well-defined fibers are made, a properly engineered setup must be in place.

3.2.10 Electrospinning modifications

Electrospinning technology has seen many advances in the recent years and researchers have been providing considerable attention to improving the setup and to solve its pressing issues. Modifications to the chamber controls, collector, nozzle styles, and how the polymer is fed are just some of the most common types of advancement in the technology. This advancement is with the goal of creating fibers that have unique properties, such as combined nano-micro scale fibers, creation of polymer composites and bi-layer nanofibers, etc.

3.2.10.1 Nozzle and collector-based modifications

Coaxial electrospinning

Coaxial electrospinning is a technique that uses a specialized nozzle and a dual-solution feed system, which allows for the injection of one solution into another at the tip of the nozzle [64–66]. The resulting individual fibers form a core–shell structure made up of two distinct polymers, usually with distinct properties. The polymers that are used to have this fiber morphology need to be immiscible so that they do not mix in the Taylor cone and during the jetting phases. If the polymers are miscible, it may result in a porous fiber morphology and the resulting fibers will blend. The use of coaxial needle allows for co-electrospinning of two polymers; the polymers do not meet until at the very end of the nozzle and are loaded into separate syringes or feedstocks. During the stretching of the polymer, the fibers solidify and lock in this core–shell structure. When electrospinning two types of polymers via this method, critical voltage, solvent evaporation rate, viscosity, and feed rates must be considered. If one of the polymers is incompatible and does not share the same parameter with the other polymer for electrospinning, it may be difficult to form a stable coaxial Taylor cone and the resulting fibers will not form a core–shell structure.

Having a core–shell fiber structure allows for utilization of more than one polymer and combines their advantageous properties into a single bulk membrane material. For example, if the core polymer has a high degradation rate, a more stable polymer can be used as the sheath polymer to prevent fast water migration towards the core polymer, thus acting as a protective layer. This is beneficial for drug delivery in order to control the release of drugs for a prolonged period instead of having a burst release, caused by the rapid degradation of the core polymer [64, 67–70]. Also, combining an elastomer with a brittle polymer is possible with this technique, as the core polymer (elastomer) can be used as a reinforcing structure for the weak polymer, to give it enough elasticity [71, 72]. The elastomer can also benefit from it as the sheath polymer can give it some yield strength (resistance to stretching). This type of polymer combination is useful in mesh implants, as nonadhering polymers are usually brittle and cannot be used as is; therefore, combining it with an elastomer can provide it with elasticity to support the tissue, without causing biological adhesion.

Core–shell electrospinning can also yield hollow fibers that are useful for various applications [64, 66, 73]. This method uses coaxial electrospinning as a preprocess where the core feed is made up of thermal degradable polymeric material and the shell is a thermally stable material (bioceramic). The bioceramic is mixed with a polymeric material and used as a sheath material feedstock. Once the coaxial electrospinning process has been carried out, the membrane is placed inside a sintering oven to thermally degrade the polymers, leaving the bioceramic with a porous hollow fiber morphology. These hollow fibers are useful in bone tissue engineering, as it can make a bioceramic with a morphology that mimics the intrinsic cortical bone structure instead of being used in granular form [74, 75].

Emulsion electrospinning

Emulsion electrospinning too is intended for making a core–shell fiber structure but without the use of a coaxial nozzle. This technique utilizes a water phase and an immiscible solvent phase that are mixed in the presence of an emulsifying agent to form the emulsion [76–79]. A surfactant is also used to stabilize the interface between the two liquids to ensure that the liquids do not mix until a voltage is applied. The common surfactants used are sodium dodecyl sulfate and Triton [80]. During the electrospinning process, the emulsion droplets within the feedstock are stretched and gradually confined, leading to their coalescence. If there are enough emulsion droplets in the feed, a continuous inner core can be formed. However, core–shell fibers from this type of electrospinning are more difficult to produce as there are more variables involved in creating a stable emulsion.

Blend electrospinning

If the goal is to blend two polymers to form a single fiber material, simply blending two polymer solutions into one feedstock is enough [81, 82]. The main consideration for this type of electrospinning is that there should be enough miscibility in the polymer solution so that phase separation is avoided. The resulting fiber will be a blend of two distinct polymers, with no chemical interaction besides some hydrogen bonding between their polymeric chains. However, this will create a random distribution of polymeric structure; thus giving a less robust and predictable property. Thus, blend electrospinning is usually carried out with polymers with similar structures and properties.

Multi-nozzle electrospinning

One or more nozzle configurations can be had to create a membrane that contains more than one type of polymeric fiber [28, 79]. This type of electrospinning can create a membrane that has a random distribution of fibers from two or more distinct polymeric feedstocks. Unlike blend electrospinning, the individual fibers from each feedstock are preserved, giving a heterogeneous polymer membrane material that contains the individual polymer's properties [83]. This strategy is also used for upscaling the electrospinning production time by having more nozzles working at the same rate [84].

Field-guided electrospinning

Moving the collector, in relation to the distance of the nozzle, has also been explored to create a membrane with varied fiber diameter and porosity level. Since the distance of the collector from the nozzle influences the fiber diameter, if the collector is moved during the electrospinning process, the resulting membrane will have a multi layered porosity level. This type of membrane is beneficial for both tissue engineering and drug delivery, as the varied fiber diameter and porosity level can induce a slower drug release and better cell attachment by mimicking the extracellular matrix environment.

Electric field manipulation electrospinning

Manipulating the electric field on the collector can give rise to a membrane that collects both random and aligned fibers. Usually, they are either aligned or random fibers are collected in a single electrospinning setup. But by manipulating the electric field, it can direct the alignment of fibers being collected [85, 86]. Here, the collector can be modified by attaching copper wires to intensify the electric field and direct a random distribution of fibers, and a cellophane tape to attenuate the electric field to align the fibers [79, 87, 88]. This field gradient directs how the fibers are being deposited on the collector, allowing for both random fibers to be deposited on single membrane. Depending on how the copper wire and cellophane tape are attached to the collector [88], various patterns can be made that can be used for tissue engineering [79].

Handheld electrospinning

Handheld electrospinning has seen recent advances, where batteries power it instead of a DC power supply to draw out the polymer solution, to form fibers [89]. The batteries are hooked to a voltage multiplier that converts electrical power from a lower voltage to a higher DC voltage, typically using a network of capacitors and diodes [89]. This allows for miniaturization of the device into something that can be used in the clinic. Handheld electrospinning uses water soluble polymers loaded in a sterile syringe [90]. The fibers are then targeted onto the affected organ that requires electrospun fiber coverage. This is highly applicable for burnt victims who require an artificial protection to protect from pathogens and preserve water loss due to exudate evaporation, due to a compromised skin barrier [90, 91]. However, this application is limited to polymers that are water soluble, as other polymeric solutions made with polar solvents may cause toxicity when directly electrospun onto the skin or organ of the patient.

Gas-assisted electrospinning

Combining gas blowing with electrospinning is another technique to produce nanoscale fibers [92, 93]. The additional airflow, placed in tangent to the nozzle, helps in further stretching of the fibers at the Taylor cone area, leading to the formation of fibers from polymers that are sensitive to high applied voltage. This strategy increases the efficiency of the process by accelerating the evaporation of highly stable solvents (e.g., water). In this way, peptide-based polymers can be electrospun without altering their native form.

Centrifugal electrospinning

This configuration uses a rotating nozzle, similar to a centrifuge, and the centrifugal force assists in stretching the polymeric fibers [28, 94]. With just pure centrifugal force, the rotation of the nozzle must be maintained at thousands of rpm, but the additional electrostatic force allows for the rotation to be reduced by 50%. This technique can increase the evaporation rate of the solvent and adds to the stretching of the polymeric fibers, allowing it to reach nanofiber dimensions lower than 100 nm. This technique also allows for the formation of aligned nanofibers and the possibility to add multiple nozzles can lower the fiber production time [95, 96].

Melt electrospinning

This method allows for highly crystalline polymers to be electrospun at a high rate. The polymer granules are melted, depending on their specific melting temperatures [57, 97]. This heating is produced by resistance heating, hot fluid circulation, air heating, or lasers. Once the polymers are melted, it is then extruded into a spinneret and subjected to an applied voltage where it is drawn out to form fibers. The diameter of the fiber produced by this method is usually larger, compared to conventional electrospinning. This is due to increased polymer melt viscosity brought about by their high molecular weight.

3.2.10.2 Nozzle-free electrospinning techniques

Bubble electrospinning

Another simple adaptation for mass production of polymeric fiber production is the use of bubble electrospinning. The concept relies on aerating a polymer solution to form bubbles on its surface [98–100]. Without the presence of an applied voltage, these bubbles will simply fizzle out. But when enough voltage is applied to counteract the surface tension and cohesion in the polymer solution, the resulting electric filed creates a tangential stress in these bubbles, making the polymer solution jet upward. The reaction then creates multiple fiber formation by making each individual bubble a Taylor cone. This technique has a significant upside in mass producing polymeric fiber membranes of thin and uniform thickness [101].

Rotating roller electrospinning

This iteration of electrospinning utilizes a rotating drum that is partially submerged on a polymeric solution [102, 103]. When the drum rotates, a thin layer of polymer solution on its surface gets electrically charged and a stream of multiple fibers is formed. The fibers jet upward and are then collected to form a nonwoven polymeric membrane. This technology is also known as nano-spider technology and is considered as one of the most innovative ideas in electrospinning. The main advantage of this technique is its high throughput and the possibility of mass-producing polymeric nanofibers at industrial scale.

Rotary cone electrospinning

A rotating conical spinneret is used to spin the polymer solution. This technique utilizes the centrifugal forces that drive the polymer solution to the edges of the cone [94, 104]. Once a voltage is applied, the polymer jets away from the edge of the rotating cone, forming multiple fibers all at once. This technique is like centrifugal electrospinning but without the use of a needle as a nozzle, and like centrifugal electrospinning, it also shares the same benefits of fast solvent evaporation and faster production rate. The production rate of this technique is about 10 g/min of polymeric solution, a rate that is significantly higher than conventional electrospinning, which usually takes 8–10 h to electrospin the same amount.

Edge electrospinning

Edge electrospinning consists of a bowl that is filled with the polymeric solution, a DC power supply, and a concentric cylindrical collector [105, 106]. The bowl is filled with a polymeric feedstock and is continuously filled to the brim. The edge of the polymer solution, where surface tension is weakest, is jetted when an applied voltage is turned on. This process produces multiple fibers at once at a high throughput. The fibers are then collected on the concentric cylindrical collector.

Ferrofluid-assisted electrospinning

A two-layer system – the lower layer being a ferromagnetic suspension and the upper layer being a polymer solution – and a permanent magnet placed at the bottom of the suspension creates spikes that align to the magnetic field lines of the magnet. These spikes also carry the polymeric solution with them and when a voltage is applied, they, in turn, draw out the polymers from the spikes created by the field lines. The spikes act as an alternative to spinneret needles and can draw out multiple fibers. The production rate is dependent on how much magnetic field lines are produced by the permanent magnet.

Droplet electrospinning

Polymeric solutions are feed to an electrically charged metal roller via droplets. The droplets are supplied by a solution distributor. Multiple solution droplets adhere to the surface of the metal roller, are electrically drawn out into fibers, and are collected on the collector placed adjacently. The rate of fiber production is dictated by how many droplets are supplied by the reservoir; thus delivering a higher production rate compared to the traditional electrospinning method.

Gas-assisted melt electrospinning

The hot air-assisted melt differential electrospinning is a variation of the melt electrospinning that utilizes the following components: high-voltage power supply, a heating system, an air pressure gun, and a copper circular ring functioning as electrode [107]. In this method, the hot gas maintains the heating on the polymer solution, allowing it to attain smaller fiber diameters at a high throughput compared to conventional melt electrospinning. The high production rate is due to the generation of multiple Taylor cones around the bottom edge of the nozzle. Like melt electrospinning, this method allows production of highly crystalline polymers that may be difficult to dissolve into a solution. However, this method is designed for high-speed production due to the use of a needle-free system.

3.3 3D bioprinting

3D printing is a technology that allows for rapid prototyping from computer-aided design (CAD) to actual plastic-made construct; this is also known as a form of additive manufacturing. This technology liberalizes the design and prototyping space and has been advancing ever since. Conventional 3D printing technology uses polymer melts as its ink and a computer-aided nozzle that can move on their x-, y-, and z-axes, as it lays down a thin layer of polymer melt and forms 3D structures based on a CAD program [108–110]. This technology has so much upside and its application to biomaterials manufacturing has seen significant advances in the recent decade [110–112].

An iteration of 3D printing is bioprinting, where the polymer melt is replaced by what is called a bioink. This bioink is made up of hydrogel materials and laced with living cells and other growth factors [113, 114]. Together, when printed, the bioink can form geometric structures that can resemble living organs in the body. This is meant to be used in tissue engineering applications to solve the longtime problem of shortage in available donated organs ready for transplant. The issue of organ shortage has been a significant problem in medicine [115], as the supply for organs donated is outpaced by their demand. In order to create enough transplantable organs, tissue engineers investigated growing tissues and, potentially, a whole organ, using 3D bioprinting. However, this is all easier said than done, as 3D printing of bioinks to form complex structures needed for an organ poses several challenges.

3.3.1 Basics of 3D bioprinting

3D bioprinting is based on the technology of 3D printing as a form of additive manufacturing, Figure 3.3 [113, 114, 116]. It comprises a software component that researchers use (mainly computer-assisted design software, CAD) to design the

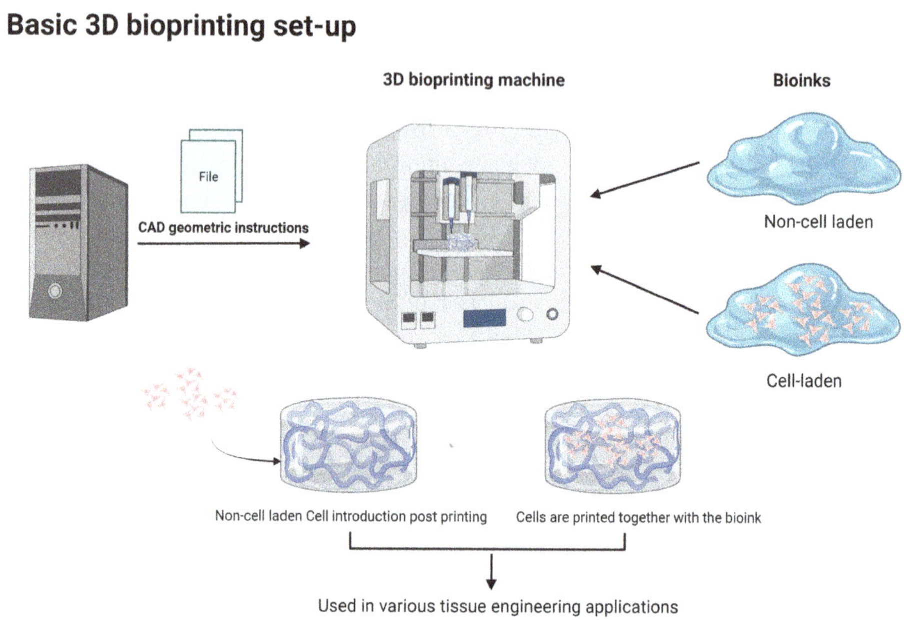

Figure 3.3: Schematics of a basic 3D bioprinting setup showing all components needed to produce a 3D geometric material used for tissue engineering applications.

geometry of the printed structure, which relies on data from imaging techniques, e.g., CT scans and MRI scans. The hardware component comprises a printhead nozzle, a bioink feedstock reservoir, and a printing stage. Certain variations of this technology have been adapted to accommodate the sensitive bioink. First, the system must be contained in a sterile environment so as to avoid contamination while printing. This is important as the living cells might get infected during printing. The printhead nozzle is moved by a step motor that can manipulate the printing direction on *x*-, *y*-, and *z*-axes. This is to ensure the layering the bioink on top of each other to form a 3D structure. The size of a printed structure is limited by these movements and by the size of the printing stage. Also, the resolution or the amount of detail on a printed structure is determined by the printhead nozzle design. The printing stage is a flat surface that is usually held on a stable position to avoid dimensional variations during printing.

The average printing time to form a structure is determined by how large the printed structure is designed to be and by the curing properties of the bioink, which in this case is the cross-linking process within the polymeric structure of the ink. The bioink should have enough mechanical properties to support itself after leaving the nozzle, otherwise the structure will collapse as the layering continues. When the printing is done, the cells that are laden inside the bioink will proliferate and migrate throughout the 3D printed structure when placed on a bioreactor. The bioreactor provides further nutrients and gas exchanges to the cells inside the structure until it is ready for transplantation.

3.3.2 Bioink

The main differentiating component between conventional 3D printing and bioprinting is the use of bioinks. Bioprinters use a variety of bioinks that can be classified into three categories; matrix, sacrificial, and support bioinks [117, 118]. Matrix bioinks are where the cells are laden and are made from naturally and synthetically derived hydrogels. Sacrificial inks are temporary support materials that are then removed once the printing is done; they do not contain any cells or components needed by the cells and are mostly used to fill in gaps in order to form channels and hollow structures in the printed structure. Support bioinks are like sacrificial inks but may be left in the structure to provide additional mechanical support in the printed structure.

Hydrogels is a class of biomaterial that is highly suited as a bioink [119, 120]. It is a highly hydrated material of hydrophilic polymers that crosslink to form 3D networks and facilitates the necessary permeability for nutrient and waste transport to the cells. In an in vitro setting, the hydrogels are able to support the cells during the printing process. This maximizes the amount of living cells that survive and proliferate once the structure is completed [114, 120–122]. It must be noted that the cells that are used for printing may be allogenic (sourced from a cell bank) or syngeneic (sourced from the same person receiving the transplant). It may be that the cells used are progenitor cells that are then induced to become adult cells of a specific organ. Adding growth factors and other differentiation molecules are also a common technique to ensure that the cells thrive and grow to specific cell lineages [116, 123].

The most common hydrogel materials for bioprinting are naturally derived due to the ease of gelation and their inherent biocompatibility toward the body and the cells being printed. Gelatin, Collagen, Hyaluronic acid, and other peptide-based polymers are common materials to form a matrix bioink [116, 124]. They possess functional groups that have natural adhesion sites for cells to thrive, as most of these materials are natural components of the extracellular matrix in the body. The preparation of these materials is quite facile as well, as they can form robust gels that have the proper rheological properties that allow it to be extruded through the nozzle of the printer head. However, these gels tend to collapse when trying to form large geometric structures [125]. Therefore, support inks are printed together with these matrix materials to act as a scaffold so that the structure being printed can hold its shape [116].

There are important factors that must be considered in choosing the right bioink for a smooth printing operation and survival of the cells in the construct being printed. These are material properties, viscosity, gelation process, and biological interaction [113, 116]. Finding the perfect balance of these properties leads to the creation of what is called "hybrid bioinks" [126], as no single material possesses the perfect balance of these properties. Polymeric structures are modified to create a bioink that is printable as also have the capability to support living cells.

3.3.2.1 Effect of polymeric properties

Material properties, such as mechanical property, degradation rate, and permeability, can affect the printability of an ink. If a material has low mechanical property it will not be able to hold its shape after extrusion and the structure will collapse. Also, the degradation rate of the gel must coincide with the proliferation rate of the cells to ensure enough scaffolding support for the cells to migrate into. Lastly, the gel must be able to facilitate the permeation of nutrients and gas and eliminate waste so as to provide the proper conditions for the cells.

3.3.2.2 Effect of viscosity

Viscosity not only affects the physical extrusion of the gel through the nozzle head, but it also determines the shear stress that the cells will experience during the printing process. Cells are highly sensitive to shear stress as their cell membranes are stretched and damaged if the viscosity is too high. This can lead to lower cell viability and affect the number of cells that can populate the printed structure. An ideal material must have a shear-thinning property, meaning that when stress is applied, the material's viscosity is reduced, allowing for a smoother flow of the material through the nozzle head and limiting the shear stresses that the cells undergo; thus increasing the viability cells after the printing is completed.

3.3.2.3 Effect of gelation process

The gelation rate of the hydrogel determines how the bioink solidifies after being extruded from the nozzle head [127]. This process is optimized to allow for a high-resolution print of the 3D structure. Before the second layer of the bioink is layered, the base layer must be able to hold the ink and hold its shape to avoid collapse. If the gelation time is too slow, it can delay the printing time; thus affecting the viability of the cells. Also, if the gelation time is too fast, it can clog the nozzle and interrupt the printing process. Therefore, an optimized gelation rate of the bioink is required to ensure an interruption-free printing process. Gelation can be achieved via the addition of cross-linkers in the bioink, to hold the hydrogel structure in place. Various kinds of cross-linkers can be used, e.g., photoinitiated, thermal, ionic, self-cross-linked, stereocomplex, click chemistry, and enzyme-catalyzed cross-linker [127, 128]. These cross-linking strategies affect the rate of gelation of the hydrogel and a strong cross-link bond between the polymeric chains must be ensured in order to hold the printed structure in place. One of the most common strategies is the use of UV cross-linkers [128, 129] as they can be easily controlled, but it must be noted that the use of UV cross-linkers at high concentrations and the exposure to prolonged UV irradiation can induce cell death. Therefore, other cross-linking strategies are also being optimized as they are more biocompatible and less harsh to cell conditions [130, 131].

3.3.2.4 Effect of biological interaction sites

Functional moieties in the matrix bioinks determine how well the cells can interact with the bioink. These functional moieties act as adhesion sites for the cells and activate internal cascades that encourage the cells to proliferate and migrate in the bioink matrix. Different peptide- and polysaccharide-based bioinks possess these adhesion sites and the cells that are laden in these bioinks can survive longer during the printing process. These active sites can also be functionalized with various growth factors and biomolecules that can further enhance the proliferation and differentiation of the cells inside the bioink. Other physicochemical properties, such as hydrophilicity and surface energy, can influence how the cells can effectively adhere and survive in the bioink. Hydrophilic polymers efficiently adsorb cell adhesion proteins and help the cells form an initial attachment to the bioink. The surface energy of the polymer also has the same effect, as the two properties are linked. High surface energy leads to increased wettability and this leads to better cell attachment, compared to materials with low surface energy properties.

3.3.3 Challenges in 3D bioprinting

At first glance, 3D bioprinting seems to be a very effective solution for creating tissues and organs on-demand. The potential of the technology creates a paradigm shift in the medical field, as there will be less reliance on donor organs to solve the transplantation demand that is putting pressure and costing lives to people who need new organs to survive. But this technology is still in its infancy and faces challenges in creating 3D structures that are like native organs [132]. The most obvious challenge is the ability of the bioprinter to create structures that are alike to what we see in a kidney [133, 134], liver [135], heart [136], pancreas, etc. [137, 138] As any anatomy and physiology student may know, these organs are very hierarchical and contain multiple levels of complexities, down to the microscopic structure, and printing precise details is deemed difficult. Next will be the selection of a suitable bioink. Bioinks not only need to be able to support the proliferation and migration of cells but they must also have the capability to mature these cells into lineages that follow the complex structure of an organ. The distribution of the cells in the printed structure must be ensured while also allowing for these cells to differentiate into various cell types to form a viable organ. Furthermore, these bioinks should be able to degrade and facilitate engraftment of the transplanted structure into the recipient's body while ensuring that the degradation products are nontoxic. A suitable bioink must also be able to print at a high resolution in a low amount of time. Up to now, the printing time of a bioprinter is hours on end, just to form a simple geometric structure. This is because the gelation or curing rate of the bioinks is relatively slow compared to conventional 3D printers. Lastly, there is still an ongoing quest to find the optimal bioink that possessed the right properties of fast curing time while being able to maintain the viability of the cells inside its matrix [136, 139, 140]. All these challenges are new avenues for the technology to expand into, and researchers worldwide are trying to advance bioprinting by solving these individual challenges, one at a time.

3.3.4 Types of 3D bioprinting

3.3.4.1 Inkjet-based bioprinting

Inkjet bioprinting is based on the traditional printing technology. This relies on precise deposition of droplets of hydrogel bioinks onto a substrate or a culture dish under a computer regulated control of the printer head [141]. This method can be further categorized into thermal and piezoelectric actuator methods, based on the droplet actuation mechanism. In thermal technology, bioink droplets are made by heating so that cavitation occurs and forces the ink out of the extrusion printer head and onto the platform or substrates. Temperature generation only occurs in microseconds, thus avoiding thermal degradation of the bioink and thermal shock to the cells. This technology is facile and is the most used method for bioprinting due to the simplicity of the ink jetting mechanism. It is also viable for thermally cross-linkable bioinks [142]. However, there are high variations in the droplet morphology that can affect the resolution of the final printed structure. This method also experiences high occurrence in nozzle blockages and can lead to disruption of the whole printing process [143]. Furthermore, the use of thermal controlled jetting places limitations on which kind of protein-based material is added to the matrix, as it can be denatured from exposure to high heat.

In piezoelectric technology, the jetting of the bioink is controlled by a piezoelectric generator attached to the nozzle head [144]. This technique does not rely on cavitation generated by temperature change in the bioink but by giving transient pressure to extrude droplets; this gives a much better droplet control in terms of size and has less chance of clogging the nozzle head. However, one downside of this method is that it can cause lysis to the cell membrane of the cells that are laden in the bioink due to the increased shear stress generated by the pressure from the piezoelectric generator. But this effect is dependent on the cell lineage and it has been found that mammalian cells, such as osteoblast, fibroblast, and chondrocytes, are more resilient to membrane deformation compared to other cell types. Thus, this type of bioprinting can be used in cartilage and bone replacement applications.

3.3.4.2 Pressure-assisted bioprinting (extrusion method)

This iteration of bioprinting relies on the extrusion of bioink using a plunger, screw, or pneumatic pressure to create 3D patterns and constructs [145, 146]. Layer-by-layer extrusion of the bioink is done by moving the nozzle, like a baker piping down icing on cake. This is a more straightforward method of laying down bioinks and is considered less harsh and more compatible to most cells and bioinks available for printing. It does not rely on any cavitation method to extrude the bioink, making it applicable to more sensitive cell types, such as stem cells. However, since it relies only on pressure to extrude the ink, it also needs a longer printing time, as it requires the hydrogel to cure before it can lay down the subsequent layers, to avoid the collapse of the structure. This means that cross-linking must be done post-extrusion and the need to incorporate UV cross-linker or other cross-linking agent may affect the viability of the cells in the long run.

3.3.4.3 Laser-assisted bioprinting

Bioprinting, using laser as the energy source, has the promise to be able to print down to the smallest scale (micro- to picometers), due to the inherent ability of light to be pulsed in those scales, just by changing the energy of the laser [144, 147]. This iteration of bioprinting comprises three parts: a pulsed laser source, a ribbon coated with liquid biological materials that are deposited on the metal film, and a receiving substrate. Here, a pulse of laser irradiates the ribbon that then causes the liquid to evaporate, which is then collected in the substrate, in a droplet form [148, 149]. Subsequent laser pulses, made in a pattern, can create geometric structures that may contain cells and proteins. The substrate contains a biopolymer or a culture medium to sustain the growth of the cells after being transferred from the ribbon. The resolution varies from pico- to micro-scale features and is affected by the thickness of the biological materials on the film, their rheological properties, the energy of the laser pulse, the wettability of the substrate, and the printing speed and organization of the structure. This technique has been proven as a viable technique to replace damaged skin by directly printing artificial skin grafts on an animal model [150]. However, its obvious limitation is the ability to print large three-dimensional (3D) structure by itself, but by combining it with other bioprinting technology, printing high-resolution structures may be at reach.

3.3.4.4 Stereolithography

Stereolithography (STL) is a way to create 3D structures by irradiating photocurable polymers solution with UV light coming from digital micromirror arrays [151]. These arrays have the capability to precisely illuminate certain sections of the polymer solution, allowing it to solidify and form 3D structures. It has been traditionally used for fabricating 3D structures from acrylics, resins, and epoxies with higher throughput, compared to conventional extrusion 3D printing. Digital light projection controls the printing process in this top-down system. In comparison with other 3D printing techniques, STL has the highest fabrication rate, and an increasing number of materials can be used in this process. Furthermore, STL as a bioprinting technology can use light-sensitive bioinks, laden with cells and biomolecules, with high precision and throughput. However, there is still a significant disadvantage for this technique, as there are very few biocompatible photo-cross-linkable polymers out there that can be used as a bioink. This technique uses UV initiators [152, 153] that, in high doses, can be toxic to biological items, such as cells and proteins. Thus, techniques using visible light spectrum has risen to avoid the use of high-energy irradiation, such as UV light. Here, a customized visible light in a STL system, including a commercial beam projector and bioinks, a mixture of PEGDA, methacrylated gelatin (GelMA), and eosin Y-based photoinitiator were used, in addition to adding an infrared (IR) filtering-water filter system, to avoid damaging the cells via IR coming from the visible light spectrum [154].

3.3.4.5 3D bioprinting in space

3D bioprinters have also reached the International Space Station (ISS), as carrying out conventional extrusion-based bioprinting in normal gravitational conditions can cause the hydrogel to collapse under its own weight [155, 156]. The use of microgravity, up in the ISS, removes the need to wait for the bioink to cure before layering subsequent layers of hydrogels. One of the reasons why 3D printing has been a challenge is because of the difficulty to create large 3D structures, because bioinks are prone to collapse under their own weight. If we can remove the weight factor in bioprinting, there will be less reliance in cross-linking the bioink using various cross-linkers that could cause toxicity to the cells. Furthermore, the printing time will be reduced as there is no need to wait for the bioink to cure. This then throws up a variety of biocompatible hydrogel materials that can be used as bioink. Most biocompatible hydrogels that are highly suited as a matrix material for cells are not viable as a bioink due to their weak cross-linking capabilities. This leads to poor structural performance as the hydrogel has the tendency to flow under the influence of gravity. Therefore, microgravity removes this flow tendency and the hydrogel bioinks retain their shape after extrusion without the need for post-processing. The microgravity also allows for the printer nozzle to be smaller and finer, giving the bioprinter a higher resolution and the ability to print microscopic details needed for a full organ. However, cells might also get affected by the lack of gravity, as some biological functions rely on physical cues caused by gravity. There is also the limitation of cost, as sending material to the ISS is still far from economical. This also means that cells from patients needing a syngeneic transplant must send their stem cells to the ISS, grow it there, and send it back to earth, severely limiting its reach to a wider population. For now, only simple geometric structures have been printed and there is still a long way to go before a full organ can be made in space. But with the advancement of spaceflight and the bioprinting technology, we can potentially have organs made from space save countless lives in the future.

3.4 3D metal and ceramic printing

Additive manufacturing, using metal or ceramic, has seen recent focus, as it has the capability to design biomaterials that can be adapted to personalized needs, allowing for efficient material use. Biomedical engineers scan an affected limb that requires implant prosthesis, which is then tailored to the exact anatomy of the recipient. Medical devices usually come prefabricated at specific designs and shapes. But 3D printing metals [157, 158] and ceramics [159, 160] have opened a new avenue for designing biomaterials made from inorganic materials, such as metal alloys and bio ceramics.

3.4.1 Selective laser sintering

Selective laser sintering (SLS) uses high power laser (e.g., a carbon dioxide laser) to fuse small particles of materials, such as metals, ceramics, glass or polymer, into a mass that has a desired 3D shape [161, 162]. The laser selectively fuses powdered material by scanning cross-sections generated from a 3D digital description of the part (e.g., from a CAD file or scan data) on the surface of a powder bed. After each cross-section is scanned, the powder bed is lowered by one-layer thickness, a new layer of material is applied on top, and the process is repeated until the part is completed. Because the density of the finished part depends on the peak laser power rather than the laser duration, an SLS machine typically uses a pulsed laser. The SLS machine preheats the bulk powder material in the powder bed, somewhat below its melting point, to make it easier for the laser to raise the temperature of the selected regions, the rest of the way to melting point. Material properties are not drastically changed by the rapid melting of the laser; it preserves the crystallinity and avoids phase transition of a material, making it suitable for varied types of materials [163–165].

Metallic and ceramic materials are made into a powder using a ball mill to form particles that are usually microns in diameter. Depending on the material being used, particle size, shape, density, roughness and porosity may vary. Metallic and bioceramic prosthesis that are fabricated using this method allow for a porous surface morphology that can increase cell attachment on its surface [164, 166]. It also allows for a more lightweight prosthesis, as biomedical engineers can optimize where to concentrate materials for load bearing and where to decrease material to save up on weight. Prosthesis for bone implants can also be tailor made for individual patients who have more specific needs. For example, cranial defects are usually irregular and must be designed depending on the geometry of the defect site. Here, biomedical engineers who are designing an implant can use 3D scan models acquired from computed tomography and convert it to a CAD file to print the desired shape of an implant. It is also useful for pediatric patients whose bones are still developing and require progressive implants and prosthesis adjustments throughout their life.

3.5 Coating methods

Coating methods are essential in biomaterial science as they are ways for functionalizing, passivation, or increase the stability of an implant, device, or drug delivery platform. In recent years, coating methods are progressively increasing to accommodate various materials. The surface properties of a biomaterial are a vital component on how they will perform and behave inside the body. Coating techniques vary depending on the substrate material that are being modified, and polymers are one of the most common materials to be used as a coating material to achieve the said goals.

Coatings, in general, can be classified into the following; stabilizing coatings and functional coatings [167–169]. To stabilize highly reactive materials, passivation or stabilizing coatings are applied to their surface to avoid material reactivity inside the body. This controls the degradation or the corrosion rate of a material to avoid the leaching of substances that can harm the body. Metal implants (especially degradable metals) use different coating methods to act as primary barriers from bodily fluids and avoid rapid degradation [167, 170]. Some metal alloys that contain metal ions harmful to the body can also be passivized by various coatings. Functional coatings, on the other hand, can give more properties to the implant or device. Most common properties are being able to carry therapeutic agents, such as antibacterial, antiviral, antiproliferative, and growth factors [171–174]. Functional coatings can also alter other surface properties of a material, such as modifying hydrophilicity or hydrophobicity, changing the surface energy, friction coefficient, and surface roughness, to modulate how biological tissues behave on the material's interface. The following will describe the different methods of coating techniques that are used in biomaterial applications.

3.5.1 Dip coating

Dip coating is a facile method of coating a material, which involves the deposition of a wet liquid film by immersion of the substrate into a solution. As the name suggests, dip coating immerses a substrate material into a precursor solution that is then lifted and dried to form a thin film coating on the substrate's surface [175, 176].

Usually, the film needs to harden and be transformed by heat treatment in order to ensure proper adhesion of the coating material to the substrate. However, this method poses several disadvantages, such as having nonuniform coating thickness due to the uncontrolled deposition of coating on the surface of the substrate [177]. The viscosity of the solution and solvent evaporation plays a role on the coating thickness. Viscous solutions tend to adhere better to the surface of a substrate due to cohesion between the molecules of the solution and the resistance to flow. After the substrate has been lifted from the immersion, it can also form a thicker coating due to more material being deposited on the surface of the substrate. Less viscous solutions can flow with less resistance; thus forming a thinner coating and adhere less to the surface of a substrate. The evaporation rate can influence the coating morphology. Choosing the right solvent allows for a more uniform coating thickness. If the evaporation rate is fast, the coating can form cracks on the coating surface, as it does not allow the solution to consolidate evenly. The angle at which the substrate is lifted from the coating solution can also influence on how thick the coating may become. Gravity can pull more coating material to the bottom part of the substrate, giving an unequal thickness. This can be overcome by rotating the substrate during the drying process, so that it evenly coats the surface of the material. Dip coating precursor solutions are mostly made up of polymers, as they possess unique adhesion properties to any substrate and facile preparation with a suitable solvent. This precursor solutions may also contain various therapeutics and can be used as a drug delivery platform.

3.5.2 Spin coating

Spin coating is a procedure used to create thin films on flat substrates by using a precursor solution and spinning it under high rotation speeds to achieve uniform thickness and with small tolerance [178–180]. This is a common method to produce microchannels on a microfluidic chip [181], as it combines with a process called soft lithography. PDMS, with photo-cross-linker, is used as a precursor coating solution, also known as photoresist; this is then applied to a wafer substrate. The PDMS is then spun and the wafer is evenly coated. Then, a mask is used that is precut with the pattern of a microchannel to create patterns on the PDMS. This process is repeated to form microstructures on the fluidic chip (e.g., channels, valves, and reservoirs). Microfluidic chips prepared in this way can be used in studying more complex biological interactions that simulates what happens inside the body. This is done by allowing multiple cell-line interactions inside the chip and by introducing other variables, such as fluid flow and biomolecule cascade interactions. This is not easily done in a normal cell culture dish, but microfluidic chips open a new avenue for researchers to create biological systems that can mimic the human body. Microfluidic chips have the promise of being able to substitute in vivo animal studies with more human-derived cell lines [182–185] (also known as organ on a chip) to be used in analyzing disease progression, cell biological interactions, stem cell differentiation, and drug analysis. This would not be possible if not for spin coating technologies, where an easy application of photoresist materials can be done with better control, as thickness can be easily adjusted based on the rotation speed of the setup.

3.5.3 Spray coating (cold spraying)

Spray coating deposits material on a substrate using a nozzle that accelerates the coating precursor (powder, liquid solution, or granules) using high pressured gas [186–188]. Solid powders (1–50 µm in diameter) are accelerated in a supersonic gas jet to velocities up to ca. 1,200 m/s. During the impact with the substrate, particles undergo plastic deformation and adhere to the surface of the substrate. This method can use varied materials, such as metals, polymers, ceramics, composite materials, and nanocrystalline powders, as coating materials. The kinetic energy of the particles, supplied by the expansion of the gas, is converted to plastic deformation energy during bonding. This method preserves the material properties as it avoids phase transformations on the materials, unlike thermal spraying techniques that use melted powders to deposit onto a substrate. This is advantageous in creating bioceramic coatings on metallic prosthesis to increase their osteoconductive properties [189, 190]. Since bioceramic properties can be altered with high temperatures, cold spraying is a viable way to keep the crystallinity of the bioceramic that is important in the osteocyte induction.

3.5.4 Electrospraying

Electrospraying uses electrodynamic forces to draw out polymeric solutions into particles that are then deposited on a substrate. Like electrospinning, electrospraying setup utilizes a syringe pump loaded with a polymeric solution and a DC power supply, and a collector. The difference is that the polymeric solution may have lower viscosity and lower molecular weight, so that instead of drawing out fibers, particles are sprayed onto the collector and deposit a thin film coating on the surface of a substrate [191, 192]. Electrospraying has distinct advantages compared to other coating methods, allowing it to produce cargo carriers for biomedical applications. Mainly, the process can be carried out at ambient conditions (temperature and pressure), which are needed for sensitive biomolecules and even living cells. Particles that are deposited can be loaded with therapeutics beforehand, adding more functionality to the coating by being a drug delivery platform. Different particle preparations, with varied morphologies, can be used for electrospraying. These morphologies, such as hollow, cup-shaped, porous, core–shell, and spherical particles can entrap various cargos with high efficiency, maximizing the therapeutic effects of a selected coating. Although the adhesion between the substrate and the coating may be less strong compared to other spraying methods due to the slower velocity of the particles, the particle coating may interact with the substrate via electrostatic forces, which is enough to ensure proper coating adhesion.

3.5.5 Electropolymerization coating

Electrocoating or electropolymerization method is used to deposit thin and uniform coatings onto a substrate using electrically charged particles out of a water suspension to coat a conductive part. In biomedical applications where polymers are used as a coating material, conductive monomers are prepared in a suspension [193, 194]. This suspension is used in an electrochemical cell and subjected to cyclic voltammetry, where the substrate acts as one of the cathodes or anodes. The free electrons forming on the surface of the substrate, once an electric current is passed, initiates the polymerization process of the monomers suspended in the suspension via oxidation and reduction processes, and this, in turn, forms a thin coating on the surface of the substrate. Over time, the resistance on the electrochemical cell increases due to the insulating effect of the thin film and this will limit the thickness of the coating. This method can be used in coating metallic implants as they can act as a great conductor and can selectively create coatings even on 3D geometric surfaces. Electropolymerization method is used to coat the surfaces of stents and other metallic to act as functional surfaces [195]. It may also be used as a passivation technique, as the polymers used in the coating process can act as a physical barrier from biological elements. Examples of polymers used in this method are polyaniline, polydopamine, polypyrole, and other phenol compounds [196–198]. The use of polymers, such as polydopamine, can increase the functionality of a metallic implant, by acting as a functional layer to conjugate drugs and nanoparticles.

3.5.6 UV-light-assisted coating

UV light can also be used as an initiator to polymerize specific monomers and deposit them on a flat surface. Patterns can be made by applying a mask on the surface of the substrate. A monomer suspension is used and the substrate is submerged into it. A UV light source is then used to irradiate the surface of the setup to initiate the polymerization process. The mask helps create a pattern that may be useful for various biomedical applications. This method has been used to create functional surfaces using phenolic materials, UV irradiation can trigger oxidative polymerization and deposition of plant-derived phenolic compounds, which opens up the possibility to create 2-D gradients and patterns of polyphenol coatings and control this polymerization temporally. Plant-derived phenolic materials used in this process include pyrogallol, tannic acid, caffeic acid, and gallic acid [199, 200], all of which contain various medicinal properties, such as antioxidant and anti-inflammatory. Usually, these phenolic compounds self-polymerize under alkaline conditions, but this method is uncontrolled and is difficult to direct onto a substrate. The UV light, besides from polymerizing the phenols, can as well excite the surface of the substrate, creating a reactive surface that attracts the monomers to adhere with strong binding energies, given that the distance of the substrate to the irradiation source is not very far. Further examples of this is when caffeic acid is used to coat the surface of biomedical-grade stainless steel and titanium to increase the biological activity of cells, and to enhance their antibacterial properties by in situ immobilization of silver nanoparticles [199]. However, this method can only be applied to flat surfaces as UV light travels in a straight line; thus limiting the coating capabilities on 3D geometric implants. This method has also been used in gold nanoparticle coating for anticancer applications where caffeic acid is used as the coating material that serves as reactive oxygen species carrying compound while being able to carry a chemotherapeutic drug (bortezomib) [199].

3.5.7 Coating method challenges

Each coating method has its own specific limitation, and no single coating method is a one size fits all solution. The coating method varies depending on the coating material, the substrate material and geometry, and the goal of the function. In order to create a truly functional coating, different techniques may be utilized but with certain compromises. Ensuring a good adhesion strength is one of the most common challenges in creating a functional or passivation coating, as the coating adhesion relies heavily on the parameters used in the process. Coatings that use ambient conditions may have less adhesion properties compared to other methods that use extreme heat or polar solvents, but they have the potential to be more functional, ultimately. Therefore, it is a compromise between functionality and adhesion, and a certain balance must be found in order to avoid coating defects that may affect the reliability of the biomaterial. Another challenge is the limitation of being able to coat biomedical devices (stents, screws, prosthesis, etc.) with complex geometric structures. It is relatively easy to create uniform coatings on 2D surfaces; coating complex devices may be difficult. Dip coating and spin coating are usually done with this kind of coating on complex geometries, but it is hard to control the variations in coating thickness and achieve uniformity. Therefore, other coating methods, such as electrospraying, electropolymerization, and cold spraying were developed. However, these methods require more preparation, utilize more energy, and need a complex setup. Finally, a quality coating method must be facile and should be capable of being used on both 3D and 2D surfaces with less energy requirement and on diverse materials.

3.6 Hydrogel preparations

Hydrogels are a class of material that is made up of crosslinked polymers with high water content (up to 95%). It is a 3D solid that results from hydrophilic polymer chains being held together by crosslinks in its intramolecular structure [201, 202]. Hydrogels are utilized for three purposes in biomedical applications. They may be used as agents for filling vacant spaces [203, 204], carriers for delivery of bioactive molecules [205–207], and 3D structures [208, 209] that act as support for cells and help in the formation of an ideal tissue. Several techniques are used in creating a hydrogel material and it all depends on the mechanism of their cross-linking. Some polymers readily form crosslinks in its polymeric structure, while some require a cross-linking agent to form intermolecular networks. The degree of cross-linking in the hydrogel network determines the rheological properties of the gel, with a higher degree of cross-linking leading to a stiffer gel, while a lower degree of cross-linking creates softer gels. Most hydrogels that are synthesized use step-growth and solution polymerization. In these techniques, the products of the hydrogel can be formed or reformed into different shapes and sizes that can be utilized in various applications.

In a co-polymerization solution or cross-linking reactions, ionic or neutrally charged monomers are mixed with a multifunctional cross-linking agent. Next, polymerization is initiated by heat, UV light [210, 211], or by a redox initiator [210, 212] that is applied in the system in the presence of the solvent. In order to remove the solvent used in the polymerization reaction, several washing steps using distilled water are required. These steps also remove unreacted monomers and cross-linking agents that may cause cytotoxic effect to the body. Cross-linking modifications can fine tune the mechanical properties and viscoelasticity for varied applications where stiffer and softer gels are required. Several cross-linking techniques are adopted for this purpose, e.g., physical cross-linking, chemical cross-linking, grafting polymerization, and radiation cross-linking. The following describes the different cross-linking methods for hydrogels meant for biomedical purposes.

3.6.1 Physical cross-linking

Physical cross-linking is relatively easy to create due to the absence of a separate cross-linking agent. This is done via the application of external physical factors, such as temperature or via ionic attraction, complex coacervation, and hydrogen bonding in the intramolecular structure of the polymers itself [213–215]. However, this type of cross-linking is not permanent and is easily reversible [216], but it is enough to be insoluble in aqueous solution, given that proper conditions are met. Hydrogels formed in this manner are easily hydrolyzed or enzymatically degraded due to the weaker links in their network.

3.6.1.1 Temperature-mediated cross-linking

Temperature plays a role on how a hydrogel is formed. In this method, hydrogels are formed by cooling polymeric solutions [217]. Examples of this is the formation of gelatin (polypeptide) [218, 219] and carrageenan (polysaccharide) [220, 221] hydrogels, where the mechanism for hydrogel formation is by helix-formation and the formation of junction zones in their polymeric networks. Both gelatin and carrageenan form entangled coils in their polymeric network when the temperature is below its melting transition temperature. These coils prevent solubilization and provide rigidity to the hydrogels. The opposite happens for some polymers, such as polyethylene oxide-polypropylene oxide and polyethylene glycol-polylactic acid, when they form hydrogel, as these polymers can form block copolymerization when they are subjected to high temperatures. These polymers can be classified as temperature responsive gelation and are useful for biomedical applications, especially in drug delivery [219]. Drugs can be entrapped in their polymeric network and released when suitable temperatures are attained, due to the reversibility of this process.

3.6.1.2 Ionic attraction-mediated cross-linking

Ionic interactions can also initiate gelation process for some polymers, such as alginate and chitosan. Polysaccharides [222–224] (e.g., alginate and chitosan) can crosslink by the addition of di- or tri-valent counter ions. This method causes the gelling of polyelectrolyte solution (e.g., Na^{+} $alginate^{-}$) with opposite charges (e.g., $Ca^{2+} + 2\ Cl^{-}$). Other examples are chitosan-polylysine hydrogel. This type of cross-linking is nonreversible but is easily hydrolyzed due to the weaker ionic interactions between the polymeric networks. This gelation process is rapid or almost instantaneous but requires that everything in the system interact with the aqueous medium (e.g., water). Thus, proper stirring is needed to ensure that the counter-ions are dissolved in the polymeric network to hold the gel structure. This type of gelation mechanism is suited for tissue engineering due to the mild conditions required for the hydrogel to form. Cells can easily tolerate the counter-ions used in this process and it does not produce immediate cytotoxic effects. However, this type of gelation process is hard to control, and the resulting gel may have unsuitable rheological properties and mechanical properties once used in vivo.

3.6.1.3 Complex coacervate-mediated cross-linking

Complex coacervate hydrogels are formed via conjugation of a poly(cation) with a poly(anion) polymer [225, 226]. An example of this conjugation is the combination of polypeptide-based polymer, which, in general, has cationic charges, while polysaccharides are anionic. When combined in an aqueous suspension, complexation between the two types of polymers occurs that results in the formation of a hydrogel. The hydrogel will then have a neutral charge if both polymers have an equal charge ratio. This results in a more neutral environment that is useful for antiadhesive applications, such as in the mitigation of biofilm formation or platelet adhesion. In drug delivery, nanogels prepared via this method can escape opsonization (particle elimination) by avoiding protein corona formation on its surface, thus improving their circulation time in the body.

3.6.1.4 Hydrogen bonding-mediated cross-linking

Hydrogen bonded hydrogels are prepared via the alteration of the aqueous solution pH of the polymers having carboxyl and hydroxyl groups [227]. An example of this cross-linking is carboxy methylcellulose [228, 229] (CMC) network, which can form hydrogels when the pH of the system is reduced to slight acidic conditions. The mechanism of this process relies on the protonation and deprotonation of carboxyl or hydroxyl groups present in the polymeric chain. This, in turn, allows for physical cross-linking of the functional groups in the polymeric chain that results in gelation. This process is reversible, depending on the pH of the system, if the hydrogel is formed by protonation. Deprotonation can then reverse the gelation process, and vice versa. This is applicable in drug delivery, especially in cancer treatment, where a pH gradient is present in the intratumoral environment [230, 231]. Once the hydrogel is protonated, it can dissociate the hydrogen bonding between the hydrogel and subsequently release any drug payload that is entrapped inside its network. Orally formulated peptides encapsulated in this type of hydrogels [232] can also benefit with this reversible gelation mechanism, as the hydrogel can protect its payload without being denatured by the acid in the gastrointestinal tract.

3.6.2 Chemical cross-linking (polymer grafting)

Chemical cross-linking is a method of producing a chemically cross-linked hydrogel using radical polymerization of low-molecular-weight monomers, branched homopolymers, or copolymers in the presence of a cross-linking agent. This method involves the grafting of monomers onto the polymer's backbone or the use of a cross-linking agent to link two polymer chains [233–236]. The cross-linking of both natural and synthetic polymers can be achieved through the reaction between their functional groups (such as COOH, OH, and NH_2) with cross-linkers. This method is known as graft polymerization and allows for the creation of complex heteropolymer hydrogels by imparting different functional groups into the polymer backbone via the attachment of a synthetic monomer onto the natural polymer, and vice versa. The resulting hydrogel may possess mechanical and biological properties that are highly valuable in biomedical applications, examples of which are shear thinning, rheological moduli modulation, and cell or growth factor adhesion sites. Conventional grafting methods use free radical polymerization of a combination of natural and synthetic monomers in the presence of chemical initiators (e.g., potassium persulfate, KPS), whereas other graft polymerization methods can be initiated using methods such as irradiation using microwaves, ultraviolet, gamma rays, and electron beams.

3.6.2.1 Microwave irradiation technique

Microwave (MW) irradiation technique uses electromagnetic waves, called microwaves, for heating, with a short wavelength ranges (between 10^{-4} to 10^{-2} m). This technique utilizes the heat generated by the water molecules present in the monomer suspension. The heat generated by the microwave can initiate the grafting of the monomers onto a polymeric chain by opening the stable bonds on both monomer and polymer molecules and allowing them to conjugate into a single polymeric chain, with side branches containing the monomer graft. The resulting heteropolymer can then form a hydrogel that contains the properties of both molecules [237, 238].

3.6.2.2 Ultraviolet, gamma ray, and electron beam irradiation technique

Ultraviolet is an electromagnetic radiation, with a wavelength shorter than MW and visible light, in the range 10–400 nm. Grafting, using UV irradiation, produces unstable excited species by knocking the free electrons on the polymer backbone via absorption of the ultraviolet radiation [239]. The free radical formed on the polymeric backbone is then attacked by the monomer, which leads to grafting [239]. The same mechanism is used by gamma ray and electron beam irradiation; however, these techniques use a higher electromagnetic wave energy, thus allowing them to be used on polymers with more stable bonds (e.g., carbon–carbon bonds) in their backbone, allowing for a wider array of polymers to be grafted, with functional side branches [240, 241].

3.6.3 Unique hydrogel iterations used in biomedicine

3.6.3.1 Cryogel formation

Cryogels (cryotropic hydrogel) are produced with a broad variety of morphologies and properties, which can be tailored for varied biomedical applications [242, 243]. Cryogels are synthesized by the cryogelation of monomers or polymeric precursors at subzero temperatures [244, 245]. Subzero temperatures can yield a supermacroporous construct that has osmotic stability and mechanical strength. The cryogelation process uses monomers and/or polymers that are mixed in an aqueous solvent and then incubated at subzero temperatures for cryopolymerization and/or gelation, with the objective of creating ice crystals within the system; the ice crystals act as a porogen (pore forming agent) during this process. After the gelation process is complete, they can then be thawed and washed at room temperature to remove unreacted compounds. The ice crystals that are then formed in the system are also melted away, leaving voids in the gel network. In the other methods, cryogels are lyophilized and ice crystals sublimated using a freeze dryer. The resulting cryogel has more flexibility and rapid reversible transformation in 3D microstructures, in response to external factors and outstanding swellability in aqueous media, in contrast to conventional hydrogel preparation [246–248]. It promotes adequate cellular migration and ingrowth, due to its supermacroporosity that can serve as a scaffold in tissue engineering. Porosity in a cryogel can facilitate adequate transport of nutrients and waste material to and from the cells seeded on them [249]. It can also hold more liquid per volume compared to traditional hydrogels, which is highly suited as an in vitro environment for cells.

3.6.3.2 Nanogel formation

Hydrogels in nanoparticle preparations (nanogel) are utilized for systemic drug delivery due to their tunable porous structure that can entrap or conjugate therapeutics at a high efficiency rate [250–252]. Nanogels range from submicron scales down to 20 nanometers, to differentiate them from microgels and in situ bulk forming hydrogels that are used for localized drug delivery. Nanoformulations have salient benefits compared to traditional formulations, as they can increase the pharmacokinetic and pharmacodynamic properties of a drug. They can potentially reduce adverse effects by increasing the biocompatibility of formulations. Since they are made up of polymeric precursors, external stimuli, such as temperature and pH gradients, can facilitate drug release by virtue of swelling, deswelling, erosion, or diffusion mechanisms. Nanogels can be prepared using emulsion polymerization method, when bifunctional monomers suspended in an oil and water phase are incorporated, and polymerization is initiated in these heterogeneous colloidal systems, with the presence of an emulsifier (An emulsifier is an additive that helps two liquids mix, e.g., water–oil emulsions) [252, 253]. Physical and chemical crosslinking are the main mechanisms in the gel formation, but the emulsification of the gel leads to its nanoscale morphology.

3.6.4 Hydrogel preparation challenges

Hydrogels are very versatile and are used heavily in biomedical research as tissue scaffolds, 3D bioprinting bioink, and as a drug delivery vehicle. However, this material has seen relatively low clinical translation due to the challenges in upscaling and mass production of a clinically safe material. Sterilization remains a challenge in producing hydrogels for clinical translation. The traditional sterilization process can alter the gelation properties of polymers and can significantly alter the material properties. Hydrogels must also be prepared almost in situ and proper aseptic handling must be ensured to avoid pathogen contamination.

Gelation rate and rheological properties must also be fine-tuned in order to be applied to the body. Rapid gelation, with very high moduli, limits its application as an injectable material and in 3D printing, due to the high probability of clogging the needle or the nozzle. Most homopolymers do not possess shear-thinning properties; thus grafting with other monomeric materials to alter its rheological properties is needed. It adds to the complexity of the procedure, reverting to the problem of upscaling production to significantly reduce the cost per weight of the material [254].

3.7 Inorganic and organic nanoparticles preparations

Nanoparticles is an emerging class of material used in various biomedical applications, from biosensing, to antibacterial agents, to drug delivery vehicle, to contrast agent, and for even novel therapies, such as hyperthermia. Materials, at nanoscale, have properties that can improve traditional therapies and technologies used in biomedicine. This improvement is because materials at the nanoscale can behave differently in comparison to bulk materials. The physical and chemical behavior of a material depends on its size due to, what is called, "the nanoeffect." The nanoeffect, to put it simply, is a result of the drastically different rules of physics that come into play at the nanoscale, as particles behave differently as the scale of their size increases or decreases. For example, large dust particles are easily collected by the body in the mucosal area of the nasal passage, whereas submicron particles can penetrate deep down to the lower respiratory tract. This is due to the ability of nanosized particles to oppose the van der Waals forces of the nasal hair and the space between the hair protrusions is too large for it to effectively filter the nano-sized dust particles. Even the microvilli of the laryngeal passage are deemed ineffective in blocking the nanoparticles from entering the lower respiratory tract. Understanding this phenomenon, biomedical researchers are using the nanoeffect to activate biological processes that usually are inaccessible with traditional materials. Nanoparticles is an umbrella term for both organic and inorganic nano-sized materials. There is a wide array of nanoparticles and they are classified based on the morphology and material they are made of. The following section of the chapter will enumerate how various nanoparticles are synthesized and how their salient properties act toward the human body.

3.7.1 Organic nanoparticles

Organic nanoparticles are made up of carbon-based materials, e.g., polymer from lipids [255], peptides [256], and saccharides [257, 258]. Different iterations of nanoparticles can be made from these raw materials and they behave differently inside the body; fundamentally, because of *t*. Their ability to circulate inside the body is defined by their surface charge, size, morphology, and molecular weight. Various techniques can influence how these nanoparticles can assemble and entrap synthetic drugs and bioactive molecules (RNA, DNA, and peptides). Most organic nanoparticles are utilized for drug delivery, for treating diseases such as cancer, genetic disorders, vaccines, and infections. They can also be used for diagnosing certain diseases, as nanoparticles can be designed to bind to specific diseased cells.

3.7.1.1 Micelle synthesis

Micelles are nanoparticles made up of amphiphilic molecules that self assemble in an aqueous solution, whereby the polar region faces the outside surface of the micelle and the nonpolar region forms the core. They are useful in drug delivery, as during this self-assembly, they can entrap drug cargo into their core and deliver it under specific biological conditions [259]. Generally, micelles have spherical structures with a hollow core. The polar shell is hydrophilic and makes the micelle water soluble that allows for intravenous administration, while the nonpolar core carries the therapeutic payload [260, 261]. The size, surface charge, dispersity, and morphology of a micelle depends on what kind of amphiphilic molecule it constitutes. Amphiphilic molecules or amphiphiles, used in micelle formations, differ in their origin; they may be a lipid-based molecule, copolymers, or even proteins. Micelles that are made up of these types of amphiphiles have different properties and morphologies. Phospholipids amphiphilic molecules are composed of a glycerol backbone, carrying a hydrophilic polar head group and two hydrophobic fatty acyl tails, while copolymer amphiphiles are generally composed of two distinct polymers conjugated via covalent or ionic bonding, which then form a hydrophobic tail and a hydrophilic end. Examples of copolymer amphiphiles are conjugations of PBO (poly(butylene oxide)), PEO (poly(ethylene oxide)), and PPO (poly(propylene oxide)) to other compounds with opposing polarity [262–265]. Copolymers can also be conjugated with natural organics, such as peptides and/or lipids, if the two components have distinct hydrophilic/phobic groups that form a single molecule. The formation of micelles are facile and do not require extreme conditions to be met once the amphiphiles are exposed to an aqueous media. When their concentration in a single solvent reaches a critical value, the critical micelle concentration (CMC), amphiphiles assemble into structures where the soluble group shields the insoluble group from the surrounding environment. And, with incorporation of therapeutic molecules in that aqueous media, micelles can entrap it with high efficiency [266–268].

3.7.1.2 Liposomes

Liposomes are spherical nano-sized particles that are made up of phospholipids or cholesterol molecules that form a bilayer structure like how a cell membrane forms. They contain a hollow core that could be used to entrap therapeutic cargoes, e.g., peptides, nucleotides (RNA, DNA), and various drug molecules [269–272]. Like micelles, liposomes can self-assemble, given proper conditions. In brief, liposome formation happens when phospholipids, being amphiphilic, impulsively form closed structures when they are hydrated in aqueous solutions [272, 273]. Such vesicles have one or more phospholipid bilayer membranes, and these structures are called lamellae. A liposome may consist of one or more lamellar layers where the polar head of the phospholipid are oriented, both in the interior and exterior of the nanoparticle. However, this structure is not limited to bilayer lamellar configuration, as self-aggregation of polar lipids is dependent on the molecular shape, temperature, and environmental and preparation conditions. This offers researchers various liposome properties modification strategies when targeting various ligands, or the therapeutic payload must be integrated in the colloidal system [274–277].

3.7.1.3 Dendrimer synthesis

Dendrimers are nanoparticles with radially symmetric molecules, homogeneous, and monodisperse structures that, typically, have a symmetric core, an inner shell, and an outer shell. The 3D structure of a dendrimer branches out radially like a tree branch, with nano-sized voids that may be used for carrying therapeutics molecules for drug delivery [278–280]. Dendrimer structure begins with a central atom or group of atoms, labeled as the core. From this central structure, the branches of other atoms called 'dendrons' grow through various chemical reaction pathways. Dendrimers are a combination of molecular chemistry and polymer chemistry. They relate to the molecular chemistry world by virtue of their step-by-step controlled synthesis, and they relate to the polymer world because of their repetitive structure of monomers. The three traditional macromolecular architectural classes (linear, cross-linked, and branched) are broadly recognized to generate rather than polydisperse products of different molecular weights. In contrast, the synthesis of dendrimers offers the chance to generate monodisperse, structure-controlled macromolecular architectures, like those observed in biological systems. Dendrimers are generally synthesized by two approaches [280]. First is called the divergent approach, with the synthesis starting from the core of the dendrimer, to which the branching arms are attached by adding building blocks in a stepwise manner. The other approach is called convergent, where the synthesis starts from the exterior, beginning with the molecular structure that becomes the outermost arm of the final dendrimer. In this strategy, the final generation number is predetermined, necessitating the synthesis of branches of a variety of requisite sizes beforehand for each generation. Dendrimers are quite flexible in their application in biomedicine. They have been explored in local delivery of non-steroidal anti-inflammatory drugs and systemic drug delivery of chemotherapeutics. Furthermore, they have been used in delivering more sensitive payloads, such as gene sequences (DNA and RNA) for gene therapy [281–285]. Dendrimer-based metal chelates can also be used as a magnetic resonance imaging contrast agent [286–288], biosensor component [289, 290], and in novel therapies, e.g., photodynamic therapy (PDT) [291].

3.7.1.4 Supramolecular nanoparticle synthesis

This is an emerging method for combining two molecular species into one supramolecule, forming a nanoparticulate formulation. The basics of this method is the utilization of functional groups that each molecule possesses and conjugating them in a way that each molecule's therapeutic efficacy is not affected [292, 293]. Examples of such a method are the use of phenolics that contain hydroxyl, catechol, and carboxyl groups, and conjugating them with a synthetic molecule that contains thiols, boronic acid, or similar groups. The conjugation between these groups is reversible by changing the pH environment of the system. In brief, the natural and synthetic components conjugate by deprotonating the said functional groups, using a weak base. The functional groups of both components then conjugate via reversible covalent bonding, which can be reversed when the supramolecules are protonated. Addition of a stabilizing trivalent ion is needed to cross-link the multiple supramolecules that have been created, to form a nanoparticle. The advantages of this method are that it can protect both molecules from other external molecules, rendering them ineffective inside the body. For example, bortezomib, a synthetic proteasome inhibitor for cancer therapy can be rendered ineffective by phenolics in the diet of a patient. This happens because the active site of bortezomib is its boronic acid functionality that is targeted by the phenolics when they meet in the blood stream. But by formulating both components in a nanoparticle formulation, it can then be uptaken by cancer cells using endocytosis pathways, subjecting the nanoparticle to an acidic environment, allowing for the reversal of the conjugation of each molecule, and enacting both their therapeutic effects. In contrast to a non-particle supramolecular formulation, the endocytosis uptake pathway is not activated; thus keeping both molecules conjugated and inactive [294, 295].

3.7.2 Inorganic nanoparticles

Inorganic nanoparticles used in biomedicine must be biocompatible, highly stable, and with clearance from the body, without causing adverse effects. Inorganic nanoparticles pertain to metallic- or ceramic-based materials and are used in a wide variety of applications, mostly in drug delivery [296], but also in bioimaging [297, 298] and novel therapies, e.g., hyperthermia [299]. This variety of functions arises from the ability of inorganic nanoparticles to be functionalized through surface modifications via the attachment of therapeutic and targeting compounds. The most common types of metallic nanoparticles used in biomedicine for drug delivery [296, 300], antibacterial agent [301, 302], and contrast imaging [303] are silver, gold, and iron nanoparticles [304], whereas ceramic-based nanoparticles, such as nanohydroxyappatite and beta-tricalcium phosphate are used for bone regeneration [304–307]. The surface properties of inorganic particles are very reactive and due to their nanosize, they can pack molecules in a highly efficient manner. Inorganic nanoparticle synthesis follows a wide range of methods and depending on the application, it may use chemical, physical, and even bio-based methods. Different methods can create different nanoparticle morphologies, such as nanowires [308, 309], nanostars [310, 311], nanospheres [312, 313], and nanosheets [314, 315]. All these morphologies can be achieved by modifying the various parameters in the synthesis process. The following section will briefly discuss these methods and how they can synthesize various types of inorganic nanoparticles.

3.7.2.1 Methods for metallic nanoparticle synthesis

Chemical reduction

Chemical reduction makes use of reagents, such as hydroquinone, dimethylamine borane, sodium borohydride, alkaline solution, and ascorbic acid, to help the reducing of metal ions or nanoparticles, with strong reducing agents yielding smaller nanoparticles compared to weak reducing agents [316–318]. An oxidation–reduction (redox) reaction is a type of chemical reaction that involves a transfer of electrons between two species. An oxidation–reduction reaction is any chemical reaction in which the oxidation number of a molecule, atom, or ion changes by gaining or losing an electron. This method is generally considered facile due to the simplicity of the required equipment (heating plate or mantle, magnetic stirrer, condenser, and glass flasks). Nanoparticles are formed when the reducing agent attacks the metal ions present in the suspension, leaving a nucleation site for like metal atoms to form a particulate morphology. Over time these particles grow, depending on the concentration of the ions present in the suspension or when the reducing agent has been exhausted. Nanoparticles can then be collected using a vacuum filter or centrifuged to let the particles precipitate.

Other oxidation–reduction methods can be achieved by using the electromagnetic spectrum (e.g., UV light, X-ray, gamma ray, microwave, or visible light) [319, 320]. Electromagnetic waveforms can excite the electrons in a suspension, and once a given molecule or atom reaches its excitation energy, the excited electrons can initiate the reduction-oxidation process, subsequently allowing the nucleation of the metal ions present in the suspension of particles to grow. A similar process can also be achieved by passing electrical current on a metal ion suspension, creating an electrochemical redox reaction and the subsequent particle formation.

Microemulsion method

Microemulsions systems are combinations of different immiscible liquids that solubilize using a surfactant, to force a homogeneous and stable state. Both organic and inorganic nanoparticles, micelles and liposomes, use this method, but in inorganic nanoparticle synthesis, the inorganic precursors are suspended in an aqueous media and they are allowed to react with materials in a nonpolar organic solvent [321–324]. Microemulsion method is the chemical reaction between two or more reactants trapped in a microemulsion system, or the reaction between a reactant and a precipitating agent. The aqueous droplets continuously collide, coalesce, and break apart, resulting in a continuous exchange of solution content. This, in turn, results in nucleation and subsequent nanoparticle formation, once the critical concentration is attained. If the concentration or the nanoparticle precursor in the system is too high or too low, it can lead to redissolution of the nanoparticle in the microemulsion system, since it cannot reach the critical size for a nanoparticle to precipitate. As the concentration of the precursor reaches the critical supersaturation value, nucleation occurs, and a stable nanoparticle is formed. This leads to a decrease in the concentration of precursors in the system, and successful nanoparticle formation occurs only between the concentrations, C_{max} and C_{min}. This step is also a limiting factor for particle size growth as more nanoparticles are formed when the C_{min} is reached. Several factors can affect the microemulsion dynamics, and these are the choice of surfactant or co-surfactant, the amount of water in the system, and reagent concentration [325]. Modifying these parameters can create different nanoparticle morphologies and properties.

Bio-based methods

Synthesis of nanoparticles using microorganisms is a relatively new method, but the potential of this method for green synthesis, as an alternative to chemical and physical methods, is outstanding. The current state of the art for inorganic nanoparticle synthesis requires the use of toxic solvents and reagents, and the waste by-products may ultimately contaminate the environment if not disposed properly. Some bacterial species are currently being investigated for their ability to reduce heavy metal ions into nanoparticles [326, 327]. This is a kind of defense mechanism they developed to mitigate the stresses when they encounter metal ions that could disrupt their biological processes, examples of which are *Pseudomonas stutzeri* [328, 329], *Pseudomonas aeruginosa* [330], *Thiobacillus ferrooxidans* [331], *T. thiooxidans* [331, 332], and *Sulfolobus acidocaldarius* [333]. There have been reports that these microbes can reduce ferric ion to its ferrous state when grown on elemental sulfur, as an energy source. *T. thiooxidans* was aerobically able to reduce ferric iron at low pH medium. The ferrous iron formed was stable to autoxidation and *T. thiooxidans* was unable to oxidize ferrous iron, but the bio reduction of ferric iron using *T. ferrooxidans* was not aerobic because of the rapid bacterial reoxidation of the ferrous iron in the presence of oxygen.

All these processes are carried out intracellularly, and different bacterial species can create a variety of size, morphological, and crystalline structure, depending on the conditions they encounter. Although the main mechanisms for controlling a precise nanoparticle property remains a challenge in this new synthesis method, once the nanoparticle is created, it can be harvested by killing the bacterial species and collecting the metal supernatant using various isolation techniques. The bacterial suspension can be heated and washed using deionized water to remove the organic components, and the remaining inorganic nanoparticles can be centrifuged or magnetically separated in the suspension.

Another green synthesis method is the use of plant phenolics as the reduction agent for metal ions. Phenolics have functional groups that can be easily deprotonized using alkali media, thus allowing metallic ions to nucleate and form into particles. The advantage of using phenolics is that they are nontoxic and require less purification and washing for the nanoparticles to become biocompatible. Furthermore, green synthesis using plant phenolics is more efficient, simpler, and economical, and can be easily scaled up to perform larger operations, in comparison to bacterial methods, as it does not require high temperatures, complex bioreactors, and does not produce toxic by-products. Lastly, biomass from plant materials is also easily accessible to most parts of the world.

3.7.2.2 Methods for ceramic nanoparticle synthesis

Sol–gel method

Synthesis of nanoceramics using sol–gel process (also known as chemical solution deposition) involves a chemical solution (sol) in liquid phase and a precursor (gel), where the two components are mixed together to form a homogeneous solution. They are then allowed to react (condensation and gelation reactions) to form a gel-like network that contains both a liquid and a solid phase [334–336]. The liquid phase is then allowed to evaporate, and the remaining solid material is then collected to be used in various biomedical applications. In nanoceramic manufacturing, the precursors are usually made up of calcium, phosphorus, and silica-based materials, with the intent of using in bone tissue engineering. Further applications of these nanoparticles have diverged into drug delivery, as mesoporous morphology can be easily achieved using the sol–gel method [337, 338]. Mesoporosity creates a high surface area-to-volume ratio and can be used to entrap therapeutics for different treatment applications [339, 340]. These mesoporous morphologies that are built from the sol–gel processes possess ordered structures of subnanometers to tens of nanometers in scale and are obtained via templating a sacrificial material on the gel network. The resulting solid phase is sintered to leave nano-sized voids and this creates the mesoporosity on the nanoparticles that are then utilized for either drug delivery or loading of growth factors for bone tissue engineering.

Sintering method

Sintering is a process where a material is subjected to heating cycles. Heat is gradually applied over a period and held for a specific amount of time, and then allowed to cool down to a specific temperature [341]. Sintering is usually done in a controlled chamber using noble gas atmosphere to prevent the oxidation of the material and, thus, complete pyrolysis. In ceramic nanoparticle synthesis, sintering usually follows the sol–gel method to deliberately remove the organic template that is used in the reaction [342]. This, in turn, will leave the ceramic material due to its high heat stability. The sintering process may also increase the density of the ceramic, making it more stable in aqueous media and prolong its stability inside the body. Other ways of doing heating cycles is by using electromagnetic radiation, such as microwave [343]. Radiation is caused by a magnetron and this vibrates the as-prepared nanopowders, heating it at a controlled rate, depending on the energy flux of the electromagnetic waves. This method allows for heat to be created directly from the material of interest, instead of having transferring heat from an external source.

Microfluidic-mediated synthesis of nanoparticles

Microfluidic systems can also be used in preparing various types of nanoparticles (polymeric, metallic, and ceramic) in a highly controlled way [344–346]. Nanoparticle sizes can be easily controlled by varying the flow parameters of different reactants inside the microfluidic channel. The salient point of this method is its ability to regulate the fluid low in microchannels, based on diffusion. Laminar flow is the common flow pattern due to the low Reynold's numbers of the reactants inside the channel. This, in turn, allows rapid mass, heat, or redox reactions in the reactants, creating a high throughput and a faster reaction process, compared to conventional methods. There are two types of microfluidic-mediated synthesis, depending on the flow pattern inside the device. These are single-phase continuous flow synthesis [347, 348] and emulsion (2-phase) microdroplets/segmented flow synthesis [349–351]. Continuous flow synthesis mix and react reagents in microchannels under diffusion-based laminar flow reaction conditions. Reaction times, temperatures, mixing efficiency, and reagent concentrations are parameters to control the particle quality. Nanoparticle synthesis takes place in micro droplet reactors to improve the mixing efficiency in microfluidic channels, and further reduces the particle size distribution. Particle size distribution is a crucial parameter on how effectively a method of synthesis can create reproducible particles at any given condition, and microfluidics can narrow down this distribution further by allowing for rapid reaction times, due to the close proximity and controlled mixing of reactants inside the microchannels. In the metallic nanoparticle synthesis using microfluidics, the nucleation process for metal ions that happens is mediated by the feed of metal ions, thus allowing for a controlled nucleation growth, based on the concentration of ions present in the continuous flow. Particle growth can be prolonged or shortened based on the flow parameters, creating a uniform particle size yield, downstream. Furthermore, this method relies on the ability of the microfluidic channel to control the temperature along the flow, rapid mixing, and allowing additional reagents to be added in order to create particles with different morphologies. Thus, a temperature control system, multiple mixing channels, and input ports might be added in the microfluidic device to allow modification on nanoparticle properties.

3.7.3 Challenges in nanoparticles synthesis

Although nanoparticles have very high upsides in biomedicine, in terms of solving numerous problems in drug delivery, biosensing, and improving biomedical device performance, it has seen relatively low clinical translation due to the challenges it faces when it comes to upscaling and reducing the cost of manufacturing. Nanoparticle synthesis used in drug delivery, for example, needs to pass numerous milestones, such as efficacy and toxicology studies, before it can reach the clinic. Synthesizing robust and identical nanoparticles, with narrow polydispersity index, has been quite a challenge due to the relatively high cost of production and low yield per synthesis. This may lead to manufacturing batch variations, which is a highly controlled regulatory standard in pharmaceuticals production. High throughput systems in the synthesis of nanoparticles is still a relatively young field and requires rigorous optimization.

There is still a large gap between academic and industrial research in the field of nanomedicine due to the key challenges in cost-effective manufacturing. Furthermore, stability of nanoparticle systems in drug formulation remains a challenge, as some classes of nanoparticles require special cold chain storage to maintain their stability, severely limiting their widespread distribution. This is especially seen in micellar, liposomal, and dendrimer nanoparticles as they have special requirements in order to maintain their morphological stability and functionality. In terms of complexity of synthesis methods, there is an inverse correlation between functionalities, such as the ability of the nanoparticle to bind to specific cells (targeting) or high drug encapsulation (porosity and surface functionality), to the simplicity on how the nanoparticles are made. Simple or facile methods usually yield just a limited functionality, whereas a highly attractive nanoparticle system, with multifunctionality, requires multiple steps in order to yield a few milligrams or grams. At the end of the day, there should be a bridge to connect this simplicity of methods to the high functionality and efficacy of a nanoparticle system for this class of biomaterials to be successfully translated into the clinic.

3.8 Bioceramic synthesis methods

In biomedicine, hydroxyapatite, beta-tricalcium phosphate, silica or Bioglass®, and composites of these are mainly usually used for bone regenerative prosthesis, fillers, and scaffolds. Depending on the synthesis methods used to create these materials, several properties, such as crystallinity, dissolution or degradation rate, mechanical properties, may be fine-tuned with regard to the final application. Bioceramics can be made with various processes, mainly wet chemical precipitation, sol–gel, hydrothermal, solid-state reactions, and conventional melting-quenching processes. Following these processes, sintering is commonly done to change the crystalline phases and the grain structure of the material. These processes will be discussed in the following section of the chapter.

3.8.1 Wet chemical precipitation synthesis of bioceramics (hydroxyapatite)

Wet chemical precipitation is used in the preparation of Hydroxyapatite (HA) due to its simplicity and the relative ease of setup. It also does not cause toxic by-products, leading to a more biocompatible yield. In brief, precursor materials, such as calcium hydroxide and phosphoric acid, are both used as calcium and phosphate source for hydroxyapatite [352–355]. Both precursors are dissolved using a suitable solvent (usually distilled water) and mixed continuously, and a neutralization reaction occurs between these precursors, yielding calcium phosphate solid particulates and more water molecules. The particulates then precipitate in the suspension, collected using a vacuum filter, and then allowed to dry in an oven. Alteration to the processing parameters, such as temperature and molar concentrations, can yield different Ca/P powder morphologies and ratios, respectively. After producing the powder, they can then be sintered using high heat (1,250 °C) under argon atmosphere, to vary the crystallinity and grain structure of the powder. This sintering process can change the density and solubility of the HA product. Wet chemical precipitation is considered. For the bioceramic called HA, there is a strict ratio requirement for the resulting powder. However, when it is set at Ca/P 1.57, 1.67, and 1.87, these are called calcium-deficient HA, stoichiometric HA, and calcium-rich HA, respectively. These types of HA have different properties, such as decomposition stability and mechanical properties. Both calcium-deficient HA 1.57 and stoichiometric HA 1.67 are highly stable under high sintering temperatures (1,300 °C), while calcium-rich HA 1.87 can decompose to calcium oxide under the same sintering conditions. Stoichiometric HA is considered to have the highest mechanical properties (hardness and fracture toughness) and density after sintering at 1,100–1,150 °C.

3.8.2 Sol–gel method in the preparation of bulk porous bioceramic

Sol–gel method has been described in the previous section of bioceramic nanoparticle synthesis and the process is similar to what has been described [356]. In brief, suitable solvents are used to dissolve the gelling precursors that contain calcium, phosphate, and/or silica source materials. The resulting solution creates a thick homogeneous gel network, with the solvent allowed to evaporate. The solid gel network can then be further processed using sintering to remove the organic gel incorporated in the network, leaving spaces that create hierarchical porosity in the inorganic ceramics. The resulting bulk bioceramic material is then used for various bone regenerative applications and, due to its porosity, allows for efficient attachment of cells and further migration deep into its geometry. The porosity also allows for increased surface area-to-volume ratio, which is beneficial for incorporating other drug or growth factors into the bioceramic's surface, making for increased bioactivity and/or biocompatibility. The porous bioceramic is also known for its flexibility, with increased deformation capacity, as it can act as a sponge-like material, highly suited for load-bearing applications. It can also further mimic the bone's natural biomechanical properties, lessening the effect of stress-shielding that may reduce the osteoblast and osteoclast activity around the material. The porous morphology efficiently allows for transfer of nutrients and waste products to and from the migrating cells, which further increases the viability and survivability of nascent cells. Hence, the sol–gel method is highly suitable for creating bioceramics for bone defect repairs, in comparison to other methods that lack the capability to create porosity in the bulk material.

3.8.3 Hydrothermal method for creating calcium oxide

Hydrothermal method explores the possibility of using natural sources of calcium from animal by-products, such as eggshells and seashells, and large animal bones from cattle and swine [357, 358]. This method allows for upcycling of these by-products into biomaterials. In this method, the animal sourced-materials are cleaned thoroughly with deionized water to remove contaminants and are boiled to allow for the removal of organic materials, such as proteins and fats that are incorporated with the animal products. Afterwards, the excess water is removed and the animal products are placed in a furnace to undergo heating cycles. The heating cycles are repeated, usually at least 3 times, to further thermally decompose the organics and remove the excess moisture in the products. The heating cycles are incrementally increased from 0 to 450 °C (stage 1), 450 to 600 °C (stage 2), and 600 to 900 °C (stage 3), with varied heat-soak times and allowed to gradually decrease in temperature, at a specific cooling rate, before the next cycle begins. The repeated heating cycles ensure that only the thermally stable components of the animal by-products remain, which are the calcium oxide. The resulting calcium oxide is then used as a precursor for making hydroxyapatite or other calcium-containing bioceramic, using other chemical techniques.

3.8.4 Solid-state reactions in the preparation of synthetic apatite

The solid-state reaction is a way of creating synthetic apatite materials by mix-grinding dry powder precursors, such as beta-tricalcium phosphate powder (TCP), calcium hydroxide ($Ca(OH)_2$), or calcium carbonate ($CaCO_3$) or a combination thereof, with natural sources [359–361]. This method can yield a calcium/phosphorus molar ratio between 1.5 and 2.0, and a particle size of less than 10 µm. The resultant powder is subjected to temperature between 600 and 1,250 °C in atmosphere or in a controlled atmospheric condition, usually with the use of noble gases, to avoid the introduction of oxygen that can decompose the material. The high temperature forces the grinded powder precursors to react with each other, by thermally decomposing the hydroxyl and carbonate species from the precursors and leaving the thermally stable elements to react, to form Ca/P powders. Other organic components are also decomposed by this step, resulting in a yield of high purity. Solid-state reaction is considered simpler compared to the conventional wet chemical precipitation method due to the noninvolvement of any aqueous or liquid solvent, removing the extra step of vacuum filtering and separate drying of the precursors. The dry-mixing method is a way to increase the throughput production of synthetic apatite, with high reproducibility, with less time, energy, and cost, giving it a high industrial value.

3.8.5 Melting-quenching method for bioglass preparation

Melt-quenching for bioactive glass preparation is a technique similar to traditional glass making, as it involves fusing a mixture of anhydrous powder material (calcium carbonate, silica dioxide, sodium carbonate, and sodium phosphate) and fusing them in a crucible using high temperatures (1,150–1,650 °C), with specific chemical compositions of 45% SiO_2, 24.5% Na_2O, 24.5% CaO, and 6% P_2O_5. The resulting melted materials are then either cast into a mold or quenched in a water bath to form what is called glass fritz [362–364]. These glass fritzes vary in particulate sizes and, subsequently, grinding using a ball mill is required to obtain uniform particulates. Grinding is usually done with acetone or any other high evaporation rate solvent to form a slurry so that the particles will not dissipate in air and increase the yield of the bioglass powder. The resulting slurry mixture is then air dried, and can be further made into other bulk preparations, such as for coatings of bone prosthesis, to increase osteoactivity or, using the sol–gel method, form a porous scaffold for bone defect repair, and even as a bone filler paste, if incorporated with a polymeric component. In casting preparation, the piece should be annealed under 500–700 °C for at least 4 h so that the internal structure of the glass is relaxed, with all internal mechanical tension and flexion being dissipated. Different ratios of precursors, annealing, and, grinding parameters can yield bioglasses with varied properties, depending on the desired application.

3.8.6 Challenges in bioceramic synthesis and manufacturing

Bioceramics have seen great success in translation to the clinic, as it is one of the oldest class of biomaterials and have been developed over a longer period, in comparison to other classes of materials. Nevertheless, its manufacturing still comes with challenges, as newer generation of bioceramics are needed to increase its functionality – not just by being osteoconductive but also by being able to perform other functions in the body. Recent applications of bioceramics now encompass drug delivery, as it can form a mesoporous morphology that is highly suited in entrapping therapeutics for cancer treatment and other disease areas. However, there is still a need to create bioceramics without the use of high temperature sintering or annealing, as it adds to the cost and time to produce the material. The sintering process ensures that the resulting bioceramic is pure, as it burns away any contaminants and unreacted precursors and solvents in the process. However, this process is time-consuming and requires a sensitive setup, which adds to the cost of material production. There are new emerging techniques that avoid further post-fabrication process steps, where the resulting bioceramic powder (e.g., hydroxyapatite) can be made ready, using SLS that allows for a rapid fusing of bioceramic precursors into 3D geometric shapes, which can be used in biomedical applications. It has possibility in personalized medicine, as the design of these geometric shapes can utilize scans from the clinic using magnetic resonance imaging or computed tomography, and render a specific shape tailored for each patient. Still, this method of additive manufacturing of bioceramics is still in its infancy and has a long way for it to be seen in the clinic.

3.9 Fabrication methods of biomedical-grade alloys

Biomedical-grade alloys are crucial in the design and development of implantable biomedical devices, such as stents, bone prosthesis, birth control implants, pacemakers leads, heart valves, etc. They are crucial in maintaining a mechanically stable device that requires long term lifespan, as these biomedical-grade alloys are known to be inert and have the capability to stay inside the body without causing any adverse effects. For an alloy to be biomedical grade, it must have a passive surface that does not cause severe local or systemic inflammatory response when interfaced with a biological element –an exemption to this are birth control intra uterine devices that stimulate an inflammation response in the fallopian tube to block the passage of both sperm and ovum. Alloys of stainless steel [365], titanium [366, 367], nickel, cobalt [368], chromium, and copper are primarily used for the said applications [369, 370]. In general, the primary process of metals includes processing the ingot to mill products in the case of wrought alloys, and casting process in the case of cast alloy. In addition, the primary products of metals can also be produced by powder metallurgy. The following section will describe the primary fabrication process of these metal ingots to form alloys that are used for various biomedical devices.

3.9.1 Thermomechanical process (TMP)

Thermomechanical process (TMP) is the most common preliminary way to convert metal ingots into mill products. TMP's main functions are to produce specific shapes of metals into slabs, blooms, and billets, and to enhance the mechanical properties of the initial raw ingot materials by grain size refinement and the production of a more uniform microstructure [371]. Forging is typically the first process prior to other TMPs. The selection of the type of TMP, after forging, depends on the mill product to be produced – billet, bar, plate, or sheet. Temperature selection varies on the type of alloy to be created [371, 372]. It is a crucial step, as some metal alloys, such as 304 stainless steel, can form instability bands when the TMP temperature are below 1,100 °C. The ideal processing method for this type of alloy must cross 1,200 °C in order to create a uniform grain structure, free from weak points [373]. This is due to the different melting and welding temperatures of whichever alloy components are selected. If the temperature is lower than the welding or melting temperatures for both alloy components, the metals can phase-separate and form instability bands that are focal points for breakage when a load or tension is applied. Other important parameters are the degree of deformation and phase composition compatibility of a specific metal. It was reported that increase in the degree of plastic deformation during the forging of dual phase Ti alloys results in lower fatigue strength due to the incompatibility of the crystallinity phases of the different titanium. It is known that different phases of metals when combined together have different ability to be deformed; therefore, different phase compositions will have different performances during and after TMP.

3.9.2 Hot iso-static pressing (HIP)

One easy way to form metallic alloys into various shapes is by the use of investment casting; however, biomedical-grade alloys, such as titanium, do not allow for an easy casting process due to their high reactivity to the atmosphere and the casting mold when they are melted. However, a technique called hot iso-stactic pressing (HIP) allows for casted metals to be processed, after casting, to consolidate their grain structure and form a dense material [374–376]. This process improves the material's mechanical properties and workability by applying a uniform pressure in all directions under high heat, which a normal investment mold does not allow. The HIP process subjects a component to both elevated temperature and isostatic gas pressure in a high-pressure containment vessel. The pressurizing component uses an inert gas so that the material does not chemically react, and maintains purity. When castings are treated with HIP, the simultaneous application of heat and pressure eliminates internal voids in the material via mechanisms of plastic deformation, creep, and diffusion bonding; subsequently improving the fatigue resistance of the alloyed material. HIP is also used as part of a sintering (powder metallurgy) process and for the fabrication of metal matrix composites, as a part of post-processing in additive manufacturing [377, 378].

3.9.3 Powder metallurgy (sintering)

This process includes blending and mixing of powdered metals that are compacted using a dye and subjecting the material to a sustained controlled heating and cooling cycle to fuse the powdered material into a consolidated form [379, 380]. This process allows porous materials to form [381]; therefore, it is suitable in manufacturing implants, such as femoral stems of total hip prostheses, and other bone interfacing devices to anchor bone tissue to the implant [382]. Porosity can be managed by controlling the sintering parameters, such as heating rate, heat-soaking temperatures, and peak temperature. As mentioned above, sintering is usually followed with a HIP process to consolidate the powder ingredients. Furthermore, improvement in the mechanical properties of implant can be achieved by conducting a HIP process after sintering, by decreasing defects such as gas or shrinkage pores [379]. One disadvantage of sintering is that it is a relatively expensive process compared to other methods due to the extended time of heating and cooling cycles, thus requiring more energy and the use of an inert atmosphere to maintain the purity of the material.

3.9.4 Superplastic deformation and isothermal forging

Superplastic deformation (SPD) is an advanced forming process of metallic implants, where a higher degree of deformation is applied to form a complex product shape, whereas a low rate forming process is required [369, 370, 383]. With this method, the use of dual phase materials is possible as long as the materials have ultra-fine grain structure. This superplasticity property comes from the dynamic recrystallization-assisted grain boundary sliding, where different rates of sliding for the different types of grain boundary are required to achieve an optimum superplasticity. For an ultra-fine grain structure to be achieved, several methods of severe plastic deformation processes are carried out [384], such as equal-channel angular pressing (ECAP) [385, 386], accumulative roll bonding (ARB) [387, 388], high-pressure torsion (HPT) [389, 390], that enable multiple layering and bonding between the different phase layer structures to cohesively bind to each other. Currently, superplasticity is used for SPD and diffusion bonding processes, which are suited for metal implants that require compliance, such as in various types of stents.

Another advanced process of creating metallic implants is by isothermal forging [370]. Here, the dies are kept at higher temperatures, thereby reducing the die-chill effect and increasing the metal flow. A relatively low strain rate condition is required to create superplasticity properties; thereby creating uniform deformation after this process. Isothermal forging creates uniform microstructure and longer lifetime of dies, and reduces the process step to obtain near net shape of the product. One disadvantage of this method is its high production cost, due to the maintenance of high temperatures during forging. Otherwise, the net shape process is directed metal deposition, which utilizes focused laser beam to melt the metal powder on the metal substrate plate [391, 392]. This process reduces the cost of production of metallic alloys, such as titanium, saving the material utilized through the process. This material saving, by focused laser melting, is quite useful in complex geometric shaped implants, as the laser can be programmed to only interact with materials of interest in the substrate plate [393].

3.9.5 Elastic metal alloy manufacturing (nitinol)

Nickel–titanium alloy has been a game changer for multiple applications, especially in biomedical devices. Nitinol has been described in the previous chapter, and fabricating it is extremely costly and difficult, due to the high reactivity of titanium. Titanium combines readily with both carbon and oxygen atom diffused in the environment, thus lowering the transformation temperature of the billet lower and not allowing the nickel atom to be incorporated in the lattice. Thus, a highly controlled smelting environment, without the presence of oxygen or carbon, must be ensured to allow for nickel to be alloyed with titanium [394–396]. There are two main methods of achieving this. First is vacuum arc remelting (VER), where it is done by striking an electrical arc between the raw material and a water-cooled copper strike plate. Melting is carried out in high vacuum and the mold itself is water-cooled copper. Second is vacuum induction melting (VEM), by using an alternating magnetic field to induce heat in the ferromagnetic metals. Both methods maintain low atmospheric pressure, removing all possible elements that can react with titanium. Other methods are also used on a boutique scale, including plasma arc melting [397, 398], induction skull melting, and e-beam melting [399, 400]. Physical vapor deposition is also used on a laboratory scale. Nitinol is easy to form while it is hot, but when it is allowed to cool down, it has the tendency to show increased frictional resistance and tool wear due to the elasticity of the alloy. Machining nitinol into a shape is also difficult due to this reason, and due to the poor thermal conductivity of the alloy, it is difficult to cool down [401]. However, other advanced machining tools, such as electrical discharge machining (EDM) and laser cutting, have made it possible to make different nitinol products, such as tubes, wires, and plates [402]. Furthermore, heat treatment of nitinol is delicate, as it has the property of being a shape memory alloy, using heat as the trigger. It is vital to fine-tune the transformation temperatures of the alloy. This is done by aging the billet so that Ni-rich phases can precipitate in the billet, depleting the matrix of nickel. When this is properly done, the resulting alloy can have an increased transformation temperature, allowing it to be formed into various shapes, without the worry of reverting into its preprogrammed shape.

3.9.6 Post-processing of metallic alloys to be used in biomedicine

After alloying metals for biomedical applications, further processing is usually required to obtain the medical-grade qualification. The first and the most common is coating the metal to increase biocompatibility or obtain bioactivity. Different polymeric coating techniques were described in the early parts of this chapter. However, there are still some techniques that remain undescribed, which are related solely for metallic implants. First is plasma spray coating [403–407], to endow reactive functional groups on the surface of the alloy. This allows for other coatings to have increased adhesion or bonding strength on the surface of the metal by virtue of hydrogen bonding. This technique is used especially in Titanium alloy for bone implants, if bioceramic coatings are needed to increase its osteogenic activity [407]. The alloy is sometimes pretreated with sand blasting to increase its roughness and surface area, to enhance the bonding strength between the bioceramic and metal. Another way of post-processing the alloy is through electropolishing, to further improve its corrosion resistance [408, 409]. This method uses an electrochemical process to remove material from a metallic piece. This, in turn, reduces the surface roughness by levelling micro-peaks and valleys, improving the surface finish. An electropolished metal has increased corrosion resistance, by reducing its surface area that interacts with a corrosive substance. One way to increase corrosion resistance using electrochemical process is by anodization [410, 411]. This process involves thickening of the natural oxide layer of the metal to protect it from corrosive substances in the body or the environment. It also enhances the aesthetic finish of the metal, improving its scratch resistance, which are beneficial for metal implants that are subjected to both constant frictional forces and load.

3.9.7 Challenges in metal fabrication for biomedical applications

Metal fabrication is one of the oldest trades in the world, and has seen tremendous advancement since the industrial revolution, when we have finally harnessed energy to melt metals to combine them with other elements, to improve their strength and endow them with properties that are conventionally not seen in metals (elasticity, increased toughness, corrosion resistance, etc.). One of the main challenges for metallic alloys is enhancing their corrosion resistance overtime and designing a method to customize implants for patients. Creating or shaping biomedical-grade alloys is deemed difficult and costly if it has to cater to specific needs of a patient. There is a growing need for specialized implants for bone regeneration, and usually, most metal implants come as prefabricated one-size-fits-all, which have the need for either reshaping of the metal implant prior implantation or sectioning parts of the bone so that the implant can fit perfectly at the implantation site. Additive manufacturing, using emerging techniques such as 3D metal printing [412–414], can be a solution to this problem, as it can manufacture customized implant shapes using scans from conventional imaging techniques and transferring those scans into a 3D model, so that a customized implant can be made. This technique also saves on cost and energy, as it only uses materials needed for the final shape of the implant, lessening waste in using raw ingots and shaping the alloy into the desired shape by forging. However, this method is still in its infancy and we still haven't seen any clinically translated 3D manufactured implants to date.

Perhaps the most important challenge for alloying is to ensure that the fatigue resistance of the resulting metal is increased and that it would survive the intended lifespan of the implant. Controlling the parameters of smelting, forging, and heat treatment ensures that the crystalline lattice of the resulting alloy fully incorporates all the required elements and the alloy is cohesively bonded to each other, avoiding weak points in the resulting alloy. However, with biomedical-grade alloys, this entails that the regular smelting and forging process is not enough and vacuum, inert gases, or high temperatures are incorporated into the fabrication design, increasing the cost of production.

3.10 Conclusions

Various synthesis and fabrication methods for different classes of biomaterials have been developed over the years. Since biomaterials is a relatively young field of science, it relies heavily on different scientific fields to improve the synthesis or fabrication methods. It combines different disciplines, from chemistry, biology, metallurgy, and engineering, where advancements in each contribute to the general field of biomaterials. The focus of biomaterialists is to produce materials with biocompatibility, bioactivity, and/or bioinertness that can be used in the development of biomedical devices, drug delivery platforms, tissue scaffolds, etc. Thus, knowing and understanding these different fields is a requisite for delivering an effective biomaterial. Improper synthesis methods can have a severe impact on the resulting material, as impurities, contaminants, and/or defects may arise and create adverse events for the end user of the biomaterial. Furthermore, understanding where to source the precursors or raw materials and how to mitigate the environmental impact of a synthesis process is crucial in order to maintain sustainability and environmental friendliness of the field. Lastly, understanding each of the challenges that the synthesis methods are facing will give rise to new and novel ways in creating sustainable, reliable, and effective biomaterials.

References

[1] Sarkar K, Gomez C, Zambrano S, Ramirez M, de Hoyos E, Vasquez H, et al. Electrospinning to forcespinning™. Materials Today. 2010;13(11):12–14.

[2] Bhardwaj N, Kundu SC. Electrospinning: A fascinating fiber fabrication technique. Biotechnology Advances. 2010;28(3):325–47.

[3] Subbiah T, Bhat GS, Tock RW, Parameswaran S, Ramkumar SS. Electrospinning of nanofibers. Journal of Applied Polymer Science. 2005;96(2):557–69.

[4] Bognitzki M, Czado W, Frese T, Schaper A, Hellwig M, Steinhart M, et al. Nanostructured fibers via electrospinning. Advanced Materials. 2001;13(1):70–72.

[5] Beachley V, Wen X. Effect of electrospinning parameters on the nanofiber diameter and length. Materials Science and Engineering: C. 2009;29(3):663–68.

[6] Jacobs V, Anandjiwala RD, Maaza M. The influence of electrospinning parameters on the structural morphology and diameter of electrospun nanofibers. Journal of Applied Polymer Science. 2010;115(5):3130–36.

[7] Haider A, Haider S, Kang I-K. A comprehensive review summarizing the effect of electrospinning parameters and potential applications of nanofibers in biomedical and biotechnology. Arabian Journal of Chemistry. 2018;11(8):1165–88.

[8] Salas C, Thompson Z, Bhattarai N. * North Carolina State University, Raleigh, NC, United States,† North Carolina A&T State University, Greensboro, NC, United States. Electrospun Nanofibers. 2016:371.

[9] Gibson P, Schreuder-Gibson H, Rivin D. Transport properties of porous membranes based on electrospun nanofibers. Colloids and Surfaces A: Physicochemical and Engineering Aspects. 2001;187:469–81.

[10] Zhang Y, Yang B, Li K, Hou D, Zhao C, Wang J. Electrospun porous poly (tetrafluoroethylene-co-hexafluoropropylene-co-vinylidene fluoride) membranes for membrane distillation. RSC Advances. 2017;7(89):56183–93.

[11] Goh Y-F, Shakir I, Hussain R. Electrospun fibers for tissue engineering, drug delivery, and wound dressing. Journal of Materials Science. 2013;48(8):3027–54.

[12] Cui Y, Huang Z, Lei L, Li Q, Jiang J, Zeng Q, et al. Robust hemostatic bandages based on nanoclay electrospun membranes. Nature Communications. 2021;12(1):1–11.

[13] Do Pham DD, Jenčová V, Kaňuchová M, Bayram J, Grossová I, Šuca H, et al. Novel lipophosphonoxin-loaded polycaprolactone electrospun nanofiber dressing reduces Staphylococcus aureus induced wound infection in mice. Scientific Reports. 2021;11(1):1–15.

[14] Jirofti N, Golandi M, Movaffagh J, Ahmadi FS, Kalalinia F. Improvement of the wound-healing process by curcumin-loaded chitosan/collagen blend electrospun

nanofibers: In vitro and in vivo studies. ACS Biomaterials Science & Engineering. 2021;7(8):3886–97.
[15] Xi K, Gu Y, Tang J, Chen H, Xu Y, Wu L, et al. Microenvironment-responsive immunoregulatory electrospun fibers for promoting nerve function recovery. Nature Communications. 2020;11 (1):1–18.
[16] Li Z, Liu Q, Wang H, Song L, Shao H, Xie M, et al. Bladder acellular matrix graft reinforced silk fibroin composite scaffolds loaded VEGF with aligned electrospun fibers in multiple layers. ACS Biomaterials Science & Engineering. 2015;1(4):238–46.
[17] Boudriot U, Dersch R, Greiner A, Wendorff JH. Electrospinning approaches toward scaffold engineering – a brief overview. Artificial Organs. 2006;30(10):785–92.
[18] Park S-H, Kim MS, Lee B, Park JH, Lee HJ, Lee NK, et al. Creation of a hybrid scaffold with dual configuration of aligned and random electrospun fibers. ACS Applied Materials & Interfaces. 2016;8(4):2826–32.
[19] Diculescu VC, Beregoi M, Evanghelidis A, Negrea RF, Apostol NG, Enculescu I. Palladium/palladium oxide coated electrospun fibers for wearable sweat pH-sensors. Scientific Reports. 2019;9(1):1–12.
[20] Shepherd LM, González E, Chen EX, Frey MW. Increasing stability of biotin functionalized electrospun fibers for biosensor applications. ACS Applied Materials & Interfaces. 2017;9 (2):1968–74.
[21] Liu Y, Hao M, Chen Z, Liu L, Liu Y, Yang W, et al. A review on recent advances in application of electrospun nanofiber materials as biosensors. Current Opinion in Biomedical Engineering. 2020;13:174–89.
[22] Han D, Gouma P-I. Electrospun bioscaffolds that mimic the topology of extracellular matrix. Nanomedicine: Nanotechnology, Biology and Medicine. 2006;2(1):37–41.
[23] Sell SA, Wolfe PS, Garg K, McCool JM, Rodriguez IA, Bowlin GL. The use of natural polymers in tissue engineering: A focus on electrospun extracellular matrix analogues. Polymers. 2010;2(4):522–53.
[24] Xu C, Inai R, Kotaki M, Ramakrishna S. Electrospun nanofiber fabrication as synthetic extracellular matrix and its potential for vascular tissue engineering. Tissue Engineering. 2004;10(7–8):1160–68.
[25] Leach MK, Feng Z-Q, Tuck SJ, Corey JM. Electrospinning fundamentals: Optimizing solution and apparatus parameters. JoVE (Journal of Visualized Experiments). 2011(47):e2494.
[26] Rodoplu D, Mutlu M. Effects of electrospinning setup and process parameters on nanofiber morphology intended for the modification of quartz crystal microbalance surfaces. Journal of Engineered Fibers and Fabrics. 2012;7(2):155892501200700217.
[27] Theron S, Zussman E, Yarin A. Experimental investigation of the governing parameters in the electrospinning of polymer solutions. Polymer. 2004;45(6):2017–30.
[28] Ding B, Wang X, Yu J. Electrospinning: Nanofabrication and Applications: William Andrew; 2018.
[29] Fong H, Chun I, Reneker DH. Beaded nanofibers formed during electrospinning. Polymer. 1999;40(16):4585–92.
[30] Liu Z, He C-H, Zhang S-Z, Zhao L. A mathematical model for the formation of beaded fibers in electrospinning. Thermal Science. 2015;19(4):1151–54.
[31] Zuo W, Zhu M, Yang W, Yu H, Chen Y, Zhang Y. Experimental study on relationship between jet instability and formation of beaded fibers during electrospinning. Polymer Engineering & Science. 2005;45(5):704–09.
[32] Li Z, Wang C. Effects of Working Parameters on Electrospinning. One-Dimensional Nanostructures: Springer; 2013. p. 15–28.

[33] Wang J, Yao S, Yu S, Huang Y, Liu W, Zhu M. Polymer templates effects on microstructure and mechanical properties of electrospun mullite nanofibers. Ceramics International. 2021.
[34] Odian G. Principles of Polymerization: John Wiley & Sons; 2004.
[35] Terayama K, Liu C-W, Higashihara T. Catalyst-transfer system in stoichiometry-independent AA+ BB-type Migita–Kosugi–Stille coupling polycondensation using ester-functionalized dibromo monomer. Polymer Journal. 2021:1–8.
[36] Sifri RJ, Padilla-Vélez O, Coates GW, Fors BP. Controlling the shape of molecular weight distributions in coordination polymerization and its impact on physical properties. Journal of the American Chemical Society. 2019;142(3):1443–48.
[37] Oelmann S, Solleder S, Meier M. Controlling molecular weight and polymer architecture during the Passerini three component step-growth polymerization. Polymer Chemistry. 2016;7(10):1857–60.
[38] McAfee T, Leonardi N, Montgomery R, Siqueira J, Zekoski T, Drenski MF, et al. Automatic control of polymer molecular weight during synthesis. Macromolecules. 2016;49 (19):7170–83.
[39] He B, Li J, Pan Z. Morphology and mechanical properties of MWNT/PMIA nanofibers by electrospinning. Textile Research Journal. 2012;82(13):1390–95.
[40] Promnil S, Numpaisal P-O, Ruksakulpiwat Y. Effect of molecular weight on mechanical properties of electrospun poly (lactic acid) fibers for meniscus tissue engineering scaffold. Materials Today: Proceedings. 2021.
[41] Härdelin L, Perzon E, Hagström B, Walkenström P, Gatenholm P. Influence of molecular weight and rheological behavior on electrospinning cellulose nanofibers from ionic liquids. Journal of Applied Polymer Science. 2013;130(4):2303–10.
[42] Mazoochi T, Hamadanian M, Ahmadi M, Jabbari V. Investigation on the morphological characteristics of nanofibrous membrane as electrospun in the different processing parameters. International Journal of Industrial Chemistry. 2012;3(1):1–8.
[43] Wu X-F, Zhou Z, Zholobko O, Jenniges JJ, Baatz B, Ahmadi M, et al. Critical condition of electrohydrodynamic jetting from a polymer-solution droplet on a conductive wire. Journal of Applied Physics. 2020;127(5):054303.
[44] Reneker DH, Yarin AL. Electrospinning jets and polymer nanofibers. Polymer. 2008;49 (10):2387–425.
[45] Okutan N, Terzi P, Altay F. Affecting parameters on electrospinning process and characterization of electrospun gelatin nanofibers. Food Hydrocolloids. 2014;39:19–26.
[46] Deitzel JM, Kleinmeyer J, Harris D, Tan NB. The effect of processing variables on the morphology of electrospun nanofibers and textiles. Polymer. 2001;42(1):261–72.
[47] De Vrieze S, Van Camp T, Nelvig A, Hagström B, Westbroek P, De Clerck K. The effect of temperature and humidity on electrospinning. Journal of Materials Science. 2009;44 (5):1357–62.
[48] Pelipenko J, Kristl J, Janković B, Baumgartner S, Kocbek P. The impact of relative humidity during electrospinning on the morphology and mechanical properties of nanofibers. International Journal of Pharmaceutics. 2013;456(1):125–34.
[49] Fashandi H, Karimi M. Pore formation in polystyrene fiber by superimposing temperature and relative humidity of electrospinning atmosphere. Polymer. 2012;53(25):5832–49.
[50] Hardick O, Stevens B, Bracewell DG. Nanofibre fabrication in a temperature and humidity controlled environment for improved fibre consistency. Journal of Materials Science. 2011;46 (11):3890–98.
[51] Cai Y, Gevelber M. The effect of relative humidity and evaporation rate on electrospinning: Fiber diameter and measurement for control implications. Journal of Materials Science. 2013;48(22):7812–26.

[52] Anindyajati A, Boughton P, Ruys A The effect of rotating collector design on tensile properties and morphology of electrospun polycaprolactone fibres. MATEC Web of Conferences: EDP Sciences; 2015. p. 02002.
[53] Motamedi AS, Mirzadeh H, Hajiesmaeilbaigi F, Bagheri-Khoulenjani S, Shokrgozar M. Effect of electrospinning parameters on morphological properties of PVDF nanofibrous scaffolds. Progress in Biomaterials. 2017;6(3):113–23.
[54] Kumar P. Effect of collector on electrospinning to fabricate aligned nano fiber 2012.
[55] Ko FK, Wan Y. Introduction to Nanofiber Materials: Cambridge University Press; 2014.
[56] Persano L, Camposeo A, Tekmen C, Pisignano D. Industrial upscaling of electrospinning and applications of polymer nanofibers: A review. Macromolecular Materials and Engineering. 2013;298(5):504–20.
[57] Hutmacher DW, Dalton PD. Melt electrospinning. Chemistry–An Asian Journal. 2011;6 (1):44–56.
[58] Cao L, Su D, Su Z, Chen X. Fabrication of multiwalled carbon nanotube/polypropylene conductive fibrous membranes by melt electrospinning. Industrial & Engineering Chemistry Research. 2014;53(6):2308–17.
[59] Huang Y, Huang Q-L, Liu H, Zhang C-X, You Y-W, Li -N-N, et al. Preparation, characterization, and applications of electrospun ultrafine fibrous PTFE porous membranes. Journal of Membrane Science. 2017;523:317–26.
[60] Balaban M, Antić V, Pergal M, Francolini I, Martinelli A, Đonlagić J. The effect of polar solvents on the synthesis of poly (urethane-urea-siloxane) s. Journal of the Serbian Chemical Society. 2012;77(10):1457–81.
[61] Nam J, Huang Y, Agarwal S, Lannutti J. Materials selection and residual solvent retention in biodegradable electrospun fibers. Journal of Applied Polymer Science. 2008;107(3):1547–54.
[62] D'Amato AR, Bramson MT, Corr DT, Puhl DL, Gilbert RJ, Johnson J. Solvent retention in electrospun fibers affects scaffold mechanical properties. Electrospinning. 2018;2(1):15–28.
[63] Balogh A, Farkas B, Verreck G, Mensch J, Borbás E, Nagy B, et al. AC and DC electrospinning of hydroxypropylmethyl cellulose with polyethylene oxides as secondary polymer for improved drug dissolution. International Journal of Pharmaceutics. 2016;505(1–2):159–66.
[64] Moghe A, Gupta B. Co-axial electrospinning for nanofiber structures: Preparation and applications. Polymer Reviews. 2008;48(2):353–77.
[65] Qu H, Wei S, Guo Z. Coaxial electrospun nanostructures and their applications. Journal of Materials Chemistry A. 2013;1(38):11513–28.
[66] Rathore P, Schiffman JD. Beyond the single nozzle: Coaxial electrospinning enables innovative nanofiber chemistries, geometries, and applications. ACS Applied Materials & Interfaces. 2020;13(1):48–66.
[67] Jiang H, Wang L, Zhu K. Coaxial electrospinning for encapsulation and controlled release of fragile water-soluble bioactive agents. Journal of Controlled Release. 2014;193:296–303.
[68] Yu B, Zhang X, Yan J, Liu D, Li L, Pei R, et al. Improved stability, antitumor effect, and controlled release of recombinant soluble trail by combining genetic engineering with coaxial electrospinning. ACS Applied Bio Materials. 2019;2(6):2414–20.
[69] Wang J, Windbergs M. Controlled dual drug release by coaxial electrospun fibers–impact of the core fluid on drug encapsulation and release. International Journal of Pharmaceutics. 2019;556:363–71.
[70] Wang C, Yan K-W, Lin Y-D, Hsieh PC. Biodegradable core/shell fibers by coaxial electrospinning: Processing, fiber characterization, and its application in sustained drug release. Macromolecules. 2010;43(15):6389–97.

[71] Li C, Li Q, Ni X, Liu G, Cheng W, Han G. Coaxial electrospinning and characterization of core-shell structured cellulose nanocrystal reinforced PMMA/PAN composite fibers. Materials. 2017;10(6):572.
[72] Kemp R. Coaxial Electrospinning for the Reinforcement of Nanofibre Mats. Stellenbosch: Stellenbosch University; 2017.
[73] Di J, Chen H, Wang X, Zhao Y, Jiang L, Yu J, et al. Fabrication of zeolite hollow fibers by coaxial electrospinning. Chemistry of Materials. 2008;20(11):3543–45.
[74] Luo Y, Lode A, Gelinsky M. Direct plotting of three-dimensional hollow fiber scaffolds based on concentrated alginate pastes for tissue engineering. Advanced Healthcare Materials. 2013;2(6):777–83.
[75] Xiao J, Wan Y, Yao F, Huang Y, Zhu Y, Yang Z, et al. Constructing 3D scaffold with 40-nm-diameter hollow mesoporous bioactive glass nanofibers. Materials Letters. 2019;248:201–03.
[76] Nikmaram N, Roohinejad S, Hashemi S, Koubaa M, Barba FJ, Abbaspourrad A, et al. Emulsion-based systems for fabrication of electrospun nanofibers: Food, pharmaceutical and biomedical applications. RSC Advances. 2017;7(46):28951–64.
[77] Zhang C, Feng F, Zhang H. Emulsion electrospinning: Fundamentals, food applications and prospects. Trends in Food Science & Technology. 2018;80:175–86.
[78] Pal P, Srivas PK, Dadhich P, Das B, Maulik D, Dhara S. Nano-/microfibrous cotton-wool-like 3D scaffold with core–shell architecture by emulsion electrospinning for skin tissue regeneration. ACS Biomaterials Science & Engineering. 2017;3(12):3563–75.
[79] Xue J, Wu T, Dai Y, Xia Y. Electrospinning and electrospun nanofibers: Methods, materials, and applications. Chemical Reviews. 2019;119(8):5298–415.
[80] Hu J, Prabhakaran MP, Ding X, Ramakrishna S. Emulsion electrospinning of polycaprolactone: Influence of surfactant type towards the scaffold properties. Journal of Biomaterials Science, Polymer Edition. 2015;26(1):57–75.
[81] Buzgo M, Mickova A, Rampichova M, Doupnik M. Blend Electrospinning, Coaxial Electrospinning, and Emulsion Electrospinning Techniques. Core-shell Nanostructures for Drug Delivery and Theranostics: Elsevier; 2018. p. 325–47.
[82] Daelemans L, Steyaert I, Schoolaert E, Goudenhooft C, Rahier H, De Clerck K. Nanostructured hydrogels by blend electrospinning of polycaprolactone/gelatin nanofibers. Nanomaterials. 2018;8(7):551.
[83] Kim IG, Lee J-H, Unnithan AR, Park C-H, Kim CS. A comprehensive electric field analysis of cylinder-type multi-nozzle electrospinning system for mass production of nanofibers. Journal of Industrial and Engineering Chemistry. 2015;31:251–56.
[84] Jiang J, Zheng G, Wang X, Li W, Kang G, Chen H, et al. Arced multi-nozzle electrospinning spinneret for high-throughput production of nanofibers. Micromachines. 2020;11(1):27.
[85] Brown DA, Kim J-H, Lee H-B, Fotouhi G, Lee K-H, Liu WK, et al. Electric field guided assembly of one-dimensional nanostructures for high performance sensors. Sensors. 2012;12(5):5725–51.
[86] You X, Ye C, Guo P. Electric field manipulation for deposition control in near-field electrospinning. Journal of Manufacturing Processes. 2017;30:431–38.
[87] Katta P, Alessandro M, Ramsier R, Chase G. Continuous electrospinning of aligned polymer nanofibers onto a wire drum collector. Nano Letters. 2004;4(11):2215–18.
[88] Nikzad MJ, Mohamadbeigi N, Sadrnezhaad SK, Mahdavi SM. Fabrication of a highly flexible and affordable transparent electrode by aligned U-shaped copper nanowires using a new electrospinning collector with convenient transferability. ACS Omega. 2019;4(25):21260–66.
[89] Xu S-C, Qin -C-C, Yu M, Dong R-H, Yan X, Zhao H, et al. A battery-operated portable handheld electrospinning apparatus. Nanoscale. 2015;7(29):12351–55.

[90] Yan X, Yu M, Ramakrishna S, Russell SJ, Long Y-Z. Advances in portable electrospinning devices for in situ delivery of personalized wound care. Nanoscale. 2019;11(41):19166–78.
[91] Haik J, Kornhaber R, Blal B, Harats M. The feasibility of a handheld electrospinning device for the application of nanofibrous wound dressings. Advances in Wound Care. 2017;6(5):166–74.
[92] Aminyan R, Bazgir S. Fabrication and characterization of nanofibrous polyacrylic acid superabsorbent using gas-assisted electrospinning technique. Reactive and Functional Polymers. 2019;141:133–44.
[93] Lauricella M, Pisignano D, Succi S. Three-dimensional model for electrospinning processes in controlled gas counterflow. The Journal of Physical Chemistry A. 2016;120(27):4884–92.
[94] Angammana CJ, Jayaram SH. Fundamentals of electrospinning and processing technologies. Particulate Science and Technology. 2016;34(1):72–82.
[95] Peng H, Liu Y, Ramakrishna S. Recent development of centrifugal electrospinning. Journal of Applied Polymer Science. 2017;134:10.
[96] Kancheva M, Toncheva A, Manolova N, Rashkov I. Advanced centrifugal electrospinning setup. Materials Letters. 2014;136:150–52.
[97] Lyons J, Li C, Ko F. Melt-electrospinning part I: Processing parameters and geometric properties. Polymer. 2004;45(22):7597–603.
[98] Liu Y, He J-H, Yu J-Y Bubble-electrospinning: A novel method for making nanofibers. Journal of Physics: Conference Series: IOP Publishing; 2008. p. 012001.
[99] He J-H, Kong H-Y, Yang -R-R, Dou H, Faraz N, Wang L, et al. Review on fiber morphology obtained by bubble electrospinning and blown bubble spinning. Thermal Science. 2012;16 (5):1263–79.
[100] Erben J, Kalous T, Chvojka J. AC bubble electrospinning technology for preparation of nanofibrous mats. ACS Omega. 2020;5(14):8268–71.
[101] Liu Y, He J-H. Bubble electrospinning for mass production of nanofibers. International Journal of Nonlinear Sciences and Numerical Simulation. 2007;8(3):393–96.
[102] Cengiz-Çallıoğlu F, Jirsak O, Dayik M. Investigation into the relationships between independent and dependent parameters in roller electrospinning of polyurethane. Textile Research Journal. 2013;83(7):718–29.
[103] Yener F, Yalcinkaya B, Jirsak O. Roller Electrospinning System: A Novel Method to Producing Nanofibers. 2013.
[104] Niu H, Lin T. Fiber generators in needleless electrospinning. Journal of Nanomaterials. 2012:2012.
[105] Begum HA, Khan K. Study on the various types of needle based and needleless electrospinning system for nanofiber production. International Journal of Text Science. 2017;6:110–17.
[106] Thoppey NM, Bochinski JR, Clarke LI, Gorga RE. Edge electrospinning for high throughput production of quality nanofibers. Nanotechnology. 2011;22(34):345301.
[107] Zhmayev E, Cho D, Joo YL. Nanofibers from gas-assisted polymer melt electrospinning. Polymer. 2010;51(18):4140–44.
[108] Ozbolat IT. 3D Bioprinting: Fundamentals, Principles and Applications: Academic Press; 2016.
[109] Ozbolat I. 3D Bioprinting: Elsevier; 2016.
[110] Lee J-Y, An J, Chua CK. Fundamentals and applications of 3D printing for novel materials. Applied Materials Today. 2017;7:120–33.
[111] Velasquillo C, Galue EA, Rodriquez L, Ibarra C, Ibarra-Ibarra LG. Skin 3D bioprinting. Applications in Cosmetology. 2013.

[112] Bedir T, Ulag S, Ustundag CB, Gunduz O. 3D bioprinting applications in neural tissue engineering for spinal cord injury repair. Materials Science and Engineering: C. 2020;110:110741.

[113] Gungor-Ozkerim PS, Inci I, Zhang YS, Khademhosseini A, Dokmeci MR. Bioinks for 3D bioprinting: An overview. Biomaterials Science. 2018;6(5):915–46.

[114] Gopinathan J, Noh I. Recent trends in bioinks for 3D printing. Biomaterials Research. 2018;22(1):1–15.

[115] Shafran D, Kodish E, Tzakis A. Organ shortage: The greatest challenge facing transplant medicine. World Journal of Surgery. 2014;38(7):1650–57.

[116] Yi H-G, Kim H, Kwon J, Choi Y-J, Jang J, Cho D-W. Application of 3D bioprinting in the prevention and the therapy for human diseases. Signal Transduction and Targeted Therapy. 2021;6(1):1–17.

[117] Prendergast ME, Solorzano RD, Cabrera D. Bioinks for biofabrication: Current state and future perspectives. Journal of 3D Printing in Medicine. 2017;1(1):49–62.

[118] Daly AC, Critchley SE, Rencsok EM, Kelly DJ. A comparison of different bioinks for 3D bioprinting of fibrocartilage and hyaline cartilage. Biofabrication. 2016;8(4):045002.

[119] Ramiah P, du Toit LC, Choonara YE, Kondiah PP, Pillay V. Hydrogel-based bioinks for 3D bioprinting in tissue regeneration. Frontiers in Materials. 2020;7:76.

[120] Kumar S, Tharayil A, Thomas S. 3D bioprinting of nature-inspired hydrogel inks based on synthetic polymers. ACS Applied Polymer Materials. 2021;3(8):3685–701.

[121] Panwar A, Tan LP. Current status of bioinks for micro-extrusion-based 3D bioprinting. Molecules. 2016;21(6):685.

[122] Yin J, Yan M, Wang Y, Fu J, Suo H. 3D bioprinting of low-concentration cell-laden gelatin methacrylate (GelMA) bioinks with a two-step cross-linking strategy. ACS Applied Materials & Interfaces. 2018;10(8):6849–57.

[123] Chimene D, Miller L, Cross LM, Jaiswal MK, Singh I, Gaharwar AK. Nanoengineered osteoinductive bioink for 3D bioprinting bone tissue. ACS Applied Materials & Interfaces. 2020;12(14):15976–88.

[124] Jian H, Wang M, Dong Q, Li J, Wang A, Li X, et al. Dipeptide self-assembled hydrogels with tunable mechanical properties and degradability for 3D bioprinting. ACS Applied Materials & Interfaces. 2019;11(50):46419–26.

[125] Groll J, Burdick JA, Cho D-W, Derby B, Gelinsky M, Heilshorn SC, et al. A definition of bioinks and their distinction from biomaterial inks. Biofabrication. 2018;11(1):013001.

[126] Van Belleghem S, Torres L Jr, Santoro M, Mahadik B, Wolfand A, Kofinas P, et al. Hybrid 3D printing of synthetic and cell-laden bioinks for shape retaining soft tissue grafts. Advanced Functional Materials. 2020;30(3):1907145.

[127] DeSimone E, Schacht K, Jungst T, Groll J, Scheibel T. Biofabrication of 3D constructs: Fabrication technologies and spider silk proteins as bioinks. Pure and Applied Chemistry. 2015;87(8):737–49.

[128] GhavamiNejad A, Ashammakhi N, Wu XY, Khademhosseini A. Crosslinking strategies for 3D bioprinting of polymeric hydrogels. Small. 2020;16(35):2002931.

[129] Ji S, Abaci A, Morrison T, Gramlich WM, Guvendiren M. Novel bioinks from UV-responsive norbornene-functionalized carboxymethyl cellulose macromers. Bioprinting. 2020;18:e00083.

[130] Lyu J, Johnson M, Creagh-Flynn J, Zhou D, Lara-Sáez I, Xu Q, et al. Instant gelation system as self-healable and printable 3D cell culture bioink based on dynamic covalent chemistry. ACS Applied Materials & Interfaces. 2020;12(35):38918–24.

[131] Huh J, Moon Y-W, Park J, Atala A, Yoo JJ, Lee SJ. Combinations of photoinitiator and UV absorber for cell-based digital light processing (DLP) bioprinting. Biofabrication. 2021;13(3):034103.

[132] Gu Y, Forget A, Shastri VP. Biobridge: An outlook on translational bioinks for 3D bioprinting. Advanced Science. 2103469.
[133] Humphreys BD. Bioprinting better kidney organoids. Nature Materials. 2021;20(2):128–30.
[134] Lawlor KT, Vanslambrouck JM, Higgins JW, Chambon A, Bishard K, Arndt D, et al. Cellular extrusion bioprinting improves kidney organoid reproducibility and conformation. Nature Materials. 2021;20(2):260–71.
[135] Sharma A, Rawal P, Tripathi DM, Alodiya D, Sarin SK, Kaur S, et al. Upgrading hepatic differentiation and functions on 3D printed silk–decellularized liver hybrid scaffolds. ACS Biomaterials Science & Engineering. 2021;7(8):3861–73.
[136] Maiullari F, Costantini M, Milan M, Pace V, Chirivì M, Maiullari S, et al. A multi-cellular 3D bioprinting approach for vascularized heart tissue engineering based on HUVECs and iPSC-derived cardiomyocytes. Scientific Reports. 2018;8(1):1–15.
[137] Zhang YS, Yue K, Aleman J, Mollazadeh-Moghaddam K, Bakht SM, Yang J, et al. 3D bioprinting for tissue and organ fabrication. Annals of Biomedical Engineering. 2017;45 (1):148–63.
[138] Jorgensen AM, Yoo JJ, Atala A. Solid organ bioprinting: Strategies to achieve organ function. Chemical Reviews. 2020;120(19):11093–127.
[139] Kumar H, Sakthivel K, Mohamed MG, Boras E, Shin SR, Kim K. Designing gelatin methacryloyl (GelMA)-Based bioinks for visible light stereolithographic 3D biofabrication. Macromolecular Bioscience. 2021;21(1):2000317.
[140] Dubbin K, Tabet A, Heilshorn SC. Quantitative criteria to benchmark new and existing bio-inks for cell compatibility. Biofabrication. 2017;9(4):044102.
[141] Li X, Liu B, Pei B, Chen J, Zhou D, Peng J, et al. Inkjet bioprinting of biomaterials. Chemical Reviews. 2020;120(19):10793–833.
[142] Ng WL, Lee JM, Yeong WY, Naing MW. Microvalve-based bioprinting–process, bio-inks and applications. Biomaterials Science. 2017;5(4):632–47.
[143] Xu H, Casillas J, Xu C. Effects of printing conditions on cell distribution within microspheres during inkjet-based bioprinting. AIP Advances. 2019;9(9):095055.
[144] Agarwal S, Saha S, Balla VK, Pal A, Barui A, Bodhak S. Current developments in 3D bioprinting for tissue and organ regeneration–a review. Frontiers in Mechanical Engineering. 2020;6:90.
[145] Papaioannou TG, Manolesou D, Dimakakos E, Tsoucalas G, Vavuranakis M, Tousoulis D. 3D bioprinting methods and techniques: Applications on artificial blood vessel fabrication. Acta Cardiologica Sinica. 2019;35(3):284.
[146] Paxton N, Smolan W, Böck T, Melchels F, Groll J, Jungst T. Proposal to assess printability of bioinks for extrusion-based bioprinting and evaluation of rheological properties governing bioprintability. Biofabrication. 2017;9(4):044107.
[147] Guillotin B, Souquet A, Catros S, Duocastella M, Pippenger B, Bellance S, et al. Laser assisted bioprinting of engineered tissue with high cell density and microscale organization. Biomaterials. 2010;31(28):7250–56.
[148] Gruene M, Unger C, Koch L, Deiwick A, Chichkov B. Dispensing pico to nanolitre of a natural hydrogel by laser-assisted bioprinting. Biomedical Engineering Online. 2011;10(1):1–11.
[149] Koch L, Brandt O, Deiwick A, Chichkov B. Laser-assisted bioprinting at different wavelengths and pulse durations with a metal dynamic release layer: A parametric study. International Journal of Bioprinting. 2017;3:1.
[150] Koch L, Kuhn S, Sorg H, Gruene M, Schlie S, Gaebel R, et al. Laser printing of skin cells and human stem cells. Tissue Engineering Part C: Methods. 2010;16(5):847–54.
[151] Melchels FP, Feijen J, Grijpma DW. A review on stereolithography and its applications in biomedical engineering. Biomaterials. 2010;31(24):6121–30.

[152] Schmidleithner C, Kalaskar DM. Stereolithography: IntechOpen; 2018.
[153] Bártolo PJ. Stereolithographic Processes. Stereolithography: Springer; 2011. p. 1–36.
[154] Wang Z, Kumar H, Tian Z, Jin X, Holzman JF, Menard F, et al. Visible light photoinitiation of cell-adhesive gelatin methacryloyl hydrogels for stereolithography 3D bioprinting. ACS Applied Materials & Interfaces. 2018;10(32):26859–69.
[155] Bicudo E, Faulkner A, Li P. Patents and the experimental space: Social, legal and geographical dimensions of 3D bioprinting. International Review of Law, Computers & Technology. 2021;35(1):2–23.
[156] Cubo-Mateo N, Podhajsky S, Knickmann D, Slenzka K, Ghidini T, Gelinsky M. Can 3D bioprinting be a key for exploratory missions and human settlements on the Moon and Mars?. Biofabrication. 2020;12(4):043001.
[157] Larsen M, Mishra R, Miller M, Dean D. Bioprinting of Bone. Essentials of 3D Biofabrication and Translation: Elsevier; 2015. p. 293–308.
[158] Wang X, Ao Q, Tian X, Fan J, Wei Y, Hou W, et al. 3D bioprinting technologies for hard tissue and organ engineering. Materials. 2016;9(10):802.
[159] Chen Z, Li Z, Li J, Liu C, Lao C, Fu Y, et al. 3D printing of ceramics: A review. Journal of the European Ceramic Society. 2019;39(4):661–87.
[160] Fahmy MD, Jazayeri HE, Razavi M, Masri R, Tayebi L. Three-dimensional bioprinting materials with potential application in pre prosthetic surgery. Journal of Prosthodontics. 2016;25(4):310–18.
[161] Kruth J-P, Wang X, Laoui T, Froyen L. Lasers and materials in selective laser sintering. Assembly Automation. 2003.
[162] Kruth JP, Mercelis P, Van Vaerenbergh J, Froyen L, Rombouts M. Binding mechanisms in selective laser sintering and selective laser melting. Rapid Prototyping Journal. 2005.
[163] Williams JM, Adewunmi A, Schek RM, Flanagan CL, Krebsbach PH, Feinberg SE, et al. Bone tissue engineering using polycaprolactone scaffolds fabricated via selective laser sintering. Biomaterials. 2005;26(23):4817–27.
[164] Yeong W, Sudarmadji N, Yu H, Chua C, Leong K, Venkatraman S, et al. Porous polycaprolactone scaffold for cardiac tissue engineering fabricated by selective laser sintering. Acta Biomaterialia. 2010;6(6):2028–34.
[165] Duan B, Wang M, Zhou WY, Cheung WL, Li ZY, Lu WW. Three-dimensional nanocomposite scaffolds fabricated via selective laser sintering for bone tissue engineering. Acta Biomaterialia. 2010;6(12):4495–505.
[166] Savalani M, Hao L, Dickens P, Zhang Y, Tanner K, Harris R. The effects and interactions of fabrication parameters on the properties of selective laser sintered hydroxyapatite polyamide composite biomaterials. Rapid Prototyping Journal. 2012.
[167] Wang G, Zreiqat H. Functional coatings or films for hard-tissue applications. Materials. 2010;3(7):3994–4050.
[168] Mahapatro A. Bio-functional nano-coatings on metallic biomaterials. Materials Science and Engineering: C. 2015;55:227–51.
[169] Roy S. Functionally graded coatings on biomaterials: A critical review. Materials Today Chemistry. 2020;18:100375.
[170] Gnedenkov SV, Sharkeev YP, Sinebryukhov SL, Khrisanfova OA, Legostaeva EV, Zavidnaya AG, et al. Functional coatings formed on the titanium and magnesium alloys as implant materials by plasma electrolytic oxidation technology: Fundamental principles and synthesis conditions. Corrosion Reviews. 2016;34(1–2):65–83.
[171] Qi H, Heise S, Zhou J, Schuhladen K, Yang Y, Cui N, et al. Electrophoretic deposition of bioadaptive drug delivery coatings on magnesium alloy for bone repair. ACS Applied Materials & Interfaces. 2019;11(8):8625–34.

[172] Ghimire A, Song J. Anti-periprosthetic infection strategies: From implant surface topographical engineering to smart drug-releasing coatings. ACS Applied Materials & Interfaces. 2021;13(18):20921–37.
[173] Müller R, Abke J, Schnell E, Macionczyk F, Gbureck U, Mehrl R, et al. Surface engineering of stainless steel materials by covalent collagen immobilization to improve implant biocompatibility. Biomaterials. 2005;26(34):6962–72.
[174] Negroiu G, Piticescu RM, Chitanu GC, Mihailescu IN, Zdrentu L, Miroiu M. Biocompatibility evaluation of a novel hydroxyapatite-polymer coating for medical implants (in vitro tests). Journal of Materials Science: Materials in Medicine. 2008;19(4):1537–44.
[175] Brinker CJ. Dip Coating. Chemical Solution Deposition of Functional Oxide Thin Films: Springer; 2013. p. 233–61.
[176] Brinker CJ, Hurd AJ. Fundamentals of sol-gel dip-coating. Journal de Physique III. 1994;4 (7):1231–42.
[177] Motealleh A, Eqtesadi S, Perera FH, Pajares A, Guiberteau F, Miranda P. Understanding the role of dip-coating process parameters in the mechanical performance of polymer-coated bioglass robocast scaffolds. Journal of the Mechanical Behavior of Biomedical Materials. 2016;64:253–61.
[178] Scriven L. Physics and applications of dip coating and spin coating. MRS Online Proceedings Library (OPL). 1988:121.
[179] Larson RG, Rehg TJ. Spin Coating. Liquid Film Coating: Springer; 1997. p. 709–34.
[180] Tyona M. A theoretical study on spin coating technique. Advances in Materials Research. 2013;2(4):195.
[181] Vecchione R, Pitingolo G, Guarnieri D, Falanga AP, Netti PA. From square to circular polymeric microchannels by spin coating technology: A low cost platform for endothelial cell culture. Biofabrication. 2016;8(2):025005.
[182] Ziółkowska K, Kwapiszewski R, Brzozka Z. Microfluidic devices as tools for mimicking the in vivo environment. New Journal of Chemistry. 2011;35(5):979–90.
[183] Jalili-Firoozinezhad S, Miranda CC, Cabral JM. Modeling the human body on microfluidic chips. Trends in Biotechnology. 2021.
[184] Bulboacă A-E, Boarescu PM, Melincovici CS, Mihu CM. Microfluidic endothelium-on-a-chip development, from in vivo to in vitro experimental models. Romanian Journal of Morphology and Embryology. 2020;61(1):15.
[185] Bhatia SN, Ingber DE. Microfluidic organs-on-chips. Nature Biotechnology. 2014;32 (8):760–72.
[186] Assadi H, Kreye H, Gärtner F, Klassen T. Cold spraying–A materials perspective. Acta Materialia. 2016;116:382–407.
[187] Grigoriev S, Okunkova A, Sova A, Bertrand P, Smurov I. Cold spraying: From process fundamentals towards advanced applications. Surface and Coatings Technology. 2015;268:77–84.
[188] Gärtner F, Stoltenhoff T, Schmidt T, Kreye H. The cold spray process and its potential for industrial applications. Journal of Thermal Spray Technology. 2006;15(2):223–32.
[189] Choudhuri A, Mohanty P, Karthikeyan J Bio-ceramic composite coatings by cold spray technology. Thermal Spray 2009 Proc Int Thermal Spray Conf: Japan Association for Earthquake Engineering; 2009.
[190] Heimann RB. Structure, properties, and biomedical performance of osteoconductive bioceramic coatings. Surface and Coatings Technology. 2013;233:27–38.
[191] Chui C-Y, Ye H. Fundamental Concepts and Insights into Electrospraying for Biomedical Applications. Biomedical Applications of Electrospinning and Electrospraying: Elsevier; 2021. p. 185–206.

[192] Williams G, Raimi-Abraham B, Luo C. Electrospinning Fundamentals. Nanofibres in Drug Delivery. London, UK: UCL Press; 2018. p. 24–59.
[193] Ling X. Formation of Polymer Coatings by Electropolymerization. 1998.
[194] Cosnier S, Karyakin A. Electropolymerization: Concepts, Materials and Applications: John Wiley & Sons; 2011.
[195] Wang J-L, Li B-C, Li Z-J, Ren K-F, Jin L-J, Zhang S-M, et al. Electropolymerization of dopamine for surface modification of complex-shaped cardiovascular stents. Biomaterials. 2014;35 (27):7679–89.
[196] Okner R, Shaulov Y, Tal N, Favaro G, Domb AJ, Mandler D. Electropolymerized tricopolymer based on N-pyrrole derivatives as a primer coating for improving the performance of a drug-eluting stent. ACS Applied Materials & Interfaces. 2009;1(4):758–67.
[197] Aguilar LE, Tumurbaatar B, Ghavaminejad A, Park CH, Kim CS. Functionalized non-vascular nitinol stent via electropolymerized polydopamine thin film coating loaded with bortezomib adjunct to hyperthermia therapy. Scientific Reports. 2017;7(1):1–11.
[198] Bhattarai DP, Shrestha S, Shrestha BK, Park CH, Kim CS. A controlled surface geometry of polyaniline-doped titania nanotubes biointerface for accelerating MC3T3-E1 cells growth in bone tissue engineering. Chemical Engineering Journal. 2018;350:57–68.
[199] Lee JY, Aguilar LE, Park CH, Kim CS. UV light assisted coating method of polyphenol caffeic acid and mediated immobilization of metallic silver particles for antibacterial implant surface modification. Polymers. 2019;11(7):1200.
[200] Aguilar LE, Lee JY, Park CH, Kim CS. Biomedical grade stainless steel coating of polycaffeic acid via combined oxidative and ultraviolet light-assisted polymerization process for bioactive implant application. Polymers. 2019;11(4):584.
[201] Ahmed EM. Hydrogel: Preparation, characterization, and applications: A review. Journal of Advanced Research. 2015;6(2):105–21.
[202] Peppas NA, Mikos AG. Preparation Methods and Structure of Hydrogels. Hydrogels in Medicine and Pharmacy: CRC Press; 2019. p. 1–26.
[203] Chan KMC, Li RH, Chapman JW, Trac EM, Kobler JB, Zeitels SM, et al. Functionalizable hydrogel microparticles of tunable size and stiffness for soft-tissue filler applications. Acta Biomaterialia. 2014;10(6):2563–73.
[204] Varma DM, Gold GT, Taub PJ, Nicoll SB. Injectable carboxymethyl cellulose hydrogels for soft tissue filler applications. Acta Biomaterialia. 2014;10(12):4996–5004.
[205] Garg T, Singh S, Goyal AK. Stimuli-sensitive hydrogels: An excellent carrier for drug and cell delivery. Critical Reviews™ in Therapeutic Drug Carrier Systems. 2013;30:5.
[206] Young S, Wong M, Tabata Y, Mikos AG. Gelatin as a delivery vehicle for the controlled release of bioactive molecules. Journal of Controlled Release. 2005;109(1–3):256–74.
[207] Milcovich G, Lettieri S, Antunes FE, Medronho B, Fonseca AC, Coelho JF, et al. Recent advances in smart biotechnology: Hydrogels and nanocarriers for tailored bioactive molecules depot. Advances in Colloid and Interface Science. 2017;249:163–80.
[208] Gao F, Xu Z, Liang Q, Liu B, Li H, Wu Y, et al. Direct 3D printing of high strength biohybrid gradient hydrogel scaffolds for efficient repair of osteochondral defect. Advanced Functional Materials. 2018;28(13):1706644.
[209] Liu C, Xu N, Zong Q, Yu J, Zhang P. Hydrogel prepared by 3D printing technology and its applications in the medical field. Colloid and Interface Science Communications. 2021;44:100498.
[210] Schacht E Polymer chemistry and hydrogel systems. Journal of Physics: Conference Series: IOP Publishing; 2004. p. 004.
[211] Petrov P, Petrova E, Tsvetanov CB. UV-assisted synthesis of super-macroporous polymer hydrogels. Polymer. 2009;50(5):1118–23.

[212] Zhu W, Ding J. Synthesis and characterization of a redox-initiated, injectable, biodegradable hydrogel. Journal of Applied Polymer Science. 2006;99(5):2375–83.
[213] Hennink WE, van Nostrum CF. Novel crosslinking methods to design hydrogels. Advanced Drug Delivery Reviews. 2012;64:223–36.
[214] Maitra J, Shukla VK. Cross-linking in hydrogels-a review. American Journal of Polymer Science. 2014;4(2):25–31.
[215] Hu W, Wang Z, Xiao Y, Zhang S, Wang J. Advances in crosslinking strategies of biomedical hydrogels. Biomaterials Science. 2019;7(3):843–55.
[216] Spoljaric S, Salminen A, Luong ND, Seppälä J. Stable, self-healing hydrogels from nanofibrillated cellulose, poly (vinyl alcohol) and borax via reversible crosslinking. European Polymer Journal. 2014;56:105–17.
[217] Tasnim T, Adkins MD, Lim T, Feng H, Magda JJ, Shea JE, et al. Thermally tunable hydrogel crosslinking mediated by temperature sensitive liposome. Biomedical Materials. 2021;16 (6):065026.
[218] Sanborn TJ, Messersmith PB, Barron AE. In situ crosslinking of a biomimetic peptide-PEG hydrogel via thermally triggered activation of factor XIII. Biomaterials. 2002;23(13):2703–10.
[219] Zhao J, Yan Y, Shang Y, Du Y, Long L, Yuan X, et al. Thermosensitive elastin-derived polypeptide hydrogels crosslinked by genipin. International Journal of Polymeric Materials and Polymeric Biomaterials. 2017;66(8):369–77.
[220] Radhakrishnan J, Subramanian A, Krishnan UM, Sethuraman S. Injectable and 3D bioprinted polysaccharide hydrogels: From cartilage to osteochondral tissue engineering. Biomacromolecules. 2017;18(1):1–26.
[221] Joshi G, Okeyoshi K, Okajima MK, Kaneko T. Directional control of diffusion and swelling in megamolecular polysaccharide hydrogels. Soft Matter. 2016;12(25):5515–18.
[222] Alvarez-Lorenzo C, Blanco-Fernandez B, Puga AM, Concheiro A. Crosslinked ionic polysaccharides for stimuli-sensitive drug delivery. Advanced Drug Delivery Reviews. 2013;65 (9):1148–71.
[223] Patil S, Jadge D. Crosslinking of polysaccharides: Methods and applications. Latest Reviews. 2008;6:2.
[224] Mizrahy S, Peer D. Polysaccharides as building blocks for nanotherapeutics. Chemical Society Reviews. 2012;41(7):2623–40.
[225] Leclercq S, Milo C, Reineccius GA. Effects of cross-linking, capsule wall thickness, and compound hydrophobicity on aroma release from complex coacervate microcapsules. Journal of Agricultural and Food Chemistry. 2009;57(4):1426–32.
[226] Hoffman AS. Hydrogels for biomedical applications. Advanced Drug Delivery Reviews. 2012;64:18–23.
[227] Zhang J, Wang N, Liu W, Zhao X, Lu W. Intermolecular hydrogen bonding strategy to fabricate mechanically strong hydrogels with high elasticity and fatigue resistance. Soft Matter. 2013;9(27):6331–37.
[228] Li J, Chen F, Lin X, Ding T. Hydrogen-bonding-assisted toughening of hierarchical carboxymethyl cellulose hydrogels for biomechanical sensing. Carbohydrate Polymers. 2021:118252.
[229] Said HM, Abd Alla SG, El-Naggar AWM. Synthesis and characterization of novel gels based on carboxymethyl cellulose/acrylic acid prepared by electron beam irradiation. Reactive and Functional Polymers. 2004;61(3):397–404.
[230] Wang Q, Zhang Y, Dai X, Shi X, Liu W. A high strength pH responsive supramolecular copolymer hydrogel. Science China Technological Sciences. 2017;60(1):78–83.

[231] Tsai -C-C, Payne GF, Shen J. Exploring pH-responsive, switchable crosslinking mechanisms for programming reconfigurable hydrogels based on aminopolysaccharides. Chemistry of Materials. 2018;30(23):8597–605.
[232] Altunbas A, Lee SJ, Rajasekaran SA, Schneider JP, Pochan DJ. Encapsulation of curcumin in self-assembling peptide hydrogels as injectable drug delivery vehicles. Biomaterials. 2011;32(25):5906–14.
[233] Bhattacharya A, Rawlins JW, Ray P. Polymer Grafting and Crosslinking: John Wiley & Sons; 2008.
[234] Parida P, Mishra SC. Synthesis, crosslinking and evaluation of graft polymer for biomedical use. 2012.
[235] Fellows CM. Mechanism and kinetics. Polymer Grafting and Crosslinking. 2009:65.
[236] Gulrez SK, Al-Assaf S, Phillips GO. Hydrogels: Methods of preparation, characterisation and applications. Progress in molecular and environmental bioengineering-from analysis and modeling to technology applications. 2011:117–50.
[237] Beşkardeş IG, Demirtaş TT, Durukan MD, Gümüşderelioğlu M. Microwave-assisted fabrication of chitosan–hydroxyapatite superporous hydrogel composites as bone scaffolds. Journal of Tissue Engineering and Regenerative Medicine. 2015;9(11):1233–46.
[238] Peng X-W, Ren J-L, Zhong L-X, Sun R-C. Synthesis and characterization of amphoteric xylan-type hemicelluloses by microwave irradiation. Journal of Agricultural and Food Chemistry. 2012;60(7):1695–702.
[239] Benson RS. Use of radiation in biomaterials science. Nuclear Instruments and Methods in Physics Research Section B: Beam Interactions with Materials and Atoms. 2002;191 (1–4):752–57.
[240] Wang H, Wen Y, Peng H, Zheng C, Li Y, Wang S, et al. Grafting polytetrafluoroethylene micropowder via in situ electron beam irradiation-induced polymerization. Polymers. 2018;10 (5):503.
[241] Sorour M, El-Sayed M, Moneem NAE, Talaat HA, Shalaan H, Marsafy SE. Characterization of hydrogel synthesized from natural polysaccharides blend grafted acrylamide using microwave (MW) and ultraviolet (UV) techniques. Starch-Stärke. 2013;65(1-2):172–78.
[242] Henderson TM, Ladewig K, Haylock DN, McLean KM, O'Connor AJ. Cryogels for biomedical applications. Journal of Materials Chemistry B. 2013;1(21):2682–95.
[243] Bakhshpour M, Idil N, Perçin I, Denizli A. Biomedical applications of polymeric cryogels. Applied Sciences. 2019;9(3):553.
[244] Zinggeler M, Schönberg J-N, Fosso PL, Brandstetter T, Rühe J. Functional Cryogel microctructures prepared by light-induced cross-linking of a photoreactive copolymer. ACS Applied Materials & Interfaces. 2017;9(14):12165–70.
[245] Kumar A, Mishra R, Reinwald Y, Bhat S. Cryogels: Freezing unveiled by thawing. Materials Today. 2010;13(11):42–44.
[246] Yetiskin B, Okay O. High-strength and self-recoverable silk fibroin cryogels with anisotropic swelling and mechanical properties. International Journal of Biological Macromolecules. 2019;122:1279–89.
[247] Raschip IE, Fifere N, Varganici C-D, Dinu MV. Development of antioxidant and antimicrobial xanthan-based cryogels with tuned porous morphology and controlled swelling features. International Journal of Biological Macromolecules. 2020;156:608–20.
[248] Dai R, Meng L, Fu Q, Hao S, Yang J. Fabrication of anisotropic silk fibroin-cellulose nanocrystals cryogels with tunable mechanical properties, rapid swelling, and structural recoverability via a directional-freezing strategy. ACS Sustainable Chemistry & Engineering. 2021;9(36):12274–85.
[249] Li J, Wang Y, Zhang L, Xu Z, Dai H, Wu W. Nanocellulose/gelatin composite cryogels for controlled drug release. ACS Sustainable Chemistry & Engineering. 2019;7(6):6381–89.

[250] Mauri E, Perale G, Rossi F. Nanogel functionalization: A versatile approach to meet the challenges of drug and gene delivery. ACS Applied Nano Materials. 2018;1(12):6525–41.
[251] Oh JK, Siegwart DJ, Lee H-I, Sherwood G, Peteanu L, Hollinger JO, et al. Biodegradable nanogels prepared by atom transfer radical polymerization as potential drug delivery carriers: Synthesis, biodegradation, in vitro release, and bioconjugation. Journal of the American Chemical Society. 2007;129(18):5939–45.
[252] Zhang H, Zhai Y, Wang J, Zhai G. New progress and prospects: The application of nanogel in drug delivery. Materials Science and Engineering: C. 2016;60:560–68.
[253] Vasudevan M, Buse E, Lu D, Krishna H, Kalyanaraman R, Shen AQ, et al. Irreversible nanogel formation in surfactant solutions by microporous flow. Nature Materials. 2010;9(5):436–41.
[254] De France KJ, Xu F, Hoare T. Structured macroporous hydrogels: Progress, challenges, and opportunities. Advanced Healthcare Materials. 2018;7(1):1700927.
[255] Mitragotri S, Stayton P. Organic nanoparticles for drug delivery and imaging. Mrs Bulletin. 2014;39(3):219–23.
[256] Jain R, Khandelwal G, Roy S. Unraveling the design rules in ultrashort amyloid-based peptide assemblies toward shape-controlled synthesis of gold nanoparticles. Langmuir. 2019;35 (17):5878–89.
[257] Seidi F, Jenjob R, Phakkeeree T, Crespy D. Saccharides, oligosaccharides, and polysaccharides nanoparticles for biomedical applications. Journal of Controlled Release. 2018;284:188–212.
[258] Yang J, Han S, Zheng H, Dong H, Liu J. Preparation and application of micro/nanoparticles based on natural polysaccharides. Carbohydrate Polymers. 2015;123:53–66.
[259] Yokoyama M. Polymeric micelles as drug carriers: Their lights and shadows. Journal of Drug Targeting. 2014;22(7):576–83.
[260] Letchford K, Burt H. A review of the formation and classification of amphiphilic block copolymer nanoparticulate structures: Micelles, nanospheres, nanocapsules and polymersomes. European Journal of Pharmaceutics and Biopharmaceutics. 2007;65 (3):259–69.
[261] Nagarajan R. Theory of Micelle Formation: Quantitative approach to Predicting Micellar properties from Surfactant Molecular Structure. Surfactant Science Series. 1997;70:1–81.
[262] Alexandridis P, Hatton TA. Poly (ethylene oxide)- poly (propylene oxide)- poly (ethylene oxide) block copolymer surfactants in aqueous solutions and at interfaces: Thermodynamics, structure, dynamics, and modeling. Colloids and Surfaces A: Physicochemical and Engineering Aspects. 1995;96(1–2):1–46.
[263] Royston GJ. Modification of Epoxy Resins with Block Copolymers of Poly (ethylene oxide) and Poly (butylene oxide): University of Sheffield; 2007.
[264] Allgaier J, Hövelmann CH, Wei Z, Staropoli M, Pyckhout-Hintzen W, Lühmann N, et al. Synthesis and rheological behavior of poly (1, 2-butylene oxide) based supramolecular architectures. RSC Advances. 2016;6(8):6093–106.
[265] Herzberger J, Niederer K, Pohlit H, Seiwert J, Worm M, Wurm FR, et al. Polymerization of ethylene oxide, propylene oxide, and other alkylene oxides: Synthesis, novel polymer architectures, and bioconjugation. Chemical Reviews. 2016;116(4):2170–243.
[266] Almeida M, Magalhães M, Veiga F, Figueiras A. Poloxamers, poloxamines and polymeric micelles: Definition, structure and therapeutic applications in cancer. Journal of Polymer Research. 2018;25(1):1–14.
[267] Zhang L, Chen Z, Yang K, Liu C, Gao J, Qian F. β-Lapachone and paclitaxel combination micelles with improved drug encapsulation and therapeutic synergy as novel nanotherapeutics for NQO1-targeted cancer therapy. Molecular Pharmaceutics. 2015;12 (11):3999–4010.

[268] Chen H-P, Chen M-H, Tung F-I, Liu T-Y. A novel micelle-forming material used for preparing a theranostic vehicle exhibiting enhanced in vivo therapeutic efficacy. Journal of Medicinal Chemistry. 2015;58(9):3704–19.
[269] Immordino ML, Dosio F, Cattel L. Stealth liposomes: Review of the basic science, rationale, and clinical applications, existing and potential. International Journal of Nanomedicine. 2006;1(3):297.
[270] Maherani B, Arab-Tehrany E, Mozafari MR, Gaiani C, Linder M. Liposomes: A review of manufacturing techniques and targeting strategies. Current Nanoscience. 2011;7(3):436–52.
[271] Shailesh S, Neelam S, Sandeep K, Gupta G. Liposomes: A review. Journal of Pharmacy Research. 2009;2(7):1163–67.
[272] Nkanga CI, Bapolisi AM, Okafor NI, Krause RWM. General perception of liposomes: Formation, manufacturing and applications. Liposomes-advances and Perspectives. 2019.
[273] Vemuri S, Rhodes C. Preparation and characterization of liposomes as therapeutic delivery systems: A review. Pharmaceutica Acta Helvetiae. 1995;70(2):95–111.
[274] Jahn A, Vreeland WN, DeVoe DL, Locascio LE, Gaitan M. Microfluidic directed formation of liposomes of controlled size. Langmuir. 2007;23(11):6289–93.
[275] Deshpande S, Brandenburg F, Lau A, Last MG, Spoelstra WK, Reese L, et al. Spatiotemporal control of coacervate formation within liposomes. Nature Communications. 2019;10 (1):1–11.
[276] Wang M, Liu Z, Zhan W. Janus liposomes: Gel-assisted formation and bioaffinity-directed clustering. Langmuir. 2018;34(25):7509–18.
[277] Yang Y, Wu Z, Wang L, Zhou K, Xia K, Xiong Q, et al. Sorting sub-150-nm liposomes of distinct sizes by DNA-brick-assisted centrifugation. Nature Chemistry. 2021;13(4):335–42.
[278] Gupta V, Nayak SK. Dendrimers: A review on synthetic approaches. Journal of Applied Pharmaceutical Science. 2015;5:03.
[279] Augustus EN, Allen ET, Nimibofa A, Donbebe W. A review of synthesis, characterization and applications of functionalized dendrimers. American Journal of Polymer Science. 2017;7 (1):8–14.
[280] Scott RW, Wilson OM, Crooks RM. Synthesis, characterization, and applications of dendrimer-encapsulated nanoparticles. The Journal of Physical Chemistry B. 2005;109 (2):692–704.
[281] Araújo R, Santos S, Igne Ferreira E, Giarolla J. New advances in general biomedical applications of PAMAM dendrimers. Molecules. 2018;23(11):2849.
[282] Svenson S, Tomalia DA. Dendrimers in biomedical applications – reflections on the field. Advanced Drug Delivery Reviews. 2012;64:102–15.
[283] Lyu Z, Ding L, Tintaru A, Peng L. Self-assembling supramolecular dendrimers for biomedical applications: Lessons learned from poly (amidoamine) dendrimers. Accounts of Chemical Research. 2020;53(12):2936–49.
[284] Moorthy H, Govindaraju T. Dendrimer architectonics to treat cancer and neurodegenerative diseases with implications in theranostics and personalized medicine. ACS Applied Bio Materials. 2021;4(2):1115–39.
[285] Kobayashi A, Yokoyama Y, Osawa Y, Miura R, Mizunuma H. Gene therapy for ovarian cancer using carbonyl reductase 1 DNA with a polyamidoamine dendrimer in mouse models. Cancer Gene Therapy. 2016;23(1):24–28.
[286] Venditto VJ, Regino CAS, Brechbiel MW. PAMAM dendrimer based macromolecules as improved contrast agents. Molecular Pharmaceutics. 2005;2(4):302–11.

[287] Han L, Li J, Huang S, Huang R, Liu S, Hu X, et al. Peptide-conjugated polyamidoamine dendrimer as a nanoscale tumor-targeted T1 magnetic resonance imaging contrast agent. Biomaterials. 2011;32(11):2989–98.

[288] Kobayashi H, Kawamoto S, Star RA, Waldmann TA, Tagaya Y, Brechbiel MW. Micro-magnetic resonance lymphangiography in mice using a novel dendrimer-based magnetic resonance imaging contrast agent. Cancer Research. 2003;63(2):271–76.

[289] Huang Y, Coman D, Hyder F, Ali MM. Dendrimer-based responsive MRI contrast agents (G1–G4) for biosensor imaging of redundant deviation in shifts (BIRDS). Bioconjugate Chemistry. 2015;26(12):2315–23.

[290] Li P, Zhang M, Sun X, Guan S, Zhang G, Baumgarten M, et al. A dendrimer-based highly sensitive and selective fluorescence-quenching sensor for Fe3+ both in solution and as film. Biosensors and Bioelectronics. 2016;85:785–91.

[291] Majoros IJ, Williams CR, Baker Jr JR. Current dendrimer applications in cancer diagnosis and therapy. Current Topics in Medicinal Chemistry. 2008;8(14):1165–79.

[292] Liang H, Zhou B, Wu D, Li J, Li B. Supramolecular design and applications of polyphenol-based architecture: A review. Advances in Colloid and Interface Science. 2019;272:102019.

[293] Myerson JW, Patel PN, Rubey KM, Zamora ME, Zaleski MH, Habibi N, et al. Supramolecular arrangement of protein in nanoparticle structures predicts nanoparticle tropism for neutrophils in acute lung inflammation. Nature Nanotechnology. 2021:1–12.

[294] Yuan H, Xu C, Zhao Y, Yu B, Cheng G, Xu FJ. Well-defined protein-based supramolecular nanoparticles with excellent MRI Abilities for multifunctional delivery systems. Advanced Functional Materials. 2016;26(17):2855–65.

[295] Chang R, Yan X. Supramolecular immunotherapy of cancer based on the self-assembling peptide design. Small Structures. 2020;1(2):2000068.

[296] Xu ZP, Zeng QH, Lu GQ, Yu AB. Inorganic nanoparticles as carriers for efficient cellular delivery. Chemical Engineering Science. 2006;61(3):1027–40.

[297] Kim D, Kim J, Park YI, Lee N, Hyeon T. Recent development of inorganic nanoparticles for biomedical imaging. ACS Central Science. 2018;4(3):324–36.

[298] Huang H-C, Barua S, Sharma G, Dey SK, Rege K. Inorganic nanoparticles for cancer imaging and therapy. Journal of Controlled Release. 2011;155(3):344–57.

[299] Iglesias GR, Jabalera Y, Peigneux A, Checa Fernández BL, Delgado ÁV, Jimenez-Lopez C. Enhancement of magnetic hyperthermia by mixing synthetic inorganic and biomimetic magnetic nanoparticles. Pharmaceutics. 2019;11(6):273.

[300] Aguilar LE, Chalony C, Kumar D, Park CH, Kim CS. Phenol-Boronic surface functionalization of gold nanoparticles; to induce ROS damage while inhibiting the survival mechanisms of cancer cells. International Journal of Pharmaceutics. 2021;596:120267.

[301] Tang Z-X, Lv B-F. MgO nanoparticles as antibacterial agent: Preparation and activity. Brazilian Journal of Chemical Engineering. 2014;31:591–601.

[302] Ma G, Chen Y, Li L, Jiang D, Qiao R, Zhu Y. An attractive photocatalytic inorganic antibacterial agent: Preparation and property of graphene/zinc ferrite/polyaniline composites. Materials Letters. 2014;131:38–41.

[303] Mulder WJ, Strijkers GJ, Van Tilborg GA, Cormode DP, Fayad ZA, Nicolay K. Nanoparticulate assemblies of amphiphiles and diagnostically active materials for multimodality imaging. Accounts of Chemical Research. 2009;42(7):904–14.

[304] Anselmo AC, Mitragotri S. A review of clinical translation of inorganic nanoparticles. The AAPS Journal. 2015;17(5):1041–54.

[305] da Silva Brum I, de Carvalho JJ, da Silva Pires JL, de Carvalho MAA, Dos Santos LBF, Elias CN. Nano-sized hydroxyapatite and β-tricalcium phosphate composite: Physico-chemical, cytotoxicity, morphological properties and in vivo trial. Scientific Reports. 2019;9(1):1–10.

[306] Dienel KE, van Bochove B, Seppälä JV. Additive manufacturing of bioactive poly (trimethylene carbonate)/β-tricalcium phosphate composites for bone regeneration. Biomacromolecules. 2019;21(2):366–75.

[307] Shu X, Liao J, Wang L, Shi Q, Xie X. Osteogenic, angiogenic, and antibacterial bioactive nano-hydroxyapatite co-synthesized using γ-polyglutamic acid and copper. ACS Biomaterials Science & Engineering. 2020;6(4):1920–30.

[308] Jones RS, Draheim RR, Roldo M. Silver nanowires: Synthesis, antibacterial activity and biomedical applications. Applied Sciences. 2018;8(5):673.

[309] Mukhtar A, Wu K, Cao X. Magnetic nanowires in biomedical applications. Nanotechnology. 2020;31(43):433001.

[310] Mousavi SM, Zarei M, Hashemi SA, Ramakrishna S, Chiang W-H, Lai CW, et al. Gold nanostars-diagnosis, bioimaging and biomedical applications. Drug Metabolism Reviews. 2020;52(2):299–318.

[311] Chirico G, Borzenkov M, Pallavicini P. Gold Nanostars: Synthesis, Properties and Biomedical Application: Springer; 2015.

[312] Allen SD, Bobbala S, Karabin NB, Scott EA. On the advancement of polymeric bicontinuous nanospheres toward biomedical applications. Nanoscale Horizons. 2019;4(2):258–72.

[313] Favi PM, Gao M, Johana Sepúlveda Arango L, Ospina SP, Morales M, Pavon JJ, et al. Shape and surface effects on the cytotoxicity of nanoparticles: Gold nanospheres versus gold nanostars. Journal of Biomedical Materials Research Part A. 2015;103(11):3449–62.

[314] Wu M, Hou P, Dong L, Cai L, Chen Z, Zhao M, et al. Manganese dioxide nanosheets: From preparation to biomedical applications. International Journal of Nanomedicine. 2019;14:4781.

[315] Li X, Shan J, Zhang W, Su S, Yuwen L, Wang L. Recent advances in synthesis and biomedical applications of two-dimensional transition metal dichalcogenide nanosheets. Small. 2017;13(5):1602660.

[316] Khan A, Rashid A, Younas R, Chong R. A chemical reduction approach to the synthesis of copper nanoparticles. International Nano Letters. 2016;6(1):21–26.

[317] Guzmán MG, Dille J, Godet S. Synthesis of silver nanoparticles by chemical reduction method and their antibacterial activity. International Journal of Chemical and Biomolecular Engineering. 2009;2(3):104–11.

[318] Rodriguez-Gattorno G, Diaz D, Rendon L, Hernandez-Segura G. Metallic nanoparticles from spontaneous reduction of silver (I) in DMSO. Interaction between nitric oxide and silver nanoparticles. The Journal of Physical Chemistry B. 2002;106(10):2482–87.

[319] Gashti MP, Almasian A. Synthesizing tertiary silver/silica/kaolinite nanocomposite using photo-reduction method: Characterization of morphology and electromagnetic properties. Composites Part B: Engineering. 2012;43(8):3374–83.

[320] Ebrahimi I, Gashti MP. Polypyrrole-MWCNT-Ag composites for electromagnetic shielding: Comparison between chemical deposition and UV-reduction approaches. Journal of Physics and Chemistry of Solids. 2018;118:80–87.

[321] Soleimani Zohr Shiri M, Henderson W, Mucalo MR. A review of the lesser-studied microemulsion-based synthesis methodologies used for preparing nanoparticle systems of the noble metals, Os, Re, Ir and Rh. Materials. 2019;12(12):1896.

[322] Yadollahpour A. Magnetic nanoparticles in medicine: A review of synthesis methods and important characteristics. Oriental Journal of Chemistry. 2015;31:271–77.

[323] Malik MA, Wani MY, Hashim MA. Microemulsion method: A novel route to synthesize organic and inorganic nanomaterials: 1st Nano Update. Arabian Journal of Chemistry. 2012;5(4):397–417.

[324] Tartaro G, Mateos H, Schirone D, Angelico R, Palazzo G. Microemulsion microstructure (s): A tutorial review. Nanomaterials. 2020;10(9):1657.

[325] Granata G, Pagnanelli F, Nishio-Hamane D, Sasaki T. Effect of surfactant/water ratio and reagents' concentration on size distribution of manganese carbonate nanoparticles synthesized by microemulsion mediated route. Applied Surface Science. 2015;331:463–71.

[326] Moballegh Nasery M, Abadi B, Poormoghadam D, Zarrabi A, Keyhanvar P, Khanbabaei H, et al. Curcumin delivery mediated by bio-based nanoparticles: A review. Molecules. 2020;25(3):689.

[327] Gonugunta P, Vivekanandhan S, Mohanty AK, Misra M. A study on synthesis and characterization of biobased carbon nanoparticles from lignin. World Journal of Nano Science and Engineering. 2012;2(3):148–53.

[328] Das S, Mandal SC. Synthesis and therapeutic activity of silver nanoparticles using bio-based protocol. The Pharma Review. 2014;12:124–28.

[329] Varshney R, Bhadauria S, Gaur MS. A review: Biological synthesis of silver and copper nanoparticles. Nano Biomedicine & Engineering. 2012;4:2.

[330] Kumar CG, Mamidyala SK. Extracellular synthesis of silver nanoparticles using culture supernatant of Pseudomonas aeruginosa. Colloids and Surfaces B: Biointerfaces. 2011;84 (2):462–66.

[331] Korbekandi H, Iravani S, Abbasi S. Production of nanoparticles using organisms. Critical Reviews in Biotechnology. 2009;29(4):279–306.

[332] Marooufpour N, Alizadeh M, Hatami M, Lajayer BA. Biological Synthesis of Nanoparticles by Different Groups of Bacteria. Microbial Nanobionics: Springer; 2019. p. 63–85.

[333] Iravani S. Bacteria in nanoparticle synthesis: Current status and future prospects. International Scholarly Research Notices. 2014:2014.

[334] Zheng K, Boccaccini AR. Sol-gel processing of bioactive glass nanoparticles: A review. Advances in Colloid and Interface Science. 2017;249:363–73.

[335] Rahman IA, Padavettan V. Synthesis of silica nanoparticles by sol-gel: Size-dependent properties, surface modification, and applications in silica-polymer nanocomposites – a review. Journal of Nanomaterials. 2012:2012.

[336] Dimitriev Y, Ivanova Y, Iordanova R. History of sol-gel science and technology. Journal of the University of Chemical Technology and Metallurgy. 2008;43(2):181–92.

[337] Baino F, Fiorilli S, Vitale-Brovarone C. Composite biomaterials based on sol-gel mesoporous silicate glasses: A review. Bioengineering. 2017;4(1):15.

[338] Deshmukh K, Kovářík T, Křenek T, Docheva D, Stich T, Pola J. Recent advances and future perspectives of sol–gel derived porous bioactive glasses: A review. RSC Advances. 2020;10 (56):33782–835.

[339] Kumar A, Mariappan C, Sarahan B. Antibacterial and structural properties of mesoporous Ag doped calcium borosilicate glass-ceramics synthesized via a sol-gel route. Journal of Non-Crystalline Solids. 2019;505:431–37.

[340] Kargozar S, Mozafari M, Hamzehlou S, Kim H-W, Baino F. Mesoporous bioactive glasses (MBGs) in cancer therapy: Full of hope and promise. Materials Letters. 2019;251:241–46.

[341] Kuang X, Carotenuto G, Nicolais L. A review of ceramic sintering and suggestions on reducing sintering temperatures. Advanced Performance Materials. 1997;4(3):257–74.

[342] Fathian Z, Maleki A, Niroumand B. Synthesis and characterization of ceramic nanoparticles reinforced lead-free solder. Ceramics International. 2017;43(6):5302–10.

[343] Kocjan A, Logar M, Shen Z. The agglomeration, coalescence and sliding of nanoparticles, leading to the rapid sintering of zirconia nanoceramics. Scientific Reports. 2017;7(1):1–8.

[344] Boken J, Soni SK, Kumar D. Microfluidic synthesis of nanoparticles and their biosensing applications. Critical Reviews in Analytical Chemistry. 2016;46(6):538–61.

[345] Song Y, Hormes J, Kumar CS. Microfluidic synthesis of nanomaterials. Small. 2008;4 (6):698–711.

[346] Kim D-Y, Jin SH, Jeong S-G, Lee B, Kang -K-K, Lee C-S. Microfluidic preparation of monodisperse polymeric microspheres coated with silica nanoparticles. Scientific Reports. 2018;8(1):1–11.

[347] Zhang D, Wu F, Peng M, Wang X, Xia D, Guo G. One-step, facile and ultrafast synthesis of phase-and size-controlled Pt–Bi intermetallic nanocatalysts through continuous-flow microfluidics. Journal of the American Chemical Society. 2015;137(19):6263–69.

[348] Gómez-de Pedro S, Puyol M, Alonso-Chamarro J. Continuous flow synthesis of nanoparticles using ceramic microfluidic devices. Nanotechnology. 2010;21(41):415603.

[349] Yang C-H, Huang K-S, Lin Y-S, Lu K, Tzeng -C-C, Wang E-C, et al. Microfluidic assisted synthesis of multi-functional polycaprolactone microcapsules: Incorporation of CdTe quantum dots, Fe 3 O 4 superparamagnetic nanoparticles and tamoxifen anticancer drugs. Lab on a Chip. 2009;9(7):961–65.

[350] Xu J, Zhang S, Machado A, Lecommandoux S, Sandre O, Gu F, et al. Controllable microfluidic production of drug-loaded PLGA nanoparticles using partially water-miscible mixed solvent microdroplets as a precursor. Scientific Reports. 2017;7(1):1–12.

[351] Abalde-Cela S, Taladriz-Blanco P, de Oliveira MG, Abell C. Droplet microfluidics for the highly controlled synthesis of branched gold nanoparticles. Scientific Reports. 2018;8 (1):1–6.

[352] Yelten-Yilmaz A, Yilmaz S. Wet chemical precipitation synthesis of hydroxyapatite (HA) powders. Ceramics International. 2018;44(8):9703–10.

[353] Ahn ES, Gleason NJ, Nakahira A, Ying JY. Nanostructure processing of hydroxyapatite-based bioceramics. Nano Letters. 2001;1(3):149–53.

[354] Ulfyana D, Anugroho F, Sumarlan S, Wibisono Y Bioceramics synthesis of hydroxyapatite from red snapper fish scales biowaste using wet chemical precipitation route. IOP Conference Series: Earth and Environmental Science: IOP Publishing; 2018. p. 012038.

[355] Nayak AK. Hydroxyapatite synthesis methodologies: An overview. International Journal of ChemTech Research. 2010;2(2):903–07.

[356] Yelten A, Yilmaz S, Oktar FN. Sol–gel derived alumina–hydroxyapatite–tricalcium phosphate porous composite powders. Ceramics International. 2012;38(4):2659–65.

[357] Cousin P, Ross R. Preparation of mixed oxides: A review. Materials Science and Engineering: A. 1990;130(1):119–25.

[358] Noviyanti AR, Akbar N, Deawati Y, Ernawati EE, Malik YT, Fauzia RP. A novel hydrothermal synthesis of nanohydroxyapatite from eggshell-calcium-oxide precursors. Heliyon. 2020;6 (4):e03655.

[359] LeGeros RZ. Biological and Synthetic Apatites. Hydroxyapatite and Related Materials: CRC press; 2017. p. 3–28.

[360] Tao S, Irvine J. Synthesis and ionic conduction of apatite-type materials. Ionics. 2000;6 (5):389–96.

[361] Zhou J, Liu Q, Xia Z. Structural construction and photoluminescence tuning via energy transfer in apatite-type solid-state phosphors. Journal of Materials Chemistry C. 2018;6 (16):4371–83.

[362] Vitale-Brovarone C, Baino F, Verné E. Feasibility and tailoring of bioactive glass-ceramic scaffolds with gradient of porosity for bone grafting. Journal of Biomaterials Applications. 2010;24(8):693–712.

[363] Brovarone CV, Verné E, Appendino P. Macroporous bioactive glass-ceramic scaffolds for tissue engineering. Journal of Materials Science: Materials in Medicine. 2006;17 (11):1069–78.

[364] Yelten A, Yilmaz S. A novel approach on the synthesis and characterization of bioceramic composites. Ceramics International. 2019;45(12):15375–84.
[365] Talha M, Behera C, Sinha O. A review on nickel-free nitrogen containing austenitic stainless steels for biomedical applications. Materials Science and Engineering: C. 2013;33 (7):3563–75.
[366] Zhang LC, Chen LY. A review on biomedical titanium alloys: Recent progress and prospect. Advanced Engineering Materials. 2019;21(4):1801215.
[367] Zhang LC, Attar H. Selective laser melting of titanium alloys and titanium matrix composites for biomedical applications: A review. Advanced Engineering Materials. 2016;18(4):463–75.
[368] Disegi JA, Kennedy RL, Pilliar R. Cobalt-base alloys for biomedical applications. ASTM Danvers. 1999.
[369] Hermawan H, Ramdan D, Djuansjah JR. Metals for biomedical applications. Biomedical Engineering-from Theory to Applications. 2011;1:411–30.
[370] Chandrasekaran M. Forging of Metals and Alloys for Biomedical Applications. Metals for Biomedical Devices: Elsevier; 2019. p. 293–310.
[371] Cojocaru V, Raducanu D, Cinca I, Vasilescu E, Drob P, Vasilescu C, et al. Improvement of the corrosion resistance and structural and mechanical properties of a titanium base alloy by thermo-mechanical processing. Materials and Corrosion. 2013;64(6):500–08.
[372] Mohammed MT, Khan ZA, Siddiquee AN. Influence of thermal and thermo-mechanical treatments on mechanical compatibility of biomedical titanium alloys: A review. Corrosion Detection in 'T'Bend Oil Pipelines Based on Fuzzy Implementation. 2012:1388.
[373] Napoli G, Di Pietro O, Stornelli G, Di Schino A. Effect of thermo-mechanical processes on microstructure evolution in austenitic stainless steels. Procedia Structural Integrity. 2019;24:110–17.
[374] Lario J, Vicente Á, Amigó V. Evolution of the microstructure and mechanical properties of a Ti35Nb2Sn alloy post-processed by hot isostatic pressing for biomedical applications. Metals. 2021;11(7):1027.
[375] Yuan B, Chung C, Zhang X, Zeng M, Zhu M. Control of porosity and superelasticity of porous NiTi shape memory alloys prepared by hot isostatic pressing. Smart Materials and Structures. 2005;14(5):S201.
[376] Kaiser R, Williamson K, O'Brien C, Ramirez-Garcia S, Browne D. Effects of hot isostatic pressing and heat treatment on cast cobalt alloy. Materials Science and Technology. 2015;31 (11):1298–304.
[377] Wu M-W, Chen J-K, Lin B-H, Chiang P-H. Improved fatigue endurance ratio of additive manufactured Ti-6Al-4V lattice by hot isostatic pressing. Materials & Design. 2017;134:163–70.
[378] Farvizi M, Javan MK, Akbarpour M, Kim H. Fabrication of NiTi and NiTi-nano Al2O3 composites by powder metallurgy methods: Comparison of hot isostatic pressing and spark plasma sintering techniques. Ceramics International. 2018;44(13):15981–88.
[379] Alshammari Y, Yang F, Bolzoni L. Mechanical properties and microstructure of Ti-Mn alloys produced via powder metallurgy for biomedical applications. Journal of the Mechanical Behavior of Biomedical Materials. 2019;91:391–97.
[380] Bolzoni L, Ruiz-Navas EM, Gordo E. Feasibility study of the production of biomedical Ti–6Al–4V alloy by powder metallurgy. Materials Science and Engineering: C. 2015;49:400–07.
[381] Li Y, Yang C, Kang L, Zhao H, Zhang W, Li Y. Biomedical porous TiNbZrFe alloys fabricated using NH4HCO3 as pore forming agent through powder metallurgy route. Powder Metallurgy. 2015;58(3):228–34.

[382] Dewidar MM, Yoon H-C, Lim JK. Mechanical properties of metals for biomedical applications using powder metallurgy process: A review. Metals and Materials International. 2006;12 (3):193–206.
[383] Khorasani AM, Goldberg M, Doeven EH, Littlefair G. Titanium in biomedical applications – properties and fabrication: A review. Journal of Biomaterials and Tissue Engineering. 2015;5 (8):593–619.
[384] Bagherifard S, Ghelichi R, Khademhosseini A, Guagliano M. Cell response to nanocrystallized metallic substrates obtained through severe plastic deformation. ACS Applied Materials & Interfaces. 2014;6(11):7963–85.
[385] Lin Z, Wang L, Xue X, Lu W, Qin J, Zhang D. Microstructure evolution and mechanical properties of a Ti–35Nb–3Zr–2Ta biomedical alloy processed by equal channel angular pressing (ECAP). Materials Science and Engineering: C. 2013;33(8):4551–61.
[386] Suresh K, Gurao N, Singh S, Suwas S, Chattopadhyay K, Zherebtsov S, et al. Effect of equal channel angular pressing on grain refinement and texture evolution in a biomedical alloy Ti13Nb13Zr. Materials Characterization. 2013;82:73–85.
[387] Rodríguez-Espinoza B, García-Pastor F, Martínez-Poveda B, Quesada A, Lopez-Crespo P. High-strength low-modulus biocompatible Nb-1Zr alloy processed by accumulative roll bonding. Materials Science and Engineering: A. 2020;797:140226.
[388] Kent D, Wang G, Yu Z, Ma X, Dargusch M. Strength enhancement of a biomedical titanium alloy through a modified accumulative roll bonding technique. Journal of the Mechanical Behavior of Biomedical Materials. 2011;4(3):405–16.
[389] Yilmazer H, Şen M, Niinomi M, Nakai M, Huihong L, Cho K, et al. Developing biomedical nano-grained β-type titanium alloys using high pressure torsion for improved cell adherence. RSC Advances. 2016;6(9):7426–30.
[390] Janeček M, Čížek J, Stráský J, Václavová K, Hruška P, Polyakova V, et al. Microstructure evolution in solution treated Ti15Mo alloy processed by high pressure torsion. Materials Characterization. 2014;98:233–40.
[391] Meacock C, Vilar R. Structure and properties of a biomedical Co-Cr-Mo alloy produced by laser powder microdeposition. Journal of Laser Applications. 2009;21(2):88–95.
[392] Zhou Y, Wei W, Yan J, Liu W, Li N, Li H, et al. Microstructures and metal-ceramic bond properties of Co-Cr biomedical alloys fabricated by selective laser melting and casting. Materials Science and Engineering: A. 2019;759:594–602.
[393] Attar H, Ehtemam-Haghighi S, Soro N, Kent D, Dargusch MS. Additive manufacturing of low-cost porous titanium-based composites for biomedical applications: Advantages, challenges and opinion for future development. Journal of Alloys and Compounds. 2020;827:154263.
[394] Sachdeva R, Miyazaki S, Dughaish Z. Nitinol as a biomedical material. Encyclopedia of Materials: Science and Technology. 2001:6155–60.
[395] Biscarini A, Mazzolai G, Tuissi A. Enhanced nitinol properties for biomedical applications. Recent Patents on Biomedical Engineering (Discontinued). 2008;1(3):180–96.
[396] Van Rompaey V, Claes G, Potvin J, Wouters K, Van de Heyning PH. Systematic review of the literature on nitinol prostheses in surgery for otosclerosis: Assessment of the adequacy of statistical power. Otology & Neurotology. 2011;32(3):357–66.
[397] Lu B, Cui X, Dong M, Ma W, Jin G. The influence of solution treatments on the microstructure, phase transformation behavior, and superelastic characteristics of nitinol synthesized by plasma arc deposition. Journal of Materials Engineering and Performance. 2020;29 (4):2491–98.
[398] Tuissi A, Rondelli G, Bassani P. Plasma arc melting (PAM) and corrosion resistance of pure NiTi shape memory alloys. Shape Memory and Superelasticity. 2015;1(1):50–57.

[399] Meisner L, Markov A, Rotshtein V, Ozur G, Meisner S, Yakovlev E, et al. Microstructural characterization of Ti-Ta-based surface alloy fabricated on TiNi SMA by additive pulsed electron-beam melting of film/substrate system. Journal of Alloys and Compounds. 2018;730:376–85.
[400] Altug-Peduk GS, Dilibal S, Harrysson O, Ozbek S. Experimental investigation on the EBM-based additively manufactured prismatic nickel–titanium SMA components. Russian Journal of Non-Ferrous Metals. 2021;62(3):357–67.
[401] Hassan M, Mehrpouya M, Dawood S. Review of the Machining Difficulties of Nickel-Titanium based Shape Memory Alloys. Applied Mechanics and Materials: Trans Tech Publ; 2014. p. 533–37.
[402] Kulkarni VN, Gaitonde V, Karnik S, Manjaiah M, Davim JP. Machinability analysis and optimization in wire EDM of medical grade NiTiNOL memory alloy. Materials. 2020;13(9):2184.
[403] Heimann RB. Plasma-Spray Coating: Principles and Applications: John Wiley & Sons; 2008.
[404] Singh H, Kumar R, Prakash C, Singh S. HA-based coating by plasma spray techniques on titanium alloy for orthopedic applications. Materials Today: Proceedings. 2021.
[405] Singh A, Singh G, Chawla V. Characterization and mechanical behaviour of reinforced hydroxyapatite coatings deposited by vacuum plasma spray on SS-316L alloy. Journal of the Mechanical Behavior of Biomedical Materials. 2018;79:273–82.
[406] Bakhsheshi-Rad H, Hamzah E, Ismail A, Aziz M, Najafinezhad A, Daroonparvar M. Synthesis and in-vitro performance of nanostructured monticellite coating on magnesium alloy for biomedical applications. Journal of Alloys and Compounds. 2019;773:180–93.
[407] Ke D, Banerjee D, Bose S. In vitro characterizations of Si4+ and Zn2+ doped plasma sprayed hydroxyapatite coatings using osteoblast and osteoclast coculture. ACS Biomaterials Science & Engineering. 2019;5(3):1302–10.
[408] Kityk A, Protsenko V, Danilov F, Pavlik V, Hnatko M, Šoltýs J. Enhancement of the surface characteristics of Ti-based biomedical alloy by electropolishing in environmentally friendly deep eutectic solvent (Ethaline). Colloids and Surfaces A: Physicochemical and Engineering Aspects. 2021;613:126125.
[409] Farabi E, Sharp J, Vahid A, Wang J, Fabijanic DM, Barnett MR, et al. Novel biodegradable Zn alloy with exceptional mechanical and in vitro corrosion properties for biomedical applications. ACS Biomaterials Science & Engineering. 2021.
[410] Sun Y, Rong Y, Zhao Y, Yao X, Hang R. The influence of substrate electropolishing on anodization behavior, corrosion resistance, cytocompatibility and antibacterial ability of NiTi alloy. Materials Letters. 2020;268:127631.
[411] Acciari HA, Palma DP, Codaro EN, Zhou Q, Wang J, Ling Y, et al. Surface modifications by both anodic oxidation and ion beam implantation on electropolished titanium substrates. Applied Surface Science. 2019;487:1111–20.
[412] Murr LE. Metallurgy principles applied to powder bed fusion 3D printing/additive manufacturing of personalized and optimized metal and alloy biomedical implants: An overview. Journal of Materials Research and Technology. 2020;9(1):1087–103.
[413] Ni J, Ling H, Zhang S, Wang Z, Peng Z, Benyshek C, et al. Three-dimensional printing of metals for biomedical applications. Materials Today Bio. 2019;3:100024.
[414] Pellizzari M, Jam A, Tschon M, Fini M, Lora C, Benedetti M. A 3D-printed ultra-low young's modulus β-Ti alloy for biomedical applications. Materials. 2020;13(12):2792.

Chapter 4
Biomaterials and the human immune system

4.1 The human immune response toward biomaterials

Biomaterials had help in the advancement of medicine in recent years by being able to improve drug delivery, increase diagnostic capabilities, and help support and regenerate damaged parts of the body. However helpful they may seem, exogenous materials are not readily accepted by the body because of the immune system. The immune system is one of the most important defenses of an organism to all foreign materials and it does not discriminate between helpful or harmful substances or materials. The immune response is a normal reaction of an organism to anything that it deems foreign or nonnative to its host [1, 2]. Its primary objective is to protect the body from pathogenic substances or organism such as toxins, bacteria, viruses, parasites, or foreign bodies that may cause disruption to the homeostasis state [3, 4]. It is considered as the last line of defense of the body as it actively scouts all of the other systems by circulating immune cells or antibodies in search of any invading pathogens or foreign materials that may have otherwise passed the primary barriers of the body such as the skin or the different mucus layers. Incapacitated or impaired immunity may be detrimental for an organism as pathogens or foreign bodies can freely invade the host's cells and cause havoc by using the cell's energy and raw materials to both replicate if it is a bacteria, virus, or parasite. Biomaterials, however useful they me be, are still considered foreign from the host's body and are considered a treat by a working immune system. Therefore, there are certain effects on a biomaterial's function once they come in contact with the immune system. In this chapter, these effects based on the fundamental workings of the immune system are discussed and how researchers are circumventing this enormous challenge to create biomaterials will not be significantly affected by the body's immunity and vice versa.

4.1.1 The basics of the human immune system

The immune system works via two means: first is the cell-mediated immune response made up of leukocytes, or white blood cells, and the second is the antibody-mediated immune response [3, 4]. Both are responsible for acting on pathogenic invaders and they work fundamentally the same by neutralizing the foreign treat but using two distinct mechanisms in doing so. Cell-mediated immunity is governed by different classes of specialized cells that are responsible for scouting and actively killing pathogens or the host's infected cells, and immune cells can also induce other biological responses such as inflammation to subdue the spread of the pathogens, and it does so by secreting a special class of biochemicals called cytokines. Cytokines are

https://doi.org/10.1515/9783110655377-004

chemical messengers made up of a large group of proteins, peptides, or glycoproteins and are responsible for mediating and regulating immunity, inflammation, and hematopoiesis. The antibody-mediated immunity works by actively binding to the foreign material or pathogenic organism so that the immune cells can recognize them or by simply overwhelming its surface for it not to infect or cause damage to the host's cells. The immune system is divided into two classifications: the innate immunity and the adaptive immunity. Innate immunity aims to keep viruses, bacteria, parasites, and other foreign particles out of your body or limit their ability to spread and move throughout the body and works immediately without the need to be activated, whereas the adaptive immunity has the capability to remember previous infections to produce antibodies and increase the production of adaptive immune cells when the same pathogen has re-entered the body (Figure 4.1).

Different classes of immune cells and antibodies are activated depending on the type of pathogen or foreign body material that had passed the primary barriers of the body and based on these classes, the body's response may vary. Immune cells are divided into different classes and they are called granulocytes (basophils, eosinophils, and neutrophils), mast cells, monocytes, dendritic cells (DCs), and natural killer cells. They are mainly responsible for neutralizing any pathogenic treat in the body and are always actively searching for these pathogens or foreign materials and dispose them using different mechanisms. There are also adaptive immune cells called B-cells and T-cells that mainly produce antibodies to neutralize infectious microbes.

4.1.2 Classes of immune cells

4.1.2.1 Granulocytes (basophil, eosinophils, neutrophils, and mast cell)

Granulocytes are a type of white blood cell mainly composed of basophil, eosinophils, and neutrophils. These cells contain small granules that help in digesting pathogenic material [5]. Basophils and eosinophils are important for host defense against parasites such as protozoa, helminths, and ectoparasites. They are also involved in allergic reactions in response to the presence of allergens such as pollen, dust, and small particulate materials [6, 7]. Neutrophils, the most numerous innate immune cells, patrol for problems by circulating in the bloodstream. They can phagocytose, or ingest, bacteria, degrading them inside special compartments called vesicles. Mast cells (also known as a mastocyte or a labrocyte) are a resident cell of connective tissue that contains many granules rich in histamine and heparin. Specifically, it is a type of granulocyte derived from the myeloid stem cell that is a part of the immune and neuroimmune systems. It is also important for defense against parasites. Mast cells are found in tissues and can mediate allergic reactions by releasing inflammatory chemicals like histamine.

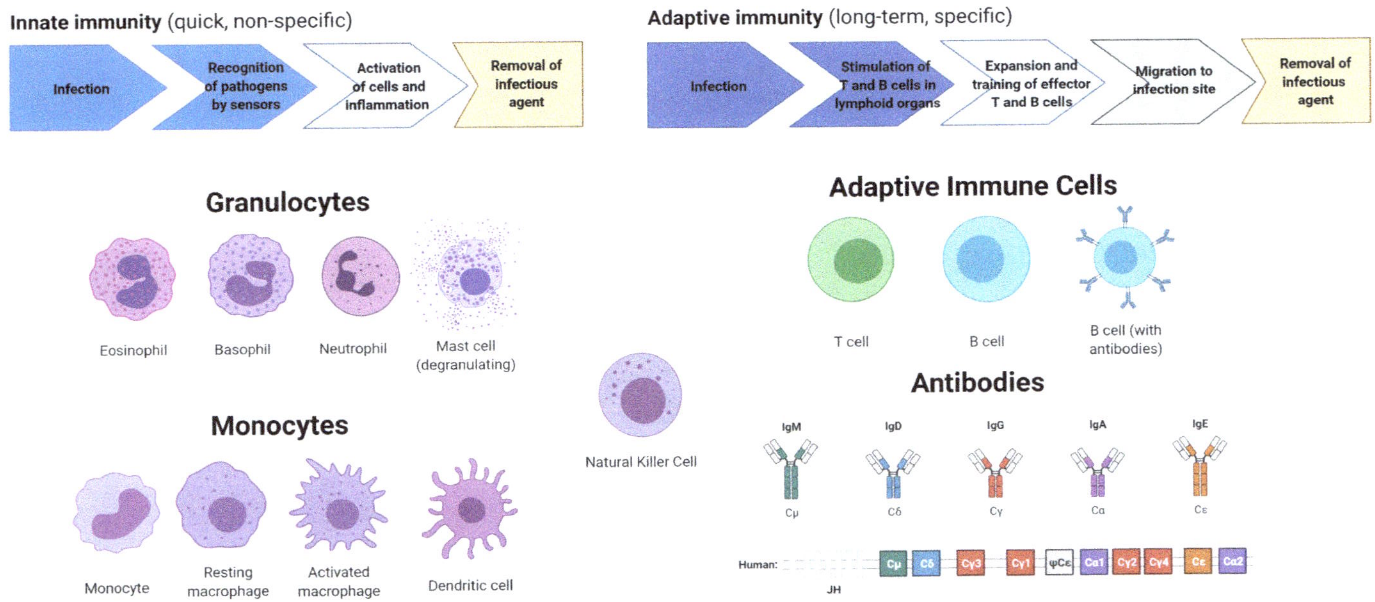

Figure 4.1: The innate and adaptive immunities and the key agents involved within these immunological processes. The innate immunity is both quick and nonspecific responders to any type of pathogens, and the adaptive immunity is both long term and specific to types of pathogens by utilizing adaptive cells and the antibodies they reproduce in cases of reinfection.

4.1.2.2 Monocytes (macrophages and dendritic cells)

Monocytes is one of the largest types of leukocytes that can differentiate into macrophages and DCs, which means that they have direct influence on the adaptive immunity as macrophages and DCs work together with B-cells and T-cells by presenting them antigen targets [8–11]. The appearance of a monocyte contains a unilobed nuclei structure and has a nongranulated cytoplasm, therefore, being classified as an agranular leukocyte. Monocytes and their macrophage and DC progeny serve three main functions in the immune system: phagocytosis, antigen presentation, and cytokine production. Macrophages are large cells that use phagocytosis as a mechanism of cell uptake where the pathogens are taken up and ingested using various enzymes and acidification. Phagocytosed pathogens are entrapped inside a vacuole and are digested inside the cytoplasm, and the resulting debris is then secreted from the cell via exocytosis. These debris become water soluble in the extracellular environment and do not cause any threat to the body. DCs are antigen-presenting cells that are responsible for communicating the antigen that it had encountered from pathogenic microbes or particles to the adaptive immune cells (B-cells and T-cells) so that these cells can seek out the microbes with the same antigen and produce the relevant antibodies to neutralize the threat. They are usually found lurking at the tissues close to the external environment such as the skin and mucus membranes and act as one of the first line of defenses whenever a pathogen crosses the primary barriers.

4.1.2.3 Natural killer cells

Natural killer cells have features of both innate and adaptive immunities. They are known for recognizing and killing virus-infected cells or tumor cells in the host's body [12–15]. They contain intracellular compartments called granules, which are filled with proteins that can create holes in the affected cell to cause apoptosis. Apoptosis is the process for programmed cell death which is a cell death pathway that does not release danger signals that can lead to greater immune activation and inflammation, unlike other cell death mechanism such as necrosis that can also damage the surrounding cells by secreting proteins and pro-inflammatory cytokines [16–19]. Though apoptosis, natural killer cells can limit the removal to only infected cells and reduce the damaged to healthy cells. Recent studies have also shown that natural killer cells have the capability to remember previous infections, like adaptive cells, and attack the same infections if it ever occurs again.

4.1.2.4 Adaptive immune cells (B-cells and T-cells)

B-cell functionalities are to present antigens to T-cells and produce antibodies to neutralize infectious pathogens [20–22]. Antibodies work by coating the surface of a pathogen to undergo three mechanisms: neutralization, opsonization, and complement activation [23–25]. Neutralization of a pathogen occurs when the antibody completely coats its surface rendering it inactive and unable to bind to host cells.

These antibodies specifically bind to the antigen present in the surface of the pathogen making it highly selective to specific viruses or other microbes. In opsonization, once the pathogen has been tagged with the antibody, immune cells like macrophages and neutrophils are alerted so that they can ingest them through phagocytosis. Whereas in compliment activation, B-cells directly lyse the pathogen using antibodies that can have cytotoxic effects on both the pathogen and infected cells [26–28]. Antibodies are expressed in two ways in a B-cell. First is through B-cell receptor (BCR), which sits on the surface of a B-cell is an antibody, meaning that the B-cell can directly bind to the pathogen with the use of antibody anchored on its surface. B-cells also secrete free antibodies to diffuse and bind to pathogens. This dual expression is important because the initial problem, for instance, a bacterium, is recognized by a unique BCR and activates the B-cell. The activated B-cell responds by secreting antibodies, essentially the BCR but in soluble form. This ensures that the response is specific against the bacterium that started the whole process. B-cells, being adaptive immune cells, mean that from the beginning of the life span, they are not active in seeking, fighting, or preventing infections. At this stage, they are called naïve B-cells as they have not been activated by DCs and have not yet instructed to make specific antibodies for the infecting pathogen. Once they are activated, these naïve B-cells differentiate into plasma cell that can create different classes of antibodies, and regulatory B-cells that once an infection has been quelled, downregulates the immune response by secreting immune-regulating cytokines. Some of the B-cells also differentiate into memory B-cells that lay dormant in the lymphatic system and memorize what antibody to produce if the host gets infected by the same pathogen in the future [28]. This response can be stimulated either naturally via exposure to active pathogen or artificially via inactive pathogen inoculation also known as vaccination [29].

T-cells are classified into two categories called CD8+ and CD4+ T-cells, based on the protein present on the cell's surface. T-cells have multiple functions, including killing infected cells and activating or recruiting other immune cells. CD8+ T-cells are also known as cytotoxic T-cells or cytotoxic lymphocytes (CTLs) and they are responsible for recognizing and removing virus-infected cells and cancer cells. CTLs have specialized compartments, or granules, containing cytotoxins that causes apoptosis to targeted infected cells or cancer cells. These granules contain highly potent cytotoxins; therefore, the release of these granules is tightly regulated by the immune system, and overexpression of these granules can also affect normal cells. The four major CD4+ T-cell subsets are TH1, TH2, TH17, and Treg, with "TH" referring to "T helper cell." TH1 coordinates immune responses against intracellular bacterial microbes and they produce and secrete molecules that alert and activate other immune cells, such as macrophages that are known for ingesting bacterial pathogens. TH2 cells are important for coordinating immune responses against extracellular parasites, such as helminths (parasitic worms), by alerting B cells, granulocytes, and mast cells. TH17 cells can produce interleukin 17 (IL-17), a signaling molecule that activates and recruits neutrophils and important for regulating immune responses. Regulatory T-cells (Tregs) monitor and inhibit the activity of other T-cells. They prevent adverse immune activation and maintain tolerance, or the prevention of immune responses against the body's own cells and antigens.

4.1.3 Communication between immune cells

Immune cells communicate in several ways, either by cell-to-cell contact or through secreted signaling molecules called cytokines and with receptors and ligands which are fundamental for cellular communication especially with immune cells. Cytokines are small proteins with diverse functions in the body [30, 31]. In the immune system, cytokines are categorized based on their function on immune cell growth, activation, and modulation of the immune response. Cytokines are classified as follows:

1) **Colony-stimulating factors** are used for cell development and differentiation of immune cells [32–34].
2) **Interferons** activate immune cells. Type I interferons mediate antiviral immune responses, and type II interferon is important for antibacterial responses [35–37].
3) **Interleukins** is a group of naturally occurring proteins (30 varieties) that provide context-specific instructions whether to activate or inhibit a specific immune response [38, 39].
4) **Chemokines** are location-specific biomolecules that attract immune cells in the site of infection. Different chemokines will direct specific immune cells to the site where they are expressed [40, 41].
5) **The tumor necrosis factor (TNF)** family of cytokines that stimulates immune-cell proliferation and activation. TNF is heavily involved in the inflammation process in order to stop the spread of an infection and increase the capillary flow on the infected area so that immune cells can easily migrate to the site [42].

On the other hand, receptors are protein structures that are expressed on the surface of a cell or in intracellular compartments which are activated by free-floating or membrane-bound ligands. The receptor–ligand communication is like a lock and key mechanism where only specific shaped proteins of ligands can activate a receptor on the surface of the cell. Ligand–receptor interaction between immune cells leads to a cascade of events inside the cell that involves networks of intracellular molecules that relay the message if a pathogen has been identified and must be engulfed or tagged with an antibody. Modulating the expression and density of various receptors and ligands, the immune cells can relay specific instructions based on the pathogenic treat at hand.

Specific receptors such as toll-like receptors (TLRs) are expressed on innate immune cells (e.g., macrophages and DCs) [43–45]. TLRs may be membrane bound or intracellular compartment bound depending on the presence of pathogen whether they are inside a cell or free floating in the extracellular space. TLRs recognize general microbial patterns, and they are essential for innate immune-cell activation and inflammatory responses.

In the adaptive immune cells, BCRs [46–48] and T-cell receptors (TCRs) [49, 50] are expressed and both are found on the cell surface, but BCRs are also secreted as antibodies to neutralize pathogens. The genes responsible for the creation of these receptors are randomly rearranged at specific cell-maturation stages, for these receptors to recognize anything and memorize previous infections so that it can act immediately when the same infection reoccurs. Therefore, memory B- or T-cells are highly specific and, upon re-exposure to their specific pathogen, can immediately induce a neutralizing immune response.

4.1.4 Immunoglobulins and their role to immune response (antibody immune response)

Immunoglobulins, also known as antibodies, are heterodimeric glycoprotein molecules produced and are structurally composed of two heavy and two light chains. They can be separated functionally into variable domains that bind antigens and constant domains that specify effector functions, such as activation of complement or binding to Fc receptors. Their role on the immune response is to recognize and bind to antigens, such as bacteria or viruses, and foreign body materials in order to neutralize them and not cause any damage in the body [51, 52]. This is also known as the antibody immune response and is an important aspect of overall immune protection of the body toward pathogens. Understanding different immunoglobulin structure and classes is important in assessing various kinds of pathogenic derived disease and is crucial in the field of biosensing. Specific immunoglobulins are typically created by the plasma cells or b-cells in response to a specific antigen presenting pathogen. Antibodies can form what is called the antibody–antigen complex [53–55], where specific antibodies attach to the antigen presented on the cell membrane of the pathogen. If pathogenic toxins are produced by infecting microbes, antibodies can also neutralize those toxins. Finally, antibodies can also attach to the receptors of viruses and bacteria for them not to infect any cells inside the body.

There are two main forms of antibodies: soluble antibodies and membrane-bound antibodies [56–58]. Soluble antibodies are free roaming in the systemic circulating while membrane-bound antibodies are bonded on the surfaces of B-cells. Soluble antibodies are usually created by plasma cells and are produced in large quantities and have shorter half-life, they are usually increased during an occurrence of an infection, while membrane-bound antibodies are attached onto memory B-cells and last longer and can easily be triggered when the same infection hits the body. There are two prototype antibodies that are produced by B-cells: IgM and IgD [59]. However, their role is to initially bind to the antigen forming an antibody–antigen complex. It is the Ig alpha and Ig beta that enable the cell to transduce the signal and respond to the presence of antigens on the cell surface. This signal transduction allows for the differentiation and proliferation of B-cells and antibody production inside the plasma cell [59, 60]. Another function of B-cells in adaptive immunity is called "class switch" that causes expression of a new antibody isotype in response to the cytokines present in the area [61, 62]. Various cytokines, such as IL-4, IL-5, IFN-gamma, and TGF-beta, are known to be responsible for class switching [63]. Class switching is important in order to respond properly to various mutations of a certain virus or a similar bacterial species infection. However, at a certain stage, the cell will lose its ability to undergo a switch to a class that has been generated before. Immune responses can vary with each antigen presented to the immune system, so quantifying specific antibody levels helps to interpret the immune response after immunization or vaccination. Human monoclonal antibody levels are used as diagnostic indicators to determine immunoglobulin-deficiency disorders [64, 65], such as autoimmune diseases and chronic inflammation that can be characterized by specific isotype deficiencies or varying concentrations of one or more immunoglobulin isotypes.

Various antibodies produced by plasma cells are classified by isotypes [66]. These antibodies differ in function and antigen responses due to their structure variation. Five major antibody classes are known: IgA, IgD, IgE, IgG, and IgM [60, 67, 68]. This classification is based on differences in amino acid sequence in the constant region (Fc) of the antibody heavy chains. IgG and IgA can be further classified into subclasses IgG1, IgG2, IgG3, IgG4, IgA1, and IgA2 [51, 69]. This is based on additional small differences in the amino acid heavy chain sequences. Based on differences in the amino acid sequence in the constant region of the light chain, immunoglobulins can be further sub-classified by determination of the type of light chain (kappa light chain or lambda light chain) [70, 71]. A light chain has two successive domains: one constant domain and one variable domain. The ratio of these two light chains differs greatly among species, but the light chains are always either kappa or lambda, but never one of each. Determination of individual subclasses is used in assessing primary immunodeficiencies or immune responses, especially if the total IgG or IgA concentration whether it is decreased or elevated and is important in diagnosing certain diseases.

4.1.5 The major histocompatibility complex and activated complement proteins (proteins role in immunity)

Proteins play a crucial role in immunity and they serve to general functions. one is the major histocompatibility complex (MHC) and the other is called activated complement proteins [72]. MHCs are proteins used by an organism's cell to identify itself as a nontreat to the immune system and the surrounding tissue by acting as carriers to present antigens on a cell's surface. There are two major types of MHC called class I and II (Figure 4.2) [73]. The MHC class I proteins are expressed by nearly all cell types, except for red blood cells and are essential for presenting viral antigens if a cell is infected [73, 74]. If a cell gets infected by a virus, it can signal the problem

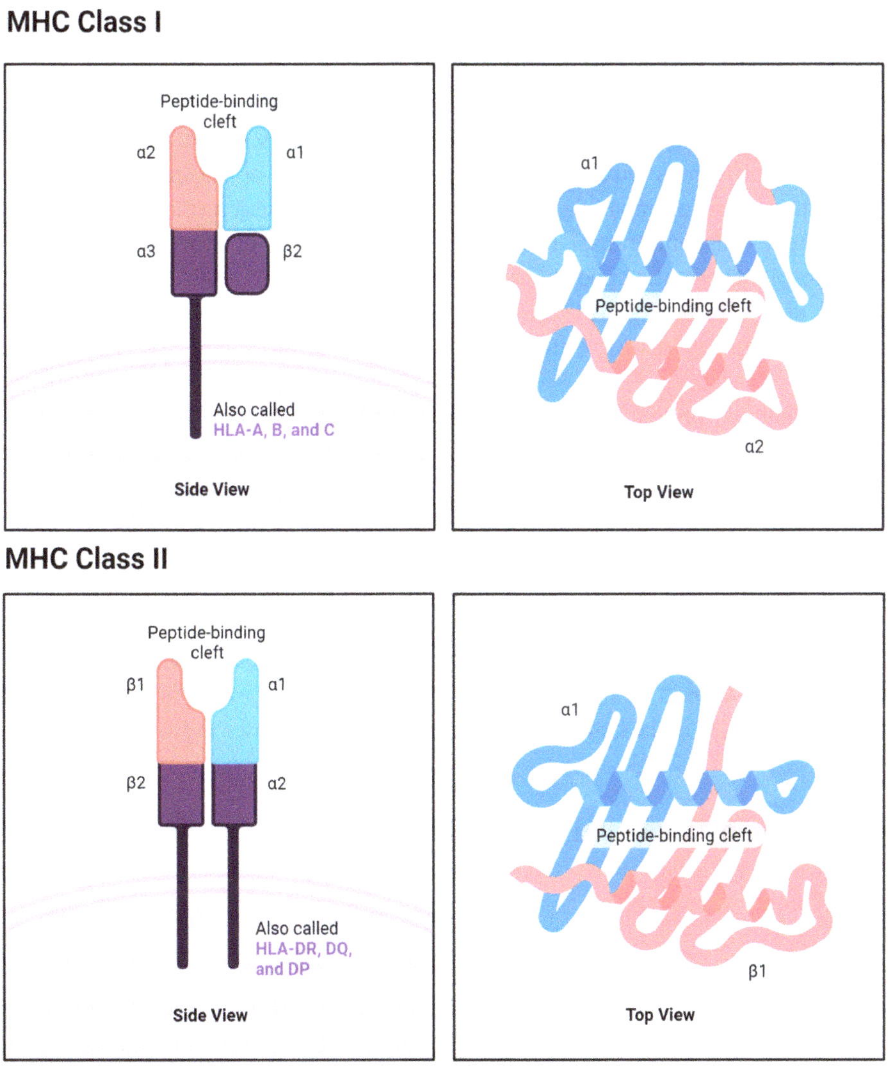

Figure 4.2: The major histocompatibility complex (MHC class 1 and 2) are both important concepts in tissue engineering as they are responsible for determining whether the implanted tissue material will be rejected or accepted. MHC class 1 and 2 are both surface proteins expressed by cells and are specific to native or nonnative cells. If the transplanted tissue has MHC incompatibility, and the immune cells can recognize it as non-self and eliminate the foreign body accordingly. This is also known as tissue implantation rejection.

through MHC class I proteins and in response, CD8+ T-cells can recognize the infection and kill the infected cell by releasing apoptotic molecules. The MHC class II proteins are mostly expressed by antigen-presenting immune cells like DCs and macrophages and are vital in communicating antigens to CD4+ T-cells. MHC class II antigens vary and include both pathogen- and host-derived molecules [73, 75, 76].

MHC proteins is the way of the body for recognizing if the cell is native or foreign to itself. Each person has a unique set of MHC proteins that are determined by their genetic makeup. Genetically similar individuals such as those seen within a family have almost similar MHC proteins. Immune cells use MHC proteins to determine whether a cell is native or foreign in organ transplantation scenario. MHC proteins of recipients and donors must match to increase the success of transplantation. If the MHC proteins are incompatible, the transplanted organ or tissue will get attacked by the immune cells, and transplant rejection occurs. In cases of bone marrow or stem cell transplantation, the opposite happens, as the donor cells attack the host's body. This is also known as graft versus host disease. Therefore, clinicians usually use strong immunosuppressing drugs to lessen the risk of transplant rejection or graft versus host disease.

Another way protein is used to clear away pathogens or dying cells and activates immune cells with the use of the complement system. A complement consists of a series of proteins found in the blood that forms a membrane-attack complex [77–79]. Complement proteins are selectively activated by enzymes when an infection occurs and once activated, complement proteins attach to a pathogen and recruit additional complement proteins that form around a pore or a hole. Complement proteins can create small holes into the pathogen by acting like a mechanical punch, creating leaks in the pathogenic microbe that leads to their death. Complement proteins like cytokines can also act as signaling molecules to alert immune cells and recruit them to the affected site in the body.

4.2 The body's general response toward biomaterials

4.2.1 Foreign body response

As mentioned in the beginning of the chapter, biomaterials are identified by the immune system as a foreign material and the immediate reaction of the body toward a nonnative material is usually hostile [80, 81]. Well-designed biomaterials that are used inside the body are expected not to elicit severe and chronic response from the immune system. Implants in general depending on which material they are made of only last in the body temporarily, for example, plastic and metal implants used for hip and knee repair need to be replaced after 20 years, whereas other materials used in stents and aesthetic implants have shorter life spans inside the body. This adds burden to the patient as they are expected to undergo another round of surgery to replace the deteriorating implanted materials and avoid further complications. All these deteriorations are primarily due to the body's immune response toward a foreign material.

The governing universal response of the body toward biomaterials made up of inert materials such as ceramics, biomedical-grade alloys, or polymers is called the foreign body immune response (Figure 4.3) [80, 81]. This initial immune response is associated with the M1 and M2 macrophages, giant cells (monocytes + macrophages), and frustrated macrophages. M1 macrophages are responsible for inflammatory response, whereas M2 microphages are associated with wound healing and tissue remodeling [82, 83]. Giant cells are the formation of fused macrophages and monocytes to increase their response to the foreign material, as they release combinations of enzymes and nitric oxide to degrade the material. These are also known as "frustrated

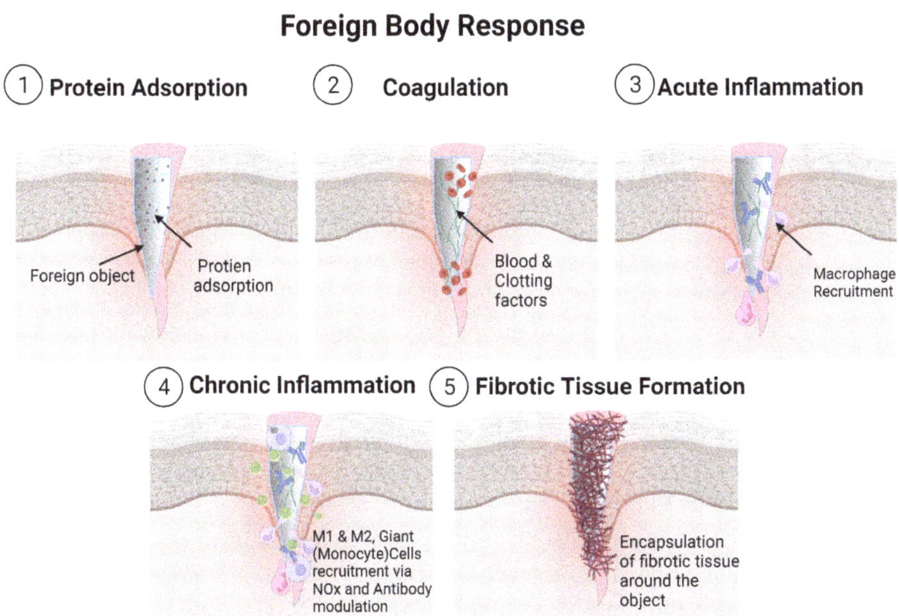

Figure 4.3: The foreign body response is the body's reaction to any foreign material that are introduced inside the body. It comes in stages depending on the severity of the response. (1) Proteins adhere onto the surface of the object in order to isolate it from the rest of the tissue and attract other immune cells to the area. (2) Blood and clotting factors are also recruited into the area to prevent further injury to the site. (3) Acute inflammatory response is triggered by the recruitment of cytokines, antibodies, and macrophages into the site. (4) If the foreign material persists, chronic inflammation starts though M1 and M2, giant monocyte recruitment and NO_x are further employed to digest the foreign material. (5) Eventually, fibrotic tissue forms around the material if it is overly persistent and the immune agents are not successful in eliminating it. The fibrotic tissue isolates the material from the rest of the body.

macrophages." If the material is small enough, they would try to phagocytose it to break them up into their basic components in their intracellular compartments. If the material size is too big to be phagocytosed or degrade using the combination of enzymes and nitric oxide, other types of immune response kicks in such as the attachment of free antibodies on the surface of the material which serves two purposes: first, they create a neutralizing barrier on the surface of the foreign material and, second, they recruit other immune cells such as neutrophils so that they can secrete their granules to degrade the material. If these responses are not enough, fibrotic cells are also recruited to encapsulate the material and create a connective tissue surrounding the foreign material and act as an impenetrable barrier to keep the host safe and prevent further material interaction with the surrounding tissue [84, 85]. In cases of tissue-derived biomaterials, such as those seen in cell-laden biomaterials or animal-derived materials such as collagen, gelatin, and hyaluronic acid, antigens are involved and therefore other immune cell types are recruited into the implantation site in addition to the abovementioned response.

4.2.2 Inflammation and wound healing process

Other host reactions are also induced when a body is implanted with a biomaterial. The immune response is just part of a sequence of reaction and to create a complete picture, the following must be discussed. First, after implantation of a biomaterial, the immediate site experiences an injury through cutting and placement of the biomaterial. The injured site experience what is called exudation where fluid, proteins, and blood cells escape from the vasculature into the surrounding tissue. Once this occurs, the biomaterial interacts with blood and might cause thrombosis and/or blood clotting. After this initial contact and if the biomaterial is hemocompatible, hemostasis happens as fibrin and other clotting factors are recruited in the site. Inflammatory products (e.g., platelets, interleukins, and cytokines) are also recruited around the implantation site. This will then trigger an acute inflammatory response where further exudation and migration of leukocytes happen due to the chemical signaling of the inflammatory products [86–88]. Here, leukocytes, specifically neutrophils, phagocytose the debris of damaged cells. After the acute inflammatory phase, macrophages, lymphocytes, and other plasma cells are recruited in the area and are involved in the chronic inflammatory step where they will try to degrade the remaining foreign body material and damaged cell debris [87, 89]. The body will now then try to heal itself by stimulating the proliferation of fibroblast cells to synthesize extracellular matrices (ECM) around the implantation area and by generating new vasculature (also known as angiogenesis). If the presence of the biomaterial does not elicit more immune response and the healing has started, regeneration of the injured tissue surrounding the biomaterial will then occur by creating a scar tissue. These scar tissues are made up of fibrotic cells [90, 91] and act as an imperfect replacement to the damaged tissue. These reactions depend on which classification of biomaterials is used in the implantation site. Some biocompatible and bioactive materials stimulate engraftment around the tissue and become part of the tissue itself, whereas bioinert materials elicit the said response in the body.

4.3 The body's response toward different classes of biomaterials and biomedical devices

Different biomaterials will elicit multiple response in the body, as some biomaterialists will design biomaterials depending on the goal of the treatment and function of the device they are developing. The following section of the chapter discusses how different classes of biomaterials interact with the body and how researchers are designing them to attenuate or increase these responses to prolong the functionality and life span of these biomedical devices (Figure 4.4).

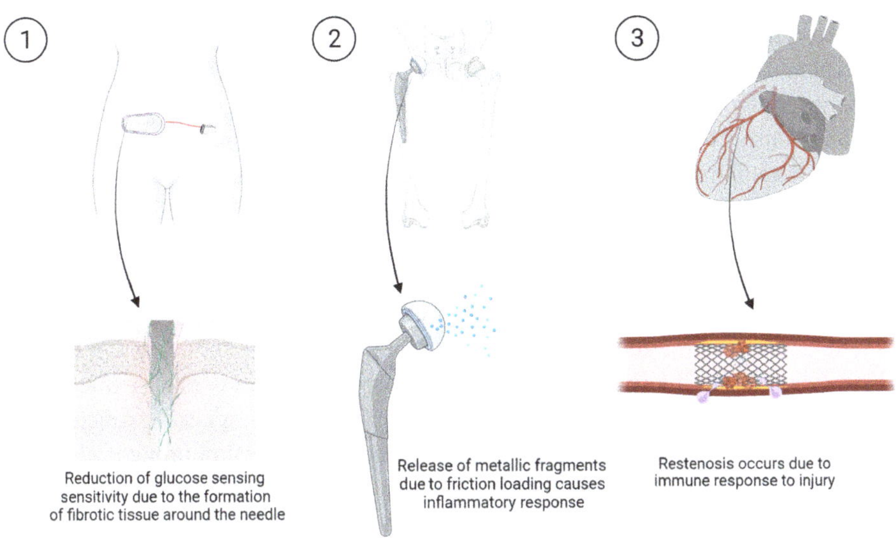

Figure 4.4: Effect of the immune system to the effectivity and longevity of biomedical devices (material response). (1) Reduction of glucose-sensing capabilities of continuous glucose monitors. (2) Inflammation due to the fragment release from metallic and ceramic prosthesis. (3) Restenosis (blockage) of stents as the immune response to injury caused by metallic stents.

4.3.1 Metallic materials and the immune system

Biomedical-grade alloys are one of the oldest and widely used materials in creating biomedical devices. They are crucial in biomedical sensing, support, and reconstruction of affected limbs. Metallic alloys are classified as inert materials and usually do not elicit severe immune response in the body. However, over time as the material experiences degradation via other means such as through friction loading, corrosion from hydrolytic, acidic, or alkaline environments, they can form fragments that may stimulate an immune response [92–94]. These are prevalent in artificial joint replacements where constant shearing and load can deteriorate the surfaces of the implant and small microscopic fragments can cause inflammation in the surrounding tissue. Furthermore, there is always a constant risk of infection corresponding to the implantation procedure of these implants that can be detrimental to the patient if left untreated. This is known as nosocomial infection brought about by contamination of the implant material with pathogenic microbes.

4.3.1.1 Artificial joint implants (titanium, stainless steel)

Artificial joint implants are made up of corrosive-resistant and mechanically robust alloy to support the constant movement and articulation of a limb. The stem of these implants is anchored inside the bone and directly interfaces with the hard tissue of the body, whereas the head or the articulating part of it directly interfaces with the other part of the implant. Bones in general have more stable environment for implants as most of them are made up of inorganic materials such as calcium and phosphate, which means that the anchorage of these implants does not elicit severe or chronic inflammatory response in the body. However, the articulating part of the implant is subjected to constant wear and this could lead to the release of ions or microscopic metallic particles in the surrounding joint. If this happens, immune cells are alerted to the area and inflammation could occur [95, 96]. This scenario can be averted by ensuring that the materials that are always in contact have strong toughness, and a combination of ceramic cup and metal head is used to reduce the friction generated when the joints articulate. Reducing the friction being generated in the sliding materials helps prevent metal ion release due to mechanical degradation [97]. Another response elicited by arthroplasty (artificial joint replacement) is called septic arthritis when nosocomial infection brought about by contamination of microbes during implantation occurs. Septic arthritis is a painful infection in a joint that can come from germs that travel through your bloodstream from another part of your body [98, 99]. To reduce this risk, antibacterial agents are usually coated onto the surface of the implant so that microbes cannot adhere onto the implant. Silver nanoparticle coating or adding antibacterial drugs onto the implant surface are the most common strategies being used by researchers to ensure that even when contamination occurs, the microbes cannot survive on the surface of the implant. New research is also being done to add anti-inflammatory coatings such as phenolics onto the surface of metals to make sure that the surrounding cells on the joints does not secrete proinflammatory cytokines that can recruit immune cells [100–102]. In other metallic bone implants, such as screws, plates, and wires, they are usually passivized by implementing oxide coatings and/or polishing them into smooth surfaces in order for the implant not to elicit severe inflammatory responses. Newer strategies in orthodontic implants also employ the addition of bioactive components on their surfaces to reduce inflammation reaction and biofilm formation [103].

4.3.1.2 Biosensing using metallic needles (insulin pumps)

Closed-loop insulin pumps are commonly referred to as "artificial pancreas" integrated systems that combine continuous or intermittent glucose reading together with the release of insulin from an external reservoir. This minimizes the need for frequent finger pricking and needle handling to dose insulin whenever needed by a diabetic patient. However, this system is known to have a temporary life span as the needle of the pumps is attacked by the innate immunity as a reaction to a foreign material [104, 105]. The usual life span of a needle in an insulin pump lasts for only 2 weeks and must be replaced with a new needle to ensure that proper glucose level sensing and insulin dosing are achieved. Insulin infusion set failure is the most common problem encountered by diabetic patients on continuous insulin therapy [106]. This occurs when infusion sets are worn for long periods of time and causing blocked or dislodged sets or insulin leakage back to the skin surface. The foreign body response is one of the main causes of this failure as the immune cells, antibodies, and fibrotic cells tend to clog the surface of the infusion needle. This failure type is considered as one of the weakest parts of continuous subcutaneous insulin infusion. Researchers are now looking for ways to solve this problem by using innovative materials to be used as an infusion set. Teflon coating on needles is one of the common and relatively successful strategy to ensure that protein and cell adhesion onto the needle is attenuated [107]. Teflon or polytetrafluoroethylene (PTFE) is a bioinert nonadhesive material that is known to have a very low dielectric constant, meaning that it reacts with very few compounds inside and outside the body. PTFE is also known to be well tolerated by the immune cells. It has the capability to be stealth-like due to its nonreactiveness. However, PTFE is nonelectroconductive meaning that it is a poor sensing material for glucose level. Thus, it must be incorporated with conductive metallic needles that can alert the immune response but in a slower and more mediated manner, prolonging the life span of the infusion needle of the insulin pump. Furthermore, insulin pumps must be anchored in the skin of the user, and the skin is known to harbor pathogens such as *Staphylococcus aureus* and *Staphylococcus epidermidis*. When a user has worn the infusion pump for prolonged periods of time, it can act as an entry point for pathogens to enter the body. The American Diabetic Association recommends changing the infusion set catheter at least every 3 days. Longer use of continuous subcutaneous insulin infusion pumps is associated with microbial colonization and may consequently lead to infection. These recommendations add cost and create inconvenience. Therefore, in order to prevent infections, significantly extend the duration between insertion site changes, researchers had also investigated incorporating antimicrobial combination of gentian violet and chlorhexidine (gendine) onto Teflon-coated infusion needles to have a synergistic biofouling and antimicrobial effect.

4.3.1.3 Vascular stents

Vascular stents are flexible metallic tubular meshes that are placed in the luminal side of a vasculature to regain patency from obstruction due to plaque build-up, thereby increasing blood flow and perfusion to an affected limb or organ. This is done by the stent by exerting radial pressure around the affected luminal side of the vasculature and press the plaque occlusion onto the sides of the lumen. This radial pressure is maintained by the structure of the stent as most stents are made up of metals such as stainless steel, chromium cobalt, and/or nickel-titanium (nitinol). The implantation procedure for stents unlike other implants is done via a minimally procedure called vascular catheterization, where a special device called percutaneous catheter enters the body via a small incision (usually along the femoral artery) and is maneuvered inside the vasculatures of the affected limb (legs) or organ (heart). Once the location of the occlusion has been identified via radiography or fluoroscopy, the catheter mechanism allows it to deploy the stent via two ways: If the stent is made of shape-memory alloys, the stent expands by itself to relieve the occlusion. The other way is through balloon expansion where a silicon balloon is used to expand the stent into its proper diameter. A vascular stent has a typical lifespan of around 5–10 years depending on the condition of the patient. One of the main reasons for a shortened device lifespan of a stent is when there is injury into the vessel created by the stent itself or the accumulation of fatty acids and blood products (plasma and erythrocytes) onto the struts of the stent. When this happens, a scenario called restenosis occurs meaning that the stent becomes occluded once again reducing blood flow and perfusion to the affected limb or organ [108, 109]. This requires further surgical intervention by removing the occluded stent and/or replacing the stent with a new one. Injury to the vessel, fatty acids, and blood products' accumulation, especially plasma, has the tendency to recruit immune cells by stimulating the surrounding endothelium to release cytokines that recruit macrophages and neutrophils to the luminal side of the stent. Once the leukocytes infiltrate the endothelial wall of the vessel, it will then release growth factors to try to heal the damage; however, this also triggers an initial inflammatory response [110–112]. The endothelium will try to envelop the stent by stimulating its proliferation process. This leads to a more severe complication called a restenotic lesion that reoccludes the vessel and making it difficult to remove the stent as it has been incorporated by the thickening of the endothelial wall of the vessel lumen. This effect is significantly observed with first generation stents, as these devices are only made up of metallic components. Newer generation of stents has the capability to avoid restenosis by adding the capabilities to release drugs that can stop the proliferation of the endothelium such as paclitaxel [113], or the accumulation of blood products on its surfaces via the use of heparin [114]. However, the controlled release of these drugs must be maintained at a therapeutic level for a period of time; thus, stents use polymeric coatings or addition of polymeric cover onto the stent strut not just as a drug delivery platform but also to mechanically block the endothelial cells from enveloping the stent struts.

4.3.2 Ceramic materials and the immune system

Ceramic materials are used mostly for bone tissue regeneration and as components in implantable bone prosthesis to increase their osteogenic activity and improved anchorage. In general, ceramics or bioceramics do not elicit immune responses as they are like natural inorganic components of the body, especially, hydroxyapatite and beta-tricalcium phosphate. They are chemically similar to the inorganic component of bone tissue, hence when an implant is coated with bioceramics, it can attenuate the immune and inflammatory response of the body. However, other studies have investigated synthetic ceramic materials. These studies have revealed that treatment with a clinical standard for ceramic implants (tricalcium phosphate–hydroxyapatite) promotes an M1 macrophage phenotype while newer ceramic materials (e.g., baghdadite and strontium–hardystonite–gahnite) induced greater expression of markers indicative of M2 macrophage phenotypes [115, 116]. There are several reasons for different macrophage phenotype activation in the presence of bioceramics: first is that the ions are being secreted by these ceramics and can cause gene expression variation of macrophages; second is the surface topography and grain sizes where macrophage phenotypes might mechanically upregulate or downregulate based on the surface roughness of the ceramic [117, 118]; third is the stiffness of the ceramic materials as cells attached preferentially to stiffer substrates and secretion of proinflammatory cytokines can be modulated based on the mechanical properties of the substrate [119]. The immunomodulatory effects of metal ions from zinc-, zirconium-, and strontium-based bioceramics are also now widely known, creating a paradigm shift away from inert bone substitutes like calcium- or phosphate-based bioceramics [120, 121].

4.3.3 Polymeric materials and the immune system

Polymers generally do not elicit immune responses in the body. However, some natural polymers can activate the body's immune response, especially those that are derived from animal sources due to the remnants of xenogeneic antigens that are not removed during the manufacturing process (Figure 4.5) [122]. Biopolymers that are applied as a scaffolding material for tissue engineering in various anatomical systems are mostly made from decellularized ECM from exogenic sources (animals).

This may lead to hyperacute rejection following organ xenotransplantation, from what is called the Gal epitope [123]. The Gal epitope is a sequence of antigen oligosaccharide "α-Gal (Galα1,3-Galβ1–4GlcNAc-R)" found in the surface of other mammals but not in humans and old-world monkeys. Humans have evolved to not express the Gal epitope due to the mutations in the α1,3-galactosyl-transferase gene, and the production of large amounts of antibodies such as IgG, IgM, and IgA as a result of constant exposure to intestinal bacteria. IgG is known to be anti-gal leading to the removal of the gal epitope in the human species. The presence of the gal epitope in biologic scaffolds can lead to calcification and degeneration of the transplanted material and efforts to attenuate this are being done by treating the scaffolding material with an enzyme called α-galactosidase to digest the oligosaccharide sequence of the gal epitope and by using transgenic herds of pigs that have been produced in which this epitope has been knocked out as a source for decellularized ECM.

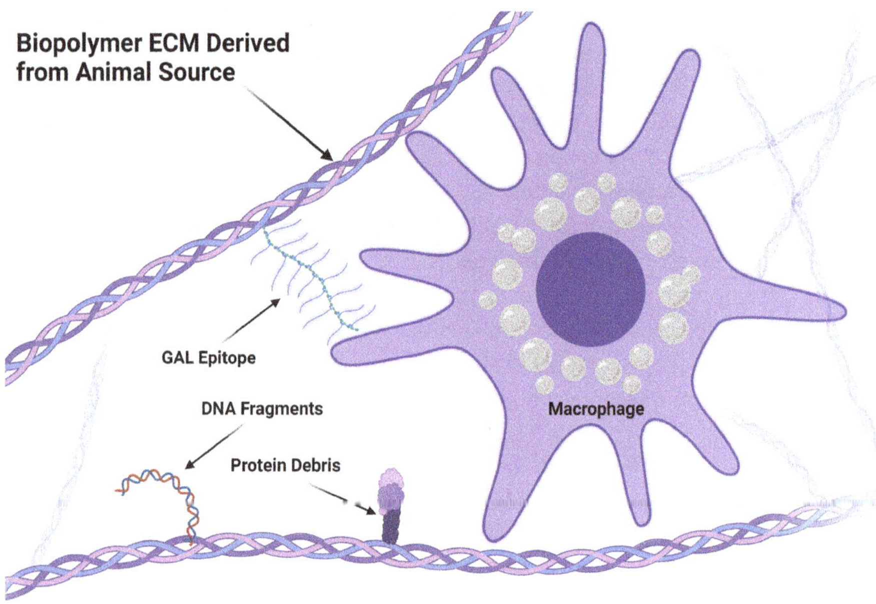

Figure 4.5: Response of the body toward animal derived (bovine and swine) biopolymer extracellular matrix materials for tissue engineering. GAL epitope, DNA fragments and protein debris that are unsuccessfully removed during the isolation and synthesis process causes macrophages to be recruited toward the material and decrease the effectivity of the implanted material.

Other antigenic factors are with remnant DNA within biologic scaffold materials [124]. DNA remnants from incomplete decellularization process of the ECM can lead to inflammatory responses following post implantation of xenogeneic ECM. However, these DNA fragments are known to be fragmented and are less likely to cause any adverse effects in tissue remodeling. Since DNA is a fragile entity and are easily fragmented over time once exposed to external environment, there are less concerns for DNA to elicit severe immune response as any remnant DNA is logically subject to the same degradation fate via enzymatic breakdown. TLRs may play an important role in this regard as they bind soluble DNA so that they can be broken down into nucleotides for future use by the cells. It has been regarded that the presence of small fragments of DNA in commercially available ECM are unavoidable, and the response they can elicit is outweighed by the positive benefits of the ECM scaffolding products. However, the host response to biologic scaffold materials composed of ECM involves both the innate and acquired immune systems, and the response is affected by device-specific variables including the intended clinical application, the source of the raw material/tissue from which the ECM is harvested, and the processing steps involved in manufacturing an approved medical device.

Bioderived ECM materials can elicit both the adaptive and innate immune responses if not completely removed with antigenic factors (e.g., DNA, surface proteins, and Gal epitope), namely the Th1 and Th2 lymphocyte responses and M1 versus M2 macrophages response, respectively. The role of T-cells, especially the Th1 and Th2 lymphocyte phenotypes, in cell-mediated immune responses to xenografts is to produce cytokines such as interleukin (IL)-2, interferon (IFN)-γ, and TNF-β leading to macrophage activation, stimulation of complement fixing Ab isotypes (IgG_{2a} and IgG_{2b} in mice), and differentiation of CD8+ cells to a cytotoxic phenotype which are all hallmarks of xenogeneic transplant rejection. Whereas the Th2 lymphocytes produce IL-4, IL-5, IL-6, and IL-10 cytokines that do not activate macrophages and that lead to production of noncomplement fixing Ab isotypes (IgG_1 in mice) and are associated with transplant acceptance. When macrophages are both recruited in the transplantation area, distinct phenotypic polarization profile can happen, like the Th1/Th2 polarization schemes for lymphocytes. When Th1 lymphocyte dominates the site, the pro-inflammatory, cytotoxic macrophage phenotype, signified as M1, is recruited to promote phagocytosis and release of enzymatic and nitric oxide degradation compounds which are associated with classic signs of inflammation, especially chronic inflammation. This contributes to transplant rejection and can hamper tissue remodeling process. But when Th2 lymphocytes are dominating the site, the anti-inflammatory macrophage phenotype, signified as M2, is recruited which promotes immunoregulation, and tissue repair, by producing high levels of IL-10 and TGF-β expression, and large amounts of arginase inhibit the release of proinflammatory cytokines, scavenge debris, promote new vasculature formation (angiogenesis), and recruit cells involved in constructive tissue remodeling.

Lastly, degradation of polymeric materials also plays a role into the immune response and overall tissue remodeling. Polymers can be cross-linked with a cross-linker and can reduce the rate of degradation of the polymeric material. However, degradation of the polymeric material must be optimized to coincide with the remodeling process of the tissue as it is a requisite for the recruitment of host cells, and ultimately to site specific tissue remodeling. If the material lingers longer than expected, it might get in between the nascent cells and surrounding tissues and limit the integration of the newly formed tissues into the host's body. The length of time that a body is exposed to foreign material with antigens certainly affects the type of immune response that will be elicited, but the effect of degradation rate upon the immune response has not been fully investigated.

4.4 Immune responses to nanomaterials

Nanoscale biomaterials such as nanoparticles and nanofibers can elicit interesting immune response in the body. The nanoscale effect can create benefits of using nanosized products in various biological and medical applications. However, the body's immune response is equipped to recognized nanoscale to microscale particles and is superbly effective in clearing them out [125–127]. This is through opsonization and phagocytosis. Nanoparticles can be tagged with what is called opsonin while in circulation and identify them as antigens for the macrophages to phagocytose them inside their intracellular space. Proteins such as opsonin bind to the surfaces of nanoparticles, and biological materials in general, immediately upon introduction into the body (Figure 4.6). The further biological response of the body is influenced by the nanoparticle–protein complex. Protein binding is one of the key elements that affect biodistribution of the nanoparticles throughout the body. But recent advances in nanoparticle engineering can be used to evade this process and to increase circulation and avoid fast clearance inside the body [128, 129]. This is allowing for nanoparticle-based delivery systems to be exploited for targeted delivery, improved circulation, and reduction of side effects of therapeutics. It is now well accepted that properties such as nanoparticle material, size, surface charge, hydrophobicity/hydrophilicity, and the steric effects of particle coating can influence the nanoparticle responses with the immune system. Nanoparticles can be coated with polymers that can evade the immune responses using poly(ethylene glycol) (PEG) or other types of polymers to provide a hydrophilic property to its surface [130–132]. However, recent studies had shown that nanoparticles coated with PEG can elicit PEG-specific antibodies, which in turn can accelerate their clearance in the body, defeating its purpose. Hence, nanoparticles should be designed based on their purpose whether to be an immunostimulant or immunosuppressing or evading system.

In oral delivery of peptides, nanoparticles had been investigated for their utilization of using the immune response to nanoparticles. It is known that crossing the mucosal lining of the gastrointestinal (GI) system is challenging as the epithelium acts as a strong barrier for particles that can act as a pathogen. Particles that are not digested by the enzymes and the acid in the GI are cleared up by the macrophages in the GI system. Intestinal lamina propria (LP) macrophages lay waiting in the intestinal lining

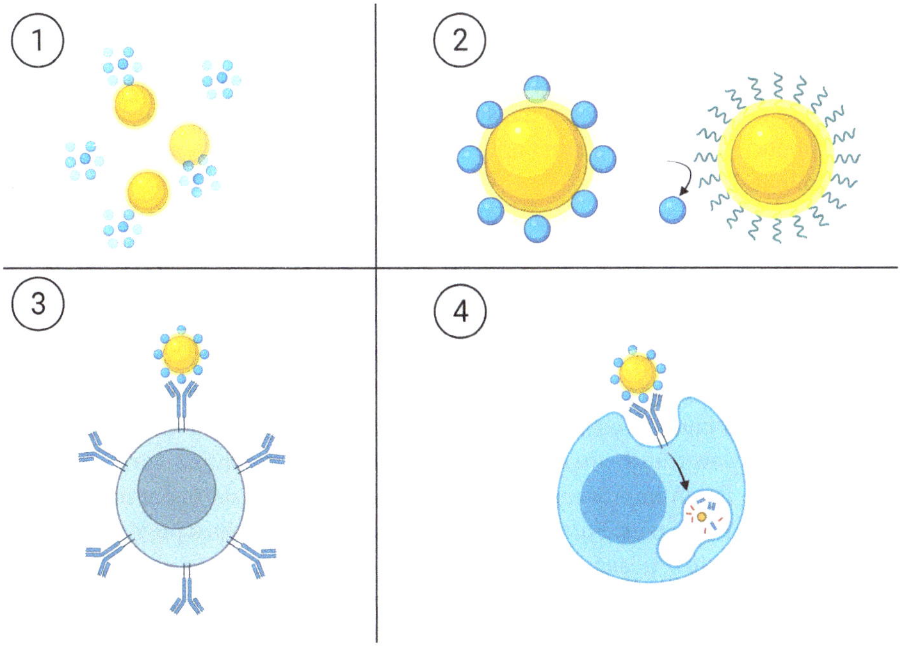

Figure 4.6: The immune response to nanoparticle biomaterials (opsonization): (1) Opsonins adhere onto the surface of nanoparticles. (2) The opsonin both neutralizes the nanoparticle and acts as a tag. PEGylated nanoparticles can repel opsonins to avoid opsonization. (3) Once tagged, the nanoparticle gets absorbed by immune cells. (4) The nanoparticle is degraded inside the endosome of the immune cell.

and are highly phagocytic and are responsible for clearing apoptotic and senescent epithelial cells. Other lymphatic tissues called Peyer's patches are small masses of tissue found throughout the ileum region of the small intestine. Also known as aggregated lymphoid nodules, they form an important part of the immune system by monitoring intestinal bacteria populations and preventing the growth of pathogenic bacteria in the intestines. These lymphatic structures are responsible for phagocytosis of nanoparticles and eliminate them in the systemic and/or the lymphatic circulation. These mechanisms can be exploited in delivering therapeutics, especially orally administered peptides. Peptides are broken down in the GI and has very low bioavailability when orally administered. However, strategies such as nanoparticle encapsulation can protect them from enzymatic and acidic degradation and increase their retention time in the GI epithelium. Once they are detected by the immune cells in the GI, they can cross this barrier and enter the lymphatic or systemic circulation, therefore increasing the bioavailability of orally administered peptides. The main translocation mechanism of these nanoparticles exploits the phagocytosis uptake of the immune cells (macrophages). However, there had been low success in using this strategy as phagocytosis using endosomal uptake subjects the nanoparticles to harsh conditions as the macrophages will degrade the components of the nanoparticles into nonharmful components. This means that the peptide payload can also be subjected into these conditions and may denature them and rendering them ineffective. Also, most nanoparticles do not necessarily get to the systemic circulation as most of them end up in the lymphatic system for elimination. Further optimization is needed to overcome these challenges.

Nanoparticles are known to stimulate the activity of the immune response. In terms of immune stimulation, different nanoparticle iterations can stimulate the adaptive or innate immunity of the body and they are generally linked to the surface charge and size of the nanoparticles. There had been reports of activation of the compliment system when nanoparticle formulated vaccines are introduced via intravenous injection and can aggravate the progression of some cancer tumors by stimulating tumor-associated immune cells and promote their conversion into a tumor-supportive phenotype. However, if the nanoparticles are introduced via intradermal or subcutaneous route, it can improve vaccine effectivity by increasing the antigenicity of conjugated weak antigens. Furthermore, nanoformulated vaccines provide an insoluble-controlled, slow release of antigens, creating a depot at the site of injection and providing protection in the destabilizing in vivo environment. One leading example of this technology that has seen clinical application is through the development of the COVID-19 vaccine, where mRNA is encapsulated in a polymeric PEGylated lipid nanoparticle drug delivery platform that protects the mRNA in vivo until it reaches the human cell and orders them to create spike proteins that are then recognized by the immune cells as antigenic and create specific antibodies to help a person fight the SARS-CoV-2 virus if they get exposed [133–136]. Lastly, immunostimulatory reactions initiated by nanoparticles are generally linked to the production of inflammatory cytokines. Several studies have linked cytokine induction by different types of nanomaterials (gold colloids, dendrimers, polymers, lipid nanoparticles, etc.) However, these nanoparticles are not directly blame for the stimulation of cytokines, but the surfactants or other antigenic components present in the formulation. Antigenicity of nanoparticles is not yet completely understood but careful analysis of formulation preparation and screening of antigenic components must be considered when formulating a nanoparticle drug delivery system in order not to induce severe reactions.

4.5 Immune modulation tools in biomaterials

Immune modulation is required in designing an effective implantable biomaterial. Several techniques are being utilized by researchers to limit the responses of the immune system toward the biomaterial being implanted in the body. In adjunct to systemic steroidal administration to lower the immune responses, several physical and chemical modifications of the biomaterial are being utilized as different material properties can modulate the behavior of proteins and cells in their interface. Besides choosing biocompatible, bioinert, or well-tolerated material, additional modifications are needed to increase the compliance of a designed biomedical implanted device to ensure that once inside the body, the immune response will elicit and not compromise the intended function of the device and will not cause any subsequent adverse effects due to rejection or severe allergic and inflammatory responses.

4.5.1 Physical modifications (surface topography and hydrophilicity)

The number one consideration of modulating or reducing the immune response is to select a material that is intrinsically inert. However, additional modifications such as creating a topography closer to biological scale in the micron and nanometric range provides a passive approach without bioactive agents to modulate macrophage behavior [137–139]. This is beneficial as a passive approach of limiting macrophage, or other cytokine adhesion to the material can increase the life span of an implanted device which can reduce the use of systemic immune-lowering drugs (steroids). Furthermore, a passive approach of modifying the physicochemical properties without resorting to addition of bioactive agents would be preferred in an implant design for regulatory considerations. This is done by creating a substrate that has an aligned micro- or nanotopography. Researchers had used aligned polymeric nanofibers using PTFE with pore sizes of 0.2, 1, and 3 µm, and aligned grated surface on poly(ε-caprolactone), poly(lactic acid), and poly(dimethyl siloxane) gratings with line width ranging from 250 nm to 2 µm [140, 141]. Their results had shown that these surface morphologies can modulate the responses of macrophages onto the material due to the reduction of the secretion of pro-inflammatory cytokines, chemokines, and reduced monocyte fusion and induced M1/M2 hybrid phenotype when macrophages are initially recruited in the site. Macrophages are responsible for foreign body responses, and their reaction to a material can further recruit other immune cells such as neutrophils and/or fibroblast cells to neutralize the material and their sensitivity to the nano- and microsurface- aligned topography can modulate the cytokine production and reduce the inflammatory response of the body onto the implanted material. The use of other geometric nano- to micropatterns and geometry such as fixed width gratings of varying depths, pillars, pits, or islands can expand the understanding of macrophage response to topography. Topography-induced change in macrophage behavior opens an opportunity to influence its phenotypic response such as cell activation, motility, maturation, and fusion in the foreign body response without the addition of bioactive cues.

Hydrophilicity of a biomaterial also plays a role in the behavior and responses of cytokines and immune cells [138, 139, 142, 143]. The material's hydrophilicity can affect the immune response by influencing the DC to activate the phenotypic type of macrophages recruited in the implantation site and decreased monocyte adhesion. Controlling of DC phenotype from proinflammatory to tolerogenic is of interest in vaccine delivery and tissue engineering/transplantation situations. Increased hydrophilicity can control the DC maturation in the implantation site consequently limiting the activation of the innate and adaptive immune responses in the body. This is because when DCs are unable to mature, it cannot present antigenic biomolecules to macrophages and adaptive cells such as T- and B-cells, thereby making the material more tolerated inside the body. There had also been reports that increasing hydrophilicity of the material can reduce the IL-6 and TNF-α production, which are both potent inflammatory cytokines.

4.5.2 Chemical modifications

Cross-linking of polymeric materials can affect the response of immune cells in the body [144, 145]. Cross-linked materials are linked to the phenotypic expression of M2 macrophages which are known to induce inflammation and reduced tissue remodeling which can be a disadvantage in tissue engineering and wound healing applications. The material's rapid degradation is desirable as it is followed by replacement with organized, site appropriate functional host tissue, which is lessened when the material has a cross-linker. This is a dilemma for biomaterialists as cross-linking of polymers can increase the biocompatibility and workability of the material. Therefore, proper selection of cross-linking method must be done to facilitate the optimal degradation rate of the polymeric material. Nontoxic cross-linking methods such as with the use of divalent or trivalent ions can facilitate a transient cross-linking in the polymeric network [146, 147]. This type of cross-linking allows for the material to be injectable or form 3D forms while having a reversible cross-linked network. This is due to the weak ionic interactions in the polymeric network which can be easily hydrolyzed by the body.

Another kind of chemical modification is the addition of inert coatings. PEG is one of the most common coating materials for limiting the immune response toward biomedical devices and drug delivery platforms (nanoparticles). This process is commonly known as PEGylation. PEG is regarded as hydrophilic and can cause retardation of protein and other antibody adhesion on the material's surface subsequently making it invisible to innate and adaptive immune response. PEG can be easily applied onto the surface of any substrate such as metals or other polymeric devices and can also be grafted chemically via ionic or covalent bonding. Furthermore, zwitterionic polymers such as poly-ethyl-2cyanoacrylate, poly-hydroxymethylacrylate, poly (carboxybetaine acrylamide), and poly[oligo(ethylene glycol) methacrylate] had been explored as alternatives to PEG as antifouling polymeric materials. Zwitterionic polymers are characterized with equal anion and cation groups on the molecular chains, which make them highly hydrophilic and antifouling [148, 149]. They can resist nonspecific protein adsorption, bacterial adhesion, and biofilm formation, which is very beneficial when tackling the problem of foreign body response and bacterial contamination on an implant [150]. Other immunomodulatory coating technologies combat the adhesion of antibodies and proteins with the use of active molecules such as osteopontin, microgels, and surfactants. All these techniques can reduce the foreign body reaction by inhibiting the adhesion of antibodies, cytokines, and other pro-inflammatory cells onto the surface of the material.

4.5.3 Active delivery of immune-modulatory compounds

A more active way of immune modulation is by incorporating molecules that are known to inhibit or reduce the activity of immune responses in a biomaterial. Adding anti-inflammatory molecules aims to modulate the magnitude and diversity of immune cell responses, or alter the phenotype of the resident cells, leading to a resolution of inflammation. This exploits drug delivery vehicles such as polymeric coatings to deliver immunomodulatory molecules passively or actively in order to influence the responding immune cells and production of cytokines in order not to create a cascade that will reject an implant or device. When a patient is receiving an implant that is known to induce immune responses such as cell-laden or ECM-derived scaffolds, it is a well-known strategy to administer systemic corticosteroid to reduce leukocyte infiltration on the biomaterial and increase the chances of implant grafting to the surrounding tissue [151]. However, this is not advisable as corticosteroids are also known to compromise the body's systemic immune responses and will allow for other pathogens to infect and cause damage to the body. Thus, localized delivery of antibiotics, nonsteroidal anti-inflammatory drugs [152, 153], antiproliferative drugs (paclitaxel, and bortezomib), monoclonal antibodies (rituximab, infliximab, adalimumab) [154, 155], and phenolics (caffeic acid, gallic acid) [155, 156] can limit the production of inflammatory cytokines such as TNF-α and various kinds of interleukins.

4.6 Conclusion

The immune response to biomaterial is a crucial consideration when designing and selecting a material for a biomedical device, drug delivery platform, or biological scaffolds. It must be considered that the body's initial response to any foreign material is hostile; therefore, proper preparation and selection must be done in order to ensure that the material does not elicit severe immune responses in the body. Making sure that there is no antigenic compounds left in material during synthesis and preparation method increases the likelihood that the biomaterial will be accepted in the body. Natural biomaterials, especially animal derived, are promising in the sense that they have biosimilar properties to our body. However, animal-derived materials, if not properly processed, can still maintain residual antigenic epitopes that will increase the rejection of a scaffold, cause allergic responses, and in worst cases severe anaphylactic shock. Properties of biomaterials such as surface topography, surface charge and hydrophilicity, degradation rate, material hardness, and chemistry can affect the immune response as these properties can influence macrophages' phenotypic fate, antibody attachment, and cytokine production around the implantation area. Lastly, the immune system in relation to biomaterials is a complex and still ongoing field of research. More effective classes of biomedical devices, drug delivery platforms, and tissue scaffolds will emerge as we understand more about the interaction of immune cells, antibodies, and cytokines toward different classes of biomaterials.

References

[1] Fu D, Wang Z, Tu Y, Peng F. Interactions between biomedical micro-/nano-motors and the immune molecules, immune cells, and the immune system: Challenges and opportunities. Advanced Healthcare Materials. 2021;10(7):2001788.

[2] Blank F, Fytianos K, Seydoux E, Rodriguez-Lorenzo L, Petri-Fink A, Von Garnier C, et al. Interaction of biomedical nanoparticles with the pulmonary immune system. Journal of Nanobiotechnology. 2017;15(1):1–9.

[3] Parham P. The Immune System: Garland Science; 2014.

[4] Crocker PR, Paulson JC, Varki A. Siglecs and their roles in the immune system. Nature Reviews. Immunology. 2007;7(4):255–66.

[5] Ribatti D, Crivellato E. Immune cells and angiogenesis. Journal of Cellular and Molecular Medicine. 2009;13(9a):2822–33.

[6] Bischoff SC. Role of mast cells in allergic and non-allergic immune responses: Comparison of human and murine data. Nature Reviews. Immunology. 2007;7(2):93–104.

[7] Galli SJ. New concepts about the mast cell. New England Journal of Medicine. 1993;328 (4):257–65.

[8] Geissmann F, Manz MG, Jung S, Sieweke MH, Merad M, Ley K. Development of monocytes, macrophages, and dendritic cells. Science. 2010;327(5966):656–61.

[9] Sprangers S, Vries T, Everts V. Monocyte heterogeneity: Consequences for monocyte-derived immune cells. Journal of Immunology Research. 2016:2016.

[10] Fani Maleki A, Rivest S. Innate immune cells: Monocytes, monocyte-derived macrophages and microglia as therapeutic targets for alzheimer's disease and multiple sclerosis. Frontiers in Cellular Neuroscience. 2019;13:355.

[11] Linton P-J, Thoman ML. Immunosenescence in monocytes, macrophages, and dendritic cells: Lessons learned from the lung and heart. Immunology Letters. 2014;162(1):290–97.

[12] Yokoyama WM, Kim S, French AR. The dynamic life of natural killer cells. Annual Review of Immunology. 2004;22:405–29.

[13] Koepsell SA, Miller JS, McKenna DH Jr. Natural killer cells: A review of manufacturing and clinical utility. Transfusion. 2013;53(2):404–10.

[14] Chiossone L, Dumas P-Y, Vienne M, Vivier E. Natural killer cells and other innate lymphoid cells in cancer. Nature Reviews. Immunology. 2018;18(11):671–88.

[15] Jost S, Altfeld M. Control of human viral infections by natural killer cells. Annual Review of Immunology. 2013;31:163–94.

[16] Reed JC. Mechanisms of apoptosis. The American Journal of Pathology. 2000;157(5):1415–30.
[17] Meier P, Finch A, Evan G. Apoptosis in development. Nature. 2000;407(6805):796–801.
[18] Elmore S. Apoptosis: A review of programmed cell death. Toxicologic Pathology. 2007;35(4):495–516.
[19] Martinez MM, Reif RD, Pappas D. Detection of apoptosis: A review of conventional and novel techniques. Analytical Methods. 2010;2(8):996–1004.
[20] Thompson AE. The immune system. Jama. 2015;313(16):1686–.
[21] den Haan JM, Arens R, van Zelm MC. The activation of the adaptive immune system: Cross-talk between antigen-presenting cells, T cells and B cells. Immunology Letters. 2014;162(2):103–12.
[22] Mekori YA, Hershko AY, Frossi B, Mion F, Pucillo CE. Integrating innate and adaptive immune cells: Mast cells as crossroads between regulatory and effector B and T cells. European Journal of Pharmacology. 2016;778:84–89.
[23] Marcucci F, Bellone M, Rumio C, Corti A. Approaches to Improve Tumor Accumulation and Interactions between Monoclonal Antibodies and Immune Cells: MAbs: Taylor & Francis; 2013. p. 34–46.
[24] Krishnamurthy A, Jimeno A. Bispecific antibodies for cancer therapy: A review. Pharmacology & Therapeutics. 2018;185:122–34.
[25] Tannenbaum S, Finko R, Cines D. Antibody and immune complexes induce tissue factor production by human endothelial cells. The Journal of Immunology. 1986;137(5):1532–37.
[26] Mizoguchi A, Bhan AK. A case for regulatory B cells. The Journal of Immunology. 2006;176 (2):705–10.
[27] Mauri C, Bosma A. Immune regulatory function of B cells. Annual Review of Immunology. 2012;30:221–41.
[28] Kurosaki T, Kometani K, Ise W. Memory B cells. Nature Reviews. Immunology. 2015;15 (3):149–59.
[29] Karlsson LC, Soveri A, Lewandowsky S, Karlsson L, Karlsson H, Nolvi S, et al. The behavioral immune system and vaccination intentions during the coronavirus pandemic. Personality and Individual Differences. 2022;185:111295.
[30] Kiermaier E, Sixt M. Fragmented communication between immune cells. Science. 2015;349 (6252):1055–56.
[31] Uhl LF, Gérard A. Modes of communication between T cells and relevance for immune responses. International Journal of Molecular Sciences. 2020;21(8):2674.
[32] Hamilton JA. Colony-stimulating factors in inflammation and autoimmunity. Nature Reviews. Immunology. 2008;8(7):533–44.
[33] Clark SC, Kamen R. The human hematopoietic colony-stimulating factors. Science. 1987;236 (4806):1229–37.
[34] Nemunaitis J. A comparative review of colony-stimulating factors. Drugs. 1997;54(5):709–29.
[35] Filippini G, Munari L, Incorvaia B, Ebers GC, Polman C, D'Amico R, et al. Interferons in relapsing remitting multiple sclerosis: A systematic review. The Lancet. 2003;361 (9357):545–52.
[36] Lengyel P. Biochemistry of interferons and their actions. Annual Review of Biochemistry. 1982;51(1):251–82.
[37] De Weerd NA, Samarajiwa SA, Hertzog PJ. Type I interferon receptors: Biochemistry and biological functions. Journal of Biological Chemistry. 2007;282(28):20053–57.
[38] Gillis S. Interleukin biochemistry and biology: Summary and introduction. 1983.
[39] Elmslie RE, Dow SW, Ogilvie GK. Interleukins: Biological properties and therapeutic potential. Journal of Veterinary Internal Medicine. 1991;5(5):283–93.

[40] Miller MD, Krangel MS. Biology and biochemistry of the chemokines: A family of chemotactic and inflammatory cytokines. Critical Reviews in Immunology. 1992;12(1–2):17–46.

[41] Miller MC, Mayo KH. Chemokines from a structural perspective. International Journal of Molecular Sciences. 2017;18(10):2088.

[42] Pontejo SM, Alejo A, Alcami A. Comparative biochemical and functional analysis of viral and human secreted tumor necrosis factor (TNF) decoy receptors. Journal of Biological Chemistry. 2015;290(26):15973–84.

[43] Uematsu S, Akira S. Toll-like receptors (TLRs) and their ligands. Toll-like Receptors (Tlrs) and Innate Immunity. 2008:1–20.

[44] Kawai T, Akira S. The role of pattern-recognition receptors in innate immunity: Update on Toll-like receptors. Nature Immunology. 2010;11(5):373–84.

[45] Takeda K, Kaisho T, Akira S. Toll-like receptors. Annual Review of Immunology. 2003;21 (1):335–76.

[46] Yam-Puc JC, Zhang L, Zhang Y, Toellner K-M. Role of B-cell Receptors for B-cell Development and Antigen-Induced Differentiation: F1000Research; 2018. p. Vol. 7.

[47] Köhler F, Hug E, Eschbach C, Meixlsperger S, Hobeika E, Kofer J, et al. Autoreactive B cell receptors mimic autonomous pre-B cell receptor signaling and induce proliferation of early B cells. Immunity. 2008;29(6):912–21.

[48] Melchers F, Ten Boekel E, Seidl T, Kong XC, Yamagami T, Onishi K, et al. Repertoire selection by pre-B-cell receptors and B-cell receptors, and genetic control of B-cell development from immature to mature B cells. Immunological Reviews. 2000;175(1):33–46.

[49] Wu L, Wei Q, Brzostek J, Gascoigne NR. Signaling from T cell receptors (TCRs) and chimeric antigen receptors (CARs) on T cells. Cellular & Molecular Immunology. 2020;17(6):600–12.

[50] Alcover A, Alarcón B, Di Bartolo V. Cell biology of T cell receptor expression and regulation. Annual Review of Immunology. 2018;36:103–25.

[51] Schroeder HW Jr, Cavacini L. Structure and function of immunoglobulins. Journal of Allergy and Clinical Immunology. 2010;125(2):S41–S52.

[52] Mix E, Goertsches R, Zett UK. Immunoglobulins – basic considerations. Journal of Neurology. 2006;253(5):v9–v17.

[53] Mariuzza R, Phillips S, Poljak R. The structural basis of antigen-antibody recognition. Annual Review of Biophysics and Biophysical Chemistry. 1987;16(1):139–59.

[54] Amit A, Mariuzza R, Phillips S, Poljak R. Three-dimensional structure of an antigen–antibody complex at 6 Å resolution. Nature. 1985;313(5998):156–58.

[55] Lafont V, Schaefer M, Stote RH, Altschuh D, Dejaegere A. Protein–protein recognition and interaction hot spots in an antigen–antibody complex: Free energy decomposition identifies "efficient amino acids". Proteins: Structure, Function, and Bioinformatics. 2007;67 (2):418–34.

[56] Krettli A, Weisz-Carrington P, Nussenzweig R. Membrane-bound antibodies to bloodstream Trypanosoma cruzi in mice: Strain differences in susceptibility to complement-mediated lysis. Clinical and Experimental Immunology. 1979;37(3):416.

[57] Manetta J, Bina H, Ryan P, Fox N, Witcher DR, Kikly K. Generation and characterization of tabalumab, a human monoclonal antibody that neutralizes both soluble and membrane-bound B-cell activating factor. Journal of Inflammation Research. 2014;7:121.

[58] Schoels M, Bombardier C, Aletaha D. Diagnostic and prognostic value of antibodies and soluble biomarkers in undifferentiated peripheral inflammatory arthritis: A systematic review. The Journal of Rheumatology. Supplement. 2011;87:20–25.

[59] Ollila J, Vihinen M. B cells. The International Journal of Biochemistry & Cell Biology. 2005;37(3):518–23.

[60] Alberts B, Johnson A, Lewis J, Raff M, Roberts K, Walter P. B Cells and Antibodies. Molecular Biology of the Cell 4th edition: Garland Science; 2002.
[61] Frasca D, Riley RL, Blomberg BB. Humoral Immune Response and B-cell Functions including Immunoglobulin Class Switch are Downregulated in Aged Mice and Humans. Seminars in immunology: Elsevier; 2005. p. 378–84.
[62] Bonilla FA, Oettgen HC. Adaptive immunity. Journal of Allergy and Clinical Immunology. 2010;125(2):S33–S40.
[63] Arai K, Nishida J, Hayashida K, Hatake K, Kitamura T, Miyajima A, et al. Coordinate regulation of immune and inflammatory responses by cytokines. Rinsho Byori. The Japanese Journal of Clinical Pathology. 1990;38(4):347–53.
[64] Husain-Syed F, Vadász I, Wilhelm J, Walmrath H-D, Seeger W, Birk H-W, et al. Immunoglobulin deficiency as an indicator of disease severity in patients with COVID-19. American Journal of Physiology-Lung Cellular and Molecular Physiology. 2021;320(4):L590–L9.
[65] Patel SY, Carbone J, Jolles S. The expanding field of secondary antibody deficiency: Causes, diagnosis, and management. Frontiers in Immunology. 2019;10:33.
[66] Sharonov GV, Serebrovskaya EO, Yuzhakova DV, Britanova OV, Chudakov DM. B cells, plasma cells and antibody repertoires in the tumour microenvironment. Nature Reviews. Immunology. 2020;20(5):294–307.
[67] Spiegelberg HL. Biological role of different antibody classes. International Archives of Allergy and Immunology. 1989;90(Suppl1):22–27.
[68] Zhang X, Calvert RA, Sutton BJ, Doré KA. IgY: A key isotype in antibody evolution. Biological Reviews. 2017;92(4):2144–56.
[69] Kitaura K, Yamashita H, Ayabe H, Shini T, Matsutani T, Suzuki R. Different somatic hypermutation levels among antibody subclasses disclosed by a new next-generation sequencing-based antibody repertoire analysis. Frontiers in Immunology. 2017;8:389.
[70] Hood L, Gray W, Sanders B, Dreyer W. Light Chain Evolution. Cold Spring Harbor Symposia on Quantitative Biology: Cold Spring Harbor Laboratory Press; 1967. p. 133–46.
[71] Abe M, Goto T, Kennel SJ, Wolfenbarger D, Macy SD, Weiss DT, et al. Production and immunodiagnostic applications of antihuman light chain monoclonal antibodies. American Journal of Clinical Pathology. 1993;100(1):67–74.
[72] Beck S, Trowsdale J. The human major histocompatibility complex: Lessons from the DNA sequence. Annual Review of Genomics and Human Genetics. 2000;1(1):117–37.
[73] Germain RN. The biochemistry and cell biology of antigen presentation by MHC class I and class II molecules: Implications for development of combination vaccines. Annals of the New York Academy of Sciences. 1995;754(1):114–25.
[74] Williams A, Peh CA, Elliott T. The cell biology of MHC class I antigen presentation. Tissue Antigens. 2002;59(1):3–17.
[75] Neefjes J, Jongsma ML, Paul P, Bakke O. Towards a systems understanding of MHC class I and MHC class II antigen presentation. Nature Reviews. Immunology. 2011;11(12):823–36.
[76] Berger AC, Roche PA. MHC class II transport at a glance. Journal of Cell Science. 2009;122(1):1–4.
[77] Sonnen AF-P, Henneke P. Structural biology of the membrane attack complex. MACPF/CDC Proteins-Agents of Defence, Attack and Invasion. 2014:83–116.
[78] Bayly-Jones C, Bubeck D, Dunstone MA. The mystery behind membrane insertion: A review of the complement membrane attack complex. Philosophical Transactions of the Royal Society B: Biological Sciences. 2017;372(1726):20160221.
[79] Morgan BP. The membrane attack complex as an inflammatory trigger. Immunobiology. 2016;221(6):747–51.

[80] Ward WK. A review of the foreign-body response to subcutaneously-implanted devices: The role of macrophages and cytokines in biofouling and fibrosis. Journal of Diabetes Science and Technology. 2008;2(5):768–77.
[81] MacLauchlan S, Skokos EA, Meznarich N, Zhu DH, Raoof S, Shipley JM, et al. Macrophage fusion, giant cell formation, and the foreign body response require matrix metalloproteinase 9. Journal of Leukocyte Biology. 2009;85(4):617–26.
[82] Vishwakarma A, Bhise NS, Evangelista MB, Rouwkema J, Dokmeci MR, Ghaemmaghami AM, et al. Engineering immunomodulatory biomaterials to tune the inflammatory response. Trends in Biotechnology. 2016;34(6):470–82.
[83] Scatena M, Eaton KV, Jackson MF, Lund SA, Giachelli CM. Macrophages: The Bad, the Ugly, and the Good in the Inflammatory Response to Biomaterials. The Immune Response to Implanted Materials and Devices: Springer; 2017. p. 37–62.
[84] Witherel CE, Abebayehu D, Barker TH, Spiller KL. Macrophage and fibroblast interactions in biomaterial-mediated fibrosis. Advanced Healthcare Materials. 2019;8(4):1801451.
[85] Rolfe B, Zhang B, Campbell G, Wang H, Mooney J, Campbell J, et al. The Fibrotic Response to Implanted Biomaterials: Implications for Tissue Engineering: IntechOpen; 2011.
[86] Shukla SK, Sharma AK, Gupta V, Yashavarddhan M. Pharmacological control of inflammation in wound healing. Journal of Tissue Viability. 2019;28(4):218–22.
[87] Jones SG, Edwards R, Thomas DW. Inflammation and wound healing: The role of bacteria in the immuno-regulation of wound healing. The International Journal of Lower Extremity Wounds. 2004;3(4):201–08.
[88] Oberyszyn TM. Inflammation and wound healing. Frontiers in Bioscience : A Journal and Virtual Library. 2007;12(8):2993–99.
[89] Koh TJ, DiPietro LA. Inflammation and wound healing: The role of the macrophage. Expert Reviews in Molecular Medicine. 2011:13.
[90] Dorrier CE, Aran D, Haenelt EA, Sheehy RN, Hoi KK, Pintarić L, et al. CNS fibroblasts form a fibrotic scar in response to immune cell infiltration. Nature Neuroscience. 2021;24 (2):234–44.
[91] Wynn TA. Cellular and molecular mechanisms of fibrosis. The Journal of Pathology: A Journal of the Pathological Society of Great Britain and Ireland. 2008;214(2):199–210.
[92] Luo Y-H, Chang LW, Lin P. Metal-based nanoparticles and the immune system: Activation, inflammation, and potential applications. BioMed Research International. 2015:2015.
[93] Ninan N, Goswami N, Vasilev K. The impact of engineered silver nanomaterials on the immune system. Nanomaterials. 2020;10(5):967.
[94] Stranavova L, Bacakova M, Novotna K, Bacakova L, Fencl J. Potential activation of the immune system on metallic materials for bone implants. Engineering of Biomaterials. 2012:15.
[95] Cadosch D, Chan E, Gautschi OP, Simmen HP, Filgueira L. Bio-corrosion of stainless steel by osteoclasts – in vitro evidence. Journal of Orthopaedic Research. 2009;27(7):841–46.
[96] Merritt K, Rodrigo JJ. Immune response to synthetic materials: Sensitization of patients receiving orthopaedic implants. Clinical Orthopaedics and Related Research (1976–2007). 1996;326:71–79.
[97] Liang H, Shi B, Fairchild A, Cale T. Applications of plasma coatings in artificial joints: An overview. Vacuum. 2004;73(3–4):317–26.
[98] Trampuz A, Zimmerli W. Diagnosis and treatment of implant-associated septic arthritis and osteomyelitis. Current Infectious Disease Reports. 2008;10(5):394–403.
[99] García-Arias M, Balsa A, Mola EM. Septic arthritis. Best Practice & Research. Clinical Rheumatology. 2011;25(3):407–21.

[100] Tan G, Xu J, Chirume WM, Zhang J, Zhang H, Hu X. Antibacterial and anti-inflammatory coating materials for orthopedic implants: A review. Coatings. 2021;11(11):1401.
[101] Shokuhfar T, Sinha-Ray S, Sukotjo C, Yarin AL. Intercalation of anti-inflammatory drug molecules within TiO 2 nanotubes. RSC Advances. 2013;3(38):17380–86.
[102] Golda-Cepa M, Chorylek A, Chytrosz P, Brzychczy-Wloch M, Jaworska J, Kasperczyk J, et al. Multifunctional PLGA/Parylene C coating for implant materials: An integral approach for biointerface optimization. ACS Applied Materials & Interfaces. 2016;8(34):22093–105.
[103] Civantos A, Martinez-Campos E, Ramos V, Elvira C, Gallardo A, Abarrategi A. Titanium coatings and surface modifications: Toward clinically useful bioactive implants. ACS Biomaterials Science & Engineering. 2017;3(7):1245–61.
[104] Pickup JC. Insulin pumps. Diabetes Technology & Therapeutics. 2019;21(S1):S-32–S-41.
[105] Herman A, De Montjoye L, Baeck M. Adverse cutaneous reaction to diabetic glucose sensors and insulin pumps: Irritant contact dermatitis or allergic contact dermatitis?. Contact Dermatitis. 2020;83(1):25–30.
[106] Rojas R, Garcia-Gabin W, Bequette BW. Multivariate statistical analysis to detect insulin infusion set failure. Proceedings of the 2011 American Control Conference: IEEE. 2011:1952–57.
[107] Knoll M, Adam S, Bahr E, Eshold J, Roß B, Steinkuhl R, et al. Minimally invasive suction sampling unit for interstitial fluid enhanced by electroosmotic mass transport. Sensors and Actuators. B, Chemical. 2002;87(1):150–58.
[108] Dibra A, Mehilli J, Braun S, Hadamitzky M, Baum H, Dirschinger J, et al. Inflammatory response after intervention assessed by serial C-reactive protein measurements correlates with restenosis in patients treated with coronary stenting. American Heart Journal. 2005;150 (2):344–50.
[109] Drachman DE, Simon DI. Inflammation as a mechanism and therapeutic target for in-stent restenosis. Current Atherosclerosis Reports. 2005;7(1):44–49.
[110] Schillinger M, Exner M, Mlekusch W, Haumer M, Ahmadi R, Rumpold H, et al.Inflammatory response to stent implantation: Differences in femoropopliteal, iliac, and carotid arteries. Radiology. 2002;224(2):529–35.
[111] Brancati MF, Burzotta F, Trani C, Leonzi O, Cuccia C, Crea F. Coronary stents and vascular response to implantation: Literature review. Pragmatic and Observational Research. 2017;8:137.
[112] Arnaoutoglou E, Kouvelos G, Koutsoumpelis A, Patelis N, Lazaris A, Matsagkas M. An update on the inflammatory response after endovascular repair for abdominal aortic aneurysm. Mediators of Inflammation. 2015:2015.
[113] Aguilar LE, Unnithan AR, Amarjargal A, Tiwari AP, Hong ST, Park CH, et al. Electrospun polyurethane/Eudragit® L100-55 composite mats for the pH dependent release of paclitaxel on duodenal stent cover application. International Journal of Pharmaceutics. 2015;478 (1):1–8.
[114] Borhani S, Hassanajili S, Tafti SHA, Rabbani S. Cardiovascular stents: Overview, evolution, and next generation. Progress in Biomaterials. 2018;7(3):175–205.
[115] Pizzoferrato A, Cenni E, Ciapetti G, Granchi D, Savarino L, Stea S. Inflammatory Response to Metals and Ceramics. Integrated Biomaterials Science: Springer; 2002. p. 735–91.
[116] Anlar HG, Bacanli M, İritaş S, Bal C, Kurt T, Tutkun E, et al. Effects of occupational silica exposure on oxidative stress and immune system parameters in ceramic workers in Turkey. Journal of Toxicology and Environmental Health. Part A. 2017;80(13–15):688–96.
[117] Wang J, Su Y, Xu L, Li D. Micro-patterned surface construction on BCP ceramics and the regulation on inflammation-involved osteogenic differentiation. Materials Science and Engineering: C. 2020;116:111220.

[118] Khang D, Liu-Snyder P, Pareta R, Lu J, Webster TJ. Reduced responses of macrophages on nanometer surface features of altered alumina crystalline phases. Acta Biomaterialia. 2009;5(5):1425–32.
[119] Graney PL, Roohani-Esfahani S-I, Zreiqat H, Spiller KL. In vitro response of macrophages to ceramic scaffolds used for bone regeneration. Journal of the Royal Society Interface. 2016;13(120):20160346.
[120] Lee J-W, Han H-S, Han K-J, Park J, Jeon H, Ok M-R, et al. Long-term clinical study and multiscale analysis of in vivo biodegradation mechanism of Mg alloy. Proceedings of the National Academy of Sciences. 2016;113(3):716–21.
[121] Han H-S, Loffredo S, Jun I, Edwards J, Kim Y-C, Seok H-K, et al. Current status and outlook on the clinical translation of biodegradable metals. Materials Today. 2019;23:57–71.
[122] Soni SS, Rodell CB. Polymeric materials for immune engineering: Molecular interaction to biomaterial design. Acta Biomaterialia. 2021.
[123] Naso F, Gandaglia A, Iop L, Spina M, Gerosa G. Alpha-Gal detectors in xenotransplantation research: A word of caution. Xenotransplantation. 2012;19(4):215–20.
[124] Badylak SF, Gilbert TW. Immune Response to Biologic Scaffold Materials. Seminars in Immunology: Elsevier; 2008. p. 109–16.
[125] Alsaleh NB, Brown JM. Immune responses to engineered nanomaterials: Current understanding and challenges. Current Opinion in Toxicology. 2018;10:8–14.
[126] Ali A, Suhail M, Mathew S, Shah MA, Harakeh SM, Ahmad S, et al. Nanomaterial induced immune responses and cytotoxicity. Journal of Nanoscience and Nanotechnology. 2016;16 (1):40–57.
[127] Hussain S, Vanoirbeek JA, Hoet PH. Interactions of nanomaterials with the immune system. Wiley Interdisciplinary Reviews. Nanomedicine and Nanobiotechnology. 2012;4(2):169–83.
[128] Anselmo AC, Gupta V, Zern BJ, Pan D, Zakrewsky M, Muzykantov V, et al. Delivering nanoparticles to lungs while avoiding liver and spleen through adsorption on red blood cells. ACS Nano. 2013;7(12):11129–37.
[129] Rodriguez PL, Harada T, Christian DA, Pantano DA, Tsai RK, Discher DE. Minimal" Self" peptides that inhibit phagocytic clearance and enhance delivery of nanoparticles. Science. 2013;339(6122):971–75.
[130] Hadjesfandiari N, Parambath A. Stealth Coatings for Nanoparticles: Polyethylene Glycol Alternatives. Engineering of Biomaterials for Drug Delivery Systems: Elsevier; 2018. p. 345–61.
[131] Kim Y-H, Min KH, Wang Z, Kim J, Jacobson O, Huang P, et al. Development of sialic acid-coated nanoparticles for targeting cancer and efficient evasion of the immune system. Theranostics. 2017;7(4):962.
[132] Hoang Thi TT, Pilkington EH, Nguyen DH, Lee JS, Park KD, Truong NP. The importance of poly (ethylene glycol) alternatives for overcoming PEG immunogenicity in drug delivery and bioconjugation. Polymers. 2020;12(2):298.
[133] Hou X, Zaks T, Langer R, Dong Y. Lipid nanoparticles for mRNA delivery. Nature Reviews Materials. 2021:1–17.
[134] Wu Z, Li T. Nanoparticle-mediated cytoplasmic delivery of messenger RNA vaccines: Challenges and future perspectives. Pharmaceutical Research. 2021;38(3):473–78.
[135] Tenchov R, Bird R, Curtze AE, Zhou Q. Lipid nanoparticles – from liposomes to mRNA vaccine delivery, a landscape of research diversity and advancement. ACS Nano. 2021.
[136] Park KS, Sun X, Aikins ME, Moon JJ. Non-viral COVID-19 vaccine delivery systems. Advanced Drug Delivery Reviews. 2021;169:137–51.

[137] Abaricia JO, Farzad N, Heath TJ, Simmons J, Morandini L, Olivares-Navarrete R. Control of innate immune response by biomaterial surface topography, energy, and stiffness. Acta Biomaterialia. 2021.

[138] Hotchkiss KM, Reddy GB, Hyzy SL, Schwartz Z, Boyan BD, Olivares-Navarrete R. Titanium surface characteristics, including topography and wettability, alter macrophage activation. Acta Biomaterialia. 2016;31:425–34.

[139] Boyan BD, Lotz EM, Schwartz Z. Roughness and hydrophilicity as osteogenic biomimetic surface properties. Tissue Engineering. Part A. 2017;23(23–24):1479–89.

[140] Fink J, Fuhrmann R, Scharnweber T, Franke RP. Stimulation of monocytes and macrophages: Possible influence of surface roughness. Clinical Hemorheology and Microcirculation. 2008;39(1–4):205–12.

[141] Yim EK, Leong KW. Significance of synthetic nanostructures in dictating cellular response. Nanomedicine. 2005;1(1):10–21.

[142] Abaricia JO, Shah AH, Musselman RM, Olivares-Navarrete R. Hydrophilic titanium surfaces reduce neutrophil inflammatory response and NETosis. Biomaterials Science. 2020;8 (8):2289–99.

[143] Dai X, Wei Y, Zhang X, Meng S, Mo X, Liu X, et al. Attenuating immune response of macrophage by enhancing hydrophilicity of Ti surface. Journal of Nanomaterials. 2015:2015.

[144] Courtman DW, Errett BF, Wilson GJ. The role of crosslinking in modification of the immune response elicited against xenogenic vascular acellular matrices. Journal of Biomedical Materials Research: An Official Journal of the Society for Biomaterials, the Japanese Society for Biomaterials, and the Australian Society for Biomaterials and the Korean Society for Biomaterials. 2001;55(4):576–86.

[145] Rojas IA, Slunt JB, Grainger DW. Polyurethane coatings release bioactive antibodies to reduce bacterial adhesion. Journal of Controlled Release. 2000;63(1–2):175–89.

[146] Yang J, Xu F, Han C-R. Metal ion mediated cellulose nanofibrils transient network in covalently cross-linked hydrogels: Mechanistic insight into morphology and dynamics. Biomacromolecules. 2017;18(3):1019–28.

[147] Ghanian MH, Mirzadeh H, Baharvand H. In situ forming, cytocompatible, and self-recoverable tough hydrogels based on dual ionic and click cross-linked alginate. Biomacromolecules. 2018;19(5):1646–62.

[148] Zheng L, Sundaram HS, Wei Z, Li C, Yuan Z. Applications of zwitterionic polymers. Reactive & Functional Polymers. 2017;118:51–61.

[149] Zhu -M-M, Fang Y, Chen Y-C, Lei Y-Q, Fang L-F, Zhu B-K, et al. Antifouling and antibacterial behavior of membranes containing quaternary ammonium and zwitterionic polymers. Journal of Colloid and Interface Science. 2021;584:225–35.

[150] Ma J, Lin W, Xu L, Liu S, Xue W, Chen S. Resistance to long-term bacterial biofilm formation based on hydrolysis-induced Zwitterion material with biodegradable and self-healing properties. Langmuir. 2020;36(12):3251–59.

[151] Kourtzelis I, Rafail S, DeAngelis RA, Foukas PG, Ricklin D, Lambris JD. Inhibition of biomaterial-induced complement activation attenuates the inflammatory host response to implantation. The FASEB Journal. 2013;27(7):2768–76.

[152] Day RO, Graham GG. Non-steroidal anti-inflammatory drugs (NSAIDs). BMJ. 2013:346.

[153] Bindu S, Mazumder S, Bandyopadhyay U. Non-steroidal anti-inflammatory drugs (NSAIDs) and organ damage: A current perspective. Biochemical Pharmacology. 2020:114147.

[154] Rispens T, de Vrieze H, de Groot E, Wouters D, Stapel S, Wolbink GJ, et al. Antibodies to constant domains of therapeutic monoclonal antibodies: Anti-hinge antibodies in immunogenicity testing. Journal of Immunological Methods. 2012;375(1–2):93–99.

[155] Saadoun D, Bodaghi B, Bienvenu B, Wechsler B, Sene D, Trad S, et al. Biotherapies in inflammatory ocular disorders: Interferons, immunoglobulins, monoclonal antibodies. Autoimmunity Reviews. 2013;12(7):774–83.
[156] Kassim M, Achoui M, Mustafa MR, Mohd MA, Yusoff KM. Ellagic acid, phenolic acids, and flavonoids in Malaysian honey extracts demonstrate in vitro anti-inflammatory activity. Nutrition Research. 2010;30(9):650–59.

Chapter 5
Biomaterials in medical device development

5.1 Biomaterials' role in medical device development

Biomedical devices had helped the field of medicine by creating more effective diagnosis, treatment, and support for various diseases. Devices in the biomedical field can be classified into three categories: Class I, II, III medical devices, with classification based primarily on risk the medical device poses. Class I medical devices are generally deemed low risk, Class II medium risk, and Class III medical devices are the highest risk (Figure 5.1) [1]. A device can be as simple as a patch to as complex as an artificial heart and vital to their development is choosing the right material to enhance or improve their performance and reduce their risk of adverse effects to the patient. Biomaterials enhance biocompatibility, bioactivity, or bio-inertness of a medical device. Material selection is a crucial factor for medical devices as they can determine the performance of the device of its intended use. For example, bandages with more breathable fabric and better adhesives are better in providing a protective barrier to a suture or wound while increasing the healing process by allowing for better gas exchange, while more advanced devices, such as pacemakers and cochlear implants, are sealed or coated by biocompatible materials to control the foreign body reaction and leakage of toxic leachable. Thus, it can be said that the advancement of biomaterials leads to the creation of more effective biomedical devices by providing properties that can improve their functionality.

5.1.1 Class I medical devices

Class I devices are low-risk devices. Examples include bandages, handheld surgical instruments, and swabs. As of now, 47% of medical devices are categorized under Class I, and 95% of these are exempt from the regulatory process due to their simplicity and low risk of adverse events toward the user in case of malfunction [1]. Class I medical devices are generally noninvasive and are used for the following: assisting healing (bandages), cell or bacteria cultures (plates, pipette, dishes), personal protective equipment (gloves, masks, face-shield), gathering vital signs (thermometer, sphygmomanometer, stethoscope), mobility assistance (wheel chairs and crutches), and laboratory consumables (swabs, beakers, sample/specimen collectors). It must be noted that medical devices should not have any electronic or digital component with software requirements to be classified under Class I. Adding an electrical/digital component renders the device to fall under Class II medical devices especially if proper algorithm design is needed to ensure proper function for the intended use is

https://doi.org/10.1515/9783110655377-005

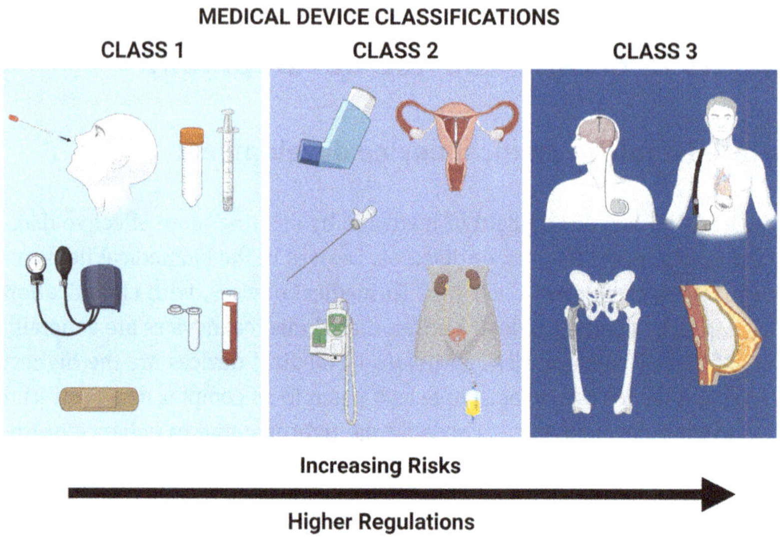

Figure 5.1: Medical device classification schematics and examples of each medical device that falls into each category. Class I medical devices contains less risks to patients and are much more widespread and diverse, it encompasses simple syringes, disposable containers, first aid and diagnostic materials, sample containers, etc. Class II medical devices contains medium risk level to patients; these include inhalers, cannulas, electronic thermometers, contraceptive implants, Foley catheters, etc. This classification is less numerous than Class I but more abundant than Class III. Class III medical devices are much less diverse and fewer in numbers and significant life-threatening risks are involved.

met. Digital thermometers, blood pressure measurement device, and electric wheelchairs fall under Class II medical devices.

5.1.2 Class II medical devices

Medium risk devices fall under the Class II category. Medium risk means that the adverse event if the devices malfunction may cause serious injuries to the end-user of the device [1]. The devices that fall under Class II category are also more invasive compared to Class I, as they are used to access deeper parts of the body or cause small puncture wounds to the skin. Class II medical devices are further divided into Class IIa and IIb. Class IIa are short-term invasive devices, e.g., injection devices, Foley catheters, needles, cannulas, nasogastric tubes, while Class IIb devices are often longer-term surgically invasive, sometimes implantable devices, e.g., intraocular lenses, birth control implants and condoms, drug dosing pens, intravenous lines and bags. These devices may cause injuries such as infection, inflammation, erythema (bruising), and/or allergic responses if not handled properly or administered incorrectly by the patient or health care workers. When a medical device has electronic or software components, it is classified as a Class II medical device, as its functionality is affected by the accuracy and robustness of the build of the electric circuitry and software algorithm programming. Thus, they can be regulated in a more stringent manner, as any malfunction or errors in the electrical and software components may lead to misdiagnosis or serious injuries.

5.1.3 Class III medical devices

Medical devices with associated high risks are classified under Class III category [1]. High risk involves life-threatening, severe injuries, and/or death in cases of a device malfunction. Ten percent of medical devices fall under Class III medical devices, are considered highly invasive, and are mostly implanted in vital organs to maintain or regain their function usually to sustain or support life. Examples of these devices are various stents, artificial heart valves, pacemakers, artificial vessels, bone implants, and artificial joint prosthesis. Other non-life-supporting Class III medical devices are used in cosmetic surgery such as silicon breast and butt implants, rhinoplasty (nose) implants, and plastic jaw fillers. Class III medical devices have the most stringent regulatory requirements as they require what is called premarket approval before being sold to the market. A premarket approval is a regulatory condition where a device must demonstrate reasonable risk and safety analysis and that using such device will ultimately benefit patient or end user compared to not using it. Therefore, a Class III medical device must undergo rigorous testing and validation in biocompatibility and robustness with proven preclinical and clinical data packages that are to be submitted before the authorities or regulatory body before getting approval for sale and distribution.

5.2 Biomaterials in medical devices (biosensors)

Biosensors are a class of medical device that is used to detect biomolecular markers, analytes, or microbes that are indicative of physiological status and conditions in the human body (Figure 5.2). The sensitivity of a biosensor can be enhanced using biomaterials that can be functionalized in order to immobilize or detect biomolecular markers with high precision and sensitivity. This will entail a more reliable diagnosis with less sampling requirements. Biosensors consist of three parts: a component that recognizes the analyte and produces a signal, a signal transducer, and an output signal. Various biomaterials are usually applied in the analyte recognition component, to improve the sensing capabilities of the device. This is a way to make a sensor more sensitive to smaller number of analytes increasing its sensitivity and improving the detection accuracy. For example, different bioreceptors can be easily entrapped within or immobilized on the surfaces of a polymeric matrix via covalent bonding on the transducer component of the sensor.

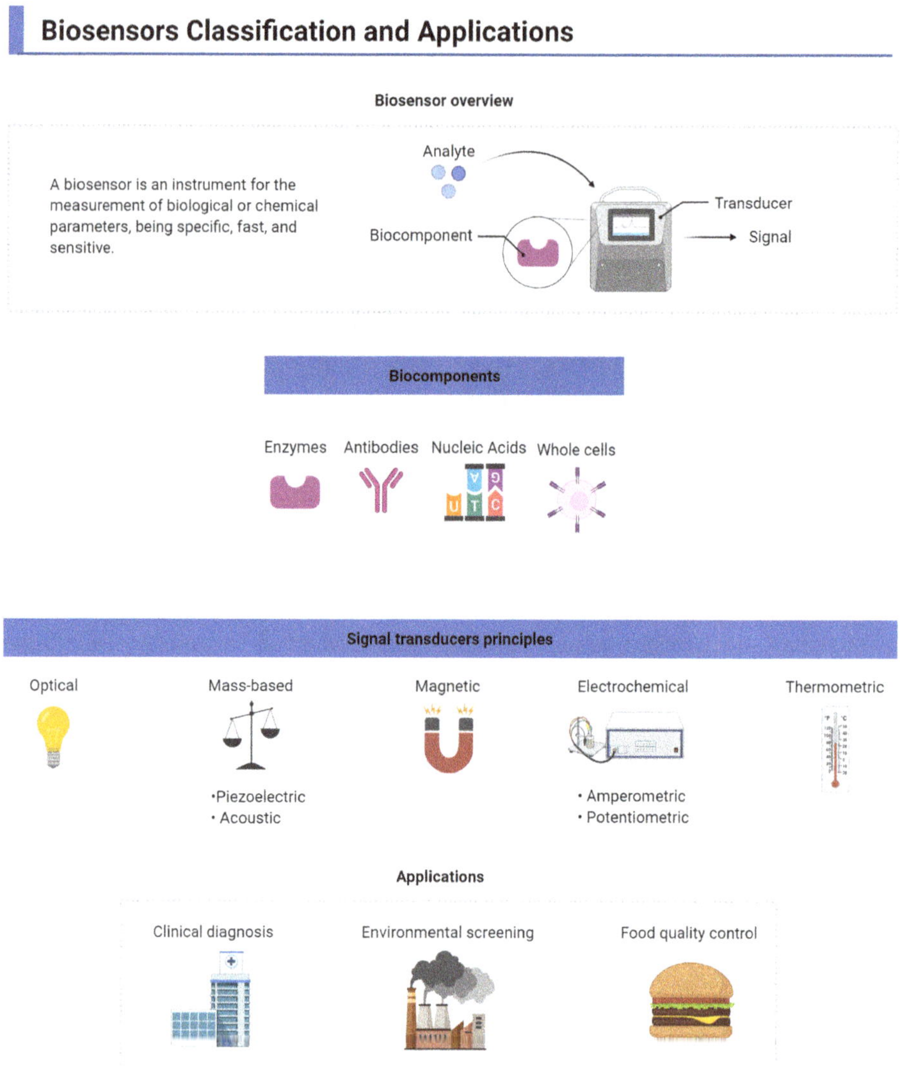

Figure 5.2: Biosensors classifications and applications. Basic components of biosensors are an analyte, a biocomponent, a transducer, and an output signal. The biocomponents (things that are being measured) could be enzyme, antibodies, nucleic acid, and whole cells. Signal transducers being used are optics, mass-based (piezo and acoustic), magnets, electrochemical, and thermometric. Changes in these transduced signals indicate the presence and levels of each biocomponent or analyte.

Biosensors are classified according to the method of biorecognition element. Based on the transducer used, biosensors are classified as an electrochemical, optical, thermal, and piezoelectric biosensor [2–4]. Furthermore, electrochemical biosensors can be classified as amperometric biosensors, meaning that the current produced during oxidation or reduction of electroactive product or reactant is measured, whereas potentiometric biosensors measure the potential of the biosensor electrode with respect to a reference electrode, and lastly, conductometric biosensors measure the change in conductance due to a specific biochemical reaction [5–7]. The most extensively researched type of biosensor are the electrochemical biosensors as they possess low detection limit, specificity, facile construction, and simple operation. Recent improvements in electronic instrumentation and electrochemical sensors are now being miniaturized and applied for in vivo monitoring or as external device for onsite clinical monitoring.

Optical biosensors, as the name suggest, rely on measurement of light absorbed or emitted due to a specific biochemical reaction of interest [8, 9]. Optical biosensors rely on multiple optical techniques such as absorption, fluorescence, luminescence, and surface plasmon resonance (SPR) [9, 10]. The most extensively researched and used optical biosensors are colorimetric biosensors due to ease of detection of the visible color change shown by these biosensors [11–13]. Thermal biosensors, on the other hand, are based on measurement on the temperature changes occurring on a specific biochemical recognition [14]. This type of biosensor utilizes the changes in enthalpy and the heat changes of a biochemical reaction which can be measured by sensitive thermistors. Lastly, piezoelectric biosensors involve measurement of mass change occurring as a result of biomolecular interaction [15, 16]. Piezoelectric materials are employed to measure the mass change by correlating with the change in oscillation frequency of the piezo crystal [17]. Commonly used materials for this type of biosensor are silica (quartz-crystals) due to their sensitivity to mass change [18]. Based on the type of biorecognition unit used, piezoelectric biosensors are further classified as enzymatic, nucleic acid–based, aptamer-based, antibody-based, whole cell–based biosensors.

5.3 Biomaterials in medical devices (vascular and nonvascular stents)

Stents are metallic tubular meshes that are placed inside the body to create patency inside an occluded bodily orifice [19]. These occlusions may be from benign or malignant tumors for nonvascular orifices [20–22], and plaque build-ups in the vasculature [23–25]. Structurally, there are no significant differences between vascular and nonvascular stents, except for the size, material, and shape. This is since the main requirement for a stent is to press the occlusion to the walls of the orifice and allow the passage of substances, food, blood, and air in the affected organ. There are multiple types of vascular and nonvascular stents. Depending on the affected organ or limb, the design considerations and biological and mechanical requirements might change, as the physiology of each body orifice varies. For nonvascular stents, meaning stents that are used outside the vasculature such as in respiratory system (trachea, and nasal stents), gastrointestinal system (esophageal, duodenal, biliary, and rectal stents), and genitourinary (ureteral and urethral stents), strong metallic mesh alloys such as stainless steel, and nickel-titanium (nitinol) are used [26–28]. Whereas for vascular stents used in treating cerebral stroke, myocardial infarction, and limb ischemia, alloys such as stainless steel [29–31] and chromium-cobalt [32, 33] are frequently used. There are inherent differences in tissue mechanical properties in these orifices that must be considered when designing stents; metal strut conformation that allows for flexibility must be used. Due to the human body's dynamic environment, the diameter of these orifices is changing depending on the activity or physiological conditions of the patient. In vascular stents, for example, stents must allow for diameter compliance in cases of vasoconstriction or vasodilation. And for nonvascular stents such as the ones used in the gastrointestinal tract due to peristalsis, the same consideration must be done with static orifices such as in respiratory and genitourinary tracts to allow for less trauma in cases of stent displacement.

5.3.1 Vascular stents

Stents were first developed to solve problems in the cardiovascular system to treat occlusion caused by plaque build-ups in the affected coronary artery, with the use of bare-metal stents (BMS) in the late 1980s as a response to limitations of plain old balloon angioplasty (POBA) [34], namely acute recoil, dissections, and need for repeat percutaneous transcatheter angioplasty (PTCA) (Figure 5.3). Since then, vascular stents had diverged from being used in cardiovascular system to cerebral vascular and large vessel (Aortic, popliteal, and femoral vessels) (Figure 5.4). The main pathology for these other vessels is somehow different as the stents are mainly applied after a diagnosis of aneurysms. Aneurysm is a vessel condition where the tunica intima, media, and externa are weakened and bulged or distends [23, 35–37]. Chronic high blood pressure can cause the vessel to rapture and could cause internal bleeding and eventual death if left untreated. Stents are placed in these aneurysm-inflicted vessels in order to maintain a laminar blood flow by creating flow diversion lessening the pressure on the vessel walls that could cause the vessel to burst [38–40]. Aneurysm coils are also placed in conjunction with a stent to ensure that blood flow and accumulation is avoided in the bulge of the aneurysm [41–43]. Here, a tightly woven mesh basket is placed inside the sac of the aneurysm. It creates a bridge across the wide neck of the aneurysm, preventing blood from getting through the spaces of the tight mesh. The blood flow continues down the artery without going into the aneurysm. These coils are usually made up of nonreactive metals such as platinum and come in a variety of shapes, sizes, and coatings that promote clotting [43–45]. In theory, coils achieve the same results as if the aneurysm has been clipped from the outside of the vessel. Other similar pathologies such as plaque build-ups can also occur in these vessels, and stenting using percutaneous intravascular catheterization intervention is also applied.

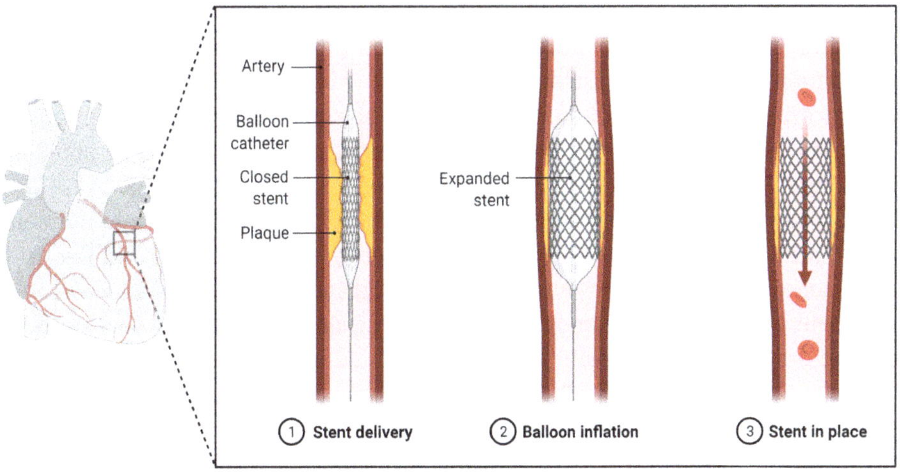

Figure 5.3: Percutaneous transcatheter angioplasty (PTCA) procedure principle.

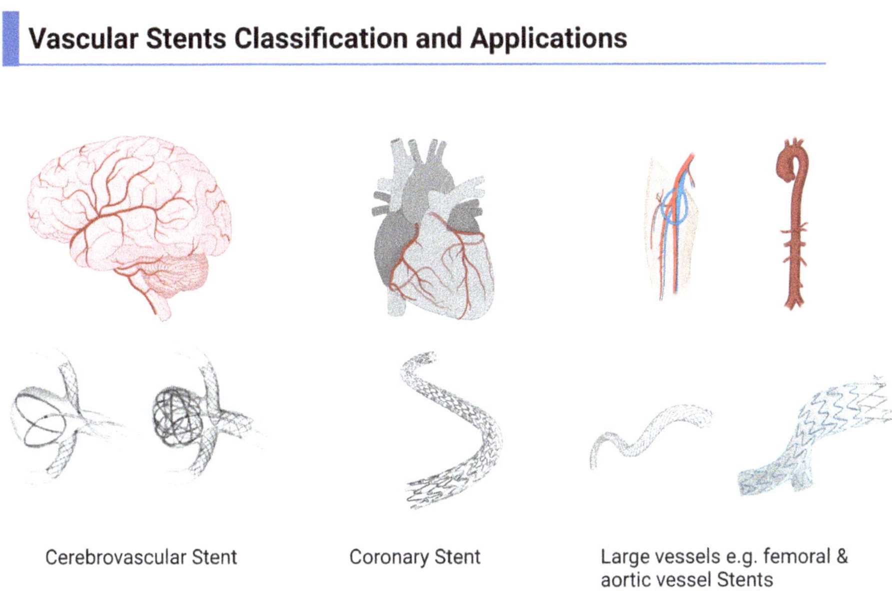

Figure 5.4: Vascular stents classifications and applications. Neurovascular or cerebrovascular stents maintain blood flow in the vasculature in the brain and could be used to prevent aneurysm ruptures. Coronary stents are used in percutaneous coronary repair to restore blood perfusion in an occluded artery in the heart. Large vessel stents (femoral and aortic) are used in cases of aortic aneurysm and atherosclerosis.

First-generation stents are made up of pure metallic alloys that expand and mechanically push and dilate the vessel to regain blood flow and maintain blood perfusion after percutaneous intervention. However, these stents have a life span of

5–10 years before reocclusion called restenosis occurs, The cause of restenosis was shown to revolve around the development of neointimal hyperplasia from vessel inflammation and smooth muscle proliferation [46–48]; thus, other biomaterials are also incorporated to expand the service life of the stent. Restenosis can be caused by various elements inside the body, from inflammatory response brought about by the accumulation of lipids that forms nascent plaques, platelet adhesion on the surface of the stent struts, or endothelial cell proliferation and migration into the stent struts; targeting either of these pathways are being investigated to avoid the risks for restenosis. This is done by attaching drugs that can inhibit those pathways. Second-generation stents, also known as drug-eluting stents (DES), were developed by adding drug delivery capabilities that releases antiproliferative and anti-blood-coagulating agents [49–51]. There are several ways drugs are attached onto the stent struts, mainly by adding a polymeric coating onto the stent struts to serve as a drug delivery platform. Polymers maintain a therapeutic level of drug release, by encapsulating or immobilizing antiproliferative drugs and release over time via various mechanisms. Drug release mechanisms used by biomaterialists and biomedical engineers vary from passive diffusion to more active release mechanisms using stimuli. This is to achieve therapeutic range of the drug payload and not cause toxicity if burst release happens. Considering the natural mechanical purpose of the stent, these polymeric drug-eluting platforms incorporated in the stent as a coating or cover must comply with the expansion of the stent during deployment and the natural dilation and constriction of the vessel. Furthermore, it must overcome hydrolytic and proteolytic degradation from the blood for a prolonged period without producing degradation products that could cause blood toxicity. Different drug immobilization techniques have been employed to load the drugs onto the stent struts. It varies from simple techniques such as dip coating [52–55], cold spraying [56–59], electrospraying [60–62], to more complex methods such as electrospinning and electro polymerization. Drugs are mixed with the polymeric solution and are coated together with the polymer and let to dry. Varying the ratio of drug to polymer weight also changes the physicochemical and mechanical properties of the coating, thus finding the optimal weight to weight ratio and proper solvent to so that the drug is homogenously distributed in the polymer. It has been a recurring problem that there is too little or too low drug loading efficiency in the polymeric coating to ensure that therapeutic levels of drug are released from the stent. Different polymers and coating techniques have been used in order that the diffusion rate of the drug is controlled in a way that at least first-order release kinetics is observed. Furthermore, stent strut thickness also plays a role in avoiding restenosis events, the use of alloys of cobalt chromium and platinum chromium allowed for manufacturing of thinner struts while maintaining adequate radial strength [32, 63]. Flow loop analysis have revealed that thinner struts reduce circulation and blood pooling inside the stent lumen when compared with similar geometric weaving but thicker strut stent, hence, reducing the risk of stent thrombosis. Thinner struts also show that the use of biocompatible polymer and improvement in coating technology with more uniform coating is feasible with less webbing and increased bonding of polymer surface to the metal surface, leading to lesser occurrences of polymer delamination. Surface irregularities on the stent strut are further avoided, thus lowering the inflammatory response and accelerating re-endothelialization on the vessel. Development of second-generation DES on this platform of a thin strut with a biocompatible polymer can create improve patient outcomes.

Vascular stent coatings can further be divided into three classes: durable, degradable, and nonpolymer (pure drug) coatings. Durable coatings are made up of polymers with nondegradable and hemocompatible materials that passively diffuse its drug payload, while degradable coatings release the drug payload via surface or bulk erosion over a period examples of polymers used in marketed vascular stents are the following (poly(styrene-b-isobutylene-b-styrene) block copolymer, PBMA: poly(n-butyl methacrylate), PVDF-HFP: poly(vinylidene fluoride)-hexafluoropropylene, PC: phosphorylcholine polymer, PLGA: poly(lactide-co-glycolide), BioLinx: hydrophobic C10-polymer/hydrophilic C19-polymer/poly(vinyl-pyrrolidone) (PVP), and PLA: polylactide). Nonpolymeric coatings are meant to use pure drugs to maximize the drug loading capability of the stent and remove the added possible toxicities related to degradation of polymers used in stent coatings. This is done by adding micro or nano porosity on the surfaces of the stent struts via various anodization methods and subsequently embedding the drugs in these porosities [64, 65]. At present the efficacy and safety of polymer-free DES in clinical practice are still in question as it might be prone for burst release of drugs that could be detrimental to the patient. However, several studies report that patients treated with polymer-free stents show similar clinical outcomes to those treated with durable and degradable polymer DES in terms of mortality, stent thrombosis, and long-term efficacy. Furthermore, in order to avoid complications of restenosis, third generation vascular stents are made with degradable struts/mesh are being developed. This entails the utilization of biodegradable metallic alloys and polymers with robust mechanical properties such as elasticity and hemocompatibility [66–68]. Biodegradable metals such as magnesium, zinc, and ferric-based alloys are one of the best candidates for third-generation stents as they can be fabricated using traditional alloying and sintering processes and be formed into meshes similar to what is being used in stainless steel–, titanium-, chromium-, cobalt-, or nitinol-based stents that are already commercialized, and also because these elements are inherently present inside the body and are considered biocompatible even after biodegradation. However, the degradation rate of these metals must still be heavily uncontrolled due to the risk of metal ion–related toxicities in the body. Thus, proper coating of polymers and/or passivation techniques are being researched to ensure that the degradation of these stents will not cause more harm to the patient [69, 70]. Polymeric-based stents are also being investigated for their used in third-generation vascular stents, as polymers have lesser toxicities related to them compared to metallic alloys that degrade inside the body [71–73]. Most bioresorbable stents are made of polylactic acid (PLA), a naturally dissolvable material that is commonly used in dissolvable implants and sutures [74, 75]. However, there are certain drawbacks of using polymer as the material for stent. This includes recoil after expansion, stent thickness limitation causing maneuverability and crossing issues, and radiopacity, making it difficult to confirm the placement of the stent inside the vessel. However, the advantage of polymeric bioresorbable stents is not through implantation, but it eliminates the cause of potential inflammation that can lead to late-stent thrombosis and restenosis in contrast to permanent metallic stents [71, 76]. Due to resorption capability of polymeric stents, it can therefore restore the natural state of vasoconstriction and vasodilatation once it has resorbed in the vessel [77]. This also allows for future intervention on the affected vessel without prior elimination of a metallic stent. It must be noted that these early polymeric bioresorbable/degradable stents have stent struts that have greater than 100 µm in thickness as to compensate with the limitations on its mechanical properties. This leads researchers to believe that thick struts are not favorable in increasing therapeutic outcomes of this generation of stents in comparison to metallic DES: Thus, newer stent designs with thinner struts are being manufactured to avoid the issue of blood pooling leading to restenosis. Lastly, while the bioresorbable stents offer potential benefits over metal stents, stent recoil has been an issue after balloon expansion [78] since polymeric stents, especially PLA, is not known to hold its shape after stress is removed. However, advances in polymer chemistry of PLA by adding other composite materials, or co-polymerizing it with other polymers might enhance its mechanical properties that could lead to the manufacturing of thinner struts with less expansion recoil.

5.3.2 Nonvascular stents

Stenting successes in creating patency in the vasculature led biomedical engineers to develop stents in nonvascular bodily orifices such as in the respiratory and gastrointestinal, genitourinary systems, and even sensory organs such as the eye. The main pathologies affecting the said systems are with the occurrence of benign or malignant growth that blockade their orifices. In the respiratory system they are the nasal, bronchial and tracheal passages. In the gastrointestinal system, it encompasses the esophagus, duodenum, colon, and biliary duct. In the genitourinary system, it is the ureters and the urethra, while in the eye, it is the choroid plexus around the cornea and the lacrimal duct. Depending on the affected orifice in the system, the stent design, size, and material considerations also vary (Figure 5.5).

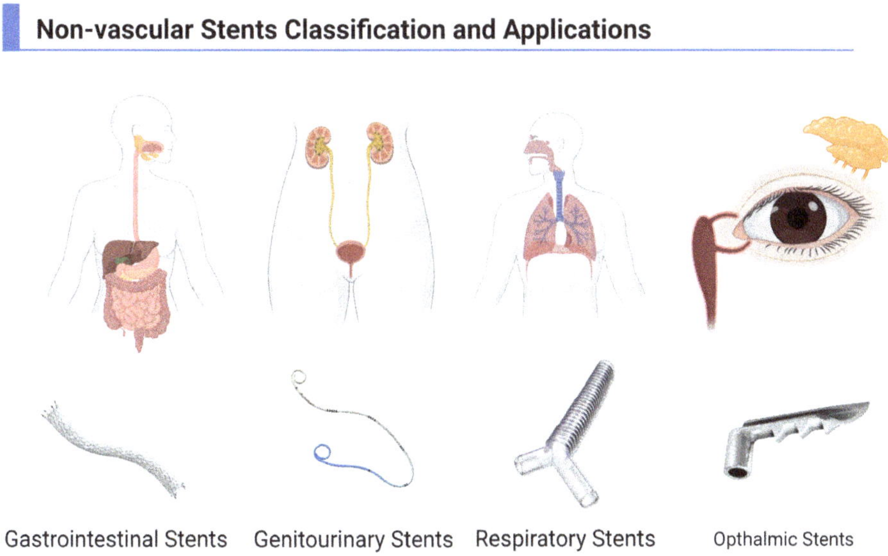

Figure 5.5: Nonvascular stents classifications and applications. Gastrointestinal stents allow food, bile, and chyme passage along the alimentary tract in cases of obstructions (e.g. cancer tumors). Genitourinary stents allow passage of urine from the kidneys toward the bladder in cases of obstructions. Respiratory stents create a wider patency in the lower respiratory tract in cases of obstruction (e.g. cancer tumors); ophthalmic stents allow flow of tears from the lacrimal glands and the flow of vitreous fluid in the cornea (glaucoma).

5.3.2.1 Gastrointestinal stents

The gastrointestinal system has seen the most variations in nonvascular stents due to the anatomical and physiological differences in the alimentary canal that can be affected by blockages that can be stented. First in the list are esophageal stents. These stents are placed usually in the inferior part of the esophagus near the sphincter that separates the stomach and the esophagus [20]. When the esophageal sphincter is compromised due to a condition called gastroesophageal reflux disease (GERD), it allows acidic gastric contents to be regurgitated into the esophagus causing chronic irritation/inflammation to the cells that line the esophagus [79, 80]. This in turn causes the cells to mutate and grow to become a malignant tumor. The growth of this tumor can block the passage of macerated food from the mouth to enter the stomach. This is presented by difficulty in swallowing (dysphagia) and will eventually cause the individual to have limited nutritional intake and wasting. In order to restore the nutritional intake of the individual suffering from esophageal cancer, clinicians place a stent to push back the tumor and allow for the passage of food into the stomach. However, the return of normal nutritional intake of the individual causes a negative feedback loop as cancer tumors are known to be veracious when it comes to stealing nutrition from the body [81]. The tumor can grow to larger proportions than before and will, in time, cause to enter the stent and block the passage of the food once again [81]. This scenario also called restenosis but found in nonvascular applications. The ingrowth of the tumor makes it also difficult to remove the stent in cases of severe restenosis. Due to this, biomedical engineers have designed polymeric covers that are to be incorporated into the stent to physically create a barrier for the tumor and the lumen of the stent where ingrowth might happen. Like vascular stents, nonvascular stents used in the GI have also seen advances in their drug-eluting capabilities. This is to check and maintain the tumor size while physically blocking the tumor from growing into the stent lumen. Drugs such as paclitaxel, 5-flourouracil, and bortezomib have been incorporated into polymeric covers that are then capable of eluting the drug in a controlled manner, whether via passive diffusion or active delivery (external or internal stimuli) [82, 83]. Technologies such as electrospinning and electropolymerization mediated coating have been explored in order to create a better drug-eluting property for the stent. Special stimuli-responsive polymers (heat, pH, enzymes) are also being used in nonvascular stenting to utilize the various triggering mechanisms for drug release [84–86]. Other types of GI stents such as the ones placed in the duodenum, pancreas, common hepatic bile duct, and the colorectal portion of the alimentary canal follow the same pathologies (cancer tumor that causes a blockage) and are seen as a remedial procedure, or as a supplementary procedure together with chemotherapy or surgery. The stents used in these parts of the GI vary in size and shape and could be made from bare metals with a polymeric coating or polymeric stent.

5.3.2.2 Genitourinary stents

Ureteral obstructions require stenting procedures to alleviate the blockage or provide a shunt for urine drainage. Ureteral stents are commonly used to remediate obstructions caused by pathologies such as nephrolithiasis (kidney stones), tumors, strictures, or retroperitoneal fibrosis [87–89]. Stents are also used in cases of prostatic obstructions or due to a complication of prostatectomy. When urine is blockaded in the ureters, it can cause severe pain and backpressure to the kidneys that can cause them to fail if not treated. In such cases, a ureteral stent is inserted into the ureter to bypass the blockage and enable urine flow from the kidneys and enter the bladder for subsequent elimination. A ureteral stent has varying diameter and length depending on the age and size of the patient, with some stents reaching 12 inches. These kinds of stents are often placed temporarily before other adjuvant treatments are done to remove the obstruction.

5.3.2.3 Pulmonary stents

Airway obstructions caused by pathologies such as cancer, tuberculosis, and other types of respiratory obstruction sometimes require stents to keep the patency of the tubular orifices in the pulmonary system (trachea-bronchial tree) open. Conditions such as central airway obstruction, chronic obstructive pulmonary disease (COPD) [90], and asthma necessitate the use of airway stents, to open narrowed airways and ease the burden of respiration. Pulmonary stents are either Y-shaped (to prop up narrowed airway bifurcation) [91, 92] or are tubular in shape [93] for relieving blockages further down the pulmonary system, depending on the pathology of the obstruction. With the increase of occurrences of respiratory illnesses in cities due to pollution, lifestyle choices such as smoking, and exposure to inflammatory chemicals, particularly in developing countries, it is predicted that the use pulmonary stents will become more common.

5.3.2.4 Ophthalmic stents

Eye disorders such as glaucoma can lead to an increase in intraocular pressure (IOP), due to fluid build-up in the eye [94, 95]. This increase in pressure is primarily due to the blockage in the eye structure called the Schlemm canal where aqueous humor exits and circulate. If left untreated this build-up in pressure can cause optic nerve compression or restriction in blood flow that could cause ischemia and eventual death of the photoreceptors in the eye which leads to reduction in visual acuity or full. There is no approved therapeutic cure for glaucoma as of now, but the use of a micro-stents helps to reduce the IOP [96, 97], so that the optic nerve is not damaged and avoid further complications. Ophthalmic stents have a maximum length of 1 mm and weighing a few micrograms; they are implanted using microsurgery and are fast becoming a preferred modality for long-term treatment of glaucoma. Furthermore, ophthalmic stents are for structurally correcting the irideocorneal angle and monocanalicular structure in cases of trauma or accidents.

Table 5.1: Materials used in nonvascular stenting.

Material and Application	Advantages	Challenges	Ref.
Polymeric (silicone) Ophthalmic stent Genitourinary	Non-biofouling Flexibility	Stent migration Poor mechanical stability	[98]
Nitinol Gastrointestinal stentPulmonary stent	Self-expanding Shape memory Ease of removal in case of restenosis	Tissue trauma due to high radial expansion and edges	[26, 69, 99, 100]
Biomedical-grade stainless steel Gastrointestinal stent	High strength Cost-effective	Imaging incompatible thrombogenic	[59, 101]
Chromium–cobalt Ureteral stent	Superelastic metal Shape memory Corrosion resistant	Tissue trauma due to high radial expansion and edges	[101]

Table 5.2: Design considerations for nonvascular stents.

Design consideration	Application	Remarks	Ref.
Strut thickness	Pulmonary stents Gastrointestinal stents Genitourinary stents	Thick struts provide better support but with reduced flexibilityThin struts provide good flexibility but reduced mechanical support	[102, 103]
Weave and braid design	Pulmonary stents Gastrointestinal stents Genitourinary stents	Weaving and braid designs can optimize the materials used for maximum radial support that the stent can provide	[104, 105]
Geometry	Gastrointestinal stents Pulmonary stents	Geometric considerations are incorporated to avoid stent migration due to constant peristalsis in the GI or movement in the bronchus to keep the stent in place.	[106]
Stent coating or cover	Pulmonary stents Gastrointestinal stents	To provide a physical barrier from ingrowth of the tumor into the stent lumen and avoid restenosis Can also provide as a drug-eluting platform	[82, 91, 107]
Drug elution	Gastrointestinal stents	To reduce/maintain tumor size while stent is in place and avoid restenosis	[108]

5.4 Biomaterial in medical devices (artificial heart valves)

Heart valves ensure that blood flow inside the heart chambers only unidirectionally so that proper blood volume is ejected per stroke (aka. Stroke volume) [109, 110]. There are four heart valves inside the human heart: first is the tricuspid valve that separates the right atrium and the right ventricle. Next is the pulmonary valve that ensures that blood enters the pulmonary artery for it to be oxygenated in the lungs. Once the blood is oxygenated, it enters the left atrium which is then separated by the mitral valve from the left ventricle. Lastly, the left ventricle then pushes the blood forward into the aortic valve to be distributed in the systemic circulation. The anatomy of these valves is almost similar with flaps made of cartilaginous-like tissues. The tricuspid and mitral valves are held together by what is called the chordae tendineae, also known as heart strings. The tricuspid, pulmonary, and aortic valves contain three flaps, while the mitral valve only has two. The function of the heart valves is vital is maintaining proper blood circulation in the body; if they are compromised due to various pathological reasons (e.g. stenosis and regurgitation) [111, 112], several complications to the body may arise. Valve stenosis is when there is thickening, narrowing, or stiffening of the heart valve rendering it unable to flex and fully close or open blocking the proper flow of the blood, while regurgitation is due to the weakening of the valves reducing its ability to fully close causing the blood to back flow. These conditions can lead to lower tissue perfusion or blood pooling in some vital organs and overexertion of the heart muscle to compensate for the reduced stroke volume. These are detrimental complications and may need surgical interventions to alleviate. In certain cases where the heart valves are totally damaged, total replacement is needed and artificial heart valve replacements are undertaken to resolve the patient and avoid heart failure due to overexertion.

To date, there are two types of artificial heart valves either being commercialized or researched upon. First is the mechanical heart valves that are made from metallic and polymeric components, and second is the tissue-engineered heart valves that are made from scaffolds laden with chondrocytes to reconstruct a healthy valve (Figure 5.6) [113, 114]. Mechanical heart valves are made from materials such as titanium and carbon. They usually consist of leaflets and a metal ring surrounded by a ring of knitted fabric, which is sewn onto the heart in place of the original valve. There are several different models available for aortic and mitral replacement surgeries [115]. The main advantage of mechanical valves is that they are very durable. However, these valves provide a surface on which blood clots can form easily. As a result, anyone who has been implanted with a mechanical valve needs to be on lifelong blood-thinning medication, such as warfarin, to prevent the development of blood clots that can cause heart attack or stroke [116, 117]. Tissue-engineered valves, also known as biological or bioprosthetic valves, are composed of animal or human tissue. The valves are derived from animal tissue such as porcine (pig) [118, 119], bovine (cow) [120], and equine (horse) [121] models, and then fixed with a preserving solution that may be mounted on a flexible frame to assist in deployment during surgery. As with mechanical valves, the bottom of a tissue valve is often surrounded by a ring of knitted fabric that is sewn onto the heart. In addition to animal-derived valves, a human tissue valve from a donor human heart, known as an allograft or homograft, may also be used as a replacement valve [122, 123]. Tissue valves can be used in open heart surgery or in a minimally invasive aortic operation known as transcatheter aortic valve implantation (TAVI) [124, 125]. Advantages of tissue valves compared to mechanical valves include the avoidance of lifelong warfarin therapy to prevent the development of blood clots. A disadvantage is their relatively poor durability compared to mechanical valves, with many requiring a re-operation in 10–20 years.

The decision to select the optimal valve type is influenced by numerous factors relating to the individual patient. Compared to tissue valves, mechanical valves increase a patient's risk of blood clot formation as well as severe bleeding from the mandatory anti-clotting medication. As a result, tissue valves are often more suitable

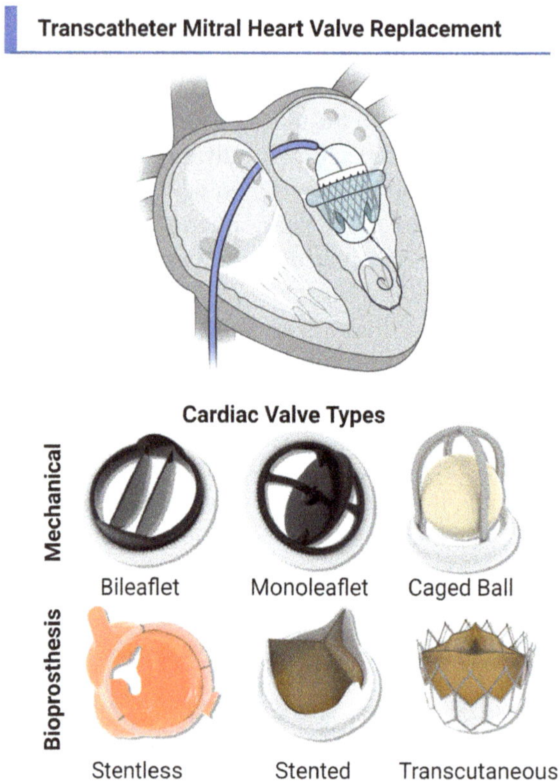

Figure 5.6: Artificial heart valve types: mechanical and bioprosthesis.

than mechanical valves for patients who have a high risk of clotting or bleeding. Women who plan to have children also fall into this category as warfarin cannot be used in pregnancy due to the risk of causing fatal malformations [126–128]. On the other hand, mechanical valves are more durable and so are often more favorable in younger patients who otherwise would likely require subsequent re-operations to replace a degenerating tissue valve. Apart from these differences, there is no significant difference in long-term survival between these two types of valves.

5.4.1 Mechanical heart valves

There are three main types of mechanical heart valves that have been developed and commercialized over the decades. The first of its kind is what is called the ball and cage valve that was developed in the 1950s, the design of which mimics a one-way ball valve where the ball moves up and down allowing for the flow of the blood to go on a single direction. It has three main components called the restraining cage, the occluder ball, and a suture ring. However, this valve has been known to create blood clots due to the material used and the friction it creates when it moves up and down. Newer generation of mechanical heart valves have used a more natural design where flaps or disks can open when blood flow is directed by the heart muscles and closes when the heart muscle relaxes. Two designs have come up from this called the bileaflet and tilting disk (monoleaflet) mechanical valves. As their name suggests the bileaflet valve consists of two flaps cut in a semi-lunar shape that opens along the valves diameter while the tilting disk only is made from a larger disk held slightly off its center to allow for the disk to tilt when blood flows. There are also other variations within these types of mechanical heart valves that aim to improve hemodynamic flow and maximize blood stroke volume per heart contraction. But still mechanical heart valves are notorious for creating blood clots that may be detrimental for its recipients; thus, patients must undergo a life-long regimen of warfarin to avoid thrombosis events. Biomaterial science has taken up the challenge to improve mechanical heart valve's hemocompatibility and avoid thrombogenic events by applying polymeric coatings on the surface of the valve materials. Polymers such as polyethylene glycol (PEG) with combination of anti-platelet and anticoagulant agents seek to improve the surface properties of the mechanical heart valves in order to avoid thrombus formation in its surface. However, this technology has not been used in clinical studies and only has shown proof of concept in animal swine models. Further investigation such as long-term antithrombogenic effects of the coating and adhesion strength must be conducted in order to say that polymeric active coatings can alleviate the problem of thrombus formation on mechanical heart valves.

5.4.2 Tissue (engineered) heart valves

Tissue valves follow a more straightforward approach when it comes to replacing the diseased heart valve [129]. They are harvested from animal sources (e.g. swine, bovine, equine) [118–121] due to mammals sharing the same cardiac anatomical structure especially when it comes to heart valves (this is also known as xenotransplantation). Also, the tissue on these valves has relatively low foreign body response because they had been decellularized and all antigen presenting components had been eliminated [130, 131]. There are two types of tissue valves: stented and stentless porcine valves. Stented tissue valves contain metallic wires that hold the valve's geometric structure and allow it to be easily implanted inside the heart [132, 133]. Stented valves are usually applied in the aortic or mitral position, examples of which are Mosaic from Medtronic and Carpentier-Edwards Perimount valves. Stented valves have leaflets made from bovine pericardium or porcine aortic valve treated with a chemical preservative and are mounted on flexible plastic or titanium metal frame (stent) covered with synthetic fabric (sewing cuff); three stent posts (struts) further support the cusps. Stentless valves, as the name suggests, forgo the use of a supporting metallic structure, thus providing less obstruction to flow and a reduced risk for endocarditis [134, 135]. Like stented valves, they are made from chemically treated porcine aortic valve and a portion of the aortic root reinforced with a pliable exterior cuff for ease of suturing. However, stentless valves are more technically demanding to implant, and they can be used only in the aortic and pulmonary positions. Tissue heart valves are less durable than their mechanical counterpart as they are made from natural animal tissues and over time constant wear can accumulate on these valves requiring further intervention after 10–20 years. Thus, this kind of valves is suited mostly for elderly patients. However, the main advantages of using tissue valves is with the removal of the use of blood-thinning agents such as warfarin as they are deemed more hemocompatible and have a better hemodynamic flow compared to mechanical valves due to the similarity in structure as to the replaced heart valves.

In order to increase the tissue valve life span, tissue-engineered heart valves with repair, remodeling, and regeneration capacity are being developed by biomedical engineers [136]. This is done via tissue engineering strategies in hopes to create a sustainable heart valve that are seeded or laden with the cells that can repair and remodel the valve in situ. There are various strategies being developed by engineers in order to do so. First is by harvesting stem cells coming from the host. These cells are then differentiated into adult valvular endothelial cells (VECs) and valvular interstitial cells (VICs). These VECs and VICs are then seeded on a decellularized scaffold, usually off-the-shelf stented or stentless valves, and are then grown to populate the entirety of the tissue valve using a bioreactor. A bioreactor is a chamber especially designed to mimic the microenvironment needed for the cells to thrive; this means supplying oxygen and elimination of carbon dioxide, and providing nutrients for the cells. This kind of TEHV is called autologous heart valve which attempts to circumvent the immunogenic response of the body toward nonnative materials. And since the cells are harvested from the same individual, the antigens the cells will present will be native to the host and will have a higher chance of implant acceptance and thus removing the need for long-term use of immunosuppressants. However, harvesting and differentiating autologous stem cells from the host is rather an expensive and time-consuming effort. Thus, other strategies using cells from another individual (allogenic) and following the same cell seeding and proliferation strategy hope to increase the life span of the valve by incorporating living cells that can repair the valve in situ. However, this requires the need for prolonged immunosuppression of the patient for the valve not to be rejected. The third strategy is by harvesting heart valve tissues from animal or human sources and decellularizing them and directly implanting the decellularized valves with the hopes that they will be populated by the host's remaining VECs and VICs. However, this strategy is expensive time consuming and ensuring that proper antigen and epitope removal during decellurization is achieved, to avoid immunogenic response and subsequent rejection. Thus, bioresorbable scaffolds using natural polymers [137] that can be found in extracellular matrix are being developed (e.g., hyaluronic acid, gelatin, and collagen) [138, 139] together with synthetic polymeric materials to form a polymeric composite that are highly biocompatible and mimics a decellularized scaffold. Lastly, a mix of all strategies where VICs and VECs that are different from an autologous stem cell source are seeded on proliferated on a bioresorbable polymeric scaffold is used to create a TEHV that is highly biocompatible to the host and does not require complex decellurization process.

There are still certain challenges about TEHV strategies. First is how to streamline the process of harvesting and differentiating autologous cells from the host and seeding them on decellularized valves or engineered polymeric scaffolds [140]. Unlike off-the-shelf tissue and mechanical valves, this new-generation TEHV requires a different regulatory standard in order to be translated into the clinic [141, 142]. The use of novel methods must be validated in order to ensure its robustness and safety once marketed as a viable alternative to commercialized heart valve devices. Sterilization procedures, toxicity assays, and consistency per batch must be standardized in the field before it can be used in the clinic. Aside from the general standards and regulations covering quality management for medical devices (ISO 13485, FDA quality system regulation and Good Manufacturing Practice), several technical standards provide guidance for the development of cardiovascular devices such as ISO 7198 (which specifies the requirements for cardiovascular implants and extracorporeal systems, vascular prostheses, and tubular vascular grafts and vascular patches), ISO 5840 (which provides general requirements for cardiac valve prostheses), and ISO 25539 (which specifies the requirements for the assessment of endovascular systems).

More technical aspects such as immunogenicity for allogenic-derived TEHV, bioresorption rate for the polymeric TEHV, remodeling rate for the transplanted cells, fibrosis and chronic inflammation occurrences, engraftment to local tissue, flow dynamics affecting cell remodeling, and other technical hurdles must be answered. Advances in computational flow modelling together with biomaterial discovery and engineering may help answer these questions [136]. Most specifically with the advent of 3D bioprinting heart valves can be easily printed from scratch and tailored specifically to the need of the patient. However, these technologies are still far from clinical applications, but researchers can use 3D printed collagen to design a human heart valve through an improved second-generation freeform, reversible embedding approach to suspend hydrogels. Recent reports have shown that tri-leaflet collagen heart valves can be printed and showed sustained functionality in a flow loop system (≤15% regurgitation) [136, 143]. Another group described a 3D-bioprinted, silicone-based heart valve featuring reinforced leaflets that were fully customizable to the patient's anatomy using a versatile multi-material additive technology [136, 144]. Computer simulations demonstrated that a bioinspired fiber design and customized leaflet shape reduced maximum stress during valve diastole, thereby improving valve durability. Additionally, in vitro hemodynamic testing showed a promisingly low regurgitation fraction of ≤ 6%. Despite the enormous advances toward the clinical translation of this technology, numerous challenges remain, such as the need for preclinical evaluation and large-scale clinical trials, before 3D printed heart valves can be used as therapeutic devices in standard clinical practice.

Advancement in TEHVs includes repair, modeling, and regenerative capacity and preclinical studies have shown that these valve models has the potential to be translated into clinic. However, numerous challenges still needs to be answered such as regulatory, logistical, and scientific concerns with regards to the clinical safety translation of TEHVs. But despite of this, decellularized homografts are still the most widely used and studied in humans and has given favorable results in longevity, valve performance, and removal of secondary surgical post-intervention. These pre-clinical evidences are strong indicators that resorbable polymer based TEHVs can break through the market in the future [136].

5.5 Biomaterial in medical devices (bone prosthesis and implants)

Movement and mobility free from pain and limitation is a hallmark of a healthy musculoskeletal system. The bones and muscles are responsible in making sure that we can articulate and move our body. Besides from movement and articulation, the musculoskeletal system, most specifically the bones, is responsible for protecting our vital organs as they house most of them inside their hard-protective structures (e.g. skull for the brain, rib cage for the lungs and heart). The bones are also mostly responsible for making new blood cells inside the soft internal tissue of all long bones called the marrow. Knowing all these functions, the importance of the musculoskeletal system is not to be overlooked, but with degeneration caused by age, trauma caused by accidents, and pathologies link to genetics, the musculoskeletal system might not be able to perform their function as intended. Thus, several devices had been developed with the aim to help alleviate musculoskeletal system's problems. First of these class of devices are bone prosthesis; these are mostly made up of metallic alloys such as titanium or stainless steel with some ceramic and/or polymeric components [145–149]. They help patients with mobility difficulty due to joint degeneration from aging or diseases such as osteoarthritis or rheumatoid arthritis by replacing the damaged joints and being anchored to the remaining healthy bone in order to restore the joint's mobility. Like prosthesis, bone implants such as screws and plates allow for reconstruction and remodeling of bones when they have been fractured by keeping them in place, while orthodontic implants allows for bone restructuring, e.g., increasing bone mass for artificial tooth implantation. Bones are moved by muscles in the body by means of contraction and relaxation of the muscle fibers. The muscles are anchored to the bones by connective tissues called tendons. Tendons, once damaged, are difficult to heal by themselves. Usually individuals who have experienced a tendon injury will have a reduced mobility because of improper healing of the tissue. Recent advances in tissue engineering have allowed for artificial polymeric materials to replace the injured tendon with the hopes to restore the individual's range of motion in the affected limb. Although bone prosthesis and implants had seen successes in being translated into clinic, there are still some setbacks that these devices encounters. First of which is the risk for bacterial colonization via the implant or prosthesis and could lead to infection for the patient. If an implant or prosthesis has been contaminated during the implantation procedure, it can create septic conditions in the body and would require further surgical intervention that adds burden to the patient. Second is to increase the bioactivity of the implant or prosthesis to induce osteogenic activity to help anchor the device in place and accelerate healing and remodeling of the hard tissue. Third is to increase biocompatibility of the implant or prosthesis by avoiding chronic inflammatory responses and increase tissue integration.

5.5.1 Bone prosthesis (artificial hip and knee prosthesis)

Joints are the location where bones come together and are vital for achieving full range of motion of the limb and provide mobility and movement for an individual. However, degeneration due to disease may cause these joints to cause loss of mobility and pain. There are different kinds of movement joints in the body depending on the location and articulation of the body part, and these are the ball and socket, hinge, condyloid, pivot, gliding, and saddle joints. Synovial joints allow for smooth movements between the adjacent bones. The joint is surrounded by an articular capsule that defines a joint cavity filled with synovial fluid. The articulating surfaces of the bones are covered by a thin layer of articular cartilage. Ligaments support the joint by holding the bones together and resisting excess or abnormal joint motions. At a synovial joint, the articulating surfaces of the bones are not directly connected, but instead encounter each other within a joint cavity that is filled with a lubricating fluid. Synovial joints allow for free movement between the bones and are the most common joints of the body. When the articular cartilage is damage due to diseases such as osteoarthritis or rheumatoid arthritis, the bones have less cushioning and may induce friction on their surfaces, and this causes severe pain and inflammation on the joint. In rheumatoid arthritis (RA), the body's immune system attacks the lining of the joint capsule, a tough membrane that encloses all the joint parts. RA is considered as an autoimmune disease and is usually treated with steroids to suppress the inflammation and overreaction of the immune system [150]. But when the lining (synovial membrane) becomes inflamed and swollen without intervention and over prolonged periods, the disease process can eventually destroy cartilage and bone within the joint, whereas osteoarthritis (OA) is a degenerative disease caused by age or genetics that worsens over time, often resulting in chronic pain [151]. Joint pain and stiffness caused by RA and OA can become severe enough to make daily tasks difficult, thus reducing the quality of life of an individual. Thus, when this happens joint replacements are the usual course of action to treat the individual. These devices or prosthesis are made up of biomedical-grade alloys, polymers, and/or ceramics to allow support and anchorage into the bone and restore articulation of the damaged joint (Figure 5.7) [145–147, 152, 153]. The most common joints that are replaced by these prostheses are the knee and hip joint. Whereas the knee, or more formally known as the tibiofemoral joint (hinge joint), is one of the largest and most complicated synovial joint in the body. The knee joins the thigh bone (femur) to the shin bone (tibia). The smaller bone that runs alongside the tibia (fibula) and the kneecap (patella) are the other bones that make the knee joint. When replacing damaged knee joint with an artificial one, the orthopedic surgeon removes and shapes the diseased part of the knee to cater to the shape of the implant. These are the distal end of the femur and the head of the tibia (lateral condyle and intercondylar eminence) which contains the articular cartilage and the medial and lateral meniscus, respectively. Knee prosthesis has two main components: first is the femoral component that is made up of metal alloys, and the tibial portion of the joint that is made of a high-grade wear-resistant plastic. The metal–plastic bearing combination is the most common type used in knee replacement implants in the clinic. Newer types of artificial knee may also be made from ceramics, with the advantage of higher wear resistance when compared to metals [154]. In an artificial knee joint, the highly polished cobalt-chromium metal moves against a very durable plastic spacer to allow

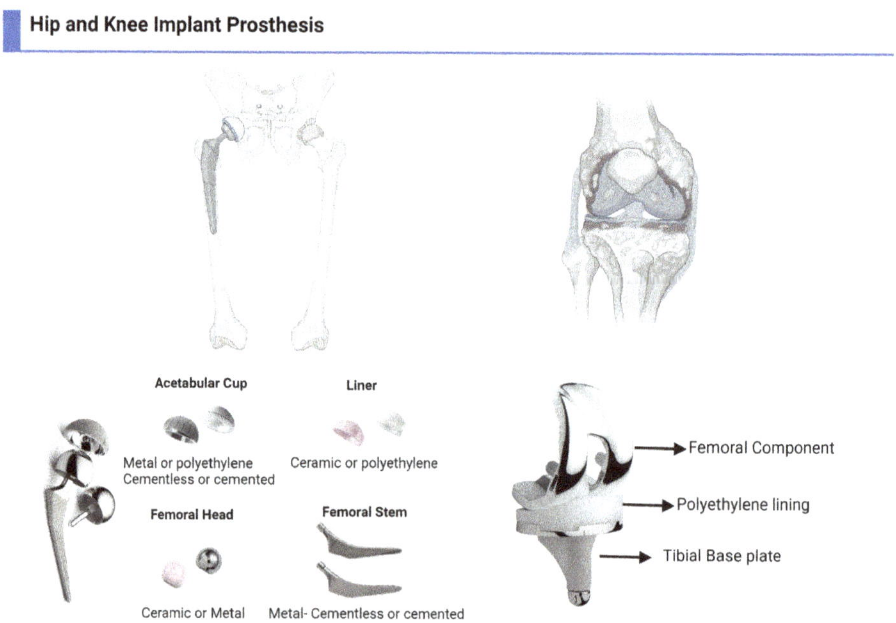

Figure 5.7: Hip and knee prosthesis parts and liner classifications. Prosthesis is implanted to replace damaged joints due to osteoarthritis or rheumatoid arthritis.

for smooth movement [155]. After implantation, the artificial knee is lubricated by the body's own synovial fluid, which is constantly replenished by living cells. The synovial lining re-forms and secretes synovial fluid and allows for a more friction less movement. However, there are still some challenges with artificial knee replacements such as toxicities from the ions released by metals under constant wear and reduction of native bone around the implant due to stress shielding [156, 157]. Thus, further improvements in coating techniques using hydroxyapatite, beta-TCP, and bioglass are needed to ensure osteointegration of the implant toward the surrounding hard tissue [158–160]. This further increases anchorage and stability of the implant subsequently avoiding the reduction of hard tissue mass around the implantation site [161–163].

There are four kinds of knee prosthesis that were developed based on the needs of an individual. It depends on what kind of material was used on the femoral and tibial ends of the implant. The most common type is the metal on plastic whereas the femoral part is made from metal alloys such as cobalt-chromium, titanium, zirconium, and nickel, and the plastic is made from high-density polyethylene. Metal on plastic is the least expensive type of implant and has the longest track record for safety and implant life span. However, one problem that can happen with plastic implants is an immune reaction triggered by tiny particles that wear away from the spacer [164]. This can cause bone to break down, leading to loosening and failure of the implant. Second is the ceramic on plastic knee prosthesis, which uses a ceramic femoral component instead of metal (or a metal component with a ceramic coating). It also rides on a plastic spacer. People who are sensitive to the nickel used in metal implants might get the ceramic type. Plastic particles from this type of implant also can lead to an immune reaction. Third is ceramic on ceramic, where both the femoral and tibial components are both made of ceramic components. Ceramic parts are the least likely to react with the body due to their biocompatibility. The inorganic components of these ceramics are known to be well tolerated in the body and are easily eliminated by the immune cells without causing severe inflammatory response [165, 166]. Also, ceramics are known to have better wear resistance compared to metal; thus, it is less prone to creating particulates inside the body. However, ceramic on ceramic prosthesis are known to make squeaking noises during articulation and movement and is not suited for active individuals as they can shatter under heavy pressure. Although they are wear-resistant from friction, ceramics are still more brittle compared to metal and plastic. Thus, this kind of prosthesis is indicated only for people who have high sensitivity to metal and plastic particles. The fourth kind is the metal-on-metal knee prosthesis where both femoral and tibial components are made of metal. Metal-on-metal implants have been used much less often in the clinic because of documented risks of trace metals leaking into the circulatory system and creating toxicities in the brain, liver, and kidneys. This arises from the friction wear from the constant articulation of the metal to metal surfaces. Metal-on-metal implants are intended to provide longer-lasting joint replacements for younger people as they remove the need to provide new prosthesis as the individual ages. But the traces of metal can cause inflammation, pain, and possibly organ damage. Metal-on-metal implants may be considered only for young, active men and women, because they may last longer than other materials.

Similarly, the hip joint is one of the most common recipients of replacement prosthesis. The hip is also one of the body's largest joints but differs from the knee as it is a ball-and-socket joint. The anatomy of this joint is formed by the acetabulum of the pelvis, which corresponds to the socket, and the ball is the femoral head, which is the upper end of the femur (thighbone). The bone surfaces of the ball and socket are covered with articular cartilage, a smooth tissue that cushions the ends of the bones and enables them to move easily. A thin tissue called the synovial membrane surrounds the hip joint. In a healthy hip, this membrane makes a small amount of fluid that lubricates the cartilage and eliminates almost all friction during hip movement. Bands of tissue called ligaments (the hip capsule) connect the ball to the socket and provide stability to the joint. During RA and OA pathologies, as discussed earlier, total hip replacement is done to remove the damaged ball-and-socket joint with prosthetic components. There are four components in a hip prosthesis: the stem, the femoral head, the liner, and the acetabular shell. The damaged femoral head is removed and replaced with a metal stem that is placed into the hollow center of the femur. The femoral stem may be either cemented or "press fit" into the bone, with the most used bone cement being an acrylic polymer called polymethylmethacrylate (PMMA) [167, 168]. A metal or ceramic ball is placed on the upper part of the stem. This ball replaces the damaged femoral head that was removed. The damaged cartilage surface of the socket (acetabulum) is removed and replaced with a metal socket. Screws or cement are sometimes used to hold the socket in place. A plastic, ceramic, or metal spacer is inserted between the new ball and the socket to allow for a smooth gliding surface. Similar to knee prosthesis, hip prosthesis can be classified depending on what kind of material was used in the articulating component of the implant. It could be metal on metal, ceramic on plastic, metal on plastic and ceramic on ceramic. Like knee prosthesis, the metal on plastic is the most common type used in the clinic but depending on the patient's age, activity level, and particular deformities of the hip, and sensitivity to metals, the surgeon might choose which kind of hip prosthesis will be used for the patient.

Hip replacements procedure may be classified into "cemented" and "cementless," depending on the type of fixation used to hold the implant in place. Cemented fixation relies on a stable interface between the prosthesis and the cement and a solid mechanical bond between the cement and the bone [169, 170]. Metal alloy stems used today rarely break but they can loosen over time. Mechanical and two biological phenomena can contribute to loosening. In the mechanical phenomena, the femoral component cracks and creates fatigue fractures in the cement which can cause the prosthetic stem to loosen and become unstable over time. This may be due to overactivity, or the patient being overweight. In the biological phenomena, the articulation of the metal ball against the polyethylene cup of the acetabular component creates polyethylene wear that forms micro particles that can trigger a biologic response that further contributes to loosening of the implant. The microscopic particulates are taken up by the cells around the joint and initiate an inflammatory response from the body. This chronic inflammatory response can cause cells to remove bits of bone around the implant, a condition called osteolysis. Due to the bone weakening from chronic inflammatory response, the stability of the implant is reduced. Bone loss can also occur around both the acetabulum and the femur, progressing from the edges of the implant due to stress shielding, a phenomenon that happens when the mechanical moduli of the implant do not correspond to the mechanical moduli of the surrounding tissue. Stress is needed to promote new bone growth, but with the metal implant shielding the surrounding bone tissue from stress, the balance between osteoblast and osteoclast activity is hampered with osteoclast activity being promoted. This will lead to implant loosening as there are less bone tissues for the implant to anchor upon.

Cementless hip prosthesis designs were introduced to attach directly to bone without the use of cement. In general, these designs are larger and longer than those used with cement [171]. They also have a surface topography that is conducive to attracting new bone growth. Most are textured or have a surface coating around much of the implant so that the new bone grows into the surface of the implant. Modern resurfacing techniques on the acetabular cup and stem components of the hip implant have porous bioactive coatings to allow bony ingrowth (osteointegration) onto the implant surface to hold the implant in place and increase bone density around the implantation area. Also, adding antibacterial agent coatings, such as silver nanoparticles, and antibacterial drugs is now being explored to combat post-implantation infection on joint implants (knee and hip) [161–163, 172, 173]. These coatings are done via the use of polymeric coatings that release the antibacterial agents in cases of cross-contamination and mishandling of the implant pre-implantation [161–163].

5.6 Biomaterial in medical devices (intraperitoneal mesh implants)

The abdominal wall consists of a complex system of several muscles which together provide protection of viscera and functions like movement of the trunk, respiration, and defecation. It is divided into posterior, lateral, and anterior walls and further divided into upper and lower quadrants. The anterior abdominal wall is composed of four muscles: the rectus abdominis, the external and internal abdominal obliquus, and the transversus abdominis. The oblique muscles rotate the trunk and allow lateral flexion of the spine. The transversus abdominis is the main muscle to retain the abdominal content, while the rectus muscle mainly takes part in ventral flexion of the spine. The rectus sheath is formed by the medial insertion of the three lateral

Table 5.3: Types of joint prosthesis based on material in contact for articulation.

Materials in contact	Advantages	Disadvantages	Ref.
Metal on plastic	Cost-effective Long track record for safety and life span	Immune reaction triggered by particles through wear of the joint	[174]
Ceramic on plastic	Reduced sensitivity to metal ions Increased wear resistance compared to metal	Particles are still generated by wear of plastic material	[175]
Ceramic on ceramic	Increased biocompatibility Ceramic parts are the least likely to react with the body and create particulates	Ceramic joint prostheses can make a squeaking noise during articulation and movement May shatter under heavy pressure into pieces that must be removed by surgery	[176, 177]
Metal on metal	Longer-lasting joint replacements for younger and active people	Traces of metal can cause inflammation, pain, and possibly organ damage	[178]

abdominal muscles. The dorsal lamina of the rectus muscle runs from the costae to the arcuate line below the umbilicus, of which no dorsal sheath exists below. The two rectus muscles are connected by the linea alba which forms the medial continuation of the rectus sheet. When this structure gets weakened a condition called a hernia occurs due to weakened muscles that may have been present since birth [179–181], or are associated with aging and repeated strains on the abdominal and groin areas. Such strain may come from physical exertion, obesity, pregnancy, frequent coughing, or straining on the toilet due to constipation [182, 183]. Hernia happens when a loop of intestine or abdominal tissue pushes through the muscle layer. A ventral hernia is a hernia that occurs at any location along the midline of the abdomen wall. There are six types of ventral hernia: inguinal hernia, femoral hernia, umbilical hernia, incisional hernia, epigastric hernia, hiatal hernia [184]. It can be diagnosed by a bulge through an opening within the abdominal muscle but small-sized hernia and hernia with special location can make the clinical diagnosis difficult. Vague symptoms may manifest if an organ is pressured or stuck. Fatty tissue usually enters a hernia first, along with or followed by the organ. In some cases, symptoms like severe pain, tenderness in the abdomen, vomiting, wind, and difficulty passing stools may develop. These last symptoms can mean that an organ or a piece of tissue trapped by the hernia is not perfectly supplied by blood which is the consequence of strangulation. A piece of herniated bowel can also cause an obstruction and prevent the bowel contents from moving through. However, the term "hernia" is used when the hernia is symptomatic, but not detectable on physical examination as other pathologies can mimic hernias like benign abdominal wall tumors (lipomas, fibroma, and hemangioma) or primary malignancies and metastases [185].

Hernia mesh, or surgical mesh, is a medical device that supports damaged tissue around hernias as it heals [184, 186–188]. Surgeons place the mesh across the area surrounding the hernia, attaching it with stitches, staples, or glue (Figure 5.8). There are several types of hernia mesh products available in the market. These include patches, plugs, and sheets. Patches are designed to go over or under the weakened or damaged tissue, whereas plugs are fitted inside the hole in the tissue, while sheets can be cut according to the geometry of the patient's hernia. Each of these hernia meshes can be further classified into the following: absorbable, nonabsorbable, synthetic [189–191], coated [192, 193] or composite [194], and animal derived. Absorbable mesh degrades and loses strength over time. It is not used to provide long-term reinforcement to the repaired hernia [195]. As the material degrades, new tissue growth is intended to provide strength to the repair. Nonabsorbable mesh is a permanent implant. It remains in the body indefinitely. Nonabsorbable mesh is supposed to provide lasting reinforcement to the repair site [196]. Hernia mesh made of synthetic materials come in woven or nonwoven sheets. The synthetic materials can be absorbable, nonabsorbable, or a combination of both. The most popular types of surgical mesh are made from polypropylene – a synthetic plastic. At least one hernia repair technique can leave mesh in contact with the intestines. This can cause the intestines to adhere to the mesh, which can result in severe complications [184, 197, 198]. Some polypropylene meshes may come coated with absorbable fatty acids

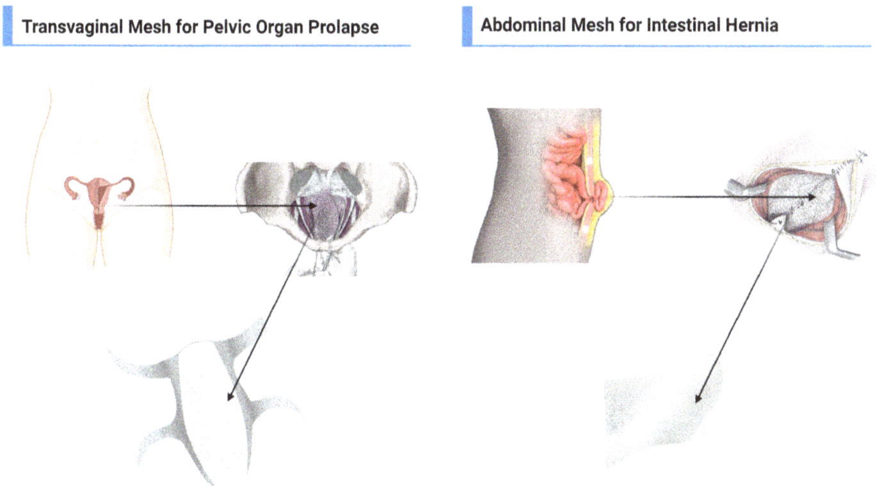

Figure 5.8: Polymeric mesh implants used inside the body (Transvaginal mesh & surgery and Abdominal mesh & surgery).

[199], cellulose, or collagen [184, 200]. Manufacturers claim these coatings prevent adhesions. Some manufacturers make hernia mesh from animal tissue. They may use the intestine or skin of animals [201]. It usually comes from a pig or cow. Manufacturers process and disinfect it. This type of mesh is also absorbable.

Recent reports of implanted mesh incorporation to the viscera of the patient had put into light the weaknesses of current material and device design. Material selection is crucial in the design and performance of a medical device. The most common material used for mesh implants (nonabsorbable) are polypropylene (PP). This polymer, although proven to be biocompatible, is not suited for such application due to the material property where it can induce adhesion of biological tissue on its surface [184, 197, 198]. Mesh-induced inflammatory responses have led to postoperative complications and inflammation and had been linked to foreign body rejection (FBR). The underlying biological response for mesh rejection has been elucidated in a study, where implantation provokes the rapid and continuous infiltration of transient macrophages. It is mainly driven by macrophages orchestrating wound healing but also granulomatous responses [202]. Chronic inflammation and subsequent rejection and damage to the surrounding tissue have been linked to PP-made meshes with representations in histological analysis containing high levels of $CD14^{+}$cells expressing HLA-DR, CD64 (FcgRI), IRF5, and CD86, indicating a myeloid cell–driven chronic inflammation in humans [184, 203].

Recent studies have tried to solve this setback and proposed different strategies using coatings such as isolated extracellular matrix proteins [204], hydrogels [205], or mesenchymal stem cells [206], in order to attenuate mesh-associated inflammation [207], while other strategies have also used nonadhering polymeric materials such as polyethyl-2-cyanoacrylate coating over polyurethane [208]. Adhesion of visceral tissue on the mesh implants due to the surface charge properties of clinically used materials (PP) is one suspected culprit for FBR of the implant. Cell adhesion on mesh implants is due to the nascent cells being produced during the wound healing process. These cells when interacting with a material can proliferate on their surface and integrate the material into the growing tissue. However, if this happens it might cause FBR due to the hydrolysis of the polymeric material into its oligomeric or monomeric forms. Hence, possible adhesion of the mesh implants must be avoided.

5.7 Conclusion

Biomaterials' role in biomedical devices is vital in progressing and improving their functions, safety, and overall performance. Selecting proper materials to be used in various medical devices can affect how the device will perform inside the body. Biocompatibility, bioinertness, and immunogenic response of the devices are dictated by the material used in medical devices and must not be overlooked when designing an effective device.

1) Biosensors can improve their sensing capabilities, making them more sensitive and capable of detecting low concentrations of biomolecules of interests, thus improving diagnosis of various health and physiological conditions.
2) Heart valves can last longer inside the body and reduce its risks of thrombogenicity by applying materials that are hemocompatible onto its flaps.
3) Vascular and nonvascular stents can also have a longer life span inside the body by avoiding restenotic events and avoid other complications such as migration.
4) Bone prosthesis performance can be improved by selecting the right contact materials to reduce friction, leachable related toxicities, and infection due to poor handling.
5) Mesh implants can be safer when proper materials are used to avoid foreign body rejection and immunogenic response that can be detrimental to the recipient.

The few examples given in this chapter are solid evidence that biomaterial development goes hand in hand with medical device development.

References

[1] Van Eck CF, Chen AF, Klatt BA, D'Antonio J, Fu F. The classification of implants: Class I, II, III. Journal of Long-term Effects of Medical Implants. 2009;19(3).

[2] Malhotra S, Verma A, Tyagi N, Kumar V. Biosensors: Principle, types and applications. International Journal of Advance Research and Innovative Ideas Education. 2017;3(2):3639–44.

[3] Alhadrami HA. Biosensors: Classifications, medical applications, and future prospective. Biotechnology and Applied Biochemistry. 2018;65(3):497–508.

[4] Kaur H, Bhosale A, Shrivastav S. Biosensors: Classification, fundamental characterization and new trends: A review. International Journal of Health Sciences and Research. 2018;8(6):315–33.

[5] Cosnier S. Electrochemical Biosensors: CRC Press; 2015.

[6] Lakard B. Electrochemical biosensors based on conducting polymers: A review. Applied Sciences. 2020;10(18):6614.

[7] Arora S, Ahmed D, Khubber S, Siddiqui S. Detecting food borne pathogens using electrochemical biosensors: An overview. IJCS. 2018;6(1):1031–39.

[8] Borisov SM, Wolfbeis OS. Optical biosensors. Chemical Reviews. 2008;108(2):423–61.

[9] Damborský P, Švitel J, Katrlík J. Optical biosensors. Essays in Biochemistry. 2016;60(1):91–100.

[10] Sohrabi H, Kholafazad Kordasht H, Pashazadeh-Panahi P, Nezhad-Mokhtari P, Hashemzaei M, Majidi MR, et al. Recent advances of electrochemical and optical biosensors for detection of C-reactive protein as a major inflammatory biomarker. Microchemical Journal. 2020:105287.

[11] Zhao VXT, Wong TI, Zheng XT, Tan YN, Zhou X. Colorimetric biosensors for point-of-care virus detections. Materials Science for Energy Technologies. 2020;3:237–49.

[12] Aldewachi H, Chalati T, Woodroofe M, Bricklebank N, Sharrack B, Gardiner P. Gold nanoparticle-based colorimetric biosensors. Nanoscale. 2018;10(1):18–33.

[13] Ahmadi Y, Soldo R, Rathammer K, Eibler L, Barišić I. Analyzing criteria affecting the functionality of G-Quadruplex-based DNA aptazymes as colorimetric biosensors and development of quinine-binding aptazymes. Analytical Chemistry. 2021;93(12):5161–69.

[14] Ramanathan K, Danielsson B. Principles and applications of thermal biosensors. Biosensors & Bioelectronics. 2001;16(6):417–23.

[15] Pohanka M. The piezoelectric biosensors: Principles and applications. International Journal of Electrochemical Science. 2017;12:496–506.

[16] Pohanka M. Overview of piezoelectric biosensors, immunosensors and DNA sensors and their applications. Materials. 2018;11(3):448.

[17] Skládal P, Dos Santos Riccardi C, Yamanaka H, da Costa PI. Piezoelectric biosensors for real-time monitoring of hybridization and detection of hepatitis C virus. Journal of Virological Methods. 2004;117(2):145–51.

[18] Kumar A. Biosensors based on piezoelectric crystal detectors: Theory and application. JOM-e. 2000;52(10):1–6.

[19] Lally C, Kelly D, Prendergast P. Stents. Wiley Encyclopedia of Biomedical Engineering. 2006.

[20] van Boeckel PG, Sijbring A, Vleggaar FP, Siersema PD. Systematic review: Temporary stent placement for benign rupture or anastomotic leak of the oesophagus. Alimentary Pharmacology & Therapeutics. 2011;33(12):1292–301.

[21] Vanderburgh L, Ho C-S. Nonvascular stents. Progress in Cardiovascular Diseases. 1996;39(2):187–200.

[22] Coté GA, Mullady DK, Jonnalagadda SS, Keswani RN, Wani SB, Hovis CE, et al. Use of a pancreatic duct stent or guidewire facilitates bile duct access with low rates of precut sphincterotomy: A randomized clinical trial. Digestive Diseases and Sciences. 2012;57(12):3271–78.

[23] Olijhoek JK, van der Graaf Y, Banga J-D, Algra A, Rabelink TJ, Visseren FL. The metabolic syndrome is associated with advanced vascular damage in patients with coronary heart disease, stroke, peripheral arterial disease or abdominal aortic aneurysm. European Heart Journal. 2004;25(4):342–48.

[24] Candemir S, White RD, Demirer M, Gupta V, Bigelow M, Prevedello L, et al. Coronary artery classification and weakly supervised abnormality localization on coronary CT angiography with 3-dimensional convolutional neural networks. arXiv preprint arXiv:191113219. 2019.

[25] Stary HC, Chandler AB, Dinsmore RE, Fuster V, Glagov S, Insull W, et al. A definition of advanced types of atherosclerotic lesions and a histological classification of atherosclerosis: A report from the committee on vascular lesions of the council on arteriosclerosis, american heart association. Circulation. 1995;92(5):1355–74.

[26] Stoeckel D, Pelton A, Duerig T. Self-expanding nitinol stents: Material and design considerations. European Radiology. 2004;14(2):292–301.

[27] Stoeckel D, Bonsignore C, Duda S. A survey of stent designs. Minimally Invasive Therapy & Allied Technologies. 2002;11(4):137–47.

[28] Nicholson T. Stents: An overview. Hospital Medicine. 1999;60(8):571–73.

[29] Migliavacca F, Petrini L, Massarotti P, Schievano S, Auricchio F, Dubini G. Stainless and shape memory alloy coronary stents: A computational study on the interaction with the vascular wall. Biomechanics and Modeling in Mechanobiology. 2004;2(4):205–17.

[30] Wang K, Shang T, Zhang L, Zhou L, Liu C, Fu Y, et al. Application of a reactive oxygen species-responsive drug-eluting coating for surface modification of vascular stents. ACS Applied Materials & Interfaces. 2021;13(30):35431–43.
[31] Talha M, Behera C, Sinha O. A review on nickel-free nitrogen containing austenitic stainless steels for biomedical applications. Materials Science and Engineering: C. 2013;33 (7):3563–75.
[32] Menown IB, Noad R, Garcia EJ, Meredith I. The platinum chromium element stent platform: From alloy, to design, to clinical practice. Advances in Therapy. 2010;27(3):129–41.
[33] Alicea LA, Aviles JI, López IA, Mulero LE, Sánchez LA. Mechanics Biomaterials: Stents. Course Materials in the Department of General Engineering, University of Puerto Rico, Mayaguez. 2004.
[34] Castaneda-Zuniga W, Formanek A, Tadavarthy M, Vlodaver Z, Edwards J, Zollikofer C, et al. The mechanism of balloon angioplasty. Radiology. 1980;135(3):565–71.
[35] Rhoton AL. Aneurysms. Neurosurgery. 2002;51(suppl_4):S1-121-S1-58.
[36] Brisman JL, Song JK, Newell DW. Cerebral aneurysms. New England Journal of Medicine. 2006;355(9):928–39.
[37] Chung B, Cebral JR. CFD for evaluation and treatment planning of aneurysms: Review of proposed clinical uses and their challenges. Annals of Biomedical Engineering. 2015;43(1):122–38.
[38] Cagnazzo F, Di Carlo D, Cappucci M, Lefevre P-H, Costalat V, Perrini P. Acutely ruptured intracranial aneurysms treated with flow-diverter stents: A systematic review and meta-analysis. American Journal of Neuroradiology. 2018;39(9):1669–75.
[39] Razavi MK, Razavi MD. Stent-graft treatment of mycotic aneurysms: A review of the current literature. Journal of Vascular and Interventional Radiology. 2008;19(6):S51–S6.
[40] Howe C, Mishra S, Kim Y-S, Chen Y, Ye S-H, Wagner WR, et al. Stretchable, implantable, nanostructured flow-diverter system for quantification of intra-aneurysmal hemodynamics. ACS Nano. 2018;12(8):8706–16.
[41] Chalouhi N, Jabbour P, Singhal S, Drueding R, Starke RM, Dalyai RT, et al. Stent-assisted coiling of intracranial aneurysms: Predictors of complications, recanalization, and outcome in 508 cases. Stroke. 2013;44(5):1348–53.
[42] Tipaldi MA, Krokidis M, Orgera G, Pignatelli M, Ronconi E, Laurino F, et al. Endovascular management of giant visceral artery aneurysms. Scientific Reports. 2021;11(1):1–6.
[43] Shapiro M, Becske T, Sahlein D, Babb J, Nelson P. Stent-supported aneurysm coiling: A literature survey of treatment and follow-up. American Journal of Neuroradiology. 2012;33(1):159–63.
[44] Krajewski S, Neumann B, Kurz J, Perle N, Avci-Adali M, Cattaneo G, et al. Preclinical evaluation of the thrombogenicity and endothelialization of bare metal and surface-coated neurovascular stents. American Journal of Neuroradiology. 2015;36(1):133–39.
[45] Henkes H, Bhogal P, Aguilar Pérez M, Lenz-Habijan T, Bannewitz C, Peters M, et al. Anti-thrombogenic coatings for devices in neurointerventional surgery: Case report and review of the literature. Interventional Neuroradiology. 2019;25(6):619–27.
[46] Kim MS, Dean LS. In-stent restenosis. Cardiovascular Therapeutics. 2011;29(3):190–98.
[47] Brower V. Stents and biology combination for restenosis. Nature Biotechnology. 1996;14(4):422–.
[48] Inoue T, Croce K, Morooka T, Sakuma M, Node K, Simon DI. Vascular inflammation and repair: Implications for re-endothelialization, restenosis, and stent thrombosis. JACC. Cardiovascular interventions. 2011;4(10):1057–66.
[49] Wessely R. New drug-eluting stent concepts. Nature Reviews. Cardiology. 2010;7(4):194–203.

[50] Roopmani P, Satheesh S, Raj DC, Krishnan UM. Development of dual drug eluting cardiovascular stent with ultrathin flexible poly (l-lactide-co-caprolactone) coating. ACS Biomaterials Science & Engineering. 2019;5(6):2899–915.
[51] Polyak B, Medved M, Lazareva N, Steele L, Patel T, Rai A, et al. Magnetic nanoparticle-mediated targeting of cell therapy reduces in-stent stenosis in injured arteries. Acs Nano. 2016;10(10):9559–69.
[52] Brinker CJ. Dip Coating. Chemical Solution Deposition of Functional Oxide Thin Films: Springer; 2013. p. 233–61.
[53] Brinker CJ, Hurd AJ. Fundamentals of sol-gel dip-coating. Journal de Physique III. 1994;4(7):1231–42.
[54] Fan Y, Li X, Yang R. The surface modification methods for constructing polymer-coated stents. International Journal of Polymer Science. 2018;2018.
[55] Mazumder M, De S, Trigwell S, Ali N, Mazumder M, Mehta J. Corrosion resistance of polyurethane-coated nitinol cardiovascular stents. Journal of Biomaterials Science. 2003;14(12):1351–62.
[56] Assadi H, Kreye H, Gärtner F, Klassen T. Cold spraying–A materials perspective. Acta Materialia. 2016;116:382–407.
[57] Grigoriev S, Okunkova A, Sova A, Bertrand P, Smurov I. Cold spraying: From process fundamentals towards advanced applications. Surface & Coatings Technology. 2015;268:77–84.
[58] Gärtner F, Stoltenhoff T, Schmidt T, Kreye H. The cold spray process and its potential for industrial applications. Journal of Thermal Spray Technology. 2006;15(2):223–32.
[59] Frattolin J, Roy R, Rajagopalan S, Walsh M, Yue S, Bertrand OF, et al. a manufacturing and annealing protocol to develop a cold-sprayed Fe-316L stainless steel biodegradable stenting material. Acta Biomaterialia. 2019;99:479–94.
[60] Chui C-Y, Ye H. Fundamental Concepts and Insights into Electrospraying for Biomedical Applications. Biomedical Applications of Electrospinning and Electrospraying: Elsevier; 2021. p. 185–206.
[61] Williams G, Raimi-Abraham B, Luo C. Electrospinning fundamentals. Nanofibres in Drug Delivery; UCL Press: London, UK. 2018;24–59.
[62] Zamani M, Prabhakaran MP, Varshosaz J, Mhaisalkar PS, Ramakrishna S. Electrosprayed montelukast/poly (Lactic-co-glycolic acid) particle based coating: A new therapeutic approach towards the prevention of In-stent restenosis. Acta Biomaterialia. 2016;42:316–28.
[63] Jorge C, Dubois C. Clinical utility of platinum chromium bare metal stents in coronary heart disease. Medical Devices (Auckland, NZ). 2015;8:359.
[64] Chen W, Habraken TC, Hennink WE, Kok RJ. Polymer-free Drug-eluting stents: An overview of coating strategies and comparison with polymer-coated drug-eluting stents. Bioconjugate Chemistry. 2015;26(7):1277–88.
[65] Cherian AM, Nair SV, Maniyal V, Menon D. Surface engineering at the nanoscale: A way forward to improve coronary stent efficacy. APL Bioengineering. 2021;5(2):021508.
[66] Hou L-D, Li Z, Pan Y, Sabir M, Zheng Y-F, Li L. A review on biodegradable materials for cardiovascular stent application. Frontiers of Materials Science. 2016;10(3):238–59.
[67] Mostaed E, Sikora-Jasinska M, Drelich JW, Vedani M. Zinc-based alloys for degradable vascular stent applications. Acta Biomaterialia. 2018;71:1–23.
[68] Gu X-N, Zheng Y-F. A review on magnesium alloys as biodegradable materials. Frontiers of Materials Science in China. 2010;4(2):111–15.
[69] Tepe G, Schmehl J, Wendel HP, Schaffner S, Heller S, Gianotti M, et al. Reduced thrombogenicity of nitinol stents – in vitro evaluation of different surface modifications and coatings. Biomaterials. 2006;27(4):643–50.

[70] Gill P, Musaramthota V, Munroe N, Datye A, Dua R, Haider W, et al. Surface modification of Ni–Ti alloys for stent application after magnetoelectropolishing. Materials Science and Engineering: C. 2015;50:37–44.
[71] Windecker S, Latib A, Kedhi E, Kirtane AJ, Kandzari DE, Mehran R, et al. Polymer-based or polymer-free stents in patients at high bleeding risk. New England Journal of Medicine. 2020;382(13):1208–18.
[72] Hobson D, Dhinakar A, Shi N, Zhang L, Wu W, Hou L, et al. Biodegradable polymeric esophagus stents. Advances in Bioinspired and Biomedical Materials. 2017;2:237–51.
[73] Chen M-C, Tsai H-W, Chang Y, Lai W-Y, Mi F-L, Liu C-T, et al. Rapidly self-expandable polymeric stents with a shape-memory property. Biomacromolecules. 2007;8(9):2774–80.
[74] Lin M, Firoozi N, Tsai C-T, Wallace MB, Kang Y. 3D-printed flexible polymer stents for potential applications in inoperable esophageal malignancies. Acta Biomaterialia. 2019;83:119–29.
[75] Siiki A, Sand J, Laukkarinen J. A systematic review of biodegradable biliary stents: Promising biocompatibility without stent removal. European Journal of Gastroenterology & Hepatology. 2018;30(8):813–18.
[76] Borhani S, Hassanajili S, Tafti SHA, Rabbani S. Cardiovascular stents: Overview, evolution, and next generation. Progress in Biomaterials. 2018;7(3):175–205.
[77] Hussain Z, Thu HE, Shuid AN, Katas H, Hussain F. Recent advances in polymer-based wound dressings for the treatment of diabetic foot ulcer: An overview of state-of-the-art. Current Drug Targets. 2018;19(5):527–50.
[78] Haude M, Erbel R, Issa H, Meyer J. Quantitative analysis of elastic recoil after balloon angioplasty and after intracoronary implantation of balloon-expandable palmaz-schatz stents. Journal of the American College of Cardiology. 1993;21(1):26–34.
[79] Menezes MA, Herbella FA. Pathophysiology of gastroesophageal reflux disease. World Journal of Surgery. 2017;41(7):1666–71.
[80] Herregods T, Bredenoord A, Smout A. Pathophysiology of gastroesophageal reflux disease: New understanding in a new Era. Neurogastroenterology & Motility. 2015;27(9):1202–13.
[81] Jang JK, Song H-Y, Kim JH, Song M, Park J-H, Kim EY. Tumor overgrowth after expandable metallic stent placement: Experience in 583 patients with malignant gastroduodenal obstruction. American Journal of Roentgenology. 2011;196(6):W831–W6.
[82] Kim S-Y, Kim M, Kim M-k, Lee H, Lee DK, Lee D-H, et al. Paclitaxel-eluting nanofiber-covered self-expanding nonvascular stent for palliative chemotherapy of gastrointestinal cancer and its related stenosis. Biomedical Microdevices. 2014;16(6):897–904.
[83] Jeong D, Lee DH, Lee DK, Na K. Nonvascular drug-eluting stent coated with sodium caprate-incorporated polyurethane for the efficient penetration of paclitaxel into tumor tissue. Journal of Biomaterials Applications. 2015;29(8):1133–44.
[84] Ma X, Xiao Y, Xu H, Lei K, Lang M. Preparation, degradation and in vitro release of ciprofloxacin-eluting ureteral stents for potential antibacterial application. Materials Science and Engineering: C. 2016;66:92–99.
[85] Rezk AI, Obiweluozor FO, Choukrani G, Park CH, Kim CS. Drug release and kinetic models of anticancer drug (BTZ) from a pH-responsive alginate polydopamine hydrogel: Towards cancer chemotherapy. International Journal of Biological Macromolecules. 2019;141:388–400.
[86] Alvarez-Lorenzo C, Concheiro A. Smart drug release from medical devices. Journal of Pharmacology and Experimental Therapeutics. 2019;370(3):544–54.
[87] Joshi H, Stainthorpe A, MacDonagh R, Keeley F, Timoney A. Indwelling ureteral stents: Evaluation of symptoms, quality of life and utility. The Journal of Urology. 2003;169 (3):1065–69.

[88] Betschart P, Zumstein V, Piller A, Schmid HP, Abt D. Prevention and treatment of symptoms associated with indwelling ureteral stents: A systematic review. International Journal of Urology. 2017;24(4):250–59.

[89] Ganatra AM, Loughlin KR. The management of malignant ureteral obstruction treated with ureteral stents. The Journal of Urology. 2005;174(6):2125–28.

[90] Iheanacho I, Zhang S, King D, Rizzo M, Ismaila AS. Economic burden of chronic obstructive pulmonary disease (COPD): A systematic literature review. International Journal of Chronic Obstructive Pulmonary Disease. 2020;15:439.

[91] Li Z-M, Lu H-B, Ren K-W, Han X-W, Wu G, Jiao D-C. Thoracic stomach–right main bronchus fistula treated with dual y-shaped covered airway stents. Clinical Radiology. 2017;72(6):517.e1–e6.

[92] Han X, Yin M, Li L, Zhu M, Ren K, Qi Y, et al. Customized airway stenting for bronchopleural fistula after pulmonary resection by interventional technique: Single-center study of 148 consecutive patients. Surgical Endoscopy. 2018;32(10):4116–24.

[93] Sumikura H, Nakayama Y, Ohnuma K, Kishimoto S, Takewa Y, Tatsumi E. In vitro hydrodynamic evaluation of a biovalve with stent (tubular leaflet type) for transcatheter pulmonary valve implantation. Journal of Artificial Organs. 2015;18(4):307–14.

[94] Nakano T, Tatemichi M, Miura Y, Sugita M, Kitahara K. Long-term physiologic changes of intraocular pressure: A 10-year longitudinal analysis in young and middle-aged Japanese men. Ophthalmology. 2005;112(4):609–16.

[95] Xu L, Wang H, Wang Y, Jonas JB. Intraocular pressure correlated with arterial blood pressure: The beijing eye study. American Journal of Ophthalmology. 2007;144(3):461–62.

[96] Kerr NM, Wang J, Barton K. Minimally invasive glaucoma surgery as primary stand-alone surgery for glaucoma. Clinical & Experimental Ophthalmology. 2017;45(4):393–400.

[97] Hooshmand J, Rothschild P, Allen P, Kerr NM, Vote BJ, Toh TY. Minimally invasive glaucoma surgery: Comparison of iStent with iStent inject in primary open angle glaucoma. Clinical & Experimental Ophthalmology. 2019;47(7):898–903.

[98] Eberhart RC, Su S-H, Nguyen KT, Zilberman M, Tang L, Nelson KD, et al. Bioresorbable polymeric stents: Current status and future promise. Journal of Biomaterials Science. 2003;14(4):299–312.

[99] Aguilar LE, GhavamiNejad A, Park CH, Kim CS. On-demand drug release and hyperthermia therapy applications of thermoresponsive poly-(NIPAAm-co-HMAAm)/polyurethane core-shell nanofiber mat on non-vascular nitinol stents. Nanomedicine: Nanotechnology, Biology and Medicine. 2017;13(2):527–38.

[100] Aguilar LE, Tumurbaatar B, Ghavaminejad A, Park CH, Kim CS. Functionalized non-vascular nitinol stent via electropolymerized polydopamine thin film coating loaded with bortezomib adjunct to hyperthermia therapy. Scientific Reports. 2017;7(1):1–11.

[101] Kapnisis K, Constantinides G, Georgiou H, Cristea D, Gabor C, Munteanu D, et al. Multi-scale mechanical investigation of stainless steel and cobalt–chromium stents. Journal of the Mechanical Behavior of Biomedical Materials. 2014;40:240–51.

[102] Menown IBA, Mamas MA, Cotton JM, Hildick-Smith D, Eberli FR, Leibundgut G, et al. Thin strut CoCr biodegradable polymer biolimus A9-eluting stents versus thicker strut stainless steel biodegradable polymer biolimus A9-eluting stents: Two-year clinical outcomes. Journal of Interventional Cardiology. 2021;2021:6654515.

[103] Nguyen DT, Smith AF, Jimenez JM. Stent strut streamlining and thickness reduction promote endothelialization. Journal of the Royal Society, Interface / the Royal Society. 2021;18(181):20210023.

[104] Liu X, Liu G, Ye P, Luo Q, Chang Z. Investigation of mechanical behaviors and improved design of V-shaped braid stents. Proceedings of the Institution of Mechanical Engineers. Part H, Journal of Engineering in Medicine. 2022;9544119221076946.
[105] van Rooij S, Sprengers ME, Peluso JP, Daams J, Verbaan D, van Rooij WJ, et al. A systematic review and meta-analysis of woven endobridge single layer for treatment of intracranial aneurysms. Interventional Neuroradiology. 2020;26(4):455–60.
[106] Ali MN, Busfield JJ, Rehman IU. Auxetic oesophageal stents: Structure and mechanical properties. Journal of Materials Science. 2014;25(2):527–53.
[107] Aguilar LE, Unnithan AR, Amarjargal A, Tiwari AP, Hong ST, Park CH, et al. Electrospun polyurethane/eudragit® L100-55 composite mats for the ph dependent release of paclitaxel on duodenal stent cover application. International Journal of Pharmaceutics. 2015;478(1):1–8.
[108] Arafat M, Fouladian P, Blencowe A, Albrecht H, Song Y, Garg S. Drug-eluting non-vascular stents for localised drug targeting in obstructive gastrointestinal cancers. Journal of Controlled Release. 2019;308:209–31.
[109] Iaizzo PA, Bianco RW, Hill AJ, Louis JDS. Heart Valves: Springer; 2013.
[110] Yoganathan AP, He Z, Casey Jones S. Fluid mechanics of heart valves. Annual Review of Biomedical Engineering. 2004;6:331–62.
[111] Perrucci GL, Zanobini M, Gripari P, Songia P, Alshaikh B, Tremoli E, et al. Pathophysiology of aortic stenosis and mitral regurgitation. Comprehensive Physiology. 2011;7(3):799–818.
[112] Valves RH. Pathophysiology of Valvular Heart Disease. Heart Failure: A Companion to Braunwald's HEART Disease E-Book. 2019;347.
[113] Alemu Y, Girdhar G, Xenos M, Sheriff J, Jesty J, Einav S, et al. Design optimization of a mechanical heart valve for reducing valve thrombogenicity – a case study with ATS valve. ASAIO Journal. 2010;56(5):389–96.
[114] Cavallo A, Gasparotti E, Losi P, Foffa I, Al Kayal T, Vignali E, et al. Fabrication and In-vitro characterization of a polymeric aortic valve for minimally invasive valve replacement. Journal of the Mechanical Behavior of Biomedical Materials. 2021;115:104294.
[115] Deviri E, Sareli P, Wisenbaugh T, Cronje SL. Obstruction of mechanical heart valve prostheses: Clinical aspects and surgical management. Journal of the American College of Cardiology. 1991;17(3):646–50.
[116] Wadhera RK, Russell CE, Piazza G. Warfarin versus novel oral anticoagulants: How to choose? Circulation. 2014;130(22):e191–e3.
[117] Lifvergren S, Chakhunashvili A, Bergman B. Improving warfarin treatment: A study using the six sigma methodology. 10th QMOD Conference Quality Management and Organiqatinal Development Our Dreams of Excellence; 18–20 June; 2007 in Helsingborg; Sweden: Citeseer; 2008.
[118] Glaser N, Jackson V, Franco-Cereceda A, Sartipy U. Survival after aortic valve replacement with bovine or porcine valve prostheses: A systematic review and meta-analysis. The Thoracic and Cardiovascular Surgeon. 2019;67(04):282–90.
[119] Jin L, Guo G, Jin W, Lei Y, Wang Y. Cross-linking methacrylated porcine pericardium by radical polymerization confers enhanced extracellular matrix stability, reduced calcification, and mitigated immune response to bioprosthetic heart valves. ACS Biomaterials Science & Engineering. 2019;5(4):1822–32.
[120] Liu J, Jing H, Qin Y, Li B, Sun Z, Kong D, et al. Nonglutaraldehyde fixation for off the shelf decellularized bovine pericardium in anticalcification cardiac valve applications. ACS Biomaterials Science & Engineering. 2019;5(3):1452–61.
[121] Harris C, Croce B, Cao C. Tissue and mechanical heart valves. Annals of Cardiothoracic Surgery. 2015;4(4):399.

[122] Lisy M, Kalender G, Schenke-Layland K, Brockbank KG, Biermann A, Stock UA. Allograft heart valves: Current aspects and future applications. Biopreservation and Biobanking. 2017;15 (2):148–57.
[123] Bonetti A, Marchini M, Ortolani F. Ectopic mineralization in heart valves: New insights from in vivo and in vitro procalcific models and promising perspectives on noncalcifiable bioengineered valves. Journal of Thoracic Disease. 2019;11(5):2126.
[124] Cribier A. Development of transcatheter aortic valve implantation (TAVI): A 20-year odyssey. Archives of Cardiovascular Diseases. 2012;105(3):146–52.
[125] Van Brabandt H, Neyt M, Hulstaert F. Transcatheter aortic valve implantation (TAVI): Risky and costly. BMJ. 2012;345.
[126] Vitale N, De Feo M, De Santo LS, Pollice A, Tedesco N, Cotrufo M. Dose-dependent fetal complications of warfarin in pregnant women with mechanical heart valves. Journal of the American College of Cardiology. 1999;33(6):1637–41.
[127] Cotrufo M, De Feo M, De Santo LS, Romano G, Della Corte A, Renzulli A, et al. Risk of warfarin during pregnancy with mechanical valve prostheses. Obstetrics and gynecology. 2002;99(1):35–40.
[128] Basude S, Hein C, Curtis S, Clark A, Trinder J. Low-molecular-weight heparin or warfarin for anticoagulation in pregnant women with mechanical heart valves: What are the risks? A retrospective observational study. BJOG: An International Journal of Obstetrics & Gynaecology. 2012;119(8):1008–13.
[129] Schmidt JB, Tranquillo RT. Tissue-engineered heart valves. Heart Valves. 2013;261–80.
[130] Waqanivavalagi SW, Bhat S, Ground MB, Milsom PF, Cornish J. Clinical performance of decellularized heart valves versus standard tissue conduits: A systematic review and meta-analysis. Journal of Cardiothoracic Surgery. 2020;15(1):1–13.
[131] Kasimir M-T, Rieder E, Seebacher G, Silberhumer G, Wolner E, Weigel G, et al. Comparison of different decellularization procedures of porcine heart valves. The International Journal of Artificial Organs. 2003;26(5):421–27.
[132] Siddiqui RF, Abraham JR, Butany J. Bioprosthetic heart valves: Modes of failure. Histopathology. 2009;55(2):135–44.
[133] Singhal P, Luk A, Butany J. Bioprosthetic heart valves: Impact of implantation on biomaterials. International Scholarly Research Notices. 2013;2013.
[134] Wilson T, Richards L, Bajwa T, Galazka P, Jahangir A, Fischer W, et al. Percutaneous removal of epicardial pacing wire: Nidus for infective endocarditis. CASE. 2021.
[135] Butany J, Ahluwalia MS, Munroe C, Fayet C, Ahn C, Blit P, et al. Mechanical heart valve prostheses: Identification and evaluation (erratum). Cardiovascular Pathology. 2003;12 (6):322–44.
[136] Fioretta ES, Motta SE, Lintas V, Loerakker S, Parker KK, Baaijens FP, et al. Next-generation tissue-engineered heart valves with repair, remodelling and regeneration capacity. Nature Reviews. Cardiology. 2021;18(2):92–116.
[137] Generali M, Dijkman PE, Hoerstrup SP. Bioresorbable scaffolds for cardiovascular tissue engineering. EMJ Interventional Cardiology. 2014;1(9):91.
[138] Nazir R. Collagen–hyaluronic acid based interpenetrating polymer networks as tissue engineered heart valve. Materials Science and Technology. 2016;32(9):871–82.
[139] Hemshekhar M, Thushara RM, Chandranayaka S, Sherman LS, Kemparaju K, Girish KS. Emerging roles of hyaluronic acid bioscaffolds in tissue engineering and regenerative medicine. International Journal of Biological Macromolecules. 2016;86:917–28.
[140] Boyd R, Parisi F, Kalfa D. State of the Art: Tissue Engineering in Congenital Heart Surgery. Seminars in Thoracic and Cardiovascular Surgery: Elsevier. 2019;807–17.

[141] Converse GL, Buse EE, Neill KR, McFall CR, Lewis HN, VeDepo MC, et al. Design and efficacy of a single-use bioreactor for heart valve tissue engineering. Journal of Biomedical Materials Research. 2017;105(2):249–59.

[142] Zhang BL, Bianco RW, Schoen FJ. Preclinical assessment of cardiac valve substitutes: Current status and considerations for engineered tissue heart valves. Frontiers in Cardiovascular Medicine. 2019;6:72.

[143] Lee A, Hudson A, Shiwarski D, Tashman J, Hinton T, Yerneni S, et al. 3D bioprinting of collagen to rebuild components of the human heart. Science. 2019;365(6452):482–87.

[144] Razavi M, Primavera R, Vykunta A, Thakor AS. Silicone-based bioscaffolds for cellular therapies. Materials Science and Engineering: C. 2021;119:111615.

[145] Ansari M, Eshghanmalek M. Biomaterials for repair and regeneration of the cartilage tissue. Bio-Design and Manufacturing. 2019;2(1):41–49.

[146] Fuchs A, Eberbach H, Izadpanah K, Bode G, Südkamp NP, Feucht MJ. Focal metallic inlay resurfacing prosthesis for the treatment of localized cartilage defects of the femoral condyles: A systematic review of clinical studies. Knee Surgery, Sports Traumatology, Arthroscopy. 2018;26(9):2722–32.

[147] Lecocq M, Linares J-M, Chaves-Jacob J, Coyle T, Roffino S, Eyraud M, et al. Total knee arthroplasty with a Ti6Al4V/PEEK prosthesis on an osteoarthritis rat model: Behavioral and neurophysiological analysis. Scientific Reports. 2020;10(1):1–15.

[148] Lemos CAA, Verri FR, de Luna Gomes JM, de Souza Batista VE, Cruz RS, Oliveira HFF, et al. Ceramic versus metal-ceramic implant-supported prostheses: A systematic review and meta-analysis. The Journal of Prosthetic Dentistry. 2019;121(6):879–86. e4.

[149] Yue S, He H, Li B, Hou T. Hydrogel as a biomaterial for bone tissue engineering: A review. Nanomaterials. 2020;10(8):1511.

[150] Guo Q, Wang Y, Xu D, Nossent J, Pavlos NJ, Xu J. Rheumatoid arthritis: Pathological mechanisms and modern pharmacologic therapies. Bone Research. 2018;6(1):1–14.

[151] Xia B, Chen D, Zhang J, Hu S, Jin H, Tong P. Osteoarthritis pathogenesis: A review of molecular mechanisms. Calcified Tissue International. 2014;95(6):495–505.

[152] Cho SM, Park J-W, Han H-S, Seok H-K, Moon M-W, Kim YC. Multifunctional composite coating as a wear-resistant layer for the bearing in total hip joint replacement. ACS Applied Materials & Interfaces. 2013;5(2):395–403.

[153] Cheng K-y, Gopal V, McNallan M, Manivasagam G, Mathew MT. Enhanced tribocorrosion resistance of hard ceramic coated Ti-6Al-4V alloy for hip implant application: In-vitro simulation study. ACS Biomaterials Science & Engineering. 2019;5(9):4817–24.

[154] Bahraminasab M, Farahmand F. State of the art review on design and manufacture of hybrid biomedical materials: Hip and knee prostheses. Proceedings of the institution of mechanical engineers, Part H. Journal of Engineering in Medicine. 2017;231(9):785–813.

[155] Braun S, Uhler M, Bormann T, Schroeder S, Jaeger S, Sonntag R, et al. Backside wear in total knee replacement: A new quantitative measurement method and a comparison of polished cobalt-chromium tibial trays with titanium tibial trays. Wear. 2021;466:203552.

[156] Sansone V, Pagani D, Melato M. The effects on bone cells of metal ions released from orthopaedic implants. A Review. Clinical Cases in Mineral and Bone Metabolism. 2013;10(1):34.

[157] Posada OM, Tate RJ, Meek R, Grant MH. In vitro analyses of the toxicity, immunological, and gene expression effects of cobalt-chromium alloy wear debris and co ions derived from metal-on-metal hip implants. Lubricants. 2015;3(3):539–68.

[158] Javadi A, Solouk A, Nazarpak MH, Bagheri F. Surface engineering of titanium-based implants using electrospraying and dip coating methods. Materials Science and Engineering: C. 2019;99:620–30.

[159] Korsacilar D, Albayrak A, Malayoğlu U. Production of a bio-compatible metallic implant via electropolymetrization method by coating a synthetic hydrogel on a titanium alloy. Journal of Biomechanics. 2011;44:17.
[160] De Giglio E, Guascito M, Sabbatini L, Zambonin G. Electropolymerization of pyrrole on titanium substrates for the future development of new biocompatible surfaces. Biomaterials. 2001;22(19):2609–16.
[161] Civantos A, Martinez-Campos E, Ramos V, Elvira C, Gallardo A, Abarrategi A. Titanium coatings and surface modifications: Toward clinically useful bioactive implants. ACS Biomaterials Science & Engineering. 2017;3(7):1245–61.
[162] Xie C-M, Lu X, Wang K-F, Meng F-Z, Jiang O, Zhang H-P, et al. Silver nanoparticles and growth factors incorporated hydroxyapatite coatings on metallic implant surfaces for enhancement of osteoinductivity and antibacterial properties. ACS Applied Materials & Interfaces. 2014;6(11):8580–89.
[163] Lyndon JA, Boyd BJ, Birbilis N. Metallic implant drug/device combinations for controlled drug release in orthopaedic applications. Journal of Controlled Release. 2014;179:63–75.
[164] Revell PA. The combined role of wear particles, macrophages and lymphocytes in the loosening of total joint prostheses. Journal of the Royal Society Interface. 2008;5 (28):1263–78.
[165] Shegarfi H, Reikeras O. Bone transplantation and immune response. Journal of Orthopaedic Surgery. 2009;17(2):206–11.
[166] Simchi A, Tamjid E, Pishbin F, Boccaccini A. Recent progress in inorganic and composite coatings with bactericidal capability for orthopaedic applications. Nanomedicine: Nanotechnology, Biology and Medicine. 2011;7(1):22–39.
[167] Vaishya R, Chauhan M, Vaish A. Bone cement. Journal of Clinical Orthopaedics and Trauma. 2013;4(4):157–63.
[168] Cawley DT, Kelly N, McGarry JP, Shannon FJ. Cementing techniques for the tibial component in primary total knee replacement. The Bone & Joint Journal. 2013;95(3):295–300.
[169] Zhou K, Yu H, Li J, Wang H, Zhou Z, Pei F. No difference in implant survivorship and clinical outcomes between full-cementless and full-cemented fixation in primary total knee arthroplasty: A systematic review and meta-analysis. International Journal of Surgery. 2018;53:312–19.
[170] Palumbo BT, Henderson ER, Edwards PK, Burris RB, Gutiérrez S, Raterman SJ. Initial experience of the journey-deuce bicompartmental knee prosthesis: A review of 36 cases. The Journal of Arthroplasty. 2011;26(6):40–45.
[171] Kim IT, Yoo JJ. Implant design in cementless hip arthroplasty. Hip & Pelvis. 2016;28(2):65–75.
[172] Romanò C, Tsuchiya H, Morelli I, Battaglia A, Drago L. Antibacterial coating of implants: Are we missing something?. Bone & Joint Research. 2019;8(5):199–206.
[173] Neut D, Dijkstra R, Thompson J, Kavanagh C, Van der Mei H, Busscher H. A biodegradable gentamicin-hydroxyapatite-coating for infection prophylaxis in cementless hip prostheses. European Cells & Materials. 2015;29:42–55.
[174] Rossi SMP, Perticarini L, Mosconi M, Ghiara M, Benazzo F. Ten-year outcomes of a nitrided Ti-6Al-4V titanium alloy fixed-bearing total knee replacement with a highly crosslinked polyethylene-bearing in patients with metal allergy. Knee. 2020;27(5):1519–24.
[175] Nakamura S, Ito H, Nakamura K, Kuriyama S, Furu M, Matsuda S. Long-term durability of ceramic tri-condylar knee implants: A minimum 15-year follow-up. The Journal of Arthroplasty. 2017;32(6):1874–79.
[176] Baruffaldi F, Mecca R, Stea S, Beraudi A, Bordini B, Amabile M, et al. Squeaking and other noises in patients with ceramic-on-ceramic total hip arthroplasty. Hip international: The

Journal of Clinical and Experimental Research on Hip Pathology and Therapy. 2020;30 (4):438–45.
[177] Banci L, Balato G, Salari P, Baldini A. Systematic review and meta-analysis of ceramic coated implants in total knee arthroplasty. Comparable mid-term results to uncoated implants. Knee Surgery, Sports Traumatology, Arthroscopy: Official Journal of the ESSKA. 2021.
[178] Megaloikonomos PD, Becher C, Van der Stok J, O'Donnell T. Femoral condyle resurfacing using an inlay metal implant: Low revision rate of 266 patients in a 5–10 years follow-up. Archives of Orthopaedic and Traumatic Surgery. Archiv fur Orthopadische und Unfall-Chirurgie. 2021.
[179] Peeters E, De Hertogh G, Junge K, Klinge U, Miserez M. Skin as marker for collagen type I/III ratio in abdominal wall fascia. Hernia. 2014;18(4):519–25.
[180] Salameh J, Talbott LM, May W, Gosheh B, Vig PJ, McDaniel DO. Role of biomarkers in incisional hernias. The American Surgeon. 2007;73(6):561–68.
[181] Fujii T, Tsutsumi S, Matsumoto A, Fukasawa T, Tabe Y, Yajima R, et al. Thickness of subcutaneous fat as a strong risk factor for wound infections in elective colorectal surgery: Impact of prediction using preoperative CT. Digestive Surgery. 2010;27(4):331–35.
[182] Mingoli A, Puggioni A, Sgarzini G, Luciani G, Corzani F, Ciccarone F, et al. Incidence of incisional hernia following emergency abdominal surgery. Italian Journal of Gastroenterology and Hepatology. 1999;31(6):449–53.
[183] Sørensen LT, Hemmingsen UB, Kirkeby LT, Kallehave F, Jørgensen LN. Smoking is a risk factor for incisional hernia. Archives of Surgery. 2005;140(2):119–23.
[184] See CW, Kim T, Zhu D. Hernia mesh and hernia repair: A review. Engineered Regeneration. 2020;1:19–33.
[185] Stojadinovic A, Hoos A, Karpoff HM, Leung DH, Antonescu CR, Brennan MF, et al. Soft tissue tumors of the abdominal wall: Analysis of disease patterns and treatment. Archives of Surgery. 2001;136(1):70–79.
[186] Kingsnorth A, LeBlanc K. Hernias: Inguinal and incisional. The Lancet. 2003;362 (9395):1561–71.
[187] LeBlanc KA, Booth WV. Laparoscopic repair of incisional abdominal hernias using expanded polytetrafluoroethylene: Preliminary findings. Surgical Laparoscopy & Endoscopy. 1993;3(1):39–41.
[188] Heniford BT, Park A, Ramshaw BJ, Voeller G. Laparoscopic ventral and incisional hernia repair in 407 patients. Journal of the American College of Surgeons. 2000;190(6):645–50.
[189] Kalaba S, Gerhard E, Winder JS, Pauli EM, Haluck RS, Yang J. Design strategies and applications of biomaterials and devices for hernia repair. Bioactive Materials. 2016;1 (1):2–17.
[190] Todros S, Pavan P, Natali A. Synthetic surgical meshes used in abdominal wall surgery: Part I – materials and structural conformation. Journal of Biomedical Materials Research. Part B, Applied Biomaterials. 2017;105(3):689–99.
[191] FitzGerald JF, Kumar AS. Biologic versus synthetic mesh reinforcement: What are the pros and cons?. Clinics in Colon and Rectal Surgery. 2014;27(04):140–48.
[192] Qiao Y, Li Y, Zhang Q, Wang Q, Gao J, Wang L. Dopamine-mediated zwitterionic polyelectrolyte-coated polypropylene hernia mesh with synergistic anti-inflammation effects. Langmuir. 2020;36(19):5251–61.
[193] Majumder A, Neupane R, Novitsky YW. Antibiotic coating of hernia meshes: The next step toward preventing mesh infection. Surgical Technology International. 2015;27:147–53.
[194] Zogbi L. The use of biomaterials to treat abdominal hernias. Biomaterials Applications for Nanomedicine. 2011;18:359–82.

[195] Charleux-Muller D, Hurel R, Fabacher T, Brigand C, Rohr S, Manfredelli S, et al. Slowly absorbable mesh in contaminated incisional hernia repair: Results of a french multicenter study. Hernia. 2021:1–9.
[196] Seiler C, Baumann P, Kienle P, Kuthe A, Kuhlgatz J, Engemann R, et al. A randomised, multi-centre, prospective, double blind pilot-study to evaluate safety and efficacy of the non-absorbable optilene® mesh elastic versus the partly absorbable ultrapro® mesh for incisional hernia repair. BMC Surgery. 2010;10(1):1–7.
[197] Earle DB, Mark LA. Prosthetic material in inguinal hernia repair: How do i choose?. Surgical Clinics of North America. 2008;88(1):179–201.
[198] Idrees S, Jindal S, Gupta M, Sarangi R. Surgical meshes–the search continues. Current Medicine Research and Practice. 2018;8(5):177–82.
[199] Schreinemacher MH, van Barneveld KW, Dikmans RE, Gijbels MJ, Greve J-WM, Bouvy ND. Coated meshes for hernia repair provide comparable intraperitoneal adhesion prevention. Surgical Endoscopy. 2013;27(11):4202–09.
[200] Lai C, Zhang S-J, Chen X-C, Sheng L-Y, Qi T-W, Yan L-P. Development of a cellulose-based prosthetic mesh for pelvic organ prolapse treatment: In vivo long-term evaluation in an ewe vagina model. Materials Today Bio. 2021;100172.
[201] Blatnik JA, Thatiparti TR, Krpata DM, Zuckerman ST, Rosen MJ, von Recum HA. Infection prevention using affinity polymer-coated, synthetic meshes in a pig hernia model. Journal of Surgical Research. 2017;219:5–10.
[202] Heymann F, von Trotha K-T, Preisinger C, Lynen-Jansen P, Roeth AA, Geiger M, et al. Polypropylene mesh implantation for hernia repair causes myeloid cell–driven persistent inflammation. JCI Insight. 2019;4:2.
[203] Qiao Y, Zhang Q, Wang Q, Lin J, Wang J, Li Y, et al. Synergistic anti-inflammatory coating “Zipped Up” on polypropylene hernia mesh. ACS Applied Materials & Interfaces. 2021;13(30):35456–68.
[204] Raj R, Shenoy SJ, Mony MP, Pratheesh KV, Nair RS, Geetha CS, et al. Surface modification of polypropylene mesh with a porcine cholecystic extracellular matrix hydrogel for mitigating host tissue reaction. ACS Applied Bio Materials. 2021;4(4):3304–19.
[205] Hu W, Zhang Z, Zhu L, Wen Y, Zhang T, Ren P, et al. Combination of polypropylene mesh and in situ injectable mussel-inspired hydrogel in laparoscopic hernia repair for preventing post-surgical adhesions in the piglet model. ACS Biomaterials Science & Engineering. 2020;6(3):1735–43.
[206] Blázquez R, Sánchez-Margallo FM, Álvarez V, Usón A, Marinaro F, Casado JG. Fibrin glue mesh fixation combined with mesenchymal stem cells or exosomes modulates the inflammatory reaction in a murine model of incisional hernia. Acta Biomaterialia. 2018;71:318–29.
[207] Liu Z, Wei N, Tang R. Functionalized strategies and mechanisms of the emerging mesh for abdominal wall repair and regeneration. ACS Biomaterials Science & Engineering. 2021.
[208] Zhao Y, Zhang Z, Wang J, Yin P, Zhou J, Zhen M, et al. Abdominal hernia repair with a decellularized dermal scaffold seeded with autologous bone marrow-derived mesenchymal stem cells. Artificial Organs. 2012;36(3):247–55.

Chapter 6
Biomaterials in cancer treatment

6.1 Cancer and its treatment modalities

Cancer is one of the most dreaded diseases in the world and has caused millions of death and suffering. It is defined by rapid and uncontrolled replication or division of abnormal cells that may cause pathological consequences in the body [1]. These cells may spread around the body a condition called "metastasis" and could cause multiple malignancies that can be detrimental to the health of a person. Cancer is not one singular disease but a group of more than 100 different and distinctive diseases and can involve any tissue of the body and have many different forms in each body area. The most common presentation of cancer is the formation of a tumor [2]. Tumor formation in the body is actually quite common, but they are usually being checked by the immune system, most especially by the immune cells (e.g., natural killer cells and/or T cells). Furthermore, not all tumors are cancerous or malignant as most of them are considered benign or noncancerous. Malignant tumors are made from abnormal cells that had accumulated mutations in their genome. The most common mutation in the cell attributed to cancer formation is with its p53 gene that tells it when to undergo apoptosis (controlled cell death) and involved in growth arrest and DNA repair [3]. However, when this gene has been damaged or mutated, the cell will then lack the genetic information to undergo the apoptotic cell cascade or repair and will continue to replicate indefinitely passing on the mutated p53 gene to its subsequent cell lineage. The accumulation of these abnormal cells may form a tumor that is malignant in nature. Mutations or damage to the cells are caused by various factors (mutagenic or carcinogenic factors), it could be environmental, lifestyle, or genetically predisposition driven. These factors or the combination of such, leads to multiple damage in the DNA causing mutations that are not beneficial to the cell. Primary examples of mutagens/carcinogens are exposure to ionizing radiation (UV light, X-ray, or Gamma rays), constant exposure to pollution and carcinogenic substances (asbestos, smoking, alcohol, drug abuse) [4–7].

Once a patient is diagnosed with cancer, clinicians usually classify the progression of the disease depending on its stage. In stage 1, cancer is characterized by the tumor being small and locally present within the organ or tissue; in stages 2 and 3, the tumor has grown into nearby tissues or lymph nodes; in stage 4, it has spread to other body parts (organs) or also known as metastatic cancer. Prognosis or how well the patient will react to treatment is defined by these stages. With stage 1 being the most responsive to treatment with generally good prognosis, in stages 2 and 3 the cancer might or might not react to treatment and require more extensive or combination treatment modalities with a prognosis of about 50%, and stage 4 is usually terminal and has lower prognosis chances of survival of less than 50%.

https://doi.org/10.1515/9783110655377-006

There are several modalities used currently in treating cancer. This all depends on how advanced the cancer stage and the location and the nature of the cancer [8]. The most common treatment modality is with chemotherapy. Chemotherapy is with the use of drugs that limit the replication of cancer cells, and/or instruct the cancer cells to undergo apoptosis [9, 10]. Radiation therapy is using targeted ionizing radiation to reduce the size of the cancer cells by damaging their intracellular proteins and/or DNA to induce cell death [11, 12]. Surgery is a more straightforward treatment of cancer where the tumor is directly resected or removed from the body [13]. Newer modalities such as immunotherapy is using the body's own immune system to target and kill malignant cells [14, 15]. Hyperthermia therapy is a well known but seldom used modality where the cancer cells are heated above the normal body temperature to induce apoptosis via heat shock [16, 17]. All of these treatment modalities can be used as a stand-alone treatment or adjuvants to each other depending on the condition of the patient and stage of the cancer. There is no silver bullet treatment for cancer as it is not a singular disease. No single drug or treatment modality is enough to cure all types of cancer. Therefore, research is still ongoing on how we can treat most cancer types. Here, biomaterials are used to advance the field of cancer treatment by reducing the severe side effects of chemotherapeutics, increase their targeting capabilities, improve diagnostic and imaging capabilities, create hyperthermic conditions inside the body in an efficient manner, used as a postoperative material to increase healing and avoid remission of cancer, as an alternative to localized brachytherapy (radiation), and as a carrier for immunotherapy stimulatory adjuvants, to cancer antigens. All of this will be discussed in detail in the subsequent parts of the chapter, to elucidate how biomaterials are being used in different treatment modalities and the current challenges each modalities are facing.

6.2 Biomaterials in cancer chemotherapy and diagnostics

6.2.1 Biomaterial drug delivery system for enhancing chemotherapy effects

Chemotherapy is a treatment modality commonly used in multiple types of cancer for reducing the proliferation of cancer cells in the body (Figure 6.1) [9, 10]. It utilizes different classes of drugs that inhibit several cell division and cell metabolic processes of rapidly dividing cells (Table 6.1). This being said, chemotherapeutics are nonselective and when administered systemically will affect most of rapidly dividing cell types, including normal cells. This will make a patient receiving chemotherapeutics experience severe side effects such as hair loss, nausea and vomiting, and general body weakness [18–20]. Because the cells responsible for hair production, gastrointestinal tract lining, and bone marrow are also rapidly dividing and are affected by the chemotherapeutics. Thus, having a drug that can target only the diseased area (cancer tumor) is highly valuable and can make the treatment more effective with less side effects. However,

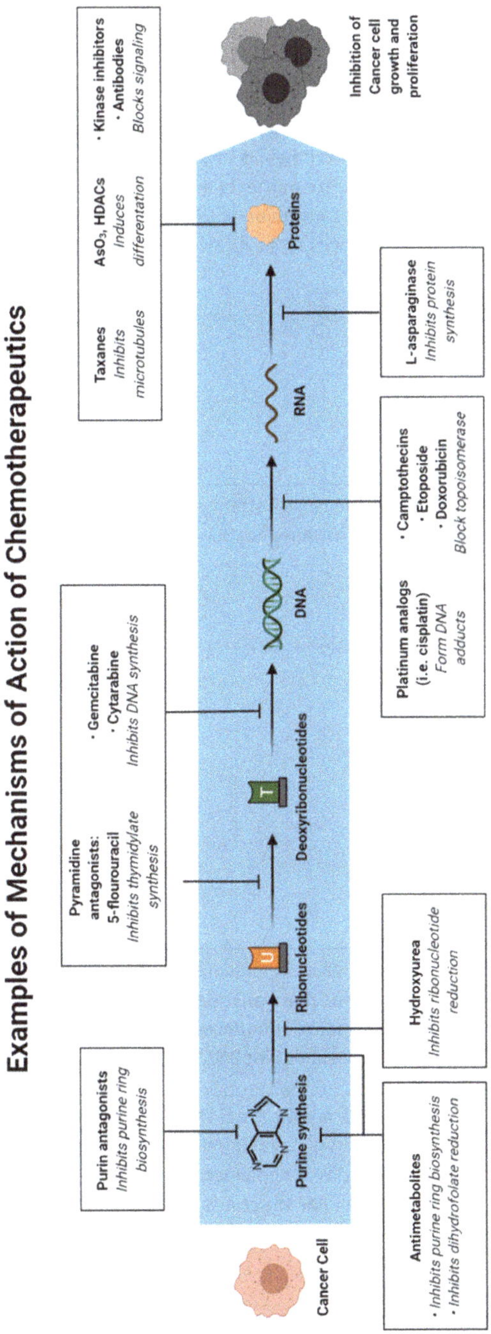

Figure 6.1: Mechanisms of different classes of chemotherapeutics to control the proliferation of cancer cells inside the body.

Table 6.1: Chemotherapeutic classes and description.

General class	Subclass and examples	Class description	Ref.
Antimetabolites	Antifolates, (methotraxate, pemetrexed) Pyrimidine antagonists, (cytarabine, 5-fluorouracil, gemcitabine) Purin antagonist, (azathioprine, fludarabine, cladribine) Ribonucleotide reductase inhibitors (hydroxyurea (hydroxycarbamide)	Inhibiting the ability of malignant cells to synthesize and metabolize nutrients and also inhibit conversion of DNA and RNA sequences for replication.	[21]
Alkylating agents	Oxazaphosphorines (cyclophosphamide) Nitrogen mustards (chlorambucil, melphalan) Imidazotetrazines (temozolamide) Nitrosureas (carmustine, streptozocin) Hydrazines (procabazine) Platinum based agents (cisplatin, carboplatin, oxaliplatin)	Crosslinks DNA strands and reduce DNA replication	[22]
Topoisomerase inhibitors	Topoisomerase I inhibitors (ironectan, topotecan) Topoisomerase II inhibitors (etoposide)	Inhibition of topoisomerase (enzymes that participate in the overwinding or underwinding of DNA) essential for DNA replication	[23]
Mitotic Inhibitors	Vinca alkaloids (vincristine, vinblastine) Taxanes (docetaxel, paclitaxel) Nontaxane (eribulin, ixabepilone, epothilone)	Inhibits mitotic processes (cell division) by breaking down, stabilizing, or polymerizing intracellular structures such as microtubules, mitotic spindles, beta-tubulin that responsible for cell division.	[24]

Table 6.1 (continued)

General class	Subclass and examples	Class description	Ref.
Antitumor antibiotics	Bleomycin Actinomycin D (dactinomycin) Anthracyclines (doxorubicin, daunorubicin, idarubicin)	Inhibits the progression of tumor forming malignancies by interfering with cell to cell interaction, DNA transcription, and formation of free radicals.	[25]
Protein Kinaseinhibitors	BCR-ABL and c-KIT tyrosine kinase inhibitors (imatinib, dasatinib, nilotinib) ALK tyrosine kinase inhibitors (alectinib, crizotinib)	Inhibition of protein and enzymes that are essential for cell metabolism	[26]
Others	Proteasome Inhibitors (bortezomib, ixazomib) PARP inhibitors (olaparib)	Inhibition of ubiquitinated apoptotic protein degradationDecreased repair of single-strand DNA breaks	[27, 28]

pure chemotherapeutics are unable to do so since they lack a targeting mechanism to allow uptake only on malignant cells. Here, drug delivery systems made up of biomaterials that are in nano and micro scales are utilized to enhance the effect of chemotherapeutic drugs.

There are several classes of biomaterials used for enhancing chemotherapeutic treatments for cancer. The first thing that comes to mind is with the use of microparticles and nanoparticles. There are multiple benefits for using nanoscale or nanoformulated chemotherapeutics for treating cancer. The size of the nanoparticles meant that it can circulate inside the body for longer periods and protecting the drug payload from other biochemical interactions inside the body. Furthermore, nanoparticulate drug delivery systems can be tailored to have capabilities to selectively target malignant cells in the body thereby enhancing their therapeutic effects and reducing side effects. This is due to the high surface to volume ratio of nanoparticles allowing it to carry therapeutics and targeting ligands at the same time. Furthermore, novel drug delivery systems are now also being tailored to possess imaging capabilities giving multifunctionality to a certain delivery system, this is also known as theranostics (therapy + diagnostics). There are multiple types of nanoparticles used in cancer therapy. Organic and inorganic nanoparticles are functionalized with targeting moieties, and imaging ligands while encapsulating or attaching covalently or noncovalently chemotherapeutic drugs.

6.2.2 Passive tumor targeting

Cancer tumors have a different physiological properties compared to normal tissues. Here nanoparticles can become an effective way of localizing chemotherapeutics due to what is called the "enhanced permeability and retention effect" (EPR, Figure 6.2) [29–31]. Cancer tumors have leaky vasculature (Figure 6.3) [32] in contrast to normal tissues since the mass grows rapidly and blood vessels need to expand quickly to accommodate the tumor's need for nutrient and oxygenation [33, 34]. This allows for the passive localization of many drugs and drug carriers due to their extravasation through leaky vasculature. This abnormal and unregulated vessel formation (angiogenesis) creates vessel walls with porosity that ranges from 40 nm to 1 μm; these leaky vessels allow relatively large nanoparticles to infiltrate into tumor masses passively without the need for a targeting ligand. Furthermore, as a rapidly growing tumor mass lacks a functioning lymphatic clearance system, the nanoparticles can enhance their accumulation in the malignant tissue. The EPR effect is the most common way for nanoparticles to target tumors and avoiding normal tissues and the majority of current nanomedicines for solid tumor treatment rely heavily on EPR effect to ensure high drug accumulation thereby improve treatment efficacy [35–37].

To allow for EPR effect to take place, passive targeting requires nanoparticles to circulate in the systemic circulation long enough for sufficient level of drug reach the target area. Therefore, strategies such as adding or modifying the surface of the nanoparticles with water soluble polymers such as polyethylene glycol (a.k.a. PEGylation) are being done [38, 39]; PEGylation is often used to make water-insoluble nanoparticles to be water-soluble in many preclinical research and also mask the nanoparticles from opsonization. One example of commercially available PEGylated nanoparticle is Doxil®, which is a PEG-coated liposomal doxorubicin used for breast cancer utilizing passive tumor accumulation [40]. Nanoparticle drug delivery systems are inherently eliminated or checked by the immune system especially the reticuloendothelial system (RES) [41–43] The reticuloendothelial system (RES) is a heterogeneous population of phagocytic cells in systemically fixed tissues that play an important role in the clearance of particles and soluble substances in the circulation and tissues. Nanoparticles that have reportedly negatively charged surfaces are known to be effectively uptaken by the RES. Thus, nanoparticles with a neutral or positive charge surfaces are expected to have a longer circulation time in the blood [44, 45].

The EPR effect, although common in most tumors, is not a perfect solution for nanoparticle utilization. It is known that the central region of metastatic or larger tumor mass does not exhibit the EPR effect [46, 47]. This is due to the extreme hypoxic condition (lack of oxygenation and faulty vasculature formation) in the area of the tumor. Hence, several adjuvant methods are used in the clinic to artificially enhance the EPR effect [48, 49]. Administering angiotensin II [48, 50] via slow infusion is used to increase systolic blood pressure and allow for more tissue perfusion in the middle of the tumor. Additionally, topical application of nitrogen oxide-

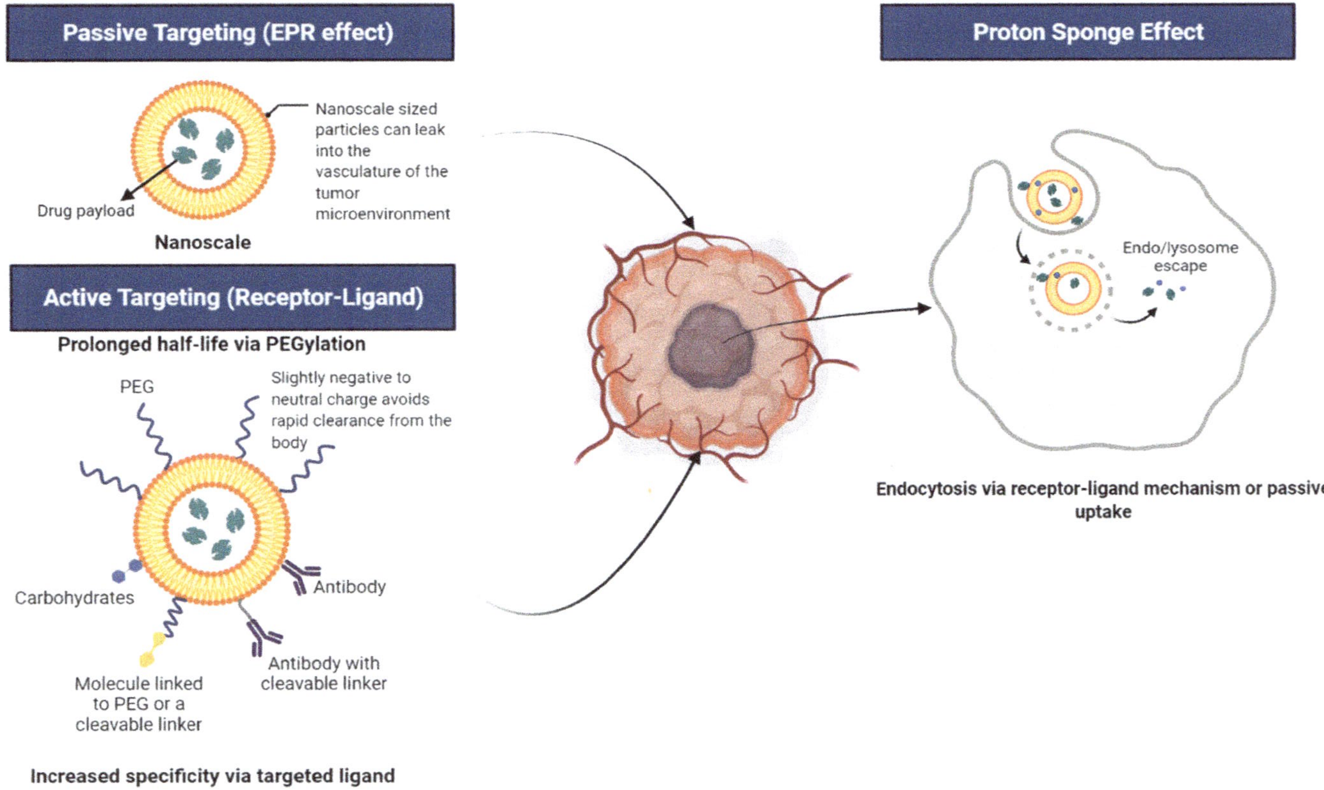

Figure 6.2: Nanoparticle mechanism on targeting (passive and active) malignant tissue and how they escape the endosome of cancer cells (proton sponge effect).

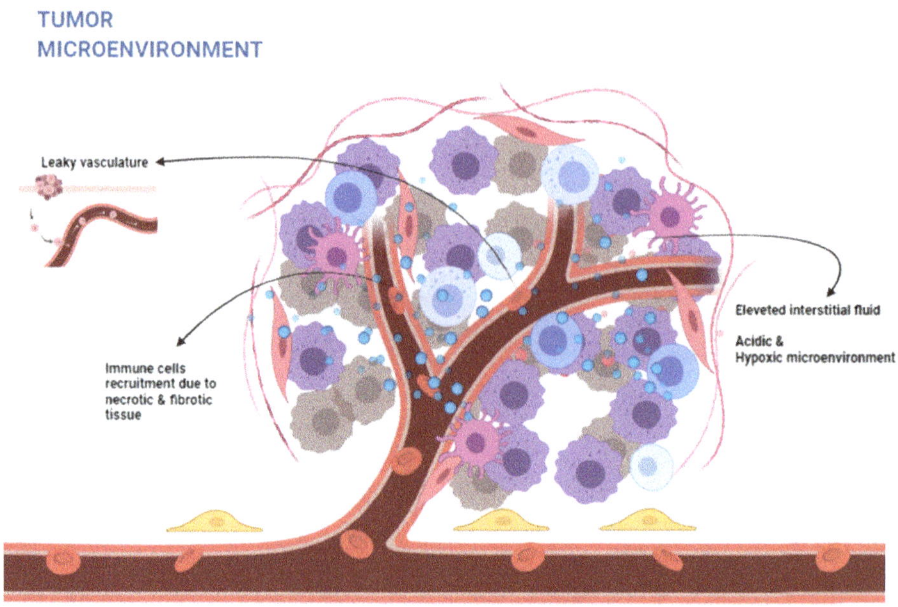

Figure 6.3: The tumor microenvironment showing leaky vasculature, elevated interstitial fluid with acidic and hypoxic environment. These unique environments can be used as triggering mechanisms for drug release and drug targeting using stimuli responsive biomaterials.

releasing agents and heat are used to expand vascular permeabilization in solid tumors [51, 52]. Passive accumulation through EPR effect is the easiest and most accepted mechanism for nanoparticle drug delivery system in treating solid tumors. However, there are several considerations in nanoparticle designs (e.g., size, molecular weight of the nanoparticles, surface charge) in order to fully realize the potential of the EPR effect. Other factors such as biocompatibility, hemocompatibility, and the ability to have prolonged plasma half-life are also vital in designing the nanoparticles for efficient passive tumor accumulation.

6.2.3 Active tumor targeting

Although the EPR effect allows for nanoparticles to accumulate in the tumor site at a higher concentration compared to normal tissues, it is not responsible for nanoparticle uptake inside malignant cells (Figure 6.2) [53]. Nanoparticle/drug cell internalization [54] is vital to ensure that the drug payload that is being carried by nanoparticulate systems are delivered efficiently to enact its therapeutic role, especially chemotherapeutic drugs that require drug activation within the intracellular structures [55]. This is especially true in delivery of nucleic acids (DNA, siRNA, miRNA) in gene therapy that require the escape of these therapeutics/molecules from endosome so they can reach desired subcellular compartments. Furthermore, EPR is nonconstant and varies in every tumor or cancer type and its effect differs from patient to patient. Thus, a more precise and active targeting is a vital feature for an effective nanoparticle treatment. Active targeting enables treatment modalities that are not achievable with EPR alone, or improve effectiveness of treatments which can be accomplished using EPR, but with less pronounced effects [56, 57]. Active targeting of nanoparticles to tumor cells, microenvironment or vasculature, uses surface modification with small molecules, antibodies, affibodies, peptides or aptamers [58, 59]. In comparison to passive targeting (EPR effect), active targeting nanoparticles carry surface ligands to facilitate selective uptake of nanoparticles to receptors present on cell or tissue of interest [58, 60]. Active targeting promotes cell uptake through receptor-mediated endocytosis [61–64]. Each cancer cell type typically presents overexpressed receptors which can be targeted by attached ligands on the nanoparticles. However, active targeting can differ from cancer cell types; thus, it is important to understand fully what kind of overexpressed receptors each cancer cell type possesses in order to choose the suitable targeting ligand to be used on a nanoparticle system. So far, most of the nanotechnology-based nanomedicines approved for clinical use or in advanced clinical trials solely rely on EPR effect. This is due to the regulatory hurdles for nanoformulated drugs to be screened thoroughly for safety and tolerability before being approved for human trials or consumption. Ligands are considered to be exotic or nonstandard therapeutic excipients that require extensive screening to ensure that they will not cause any long-term or short-term adverse effects. However, with the progression of the field of nanomedicine, it is expected that next-generation nanotherapies may possess active targeting ligands that enable or enhance intracellular uptake, intracellular trafficking, and penetration of physiological barriers which are still a major hurdle in cancer treatment.

6.2.4 The proton sponge effect

Once the nanoparticle is uptaken by malignant cells, one major hurdle must be overcome to ensure that the drug or therapeutic molecule payload is properly delivered. Nanoparticles that are uptaken by malignant cells are typically enveloped by a structure called the endosome (Figure 6.2) [65, 66]. Particle uptake is governed mostly by receptor-mediated endocytosis where an active targeting ligand creates a link with a receptor bound in the cell membrane of the malignant cells [67–69]. Once this has been done, the particle is enveloped by a structure called endosome. Endosomes provide an environment for materials to be sorted before they reach the degradative lysosome [70]. The endosomal microenvironment is acidic (pH 6.5 to 4.0) in comparison to intracellular pH of 7.0 [71]. Usually, particles entrapped inside endosomes are maintained inside their lumen and have difficulty escaping them. This, in turn, reduces the therapeutic effects of the nanoparticle as it is being trafficked inside intracellular compartments that neutralize it. In order to maintain the nanoparticle's therapeutic efficiency, it must escape the endosomal lumen and deliver its payload in the intracellular space so that the drug can enact its mechanism of action [72, 73]. The proton sponge effect (SPE) is one route researchers used to overcome endosomal trafficking [74, 75]. SPE happens when nanoparticles recruits water molecules inside the luminal side of the endosome by virtue of attracting hydrogen and chloride ions. The ionic difference between the intraluminal side of the endosome and the intracellular space attracts water molecules to rush into the endosomal space to maintain electrical neutrality; this process leads to endosomal swelling and rupture [76, 77] making the nanoparticles free to release its therapeutic payload into the intracellular organelles or compartments of interest. Nanoparticles are designed to react with the acidic conditions of the endosomal lumen by containing cationic charges [78, 79]. Cationic polymers and polyplexes are known to heavily recruit hydrogen and chloride ions once they are enveloped inside the endosome. Another way is to make achieve SPE is to make the nanoparticle acidic pH-responsive, pH-responsive materials can also attract hydrogen and chloride ions into the endosomal lumen by having a reversible charge variations [80]. Usually, pH-responsive polymers have positive and negatively charged moieties that are conjugated in neutral or alkaline pH; this can be reversibly conjugated when exposed to acidic environment. This dissociation of charged moieties can expose positively charged end groups that may then recruit negatively charged ions and water molecules inside the endosomal compartment leading it to burst. SPE is still a heavily debated hypothesis for increasing therapeutic effects of nanoparticle drug delivery systems [74]. As endosomal trafficking is a highly complex mechanism [81, 82], it is not still fully understood whether what other nanoparticle properties may induce SPE.

6.2.5 Nanoparticle transport across physiological barriers

Another hurdle for nanoparticle drug delivery systems in cancer therapy are the physiological barriers found in the body. There are certain physiological barriers which main function is to protect an organ from exogenous materials that may cause adverse effects on the physiological functions of the body [83]. One example of this is the blood–brain barrier (BBB) [84–86], the BBB is a selective physiological barrier that protects the brain from pathogenic factors. The BBB only allows water molecules, gases such as oxygen, lipid-soluble molecules, and other necessary substances like glucose to translocate toward the brain. BBB is composed of blood vessels that tightly regulate the movement of the said substances. Treating brain tissue malignancies is deemed challenging due to this barrier as systemic administration of nanoparticles is typically regulated by the BBB. Hence, special design considerations for nanoparticle systems must be incorporated in order to cross the BBB. This is usually done by active targeting receptors that are located in the BBB in order to translocate nanoparticles into the central nervous system. Receptor-mediated transcytosis (RMT) is one such class of transport system [61, 87]. Endogenous peptides use RMT to translocate themselves across the BBB. Nanoparticles functionalized with ligands akin to attaching to the receptors present along the endothelial cells in the BBB can pass through the barrier allowing them to penetrate the tissues in the brain. Examples of identified receptor ligands that can be used nanoparticle drug delivery systems in cancer treatment are the following; Transferrin receptor (TfR) binding to OX26 conjugated liposomes [88–90]. Human TfR, a transmembrane glycoprotein, is composed of two 90 kDa subunits that are joined by disulfide bonds with each subunit binding one Tf molecule while OX26 is a monoclonal antibody (mABs). The OX26 brain drug delivery system has been validated quite thoroughly, using covalent linkage/disulfide linkage or streptavidin/biotin (SA/B) linkage strategy. Insulin receptors are also abundant along the BBB; the brain requiring high caloric intake needs insulin to facilitate the utilization of glucose in the body. IGF or immunoglobulin F can also be found overexpressed in the BBB with its mABs (IGF-II) and mannose-bearing glycoproteins. Furthermore, low-density lipoproteins or LDL-receptor-related proteins (LRPs) [91] have been reported to mediate transport of various ligands, including two members of the Tf family, lactoferrin and melanotransferrin, across the BBB. And lastly, cell-penetrating peptides (CPPs) [92, 93] are short amino acid sequences (10–27 amino acid residues) that possess several positive charges. They are powerful tools for transporting proteins and other molecules into cells and across the BBB. Some CPPs, for example, the transcription-activating factor Tat, penetratin and the Syn-B vector, are derived from natural sequences [94]. If the nanoparticle is functionalized with ligands that target the said receptors, it may help its translocation into the brain tissue and help in treating central nervous system malignancies.

Cancer tumors contain natural barriers that nanoparticle drug delivery systems must overcome to ensure therapeutic efficiency. These include the tumor stroma (e.g., biological barriers) and tumor endothelium barriers (e.g., functional barriers) [95]. Biological barriers restrict the movement of nanoparticles along the intratumoral environment while functional barriers disrupt the transport of intact nanoparticles the tumor mass: These include elevated interstitial fluid pressure and acidic environment found in the tumoral environment. Thus, it is important to design nanoparticle drug delivery systems that can circumvent the said barriers. It is important to understand tumor physiology in order to a treatment suited that can elicit proper chemotherapeutic response. The tumor microenvironment (TME) [96, 97] is a heterogeneous system composed of disarrayed vasculature, scarred and necrotic tissues mixed with immune cells all mixed in an extracellular matrix. The cell density inside the TME is very high and creates elevated interstitial fluid pressure that can be disruptive for free diffusion of nanoparticles. Thus, nanoparticle zeta potential is very important in characterizing its behavior in varying pressure gradients to ensure that they can be stable and disperse avoiding agglomeration and stunted diffusion along the tumor mass. Strategies to reduce interstitial fluid pressure to improve tumor penetration include adjuvant pharmacological and external interventions to normalize vasculature within TME [98]. This is done by administering hypertonic solutions intratumorally or with the use of heat using radiofrequency and high-intensity focused ultrasound. Furthermore, the TME is found to be more acidic compared to normal physiological tissues due to the overproduction of lactic acid since a tumor mass can become hypoxic and starts to undergo anaerobic respiration. This acidity tends to destabilize nanoparticle drug delivery systems making drug delivery unpredictable and inconsistent [99, 100]. However, newer strategies are being employed by creating pH-responsive nanoparticles that can alter their surface charge when exposed to acidic pH [99, 101]. This stimuli responsiveness utilizes the acidic conditions of the TME in order to release its drug payload in the ECM making this functional barrier a positive agent in delivering chemotherapeutics.

6.2.6 Theranostics

An emerging new chemotherapeutic trend combines the ability to deliver drugs to malignant cells while allowing it to be monitored or locate their distribution and dispersion in the tumor using conventional imaging techniques such as X-ray, magnetic resonance imaging (MRI), fluorescence imaging, and ultrasound. This emerging field in cancer therapy is called theranostics, a combination of the words "therapy" and "diagnostics" [102, 103]. Theranostics relies on the field of nanotechnology, most specifically nanoparticles such as gold [104], iron [105, 106], quantum dots [107–109], and nanobubbles [110, 111]. Nanoparticles allow for chemotherapeutics to be carried toward malignant cells via passive or active targeting but depending on the material used, the nanoparticle can also be utilized as an imaging agent. This are done by the following mechanisms; inorganic nanoparticles such as gold- or iron-based nanoparticles can be imaged using high-energy electromagnetic radiation (X-ray) or magnetic field (MRI) as they can attenuate or enhance the transmission signal of the imaging technique, respectively; quantum dots have the capability to be functionalized with fluorescence molecule or even act alone as the fluorescence agent as it can emit light when excited with the proper wavelength of electromagnetic radiation; Nanobubbles can be imaged by being hyperechoic when exposed to ultrasound. This may give clinicians the idea for the location and size of the tumor showing details of where it has been uptaken or localized elsewhere inside the body.

6.3 Biomaterials used in surgical intervention of cancer

Surgical intervention is a standard treatment for cancer exhibiting solid malignancies. This could mean resecting parts or total removal of an affected organ. Resecting or total removal will depend on which part of the body is afflicted with cancer, for nonvital organs such as the breast and genitals, bones, and sensory (eyes) total removal may be advised by the oncologist, while for vital organs, resection is usually indicated. The surgical intervention plan may also vary depending on the stage of cancer. Some cancers that are too far advanced or with tumor size that are too large and heavily vasculated are usually contraindicated for surgery as it poses high risks for the patient's life. For stage 4 cancer, typically palliative surgery is done to relieve symptoms and help the patient live a somewhat comfortable end of life [112–114]. While for stages of cancer that can still be managed and has not yet metastasized, a course of chemotherapy or radiation therapy is done to reduce the size of the tumor before undergoing surgery. Surgery can also precede adjuvant therapies (chemo and radio) to ensure that there are no remaining malignant cells in the body post removal of the tumor. There are several types of surgery in cancer treatment that are helpful to people with cancer. Some surgeries are used in combination with other types of treatment. These types of surgeries are described in Table 6.2.

Table 6.2: Types of surgery used in cancer treatment and management.

Surgery type	Description	Ref
Curative	Curative surgery is the primary form of surgical intervention for cancer treatment with the intention to remove all tumor mass or nonvital affected organ. This is usually followed up by adjuvant therapies such as radiation or chemotherapy.	[115]
Preventive	Preventive surgery is a form of prophylactic treatment that removes potential or suspected malignant formation. Example of this is the removal of polyps in the colon or radical mastectomy of two breasts, even though only one breast is affected by cancer, or total hysterectomy for ovarian cancer.	[116]
Diagnostic	Diagnostic surgery is performed to determine if the mass in the body is malignant or benign. Example of this is bone marrow biopsy or liver biopsy.	[117]
Staging	Staging surgery is meant to uncover the extent of cancer. Example of this is laparoscopy (a viewing tube with a lens or camera is inserted through a small incision to examine the inside of the body and to remove tissue samples) is an example of a surgical staging procedure.	[118]
Debulking	Debulking surgery removes a portion, of a cancerous tumor. This is done when the tumor is heavily vascularized or too large and meant to reduce its mass prior to adjuvant therapies, e.g., chemotherapy and radiotherapy.	[119]

Table 6.2 (continued)

Surgery type	Description	Ref
Palliative	Palliative surgery is used to treat cancer at advanced stages. It is meant to help patient live a comfortable end of life or correct problems that cancer treatment may have caused.	[114]
Supportive	Supportive surgery is does not work to cure cancer but helps other cancer treatments work effectively. Example of this is the insertion of a catheter to help with chemotherapy.	[120]
Restorative	Restorative surgery is meant to restore physical appearances for either aesthetic of functional purposes. For example, women with breast cancer sometimes need breast reconstruction surgery to restore the physical shape of the affected breast(s), or adding prosthetic limbs for amputated leg or arm afflicted by bone cancer.	[121]

Although surgery can remove the majority of the tumor, clinicians are still not 100% sure if there will be remission of cancer in the resected part of the body. This is especially true in cancers located in vital organs; thus, biomaterials play an important role to ensure to increase the chances of remission in the site where the cancer tumor is removed or resected. One example of this is in brain tumor surgery. Glioma in the brain once resected usually leaves a void that must be filled to retain cerebral pressure within normal values (Figure 6.4). The materials used in this post surgical intervention are made from bioresorbable materials of 1,3-Bis(4-carboxyphenoxy) propane and sebacic acid (PA-CPP-SA) and commercially named Gliadel® [122–124]. It is formed into erodable wafers that are implanted directly to the brain where a segmented tumor was previously located. This helps in maintaining volume and absorbing exudates that form from the resected site. Addition of drug (carmustine) [125, 126] in these materials also forms a localized drug delivery platform that ensures that malignant cells that are not removed during the surgery are eliminated and increases the chances for remission. The wafers will be fully absorbed by the body in 4 weeks time and do not incur blockage with the flow of the cerebrospinal fluid.

Other postsurgical biomaterials being used to mitigate the recurrence of resected tumors are applied in breast cancer surgery. Breast cancer is one of the leading causes of death in women and is usually intervened with a surgical procedure called mastectomy. Mastectomy can be partial or complete removal of breast tissue and surrounding lymph nodes and a form of curative surgery with the intent to halt the spread of the cancer cells to other vital organs [127, 128]. This also ensures that the remission chances is increased as there could be no possible breast tissue for cancer to grow into. Here, nanofibers are researched upon for their application in delivering chemotherapeutics locally and be as a scaffold for reconstructing toward the remaining breast tissue [129, 130]. This kind of material arises from the need to

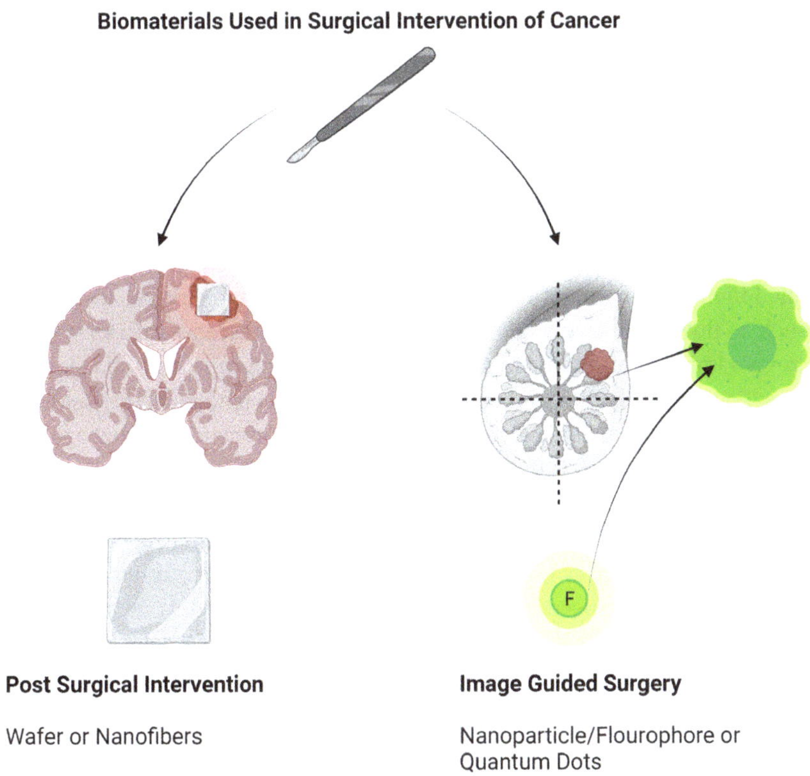

Figure 6.4: Biomaterials used in surgical intervention of cancer, wafer and nanofibers for post-surgical intervention of intracranial tumor to facilitate remodeling and reduction of recurrence of tumor growth. Wafers are loaded with chemotherapeutics that is released over time. Image-guided surgery with fluorophore-loaded nanoparticle/quantum dot tagging that allows real-time visualization of cancer cells through fluorescence.

regain breast tissue after it has been resected to avoid the spread of malignant cells to the surrounding breast mass. Partial mastectomy is usually done for breast cancer that is caught at an early stage. However, this poses an issue that there may be remaining malignant cells that are not taken out by the procedure. In order to ensure that the resected area will be completely free from malignant cells, the nanofibers can act as a drug release platform that can be implanted and release via stimuli induction an array of chemotherapeutic drugs to be uptaken by remaining malignant cells [131–133]. After the drug payload has been exhausted, the remaining nanofibers can act as a scaffolding material for tissue regeneration as it can mimic the architecture of the extracellular matrix of the breast tissue. Over time, the resected area will regain some of its mass maintaining an aesthetic look for the patient.

Another use of biomaterials in surgical intervention for cancer is through image-guided tumor surgery (Figure 6.4) [134]. As mentioned above, surgical removal of tumors may be challenging and reduces the chances of remission if the surgeons leave behind malignant cells that are invisible with using just the naked eye. Here, nanoparticles are used to tag cells that highly express receptors or antigens that are distinct for malignant cells [135–137]. As described in the earlier section of the chapter, nanoparticles can be used to selectively bind to cancer cells due what is called the EPR effect and via active targeting using ligand-receptor-mediated endocytosis. Nanoparticles can be decorated with fluorescent dyes that illuminate when excited with the proper wavelength of light [138, 139]. For instance, fluorescent dyes utilize what is called immunofluorescence which is commonly used in biological assays and imaging such as flow cytometry, ELISA, and confocal or fluorescence microscopy are used for in vivo imaging of tumors in animals as proof of concept. This concept utilizes the specificity of antibodies to their antigen allowing them to be highly selective only to target cells (cancer cells) to give clinicians a better distinction of what are normal cells to malignant tissue. It is done by conjugating an antibody molecule together with the fluorophore dye and attaching it to a nanoparticle system [140], whether it is inorganic- or organic-based, and administered toward the body. This allows the dyes to be used not only in in vitro or histological settings but also in vivo systems as well. For example, nanoparticles used for imaging are with quantum dots or other inorganic metals such as gold. When administered in vivo, the nanoparticles can be selectively uptaken by malignant cells allowing them to be seen when exposed to suitable light probe. Different light sources are used in this application. For, fluorophore functionalized nanoparticles, blue (470 nm), red (665–700 nm), and green (550 nm) wavelengths of light are typically used to excite the fluorophore and emit light slightly longer than the initial excitation wavelength (typically around 50 nm). However, fluorophore dyes are known to be unstable, thus limiting their applications for in vivo delivery [141]. Thus, other inorganic nanoparticle systems such as quantum dots [142, 143] and gold materials [144] are also being explored for a more stable, real-time probing of cancer cells and help visualize all malignancies during the operation.

Quantum dots range between 2 and 8 nm in size and have broad absorption band with narrow and symmetric emission band (full width at half-maximum ~25 · 40 nm). Quantum dots typically emit at large range of wavelength varying from visible to NIR spectral range (5, 54, 55). This is due to a radiative recombination of an exciton (an electron–hole pair), which is has a long lifetime (56) (>10 ns) leading to the emission of a photon in a narrow and symmetric energy band. In contrast to conventional fluorescent dyes (fluorophores), quantum dots have several attractive optical features that are more favorable for long-term, multi-target, and highly sensitive bioimaging applications in vivo. Since quantum dots have a large molar extinction coefficient (also known as attenuation coefficient) value compared to organic fluorescent compounds in the order of $0.5–5 \times 10^6$/M cm, they are capable of absorbing excitation photons very efficiently (10–50 times faster than organic dyes). The quantum dot's higher rate of absorption is directly correlated to the brightness of light that it emits with approximately 10–20 times brighter than organic dyes, allowing for highly sensitive fluorescence imaging. Furthermore, the stability of quantum dots is better than fluorescent dyes as they are photostable and do not oxidize in presence of singlet oxygen species [145, 146] and can be administered inside the body. However, further modifications must be needed for a quantum dot system to be safe in vivo. Although quantum dots have superb bioimaging capabilities and stability, they must be surface-modified to reduce their toxicity. Thus, another form of inorganic nanoparticle that can be used in image-guided surgery is the use of gold nanoparticles (AuNP). Gold nanoparticles can be excited with 310 nm of and emit at 426 nm. Although they are considered inferior to quantum dots, they scatter light strongly and therefore photobleaching does not occur with gold nanoparticles and, unlike quantum dots, gold nanoparticles are likely nontoxic.

6.4 Biomaterials used in hyperthermia treatment of cancer

Hyperthermia therapy (HT) [147, 148] is considered as one of the oldest treatments for cancer and as an adjuvant for chemotherapy and radiotherapy to improve treatment outcomes [149–151]. It is defined by elevating the body temperature close to 43–45 °C to initiate apoptosis in malignant tissues (cancer tumor, Figure 6.5). Apoptosis is achieved because tumor tissues have inefficient cooling capabilities due to rapid vascularization (angiogenesis) [152, 153]. When a is tumor is heated to elevated temperatures, the inefficient cooling happens due to this randomized vasculature and structure of the cells, as blood and tissue cannot dissipate the heat along the rest of body. This also allows for fluid to rush into the internal site of the tumor due to the expansion of the cells making other therapies such as chemotherapy more effective. HT can be achieved via various methods such as submersion to warm water bath, radiant heat through heat lamps, or by using hot blankets, inductive loops, or thermal chambers. This is known as whole body hyperthermia (WBH) which elevates the body temperature above normal (36.5–37.5) [154]. The effects of whole body hyperthermia in cancer treatment is to increase blood perfusion in and around the tumor area to render primary therapies such as chemo or radiation therapies more effective. Hyperthermia is known as an adjuvant therapy for these conventional treatments and could provide more favorable prognosis for the patient.

Conventionally, the patient that undergoes WBH needs to be sedated to last throughout the treatment since prolonged hyperthermia can be an uncomfortable and intolerable experience [155, 156]. Thus, regional or localized hyperthermia therapy was developed in order to spare body parts that do not require elevated temperatures for the treatment to be effective. Regional or partial HT is a method for heating large parts of the body and there are three key methods to achieve it: First is with peripheral heat applicators. This heats tumors using coherent arrays of dipole antenna couples placed in a loop design around the patient [157–159]. The

Whole Body Hyperthermia	Regional Hyperthermia	Local Hyperthermia
Heat Blanket Heat Bath Heat Lamps Inductive Loops Thermal Chambers	Peripheral heat applicators Thermal perfusion Constant hyperthermic peritoneal perfusion	Microwave antenna Focused Ultrasound Resisting wire embeds Nanoparticle and Thermalseeds

Figure 6.5: Hyperthermia modalities for treating cancer.

antennas produce microwave or radiofrequency energy to be absorbed in the target tissue. Second is with thermal perfusion where part of the patient's blood is fed into an external heater using a circulating pump and returned to the limb or organ, along with chemotherapeutic drugs. This is also called hyperthermic intraperitoneal chemotherapy (HIPEC) [160, 161]. Third is constant hyperthermic peritoneal perfusion (CHPP) [162] in which the abdominal cavity is bathed in warm fluid that contains anticancer drugs. Local hyperthermia is done via microwave heating, focused ultrasound, resisting wire embeds, and infrared radiators directly applied to the site of the tumor. Superficial placed malignancies such as melanoma are easily treated with hyperthermia but deeper-seated malignancies poses several challenges on how to give the proper thermal dose. Thermal dosage is crucial in achieving therapeutic effects for hyperthermia, both for local and for whole body. If the tumor is underheated (<40–43 °C) it may pose resistance for future hyperthermia therapy by creating heat shock proteins (HSP) [163, 164] within the malignant cells. HSP are a family of proteins that are upregulated when a cell experiences stressful conditions not just including thermal shock but also inflammation, UV irradiation, wound healing, and tissue remodeling. However, in this context HSP can make cancer cells more resilient to thermal shock that hyperthermia induces as they can be accustomed under stressful conditions if the thermal dosage is insufficient.

Here, biomaterials can serve as means to deliver therapeutic heating toward malignant cells while also serving as a delivery platform for chemotherapeutics (Figure 6.6). New experimental means of delivering heat to affected limbs and organs are being achieved using developed biomaterials such as nanoparticles, nanofibers, metallic implants, and with the development of external devices such as magnetic field oscillators, and infrared lasers. Contemporary research in experimental hyperthermia therapy uses these two main external stimuli sources to induce heat on biomaterials in order for it to transfer thermal energy to malignant cells that could lead to apoptosis. The advantage of using biomaterials for hyperthermia therapy is that it leads to multifunctionality.

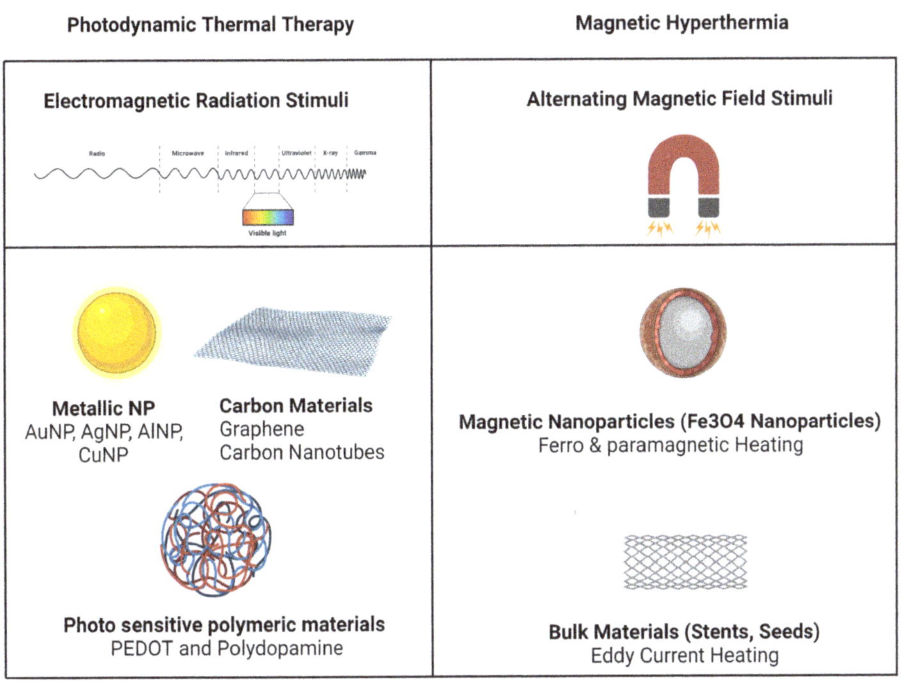

Figure 6.6: Biomaterials used for two types of hyperthermia modalities (photodynamic and magnetic). Photodynamic therapy uses specific wavelength of light (near-infrared light) and magnetic hyperthermia uses alternating magnetic field.

Table 6.3: Hyperthermia methods in cancer treatment.

Method	Mechanism	Material	Ref.
Whole body hyperthermia (WBH)	Elevating the whole body temperature to above 41–43 °C via radiative heat or extracorporeal technologies	Water bath and radiant heat with ultraviolet radiation. In radiant WBH, heat is superficially applied to the whole body using hot blankets, inductive loops, or thermal chambers Extracorporeal WBH is obtained by extraction of extracorporeally heated blood. A circuit of blood is shaped beside the body forming an extracorporeal ring. The circulating blood is crossed through a water bath or hot air to heat it and then infused into the main vein. The anticipated body temperature is adjusted and measured by altering the current volume of the warmed injected blood.	[154]
Regional hyperthermia	Part of the patient's body is heated via microwave or radio (RF) heating	Coherent arrays of dipole antenna couples placed in a loop design around the patient. The antennas produce microwave or radiofrequency energy to be absorbed in the target tissue. With such apparatus, target tissue can be heated to 41–42 °C.	[17]
Local hyperthermia	Localized (under tissue level) heating via bioheat transfer of thermal generating material or via microwaves, radio waves, or ultrasound	Biomaterial-mediated heat generation via magnetic nanoparticles, conductive materials, and photosensitive materials.Focused ultrasound, micro and radio waves can be delivered toward the tumor and affected tissue and generate bioheat transfer.	[149]

6.4.1 Magnetic hyperthermia (nanoparticle heating)

Magnetic hyperthermia is an experimental treatment for cancer, based on the fact that magnetic nanoparticles can convert an external electromagnetic energy alternating high-frequency field to heat [165–167]. The first thing that comes to mind is the use of nanosized ferromagnetic and paramagnetic materials such as iron oxides (Fe_3O_4 and Fe_2O_3) [168–170], also known as magnetic nanoparticles (MNP) [171]. MNPs are known to be biocompatible and easy to functionalize with polymeric coatings, and chemotherapeutic drugs making it suitable not only as a drug carrier but also to deliver heat to malignant cells. The way it do so is with the use of an external oscillating/alternating magnetic field (AMF). Alternating magnetic field (AMF)-assisted thermal therapy has received widespread attention for tumor treatment recently. In thermal therapy, the localized MNPs in cancerous tissue generate abundant heat via a high-frequency (300–1,100 kHz) AMF for an irreversible thermal injury [172, 173]. MNPs generate this heat via Néel–Brownian motions and/or hysteresis loss of magnetic nanoparticles using alternating magnetic fields [174, 175].

The heating power and efficiency of these MNP are dependent on particle properties and treatment settings. First, there are differences on how efficient a nanoparticle can generate heat. Larger sized nanoparticles (>50 nm) can possess multiple magnetic domains which have to be aligned first in order to react with the external magnetic field and generate heat [176]. This are usually found in paramagnetic of ferromagnetic materials, whereas a class of magnetic nanoparticles called superparamagnetic NPs (SMNP) are considered to have the most efficient heating properties as they can generate heat almost instantaneously due to their high susceptibility to an external magnetic field. This susceptibility comes from their properties of having a single magnetic domain compared to paramagnetic and ferromagnetic materials. As we go down in nanoparticle size (50 to 15 nm), the magnetic domains of each particle become singular. This single magnetic domain means that they can react to an external magnetic field faster with lower intensity needed from the external source. Thus, less power input and lower frequency requirements. Furthermore, the relatively low heating rate of conventional magnetic hyperthermia using paramagnetic and ferromagnetic NPs, for example, often requires a high concentration of nanoparticles to be injected into the human body, which not only would result in potential toxicity, but with the use of SPMNs the required concentration of nanoparticles can be reduced while still being able to achieve therapeutic ranges of thermal dose [170, 177, 178].

In terms of biological response, smaller nanoparticles mean that they can fenestrate or exploit the EPR effect in the tumor more efficiently and carry more therapeutic payload due to the increased volume to surface area ratio. The use of nanoparticle hyperthermia also includes the convenience of having a singular platform for hyperthermia and chemotherapy. As mentioned above, the surfaces of MNP/SMNPs can be easily functionalized via conventional coating methods; this is usually done via a two-step synthesis. First, the inorganic nanoparticle is synthesized either via co-precipitation [179, 180] or via thermal decomposition [181, 182] methods. Then, the emulsion-solvent [183, 184] extraction/evaporation [185] method. The surface properties of MNPs are highly reactive toward polymers that can serve as drug conjugation sites. The polymeric coating can also serve as a biocompatibility enhancer as it can create an interface toward biological substances (e.g., proteins, blood components, and enzymes) and the inorganic component of the nanoparticle system. Lastly, the ability of the MNPs to be influenced by an external magnetic field means that they can also be driven intently to concentrate inside or around the tumor using magnetic actuation and control in a 3-axis Helmholtz electromagnetic coil setup; although still very early in development, the potential of controlling the MNPs inside the body using a controlled magnetic field means that there can be enhanced concentration of relatively small amounts of MNPs and enhanced targeting capabilities only for selected sites.

6.4.2 Magnetic hyperthermia (bulk heating)

Other forms of magnetic hyperthermia have also been developed using bulk conductive materials such as nitinol alloy or stainless steel [186, 187]. The mechanism for heating these materials under alternating magnetic field depends on eddy current heating also known as induction heating. Eddy currents are small localized electrical current loops generated when a conductive material is exposed to an alternating magnetic field. The eddy currents flowing in the material creates ohmic resistance and generates heat in consequence, this is also known as joule/ohmic heating. Given the proper frequency of oscillation on the magnetic field, heating of these bulk materials can be controlled for therapeutic range of hyperthermia. Preclinical proof of concept studies had been done using thermoseeds and chemothermal seeds made up of stainless steel and nitinol, respectively. Other forms of bulk induction hyperthermia using nitinol stents [188, 189] have also been investigated to deliver chemotherapeutic drugs while giving hyperthermia in malignancies found in the gastrointestinal tract (esophagus and duodenum).

A study conducted in rabbit liver used thermoseeds made from biomedical-grade stainless steel to reduce tumor size [190, 191]. Since stainless steel is conductive in nature, once exposed to AMF it while create eddy current heating and transfer it to the surrounding tissue. The thermoseeds are placed similar to brachytherapy seeds known to deliver localized radiation doses toward and around the tumor site. But instead of delivering radiation which can be uncontrolled and may cause cell death even in normal cells, thermoseeds can deliver therapeutic heating via bioheat transfer. In contrast to nanoparticle hyperthermia, the immediate heating of these thermoseeds can be beneficial in reaching a complete and uniform isotherm of 43–45 °C in the tissue. The temperature reached by a thermoseed depends on multiple variables, such as the strength and frequency of the electromagnetic field, the orientation of the thermoseed within the coil, its proximity to other thermoseeds, the permeability and electrical conductivity of the thermoseed, the local blood perfusion rate, and the thermoseed's thermophysical properties. By adjusting some of these variables, it is possible to achieve a uniform isotherm of hyperthermia within the target tissue. An improvement on this idea arises when it thermal seeds are imbued with drug delivery capabilities. Another proof of concept study is where hyperthermia and chemotherapy are combined to decrease tumor progression in a nude mice model. The chemothermal seeds are made from nitinol tubes and were functionalized using electropolymerization with polydopamine to act as a pH-dependent drug-releasing platform to release bortezomib [192–194]. The inside of the chemothermal seeds was also added with a pH-soluble hydrogel made up of poly(dopamine-co-acrylamide)-compl-poly(boronic acid-co-acrylamide) loaded with 5-fluorouracil to give dual drug synergism. The results show that the tumor progression was halted on the group that received subsequent hyperthermia with dual drug therapies. This further solidifies the idea of hyperthermia being beneficial in enhancing the effect of chemotherapeutic drugs compared to using each single treatment modality alone.

Since other implanted biomedical devices such as the ones found in the gastrointestinal malignancies (esophagus, duodenum, biliary, and rectum) are made up of conductive materials, once subjected to AMF, direct hyperthermia can be enacted over the tumor and can be used to reduce its progression. In another example where this has been investigated, esophageal and duodenal stents were combined again with chemotherapy as they made from nitinol. In one research an esophageal stent with the capability to deliver two types of drugs (paclitaxel and 5-fluorouracil) has been developed [195–197]. The induced hyperthermia generated from the stent caused the thermoresponsive polymeric stent cover to deliver its drug payload on an on-demand manner. The stent also has the capability to either release just a single drug if the temperature is only set at 40–43 °C (paclitaxel) and release both paclitaxel and 5-flourouracil when the temperature reach above 45 °C. The stent cover is made from an electrospun core-shell of poly-NIPAAm-co-HMAAm (shell) and polyurethane (core). The polyurethane added mechanical stability for the polymeric cover while the poly-NIPAAm-co-HMAAm (shell) acted as the thermoresponsive part. The ability to either one or two drugs depended on the ration of NIPAAm and HMAAm during the polymerization process, where a 9:1 ratio of NIPAAm to HMAAm yeilded an LCST point of 40–43 °C while an 8:2 ratio brought about a higher LCST point at above 45 °C. This is beneficial as there are two types of cancer typically developed in the esophagus. The first one is called squamous cell carcinoma that affects the cells that lines the esophagus [198], while the other more aggressive type is called adenocarcinoma that affects the glandular cells near the lower esophageal sphincter [199]. If the patient has the squamous cell type of cancer, the physician can opt for lower hyperthermia therapy with single drug release of PTX, but if the patient has the more aggressive type of adenocarcinoma, a dual drug release and higher temperature of hyperthermia can be done using the same stent. A similar approach has also been done with a duodenal stent, where the surface of the nitinol struts was functionalized with electropolymerized dopamine to act as a pH-dependent releasing platform for bortezomib and the addition of hyperthermia adds to the anticancer effects of the drug [200].

Most clinically approved nonvascular stents that are used in the gastrointestinal and respiratory systems are ferromagnetic and have direct contact with a malignancy, thereby making bulk induction hyperthermia a convenient way to introduce hyperthermia found in these systems. And the addition of chemotherapeutic delivery can enhance the tumor-inhibiting effects of both treatment modalities. Furthermore, heating bulk material is much easier compared to nanoparticle-mediated hyperthermia. There is less need for administering high concentrations of nanoparticles, and the heat can easily be concentrated only where it is needed. However, controlling a proper isotherm may pose a problem as bulk heating can reach up to very high temperatures if the AMF intensity is left uncontrolled. It may lead to tissue ablation of normal tissues that may cause adverse effects on subsequent treatments. But advances in biomaterial metallurgy can be used as nitinol can be programmed to have lower curie temperature in order to make it safer when overheating is achieved.

6.4.3 Photodynamic hyperthermia therapy

The use of electromagnetic radiation and converting it to heat for treating various diseases especially cancer has been literally put into light as new emerging biomaterials can transform with high-efficiency specific wavelength of light into heat that can be used for hyperthermia therapy. Photothermal therapy (PPT) [201] is an extension of photodynamic therapy (PDT) [202] and a novel means of delivering thermal dose to cancer using a photosensitizer [203]. Unlike PDT that requires oxygen to interact with target cells or tissues, PTT can be activated in hypoxic conditions which are prevalent in cancer environment. The main mechanism for heat generation is through the excitation of the photosensitive material with a specific band of light and causing it to vibrate which in turn generates heat which can kill cancer cells. PTT-sensitive materials are known to be excited using longer wavelength of light (infrared to near-infrared) which is less ionizing compared to higher-energy wavelength of light (ultraviolet), therefore causing less harm to normal cells.

However, photothermal therapy has some inherent limitations [204] as compared to magnetic hyperthermia as it relies on electromagnetic radiation (light) which has a lower tissue penetration capabilities, in contrast to an alternating magnetic field which has a higher tissue permeation inside the body. PTT is conducted with the use of near-infrared (NIR) light [205], and the penetration depth of this light onto biological tissue is demonstrated to range from 0.5–5.0 mm. NIR can be further classified into two windows (NIR-I and NIR-II). In NIR-I the penetration depth is at least ~1 mm, and approach at 2 mm before it gets absorbed by the tissue and other biological components. This is also known as the first window near-infrared light with wavelengths ranging from 650 to 1,350 nm. However, its penetration depth can be further enhanced by using a second window NIR (NIR-II) at ranges between 1,000 and 1,700 nm. This limitation, however, does not impede potential applications of PTT as a cancer treatment as the materials used in this application are deemed more biocompatible than magnetic nanoparticles.

6.4.4 Metallic nanoparticle-mediated PTT

Metallic nanoparticles (AuNP, AgNP, AlNP, and CuNP) are known to be a photosensitive material. The free electrons of these metallic NPs when exposed to NIR light can oscillate coherently with the electromagnetic radiation waveform. More specifically, when NIR light is shone on these NPs, it forms a dipole oscillation along the direction of the electric field and when the oscillation reaches its maximum, this frequency is called the surface plasmon resonance (SPR) [206, 207].While all of these types of metallic NPs can undergo SPR and generate heat, the most suited for cancer PTT is AuNP. AuNP in particular is known for its bioinertness and ease of functionalization via chemosorption. Thus, similar to magnetic nanoparticles, they can be loaded with ligands and chemotherapeutic drugs to specifically target malignant cells. However, its SPR to NIR depends in its shape and size, and changes in their dimensions such as in their length and width or their aspect ratio may significantly affect their heat generation [208, 209]. It has been found that Au nanorods have excellent SPR with NIR-I and II, thereby making it ideal for PTT.

6.4.5 Carbon-based nanomaterial-mediated PTT

Pure carbon-based materials such as graphene and graphene oxide can also be used as a photosensitizer when excited with NIR laser with wavelength of 808 nm. Graphene and graphene oxide also exhibit surface plasmon resonance similar to metallic nanoparticles. It may also be functionalized with various peptides, ligands, and drugs via covalent and noncovalent bonding, and van der Waals interaction [210, 211]. However, there are still some debates whether using graphene intravenously is safe; thus, further safety studies must be done before it can be explored as a nano-drug carrier with PTT capabilities. It must be noted, however, that graphene and graphene oxide can be co-electrospun with biocompatible polymers and still exhibit their SPR. Thus, it can be used on superficial malignancies such as melanoma or in breast cancer wound dressing to deliver thermal dosage to enhance radiation and chemo treatments.

6.4.6 Polymeric material–mediated PTT

Some class of photosensitive polymeric-based materials also exhibits conversion of light energy to thermal energy, albeit over different mechanism. The nonradiative process for heat generation of polymers is different from that of metals which is related with surface plasmon resonance. Specifically, when the polymer is excited by the suitable wavelength of light, it can be activated to the excited state (S1). The excited state decays back to the ground state mainly through three processes: emitting a photon (i.e., fluorescence), intersystem crossing (which may further lead to phosphorescence or triplet–triplet annihilation with other species), and nonradiative relaxation. The nonradiative excitation pathway also manifests as heat, but this pathway is usually competitive with intersystem crossing overtaking the vibrational relaxation pathway resulting in reduced heat generation. Thus, light emitting and intersystem crossing must be inhibited. This is done by creating closed stacking of the molecular/polymeric sensitizers increasing intermolecular collisions [212, 213]. Such collisions efficiently quench the molecular fluorescence and intersystem crossing, giving rise to much increased yield of nonradiative relaxation. Polymers such as polyaniline (PANI) [214], polypyrole (PPY) [215, 216], polythiophene (PTh) [212], poly (3,4-ethylenedioxythiophene):poly(4-styrenesulfonate) (PEDOT) [217] and polydopamine (PDA) [217–219] have been reported to exhibit photothermal conversion of NIR light at wavelengths of 808 nm. Since they are polymeric materials, they can be fabricated into multiple forms such as nanoparticles, nanofibers, nanocoatings, and conjugated with chemotherapeutics and targeting ligands.

6.5 Biomaterials used in immunotherapy of cancer

The immune system is now being utilized as a new treatment modality for cancer therapy. This is called immunotherapy where the person's own immune system is used to attack cancer cells. This is done by boosting or programming it to recognize malignant cells as the body's immune system has the capability to eliminate harmful pathogens (Figure 6.7). However, malignant cells are usually left undetected due to their normal origins and the immune system do not recognize them as foreign or pathogenic. Malignancies usually happen when sufficient damage and mutations occur; thus, the antigen they present are not scouted by the immune cells and removed accordingly. Furthermore, malignant cells can avoid detection by secreting substances that keep them invisible to immune cells and even if malignant cells get detected by immune cells, the response of the immune system [220, 221] is not strong enough to completely eliminate them all. Hence, researchers have found ways of help the immune system to recognize cancer cells and create a strong and aggressive response towards them. This is extremely beneficial as immunotherapy does not have severe side effects as compared to conventional treatment for cancer as the body naturally eliminates cancer without or with minimal aid of synthetic substances or ionizing radiation.

6.5.1 Implantable and injectable biomaterials for immunotherapy

Biomaterial science is now also helping the new modality of immunotherapy to be realized in the clinic and how to more efficiently deliver immunomodulatory

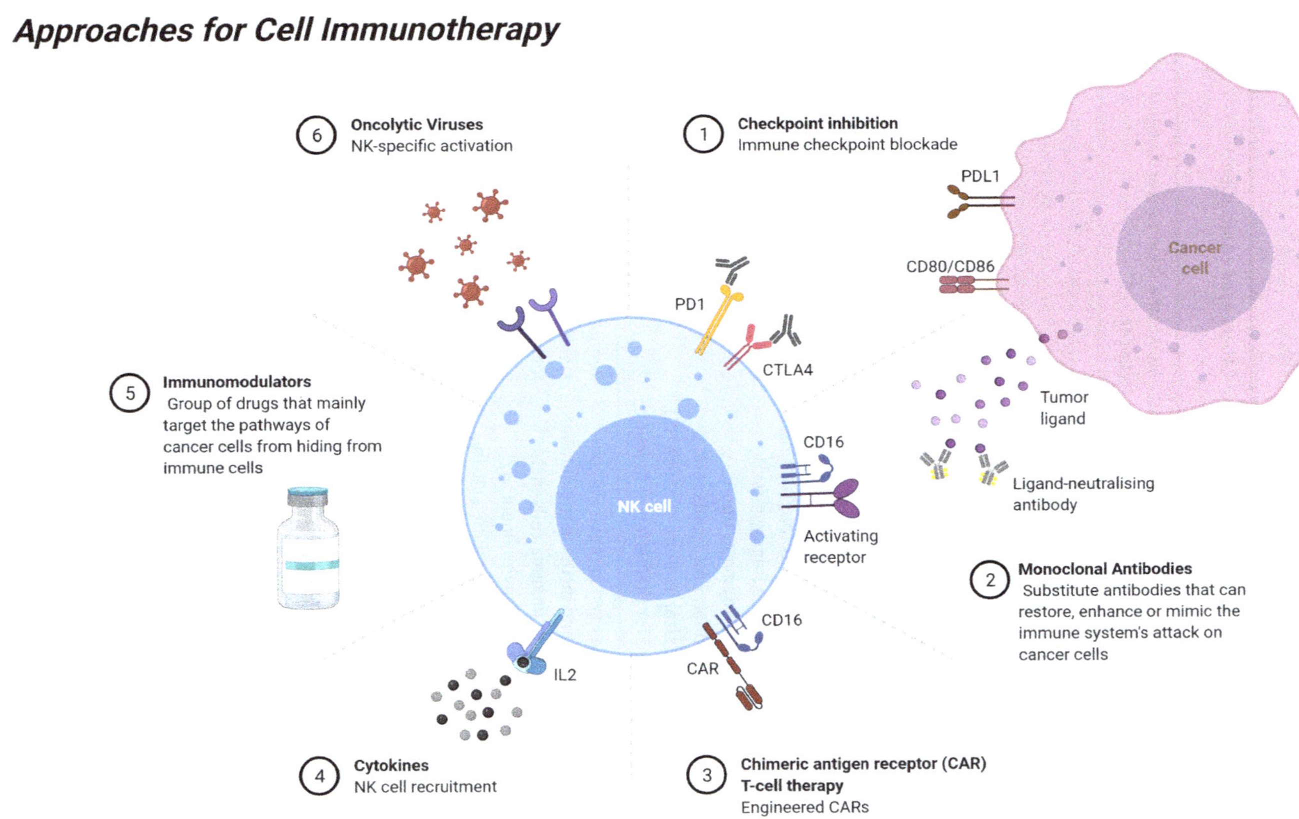

Figure 6.7: Approaches for cancer cell immunotherapy. Different immunotherapeutics can be used to trigger an immune response towards cancer cells.

Table 6.4: Types of immunotherapy used for cancer treatment.

Immunotherapy type	Description	Cancer-type application	Ref.
Checkpoint inhibitors	Drugs are used to block checkpoint proteins from binding with their partner proteins. This prevents the "off" signal from being sent, allowing the T cells to kill cancer cells. Examples of these drugs are (anti-CTLA-4) ipilimumab, nivolumab, and pembrolizumab (anti-PD-1).	Melanoma (skin cancer) Lung cancer	[222]
Chimeric antigen receptor (CAR) T-cell therapy	CAR T-cells mediate MHC-unrestricted tumor cell killing by enabling T cells to bind target cell surface antigens through a single-chain variable fragment (scFv) recognition domain. Upon engagement, CAR T-cells form a nonclassical immune synapse (IS), required for their effector function.	Large B-cell lymphoma (DLBCL), follicular lymphoma, mantle cell lymphoma, multiple myeloma, and B-cell acute lymphoblastic leukemia (ALL)	[223]
Cytokines	Cytokines can improve antigen priming and increase the number of effector immune cells in the tumor microenvironment and enhance their cytolytic activity. The two main types of cytokines used to treat cancer are interleukins and interferons.	Renal cell carcinoma, metastatic melanoma, hairy cell leukemia, follicular lymphoma, Kaposi's sarcoma	[224]
Immunomodulators	Immunomodulators are a group of drugs that mainly target the pathways of cancer cells from hiding from immune cells. It works by modulating proteins involved FGF-dependent angiogenesis in vivo and exhibits antineoplastic activity	Multiple myeloma, mantle cell lymphoma, follicular lymphoma, and marginal zone lymphoma	[225]
Vaccines	Vaccines help prime immune cells to identify malignant cells by presenting them with antigens inherent to specific types of cancers.	Cervical cancer (commercialized) Various types of cancer (under development)	[226]

Table 6.4 (continued)

Immunotherapy type	Description	Cancer-type application	Ref.
Monoclonal antibodies	Monoclonal antibodies are laboratory-produced molecules engineered to serve as substitute antibodies that can restore, enhance or mimic the immune system's attack on cancer cells. They are designed to bind to antigens that are generally more numerous on the surface of cancer cells than healthy cells. Examples are ibritumomab tiuxetan, obinutuzumab, ofatumumab, and rituximab.	Lymphoma, multiple myeloma	[227]
Oncolytic viruses	Oncolytic viruses infect and help destroy cancer cells by directly stimulating the production of immune cells when directly injected in the tumor microenvironment. They also increase the effect of other immunotherapy (checkpoint inhibitors).	Melanoma	[228]

compounds to the body and target site [229, 230]. There are several ways biomaterials are utilized in immunotherapy (Figure 6.8). First is with local delivery of immunomodulatory compounds (cytokines, drugs, Mab, etc.) with the use of implantable biomaterials [231, 232]. In this context, the implanted biomaterial is placed subcutaneously via minor surgical procedure. The immunomodulatory compounds are released in a controlled manner from this implant. The implants are made from biocompatible polymers (e.g., PLGA, alginate, gelatin, and hyaluronic acid) that release the immunomodulatory compound to stimulate or prime the immune system to be more vigilant and active in seeking cancer cells. This need for immunomodulatory implants arises from the high cost and complex procedure of dendritic cell (DC) programming ex vivo. In practice DCs [233] are harvested from a patient's body and are presented with antigens that are unique for the specific cancer cell type; this in turn activates the DCs' maturation phase and present tumor antigens to naïve T-cells once they are reintroduced in the body. The naïve T-cells then mature into tumor specific T-cells and target the cancer tumor. As mentioned, this process is highly complex and time-consuming, and the survival rate of the DCs are relatively low. Thus, in vivo DC maturation and programming are very helpful in simplifying this process. In one example, granulocyte-macrophage

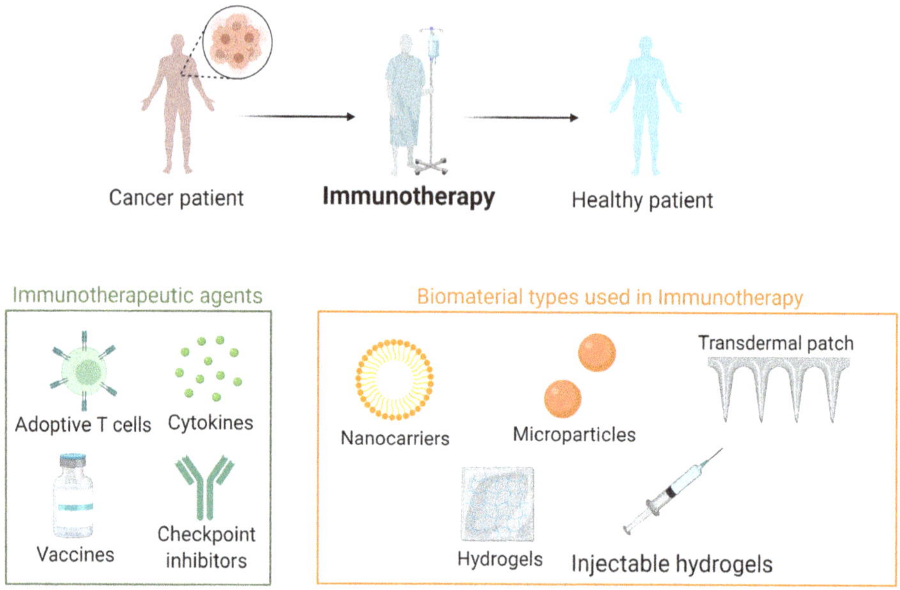

Figure 6.8: Biomaterials used in cancer immunotherapy. These are materials used in inducing immune response to cancer cells by delivering immunotherapeutic agents inside the body.

colony stimulating factor (GM-CSF) [234] has been encapsulated in a polylactide-co-glycolide scaffold with highly porous morphology synthesized using gas foaming process [235, 236]. GM-CSF is known to be able to stimulate the expansion and of DC, macrophages, and neutrophils and can indirectly regulate T-cell proliferation and activation in vivo. The successful encapsulation and release of GM-CSF in an implantable polymeric scaffold removes the need for harvesting DC's and reintroducing them inside the body. The study has shown that the GM-CSF that has been released from the scaffold successfully recruited and created matured DCs that can migrate to the lymph nodes to present tumor antigens to naïve T-cells. In their result, where they compared the number of DCs recruited by the scaffold in vivo and by a vaccine introduced ex vivo, the amount of DCs are roughly the same but with the implant creating a more streamlined process overall. Injectable polymeric materials such as hydrogels [237] can also be utilized as a controlled release platform for immunomodulatory compounds. The advantage for using injectable biomaterials is the removal for surgical implantation. This creates better compliance and ease of administration for the patient and for the clinician, respectively. These injectable hydrogels can be made from shear thinning materials or with in situ hydrogels made from alginate [238]. The gelation mechanics can be tuned to occur in vivo with the presence of Ca^{2+} ions. Over time, the hydrogel will be hydrolyzed once their immunomodulatory compounds have been exhausted. One factor to consider when designing an immunomodulatory implant and injectable biomaterial is that processing parameters should be done using mild conditions. Immunomodulatory agents such as growth factors, cytokines, and other biologicals are highly sensitive to harsh processing conditions since they are made with proteins. Furthermore, the effect of the polymeric material in the maturation of the DC should be minimal to nonexistent as the DCs should be only influenced by the immunomodulatory agent and not by the leachables or any epitope that may be incorporated in the polymeric structure.

6.5.2 Transdermal biomaterials for immunotherapy

Transdermal delivery platforms can be utilized to further enhance the compliance and ease of administration for immunomodulatory compounds [239–241]. This is inherently easier when it comes to administration, but it comes with several challenges. Transdermal delivery utilizes the skin as a point of entry for immunomodulatory agents; thus, the problem of low permeability for the active agent must be answered if it is intended to be used in other forms of cancer. But in terms of treating melanoma (aka skin cancer) transdermal delivery will be sufficient enough to deliver immunomodulatory agents towards the target site due to its proximity [242]. In a proof of concept study done on mice models, transdermal delivery of anti-PD1 using microneedles was successfully demonstrated to treat melanoma [243]. The microneedle array carries porous nanoparticles loaded with the immune checkpoint inhibitor agent (anti-PD1), which is then placed over the melanoma tumor. The microneedles facilitated the point of entry for the nanoparticles and after dissolution released the nanoparticles which are pH sensitive. This pH sensitivity allows for a controlled released on the tumor microenvironment (TME) as it is a well-established fact that TME has lower pH compared to normal physiological tissues. The immunomodulatory agent can now inhibit the stealth properties of the tumor and allow the immune cells to attack the cancer cells. In their results, the group which got the transdermal delivery of anti-PD-1 had a better survival rate compared to the group that received an intravenous administration of the same agent. This method of delivering immunomodulatory compounds promises a significant advancement in melanoma treatment. Creating an easily applicable biomaterial to treat melanoma will be as easy as treating any other kind of skin condition. However, it must be noted that this method works better in conjunction with other therapies or with other classes of immune checkpoint inhibitors.

6.5.3 Nanoparticle technology in cancer immunotherapy

Nanomedicine, most specifically nanoparticle drug delivery, is also being utilized heavily in multiple immunotherapy sub-modalities for the delivery of vaccine, checkpoint inhibitors, and cytokines. This is due to the ability of micelles, liposomes, polymeric, and lipid-based nanoparticles to protect the sensitive immunomodulatory payloads and target specific cell lines (immune or malignant cells) via passive (EPR effect) and active targeting (ligand-receptor mechanism). Nanoparticles can not only target specific cells; they can also increase half-life and stability of immunomodulatory agents and can also deliver the therapeutic payload intracellularly, making them attractive for targeting immune cells and as vector for mRNA vaccines. Another salient point for using nanoparticles in immunotherapy can be explained by the limitations of local delivery methods such as intratumoral injections of biomaterials or implantable scaffolds as described in the previous section. In more inaccessible tumors found in the body, the need for a systemic delivery system is being realized.

Nanoparticle-based approaches were explored to target immunomodulatory agents directly toward T-cells. A research using nanoparticles of PLGA-PEG containing SD-208, a TGF-BetaR1 inhibitor, was used to restore T-cell function and recruit lymphocytes towards non-inflamed tumor environment [244–246]. The results have shown longer survival rate in mouse colorectal cancer with targeted delivery of the TLR7/TLR8 agonist increasing the proportion of tumor-infiltrating CD8+ T-cells and sensitized tumors to anti-PD-1 therapy. This method of directly targeting immune cells in the circulatory system can enhance localization of tumor infiltrating immune cells in the tumor, in comparison to direct localization or targeting of malignant cells, since nanoparticles can have inefficient infiltrating capabilities due to biological and functional barriers present in most cancers.

Another example of directly targeting immune cells using a cytokine for inducing cancer apoptosis is through liposomes functionalized with vascular adhesion receptor E-selectin and immune cytokine TRAIL (aka TNFSF10). E-selectin ligand can selectively bind in both circulating cancer cells and immune cells, while TRAIL can induce cancer apoptosis. TRAIL is known to have a low in vivo half-life but being bonded to immune cells increases their circulation time and allows for a higher apoptosis response in solid cancer tumor models. This results in increased immune cells migrating into the tumor and releasing cytokines and thereby increasing the antitumor response. This strategy is also beneficial to targeting metastatic malignant cells that are already circulating inside the body as these nanoparticle-coated immune cells were initially utilized to target and kill circulating tumor cells in the bloodstream.

Nanoparticle drug delivery system is also being utilized in developing vaccines for cancer prevention with mRNA technology. Messenger ribonucleic acids (mRNA) are small fragments of genetic code that instructs the cells to produce proteins; as a vaccine, it is meant to instruct cells to produce antibodies to fight off or target certain pathogens (e.g., viruses and bacteria). As a cancer vaccine, mRNA instructs T-cells to recognize mutations found in cancer cells and help the body eliminate them before they form tumors. This can help prevent progression of the disease or help increase the remission of cancer symptoms. In more optimistic view of the treatment, it can help prevent the occurrence of cancer for those people who have higher genetic predisposition or risks in developing it. This is a significant paradigm shift in cancer treatment and prevention as mRNA technology may help reduce the burden on healthcare and patient who are suffering from certain kinds of cancers. However, delivery of mRNA can be challenging and is now being explored in two ways. The first one is through viral vectors and the other is through lipid-based nanoparticle delivery. Viral vector delivery of mRNA has its salient points as it can be very selective to specific cells, as it can utilize the inherent virulence of the virus in targeting cells inside the body. Examples of viral vectors being investigated for delivery of mRNA and other nucleic acids are lentiviruses, adeno-associated viruses, and the Sendai virus [247]. However, the use of viral delivery systems is challenging as it can induce unwanted immune responses in the body, as it is still present antigens that may be scouted by the immune system as harmful. As an alternative, nonviral vectors such as lipid nanoparticles are being utilized as a preferred more biocompatible means to deliver mRNA to T-cells. In a recent study, a lipid-based nanoparticle mRNA vaccine was designed to target dendritic cells in vivo without the need for targeting ligands. This platform comprises several cationic lipids, DOTMA (1,2-di-*O*-octadecenyl-3-trimethylammonium-propane) and DOTAP (1,2-dioleoyl-3-trimethylammonium propane) [244, 248], as well as the zwitterionic lipid DOPE (1,2-dioleoyl-sn-glycero-3-phosphoethanolamine) [248], which form complexes with anionic mRNA via electrostatic interaction and hydrogen bonding. Their results show an improved intravenous mRNA administration to dendritic cells in the spleen and lymph tissues on a mice model, and these dendritic cells are able to express tumor antigens, present them to naive T-cells and mediate potent rejection of tumors in murine models of melanoma and lung and colorectal cancer. Personalized vaccines are also a new promising way of using mRNA for treatment of cancer. mRNA vaccines can be made using a generic process allowing for quick production of patient -specific neoantigens that are identified by analyzing the tumor exome. Furthermore, mRNAs can also be used in cellular therapies to transfect patient-derived cells in vitro and reintroduce the programmed cells back into the patient.

6.6 Conclusion

Cancer has always been a challenging disease to overcome but we are now starting to understand more on how to prevent, treat, and cure it. Advances in therapeutic modalities are being driven by basic science in understanding cell biochemical processes of cancer, the tumor microenvironment and tissue physiological conditions. New bioactive anticancer compounds whether it may be synthetic or biological are being discovered that helps in alleviating the burdens of cancer in our society. It is just a matter of how effectively these compounds are being delivered, removal of severe adverse effects, increasing therapeutic effects, and overall safety. Biomaterials have contributed well in making sure these goals of treatment are met and improved. Biomaterial science has brought about nanoparticles for targeted delivery for chemotherapeutic and immunotherapy compounds, materials for increasing surgical effectivity and removal of recurrence of cancer, advancement of adjuvant therapies such as hyperthermia therapy and combining it with chemotherapy, and discovery of combination therapies such as "theranostics." The future of cancer treatment is a collaborative effort and biomaterial science has always been a great tool to meet the challenges of this dreaded disease.

References

[1] Pantel K, Alix-Panabières C, Riethdorf S. Cancer micrometastases. Nature Reviews Clinical Oncology. 2009;6(6):339–51.

[2] Cooper GM, Hausman RE. The development and causes of cancer. The Cell: A Molecular Approach. 2000:2.

[3] Daya-Grosjean L, Dumaz N, Sarasin A. The specificity of p53 mutation spectra in sunlight induced human cancers. Journal of Photochemistry and Photobiology B: Biology. 1995;28 (2):115–24.

[4] Seo K-Y, Jelinsky SA, Loechler EL. Factors that influence the mutagenic patterns of DNA adducts from chemical carcinogens. Mutation Research/Reviews in Mutation Research. 2000;463(3):215–46.

[5] Mena S, Ortega A, Estrela JM. Oxidative stress in environmental-induced carcinogenesis. Mutation Research/Genetic Toxicology and Environmental Mutagenesis. 2009;674 (1–2):36–44.

[6] Wogan GN, Hecht SS, Felton JS, Conney AH, Loeb LA. Environmental and Chemical Carcinogenesis. Seminars in Cancer Biology: Elsevier; 2004. p. 473–86.

[7] Smith KC. Spontaneous mutagenesis: Experimental, genetic and other factors. Mutation Research/Reviews in Genetic Toxicology. 1992;277(2):139–62.

[8] Siegel R, DeSantis C, Virgo K, Stein K, Mariotto A, Smith T, et al. Cancer treatment and survivorship statistics. CA: A Cancer Journal for Clinicians. 2012;62(4):220–41.

[9] Chu E, Sartorelli A. Cancer chemotherapy. Basic & Clinical Pharmacology. 2004;9:898–930.

[10] Perry MC. The Chemotherapy Source Book: Lippincott Williams & Wilkins; 2008.

[11] Jaffray DA, Gospodarowicz MK. Radiation therapy for cancer. Disease Control Priorities. 2015;3:239–47.

[12] Baskar R, Lee KA, Yeo R, Yeoh K-W. Cancer and radiation therapy: Current advances and future directions. International Journal of Medical Sciences. 2012;9(3):193.

[13] Coffey JC, Wang J, Smith M, Bouchier-Hayes D, Cotter T, Redmond H. Excisional surgery for cancer cure: Therapy at a cost. The Lancet Oncology. 2003;4(12):760–68.

[14] Couzin-Frankel J. Cancer Immunotherapy. American Association for the Advancement of Science; 2013.

[15] Farkona S, Diamandis EP, Blasutig IM. Cancer immunotherapy: The beginning of the end of cancer?. BMC Medicine. 2016;14(1):1–18.

[16] Sethi M, Chakarvarti S. Hyperthermia techniques for cancer treatment: A review. International Journal of PharmTech Research. 2015;8(6):292–99.

[17] Longo TA, Gopalakrishna A, Tsivian M, Van Noord M, Rasch CR, Inman BA, et al. A systematic review of regional hyperthermia therapy in bladder cancer. International Journal of Hyperthermia. 2016;32(4):381–89.

[18] Monsuez -J-J, Charniot J-C, Vignat N, Artigou J-Y. Cardiac side-effects of cancer chemotherapy. International Journal of Cardiology. 2010;144(1):3–15.
[19] Love RR, Leventhal H, Easterling DV, Nerenz DR. Side effects and emotional distress during cancer chemotherapy. Cancer. 1989;63(3):604–12.
[20] Paus R, Haslam IS, Sharov AA, Botchkarev VA. Pathobiology of chemotherapy-induced hair loss. The Lancet Oncology. 2013;14(2):e50–e9.
[21] Konstantinov SM, Berger MR. Antimetabolites. In: Offermanns S, Rosenthal W, editors. Encyclopedia of Molecular Pharmacology. Berlin, Heidelberg: Springer Berlin Heidelberg; 2008. p. 147–52.
[22] Konstantinov SM, Berger MR. Alkylating agents. In: Offermanns S, Rosenthal W, editors. Encyclopedia of Molecular Pharmacology. Berlin, Heidelberg: Springer Berlin Heidelberg; 2008. p. 53–57.
[23] Danks MK. Topoisomerase enzymes as drug targets. In: Schwab M, editor. Encyclopedic Reference of Cancer. Berlin, Heidelberg: Springer Berlin Heidelberg; 2001. p. 900–03.
[24] van Vuuren RJ, Visagie MH, Theron AE, Joubert AM. Antimitotic drugs in the treatment of cancer. Cancer Chemotherapy and Pharmacology. 2015;76(6):1101–12.
[25] Gause GF, Dudnik YV. Antitumor antibiotic carminomycin: Mechanism of action. In: Hellmann K, Connors TA, editors. Chemotherapy: Cancer Chemotherapy II. Boston, MA: Springer US; 1976. p. 169–73.
[26] Gerits N, Moens U. Protein kinase inhibitors. In: Offermanns S, Rosenthal W, editors. Encyclopedia of Molecular Pharmacology. Berlin, Heidelberg: Springer Berlin Heidelberg; 2008. p. 1008–12.
[27] Oerlemans R, Berkers CR, Assaraf YG, Scheffer GL, Peters GJ, Verbrugge SE, et al. Proteasome inhibition and mechanism of resistance to a synthetic, library-based hexapeptide. Investigational New Drugs. 2018;36(5):797–809.
[28] Sunada S, Miki Y. MPARP inhibitors: Echanism of action. In: Nakamura S, Aoki D, Miki Y, editors. Hereditary Breast and Ovarian Cancer: Molecular Mechanism and Clinical Practice. Singapore: Springer Singapore; 2021. p. 281–92.
[29] Nel A, Ruoslahti E, Meng H. New insights into "permeability" as in the enhanced permeability and retention effect of cancer nanotherapeutics. ACS Publications. 2017:9567–69.
[30] Kalyane D, Raval N, Maheshwari R, Tambe V, Kalia K, Tekade RK. Employment of enhanced permeability and retention effect (EPR): Nanoparticle-based precision tools for targeting of therapeutic and diagnostic agent in cancer. Materials Science and Engineering: C. 2019;98:1252–76.
[31] Yhee JY, Son S, Son S, Joo MK, Kwon IC. The EPR effect in cancer therapy. Cancer Targeted Drug Delivery: Springer. 2013:621–32.
[32] Dvorak H, Nagy J, Dvorak A. Structure of solid tumors and their vasculature: Implications for therapy with monoclonal antibodies. Cancer Cells (Cold Spring Harbor, NY: 1989). 1991;3(3):77–85.
[33] Jain R. Measurements of tumor vascular leakiness using DCE in brain tumors: Clinical applications. NMR in Biomedicine. 2013;26(8):1042–49.
[34] Dvorak HF, Weaver VM, Tlsty TD, Bergers G. Tumor microenvironment and progression. Journal of Surgical Oncology. 2011;103(6):468–74.
[35] Nakamura H, Jun F, Maeda H. Development of next-generation macromolecular drugs based on the EPR effect: Challenges and pitfalls. Expert Opinion on Drug Delivery. 2015;12(1):53–64.
[36] Acharya S, Sahoo SK. PLGA nanoparticles containing various anticancer agents and tumour delivery by EPR effect. Advanced Drug Delivery Reviews. 2011;63(3):170–83.

[37] Park J, Choi Y, Chang H, Um W, Ryu JH, Kwon IC. Alliance with EPR effect: Combined strategies to improve the EPR effect in the tumor microenvironment. Theranostics. 2019;9(26):8073.
[38] Hatakeyama H, Akita H, Harashima H. The polyethyleneglycol dilemma: Advantage and disadvantage of PEGylation of liposomes for systemic genes and nucleic acids delivery to tumors. Biological and Pharmaceutical Bulletin. 2013;36(6):892–99.
[39] Awad NS, Paul V, Mahmoud MS, Al Sawaftah NM, Kawak PS, Al Sayah MH, et al. Effect of pegylation and targeting moieties on the ultrasound-mediated drug release from liposomes. ACS Biomaterials Science & Engineering. 2019;6(1):48–57.
[40] Chen E, Chen B-M, Su Y-C, Chang Y-C, Cheng T-L, Barenholz Y, et al. Premature drug release from polyethylene glycol (PEG)-coated liposomal doxorubicin via formation of the membrane attack complex. ACS Nano. 2020;14(7):7808–22.
[41] Iio M, Wagner HN. Studies of the reticuloendothelial system (RES). I. Measurement of the phagocytic capacity of the RES in man and dog. The Journal of Clinical Investigation. 1963;42(3):417–26.
[42] Wagner HN, Iio M, Hornick RB. Studies of the reticuloendothelial system (RES). II. Changes in the phagocytic capacity of the RES in patients with certain infections. The Journal of Clinical Investigation. 1963;42(3):427–34.
[43] Wagner HN, Iio M. Studies of the reticuloendothelial system (RES). III. Blockade of the RES in man. The Journal of Clinical Investigation. 1964;43(8):1525–32.
[44] Liu L, Hitchens TK, Ye Q, Wu Y, Barbe B, Prior DE, et al. Decreased reticuloendothelial system clearance and increased blood half-life and immune cell labeling for nano-and micron-sized superparamagnetic iron-oxide particles upon pre-treatment with Intralipid. Biochimica Et Biophysica Acta (Bba)-general Subjects. 2013;1830(6):3447–53.
[45] Li S-D, Huang L. Nanoparticles evading the reticuloendothelial system: Role of the supported bilayer. Biochimica Et Biophysica Acta (Bba)-biomembranes. 2009;1788(10):2259–66.
[46] Maeda H. Toward a full understanding of the EPR effect in primary and metastatic tumors as well as issues related to its heterogeneity. Advanced Drug Delivery Reviews. 2015;91:3–6.
[47] Fang J, Nakamura H, Maeda H. The EPR effect: Unique features of tumor blood vessels for drug delivery, factors involved, and limitations and augmentation of the effect. Advanced Drug Delivery Reviews. 2011;63(3):136–51.
[48] Wu J. The enhanced permeability and retention (EPR) effect: The significance of the concept and methods to enhance its application. Journal of Personalized Medicine. 2021;11(8):771.
[49] Greish K. Enhanced permeability and retention (EPR) effect for anticancer nanomedicine drug targeting. Cancer Nanotechnology: Springer. 2010:25–37.
[50] Fang J, Sawa T, Maeda H. Factors and mechanism of "EPR" effect and the enhanced antitumor effects of macromolecular drugs including SMANCS. Polymer Drugs in the Clinical Stage. 2004:29–49.
[51] Li R, Zheng K, Yuan C, Chen Z, Huang M. Be active or not: The relative contribution of active and passive tumor targeting of nanomaterials. Nanotheranostics. 2017;1(4):346.
[52] Manoharan D, Li W-P, Yeh C-S. Advances in controlled gas-releasing nanomaterials for therapeutic applications. Nanoscale Horizons. 2019;4(3):557–78.
[53] Costa EC, Gaspar VM, Marques JG, Coutinho P, Correia IJ. Evaluation of nanoparticle uptake in co-culture cancer models. PLoS One. 2013;8(7):e70072.
[54] Chen WC, Zhang AX, Li S-D. Limitations and niches of the active targeting approach for nanoparticle drug delivery. European Journal of Nanomedicine. 2012;4(2–4):89–93.
[55] Liu F-S. Mechanisms of chemotherapeutic drug resistance in cancer therapy – a quick review. Taiwanese Journal of Obstetrics and Gynecology. 2009;48(3):239–44.

[56] Schleich N, Po C, Jacobs D, Ucakar B, Gallez B, Danhier F, et al. Comparison of active, passive and magnetic targeting to tumors of multifunctional paclitaxel/SPIO-loaded nanoparticles for tumor imaging and therapy. Journal of Controlled Release. 2014;194:82–91.
[57] Bertrand N, Wu J, Xu X, Kamaly N, Farokhzad OC. Cancer nanotechnology: The impact of passive and active targeting in the era of modern cancer biology. Advanced Drug Delivery Reviews. 2014;66:2–25.
[58] Bazak R, Houri M, El Achy S, Kamel S, Refaat T. Cancer active targeting by nanoparticles: A comprehensive review of literature. Journal of Cancer Research and Clinical Oncology. 2015;141(5):769–84.
[59] Yoo J, Park C, Yi G, Lee D, Koo H. Active targeting strategies using biological ligands for nanoparticle drug delivery systems. Cancers. 2019;11(5):640.
[60] Ling D, Hackett MJ, Hyeon T. Surface ligands in synthesis, modification, assembly and biomedical applications of nanoparticles. Nano Today. 2014;9(4):457–77.
[61] Zaki NM, Tirelli N. Gateways for the intracellular access of nanocarriers: A review of receptor-mediated endocytosis mechanisms and of strategies in receptor targeting. Expert Opinion on Drug Delivery. 2010;7(8):895–913.
[62] Gonda A, Kabagwira J, Senthil GN, Wall NR. Internalization of exosomes through receptor-mediated endocytosis. Molecular Cancer Research. 2019;17(2):337–47.
[63] Qian ZM, Li H, Sun H, Ho K. Targeted drug delivery via the transferrin receptor-mediated endocytosis pathway. Pharmacological Reviews. 2002;54(4):561–87.
[64] Gao H, Shi W, Freund LB. Mechanics of receptor-mediated endocytosis. Proceedings of the National Academy of Sciences. 2005;102(27):9469–74.
[65] Cho K, Wang X, Nie S, Shin DM. Therapeutic nanoparticles for drug delivery in cancer. Clinical Cancer Research. 2008;14(5):1310–16.
[66] Donahue ND, Acar H, Wilhelm S. Concepts of nanoparticle cellular uptake, intracellular trafficking, and kinetics in nanomedicine. Advanced Drug Delivery Reviews. 2019;143:68–96.
[67] Ding H-M, Ma Y-Q. Role of physicochemical properties of coating ligands in receptor-mediated endocytosis of nanoparticles. Biomaterials. 2012;33(23):5798–802.
[68] Vácha R, Martinez-Veracoechea FJ, Frenkel D. Receptor-mediated endocytosis of nanoparticles of various shapes. Nano Letters. 2011;11(12):5391–95.
[69] Decuzzi P, Ferrari M. The role of specific and non-specific interactions in receptor-mediated endocytosis of nanoparticles. Biomaterials. 2007;28(18):2915–22.
[70] Yim WW-Y, Mizushima N. Lysosome biology in autophagy. Cell Discovery. 2020;6(1):1–12.
[71] Ko M, Quiñones-Hinojosa A, Rao R. Emerging links between endosomal pH and cancer. Cancer and Metastasis Reviews. 2020;39(2):519–34.
[72] Smith SA, Selby LI, Johnston AP, Such GK. The endosomal escape of nanoparticles: Toward more efficient cellular delivery. Bioconjugate Chemistry. 2018;30(2):263–72.
[73] Parodi A, Evangelopoulos M, Arrighetti N, Cevenini A, Livingston M, Khaled SZ, et al. Endosomal escape of polymer-coated silica nanoparticles in endothelial cells. Small. 2020;16(36):1907693.
[74] Vermeulen LM, De Smedt SC, Remaut K, Braeckmans K. The proton sponge hypothesis: Fable or fact?. European Journal of Pharmaceutics and Biopharmaceutics. 2018;129:184–90.
[75] Behr J-P. The proton sponge: A trick to enter cells the viruses did not exploit. CHIMIA International Journal for Chemistry. 1997;51(1–2):34–36.
[76] Wojnilowicz M, Glab A, Bertucci A, Caruso F, Cavalieri F. Super-resolution imaging of proton sponge-triggered rupture of endosomes and cytosolic release of small interfering RNA. ACS Nano. 2018;13(1):187–202.
[77] Martens TF, Remaut K, Demeester J, De Smedt SC, Braeckmans K. Intracellular delivery of nanomaterials: How to catch endosomal escape in the act. Nano Today. 2014;9(3):344–64.

[78] Shi B, Zhang H, Bi J, Dai S. Endosomal pH responsive polymers for efficient cancer targeted gene therapy. Colloids and Surfaces B: Biointerfaces. 2014;119:55–65.

[79] Zhao H, Li Q, Hong Z. Paclitaxel-loaded mixed micelles enhance ovarian cancer therapy through extracellular pH-triggered PEG detachment and endosomal escape. Molecular Pharmaceutics. 2016;13(7):2411–22.

[80] Kanamala M, Wilson WR, Yang M, Palmer BD, Wu Z. Mechanisms and biomaterials in pH-responsive tumour targeted drug delivery: A review. Biomaterials. 2016;85:152–67.

[81] Weeratunga S, Paul B, Collins BM. Recognising the signals for endosomal trafficking. Current Opinion in Cell Biology. 2020;65:17–27.

[82] Elkin SR, Lakoduk AM, Schmid SL. Endocytic pathways and endosomal trafficking: A primer. Wiener Medizinische Wochenschrift. 2016;166(7):196–204.

[83] Bertrand N, Leroux J-C. The journey of a drug-carrier in the body: An anatomo-physiological perspective. Journal of Controlled Release. 2012;161(2):152–63.

[84] Bicker J, Alves G, Fortuna A, Falcão A. Blood–brain barrier models and their relevance for a successful development of CNS drug delivery systems: A review. European Journal of Pharmaceutics and Biopharmaceutics. 2014;87(3):409–32.

[85] Smith QR. A review of blood-brain barrier transport techniques. The Blood-Brain Barrier. 2003:193–208.

[86] Davson H. Review lecture. The blood-brain barrier. The Journal of Physiology. 1976;255(1):1–28.

[87] Thuenauer R, Müller SK, Römer W. Pathways of protein and lipid receptor-mediated transcytosis in drug delivery. Expert Opinion on Drug Delivery. 2017;14(3):341–51.

[88] Ulbrich K, Hekmatara T, Herbert E, Kreuter J. Transferrin-and transferrin-receptor-antibody-modified nanoparticles enable drug delivery across the blood–brain barrier (BBB). European Journal of Pharmaceutics and Biopharmaceutics. 2009;71(2):251–56.

[89] Johnsen KB, Moos T. Revisiting nanoparticle technology for blood–brain barrier transport: Unfolding at the endothelial gate improves the fate of transferrin receptor-targeted liposomes. Journal of Controlled Release. 2016;222:32–46.

[90] Daniels TR, Bernabeu E, Rodríguez JA, Patel S, Kozman M, Chiappetta DA, et al. The transferrin receptor and the targeted delivery of therapeutic agents against cancer. Biochimica Et Biophysica Acta (Bba)-general Subjects. 2012;1820(3):291–317.

[91] May P, Woldt E, Matz RL, Boucher P. The LDL receptor-related protein (LRP) family: An old family of proteins with new physiological functions. Annals of Medicine. 2007;39(3):219–28.

[92] Milletti F. Cell-penetrating peptides: Classes, origin, and current landscape. Drug Discovery Today. 2012;17(15–16):850–60.

[93] Kerkis A, Hayashi MA, Yamane T, Kerkis I. Properties of cell penetrating peptides (CPPs). IUBMB Life. 2006;58(1):7–13.

[94] Hervé F, Ghinea N, Scherrmann J-M. CNS delivery via adsorptive transcytosis. The AAPS Journal. 2008;10(3):455–72.

[95] Motz GT, Santoro SP, Wang L-P, Garrabrant T, Lastra RR, Hagemann IS, et al. Tumor endothelium FasL establishes a selective immune barrier promoting tolerance in tumors. Nature Medicine. 2014;20(6):607–15.

[96] Hinshaw DC, Shevde LA. The tumor microenvironment innately modulates cancer progression. Cancer Research. 2019;79(18):4557–66.

[97] Nicholas NS, Apollonio B, Ramsay AG. Tumor microenvironment (TME)-driven immune suppression in B cell malignancy. Biochimica Et Biophysica Acta (Bba)-molecular Cell Research. 2016;1863(3):471–82.

[98] Yang H, Tong Z, Sun S, Mao Z. Enhancement of tumour penetration by nanomedicines through strategies based on transport processes and barriers. Journal of Controlled Release. 2020.

[99] He X, Li J, An S, Jiang C. pH-sensitive drug-delivery systems for tumor targeting. Therapeutic Delivery. 2013;4(12):1499–510.

[100] Chen B, Dai W, He B, Zhang H, Wang X, Wang Y, et al. Current multistage drug delivery systems based on the tumor microenvironment. Theranostics. 2017;7(3):538.

[101] Uthaman S, Huh KM, Park I-K. Tumor microenvironment-responsive nanoparticles for cancer theragnostic applications. Biomaterials Research. 2018;22(1):1–11.

[102] Shanbhag PP, Jog SV, Chogale MM, Gaikwad SS. Theranostics for cancer therapy. Current Drug Delivery. 2013;10(3):357–62.

[103] Kim H, Kwak G, Kim K, Yoon HY, Kwon IC. Theranostic designs of biomaterials for precision medicine in cancer therapy. Biomaterials. 2019;213:119207.

[104] Shao J, Griffin RJ, Galanzha EI, Kim J-W, Koonce N, Webber J, et al. Photothermal nanodrugs: Potential of TNF-gold nanospheres for cancer theranostics. Scientific Reports. 2013;3(1):1–9.

[105] Rajkumar S, Prabaharan M. Multi-functional nanocarriers based on iron oxide nanoparticles conjugated with doxorubicin, poly (ethylene glycol) and folic acid as theranostics for cancer therapy. Colloids and Surfaces B: Biointerfaces. 2018;170:529–37.

[106] Ansari MO, Ahmad MF, Shadab G, Siddique HR. Superparamagnetic iron oxide nanoparticles based cancer theranostics: A double edge sword to fight against cancer. Journal of Drug Delivery Science and Technology. 2018;45:177–83.

[107] Hassan M, Gomes VG, Dehghani A, Ardekani SM. Engineering carbon quantum dots for photomediated theranostics. Nano Research. 2018;11(1):1–41.

[108] Tade RS, Patil PO. Theranostic prospects of graphene quantum dots in breast cancer. ACS Biomaterials Science & Engineering. 2020;6(11):5987–6008.

[109] Wang Y, Wang Y, Chen G, Li Y, Xu W, Gong S. Quantum-dot-based theranostic micelles conjugated with an anti-EGFR nanobody for triple-negative breast cancer therapy. ACS Applied Materials & Interfaces. 2017;9(36):30297–305.

[110] Wang S, Li F, Qiao R, Hu X, Liao H, Chen L, et al. Arginine-rich manganese silicate nanobubbles as a ferroptosis-inducing agent for tumor-targeted theranostics. Acs Nano. 2018;12(12):12380–92.

[111] Li Y, Wan J, Zhang Z, Guo J, Wang C. Targeted soft biodegradable glycine/PEG/RGD-modified poly (methacrylic acid) nanobubbles as intelligent theranostic vehicles for drug delivery. ACS Applied Materials & Interfaces. 2017;9(41):35604–12.

[112] Hope JM, Pothuri B. The role of palliative surgery in gynecologic cancer cases. The Oncologist. 2013;18(1):73.

[113] Miner TJ, Jaques DP, Karpeh MS, Brennan MF. Defining palliative surgery in patients receiving noncurative resections for gastric cancer. Journal of the American College of Surgeons. 2004;198(6):1013–21.

[114] Cohen JT, Miner TJ. Patient selection in palliative surgery: Defining value. Journal of Surgical Oncology. 2019;120(1):35–44.

[115] Sato S, Kunisaki C, Tanaka Y, Sato K, Miyamoto H, Yukawa N, et al. Curative-intent surgery for stage IV advanced gastric cancer: Who can undergo surgery and what are the prognostic factors for long-term survival? Annals of Surgical Oncology. 2019;26(13):4452–63.

[116] Akin E, Gonullu E, Altintoprak F. Prophylactic resections for genetic predisposition of colon and rectum. In: Dilek ON, Uranues S, Latifi R, editors. Prophylactic Surgery. Cham: Springer International Publishing; 2021. p. 201–12.

[117] Underwood JCE. Reporting and Classification of Biopsy Diagnoses. Introduction to Biopsy Interpretation and Surgical Pathology. London: Springer London; 1981. p. 131–38.

[118] Picci P. Staging. In: Picci P, Manfrini M, Donati DM, Gambarotti M, Righi A, Vanel D, editors. Diagnosis of Musculoskeletal Tumors and Tumor-like Conditions: Clinical, Radiological and Histological Correlations – The Rizzoli Case Archive. Cham: Springer International Publishing; 2020. p. 31–33.

[119] Zheng H, Gao Y-N. Primary debulking surgery or neoadjuvant chemotherapy followed by interval debulking surgery for patients with advanced ovarian cancer. Chinese Journal of Cancer Research. 2012;24(4):304–09.

[120] Metzger U, Weder W, Mona D, Largiadèr F. Palliative and Supportive Surgery for Gastrointestinal Cancer. Berlin, Heidelberg: Springer Berlin Heidelberg; 1988. p. 143–47.

[121] Koshima I. History of reconstructive microsurgery: From myth to reality. In: Maruccia M, Giudice G, editors. Textbook of Plastic and Reconstructive Surgery: Basic Principles and New Perspectives. Cham: Springer International Publishing; 2022. p. 77–86.

[122] Lendlein A, Sisson A. Handbook of Biodegradable Polymers: Isolation, Synthesis, Characterization and Applications: John Wiley & Sons; 2011.

[123] Śmiga-Matuszowicz M, Korytkowska-Wałach A, Nowak B. Isosorbide-based polysebacates as polymeric components for development of in situ forming implants. Polymers for Advanced Technologies. 2019;30(4):1072–82.

[124] Chesterman J, Zhang Z, Ortiz O, Goyal R, Kohn J. Biodegradable polymers. Principles of Tissue Engineering. 2020:317–42.

[125] Panigrahi M, Das P, Parikh P. Brain tumor and Gliadel wafer treatment. Indian Journal of Cancer. 2011;48(1):11.

[126] Bregy A, Shah AH, Diaz MV, Pierce HE, Ames PL, Diaz D, et al. The role of Gliadel wafers in the treatment of high-grade gliomas. Expert Review of Anticancer Therapy. 2013;13(12):1453–61.

[127] Do Hammer C, Fanning A, Crowe J. Overview of breast cancer staging and surgical treatment options. Cleveland Clinic Journal of Medicine. 2008;75:S11.

[128] Loukas M, Tubbs RS, Mirzayan N, Shirak M, Steinberg A, Shoja MM. The history of mastectomy. The American Surgeon. 2011;77(5):566–71.

[129] Akpan U, Pellegrini M, Obayemi J, Ezenwafor T, Browl D, Ani C, et al. Prodigiosin-loaded electrospun nanofibers scaffold for localized treatment of triple negative breast cancer. Materials Science and Engineering: C. 2020;114:110976.

[130] Sims-Mourtada J, Niamat RA, Samuel S, Eskridge C, Kmiec EB. Enrichment of breast cancer stem-like cells by growth on electrospun polycaprolactone-chitosan nanofiber scaffolds. International Journal of Nanomedicine. 2014;9:995.

[131] Sasikala ARK, Unnithan AR, Thomas RG, Ko SW, Jeong YY, Park CH, et al. Multifaceted implantable Anticancer device for potential postsurgical breast cancer treatment: A single platform for synergistic inhibition of local regional breast cancer recurrence, surveillance, and healthy breast reconstruction. Advanced Functional Materials. 2018;28(8):1704793.

[132] Singh SK, Singh S, Lillard JW Jr, Singh R. Drug delivery approaches for breast cancer. International Journal of Nanomedicine. 2017;12:6205.

[133] Tiwari AP, Bhattarai DP, Maharjan B, Ko SW, Kim HY, Park CH, et al. Polydopamine-based implantable multifunctional nanocarpet for highly efficient photothermal-chemo therapy. Scientific Reports. 2019;9(1):1–13.

[134] Chen C, Tang Y, Ding D. Intramolecular motion-associated biomaterials for image-guided cancer surgery. Smart Materials in Medicine. 2020;1:24–31.

[135] Lee HJ, Ehlerding EB, Jiang D, Barnhart TE, Cao T, Wei W, et al. Dual-labeled pertuzumab for multimodality image-guided ovarian tumor resection. American Journal of Cancer Research. 2019;9(7):1454.

[136] Marotta CB, Haber T, Berlin JM, Grubbs RH. Surgery-guided removal of ovarian cancer using up-converting nanoparticles. ACS Applied Materials & Interfaces. 2020;12(43):48371–79.

[137] Yoshii Y, Tashima H, Iwao Y, Yoshida E, Wakizaka H, Akamatsu G, et al. Immuno-OpenPET: A novel approach for early diagnosis and image-guided surgery for small resectable pancreatic cancer. Scientific Reports. 2020;10(1):1–11.
[138] Zargoosh K, Aliabadi HM. SrAl2O4: Eu2+: Dy3+/WO3/polyester nanocomposite as a highly efficient and environmentally friendly photocatalyst for removal of dyes from industrial wastes. Environmental Nanotechnology, Monitoring & Management. 2019;12:100273.
[139] Lv Y, Liu P, Ding H, Wu Y, Yan Y, Liu H, et al. Conjugated polymer-based hybrid nanoparticles with two-photon excitation and near-infrared emission features for fluorescence bioimaging within the biological window. ACS Applied Materials & Interfaces. 2015;7(37):20640–48.
[140] Jenkins R, Burdette MK, Foulger SH. Mini-review: Fluorescence imaging in cancer cells using dye-doped nanoparticles. RSC Advances. 2016;6(70):65459–74.
[141] Patsenker L, Gellerman G. Fluorescent reporters for drug delivery monitoring. Israel Journal of Chemistry. 2020;60(5–6):504–18.
[142] Resch-Genger U, Grabolle M, Cavaliere-Jaricot S, Nitschke R, Nann T. Quantum dots versus organic dyes as fluorescent labels. Nature Methods. 2008;5(9):763–75.
[143] Altınogˇlu EI, Russin TJ, Kaiser JM, Barth BM, Eklund PC, Kester M, et al. Near-infrared emitting fluorophore-doped calcium phosphate nanoparticles for in vivo imaging of human breast cancer. ACS Nano. 2008;2(10):2075–84.
[144] Daraee H, Eatemadi A, Abbasi E, Fekri Aval S, Kouhi M, Akbarzadeh A. Application of gold nanoparticles in biomedical and drug delivery. Artificial Cells, Nanomedicine, and Biotechnology. 2016;44(1):410–22.
[145] Kholikov K, Ilhom S, Sajjad M, Smith ME, Monroe JD, San O, et al. Improved singlet oxygen generation and antimicrobial activity of sulphur-doped graphene quantum dots coupled with methylene blue for photodynamic therapy applications. Photodiagnosis and Photodynamic Therapy. 2018;24:7–14.
[146] Zhang J, Wu S, Lu X, Wu P, Liu J. Lanthanide-boosted singlet oxygen from diverse photosensitizers along with potent photocatalytic oxidation. ACS Nano. 2019;13 (12):14152–61.
[147] Mahmoudi K, Bouras A, Bozec D, Ivkov R, Hadjipanayis C. Magnetic hyperthermia therapy for the treatment of glioblastoma: A review of the therapy's history, efficacy and application in humans. International Journal of Hyperthermia. 2018;34(8):1316–28.
[148] Mallory M, Gogineni E, Jones GC, Greer L, Simone CB II. Therapeutic hyperthermia: The old, the new, and the upcoming. Critical Reviews in Oncology/hematology. 2016;97:56–64.
[149] Huilgol NG, Gupta S, Sridhar C. Hyperthermia with radiation in the treatment of locally advanced head and neck cancer: A report of randomized trial. Journal of Cancer Research and Therapeutics. 2010;6(4):492.
[150] Behrouzkia Z, Joveini Z, Keshavarzi B, Eyvazzadeh N, Aghdam RZ. Hyperthermia: How can it be used?. Oman Medical Journal. 2016;31(2):89.
[151] Mahmood J, Shukla HD, Soman S, Samanta S, Singh P, Kamlapurkar S, et al. Immunotherapy, radiotherapy, and hyperthermia: A combined therapeutic approach in pancreatic cancer treatment. Cancers. 2018;10(12):469.
[152] Mancuso MR, Davis R, Norberg SM, O'Brien S, Sennino B, Nakahara T, et al. Rapid vascular regrowth in tumors after reversal of VEGF inhibition. The Journal of Clinical Investigation. 2006;116(10):2610–21.
[153] Shen Y, Liu P, Zhang A, Xu LX. Study on tumor microvasculature damage induced by alternate cooling and heating. Annals of Biomedical Engineering. 2008;36(8):1409.
[154] Lassche G, Crezee J, Van Herpen C. Whole-body hyperthermia in combination with systemic therapy in advanced solid malignancies. Critical Reviews in Oncology/hematology. 2019;139:67–74.

[155] Ostberg JR, Taylor SL, Baumann H, Repasky EA. Regulatory effects of fever-range whole-body hyperthermia on the LPS-induced acute inflammatory response. Journal of Leukocyte Biology. 2000;68(6):815–20.
[156] Zhao C, Dai C, Chen X. Whole-body hyperthermia combined with hyperthermic intraperitoneal chemotherapy for the treatment of stage IV advanced gastric cancer. International Journal of Hyperthermia. 2012;28(8):735–41.
[157] Lindner LH, Issels RD. Hyperthermia in soft tissue sarcoma. Current Treatment Options in Oncology. 2011;12(1):12–20.
[158] Habash RW, Bansal R, Krewski D, Alhafid HT. Thermal therapy, part 2: Hyperthermia techniques. Critical Reviews™ in Biomedical Engineering. 2006;34:6.
[159] Kok HP, Cressman EN, Ceelen W, Brace CL, Ivkov R, Grüll H, et al. Heating technology for malignant tumors: A review. International Journal of Hyperthermia. 2020;37(1):711–41.
[160] Huo Y, Richards A, Liauw W, Morris D. Hyperthermic intraperitoneal chemotherapy (HIPEC) and cytoreductive surgery (CRS) in ovarian cancer: A systematic review and meta-analysis. European Journal of Surgical Oncology (EJSO). 2015;41(12):1578–89.
[161] Bakrin N, Classe J, Pomel C, Gouy S, Chene G, Glehen O. Hyperthermic intraperitoneal chemotherapy (HIPEC) in ovarian cancer. Journal of Visceral Surgery. 2014;151(5):347–53.
[162] Park BJ, Alexander HR, Libutti SK, Wu P, Royalty D, Kranda KC, et al. Treatment of primary peritoneal mesothelioma by continuous hyperthermic peritoneal perfusion (CHPP). Annals of Surgical Oncology. 1999;6(6):582–90.
[163] Pelz JO, Vetterlein M, Grimmig T, Kerscher AG, Moll E, Lazariotou M, et al. Hyperthermic intraperitoneal chemotherapy in patients with peritoneal carcinomatosis: Role of heat shock proteins and dissecting effects of hyperthermia. Annals of Surgical Oncology. 2013;20 (4):1105–13.
[164] Calderwood SK, Theriault JR, Gong J. How is the immune response affected by hyperthermia and heat shock proteins?. International Journal of Hyperthermia. 2005;21(8):713–16.
[165] Périgo EA, Hemery G, Sandre O, Ortega D, Garaio E, Plazaola F, et al. Fundamentals and advances in magnetic hyperthermia. Applied Physics Reviews. 2015;2(4):041302.
[166] Ortega D, Pankhurst QA. Magnetic hyperthermia. Nanoscience. 2013;1(60):e88.
[167] Giustini AJ, Petryk AA, Cassim SM, Tate JA, Baker I, Hoopes PJ. Magnetic nanoparticle hyperthermia in cancer treatment. Nano Life. 2010;1(01n02):17–32.
[168] Khlusov IA, Zagrebin L, Shestov S, Itin V, Sedoi V, Feduschak T, et al. Colony-forming activity of unipotent hemopoietic precursors under the effect of nanosized ferrites in a constant magnetic field in vitro. Bulletin of Experimental Biology and Medicine. 2008;145(1):151–57.
[169] Marinca T, Chicinaş I, Isnard O, Neamţu B. Nanocrystalline/nanosized Fe_3O_4 particles obtained by heat treatment and mechanical milling. Optoelectronics and Advanced Materials-Rapid Communications. 2015;9(5–6):730–33.
[170] Rajan A, Sahu NK. Review on magnetic nanoparticle-mediated hyperthermia for cancer therapy. Journal of Nanoparticle Research. 2020;22(11):1–25.
[171] Shubayev VI, Pisanic TR II, Jin S. Magnetic nanoparticles for theragnostics. Advanced Drug Delivery Reviews. 2009;61(6):467–77.
[172] Stocke NA, Sethi P, Jyoti A, Chan R, Arnold SM, Hilt JZ, et al. Toxicity evaluation of magnetic hyperthermia induced by remote actuation of magnetic nanoparticles in 3D micrometastasic tumor tissue analogs for triple negative breast cancer. Biomaterials. 2017;120:115–25.
[173] Fathi Karkan S, Mohammadhosseini M, Panahi Y, Milani M, Zarghami N, Akbarzadeh A, et al. Magnetic nanoparticles in cancer diagnosis and treatment: A review. Artificial Cells, Nanomedicine, and Biotechnology. 2017;45(1):1–5.
[174] Xu H, Pan Y. Experimental evaluation on the heating efficiency of magnetoferritin nanoparticles in an alternating magnetic field. Nanomaterials. 2019;9(10):1457.

[175] Wu K, Su D, Liu J, Saha R, Wang J-P. Magnetic nanoparticles in nanomedicine: A review of recent advances. Nanotechnology. 2019;30(50):502003.

[176] Martinez-Boubeta C, Simeonidis K, Makridis A, Angelakeris M, Iglesias O, Guardia P, et al. Learning from nature to improve the heat generation of iron-oxide nanoparticles for magnetic hyperthermia applications. Scientific Reports. 2013;3(1):1–8.

[177] Danaei M, Dehghankhold M, Ataei S, Hasanzadeh Davarani F, Javanmard R, Dokhani A, et al. Impact of particle size and polydispersity index on the clinical applications of lipidic nanocarrier systems. Pharmaceutics. 2018;10(2):57.

[178] Figuerola A, Di Corato R, Manna L, Pellegrino T. From iron oxide nanoparticles towards advanced iron-based inorganic materials designed for biomedical applications. Pharmacological Research. 2010;62(2):126–43.

[179] Nosrati H, Salehiabar M, Manjili HK, Danafar H, Davaran S. Preparation of magnetic albumin nanoparticles via a simple and one-pot desolvation and co-precipitation method for medical and pharmaceutical applications. International Journal of Biological Macromolecules. 2018;108:909–15.

[180] Cherian E, Rajan A, Baskar G. Synthesis of manganese dioxide nanoparticles using co-precipitation method and its antimicrobial activity. International Journal of Modern Science and Technology. 2016;1(01):17–22.

[181] Razmara Z, Rezvani AR, Saravani H. Fischer–Tropsch reaction over a Co 2-Ni-Mn/SiO 2 nanocatalyst prepared by thermal decomposition of a new precursor. Chemical Papers. 2017;71(4):849–56.

[182] Pascu O, Carenza E, Gich M, Estradé S, Peiró F, Herranz G, et al. Surface reactivity of iron oxide nanoparticles by microwave-assisted synthesis; comparison with the thermal decomposition route. The Journal of Physical Chemistry C. 2012;116(28):15108–16.

[183] Kızılbey K. Optimization of rutin-loaded PLGA nanoparticles synthesized by single-emulsion solvent evaporation method. Acs Omega. 2019;4(1):555–62.

[184] Jenjob R, Phakkeeree T, Seidi F, Theerasilp M, Crespy D. Emulsion techniques for the production of pharmacological nanoparticles. Macromolecular Bioscience. 2019;19 (6):1900063.

[185] Ganesan V, Hariram M, Vivekanandhan S, Muthuramkumar S. Periconium sp. (endophytic fungi) extract mediated sol-gel synthesis of ZnO nanoparticles for antimicrobial and antioxidant applications. Materials Science in Semiconductor Processing. 2020;105:104739.

[186] Fayazzadeh S, Khodaei M, Arani M, Mahdavi S, Nizamov T, Majouga A. Magnetic properties and magnetic hyperthermia of cobalt ferrite nanoparticles synthesized by hydrothermal method. Journal of Superconductivity and Novel Magnetism. 2020;33(7):2227–33.

[187] Yang N, Gong F, Cheng L, Lei H, Li W, Sun Z, et al. Biodegradable magnesium alloy with eddy thermal effect for effective and accurate magnetic hyperthermia ablation of tumors. National Science Review. 2021;8(1):nwaa122.

[188] Park J-H, Kim MT, Kim KY, Bakheet N, Kim T-H, Jeon JY, et al. Local heat treatment for suppressing gastroduodenal stent-induced tissue hyperplasia using nanofunctionalized self-expandable metallic stent in rat gastric outlet model. ACS Biomaterials Science & Engineering. 2020;6(4):2450–58.

[189] Aguilar LE, GhavamiNejad A, Park CH, Kim CS. On-demand drug release and hyperthermia therapy applications of thermoresponsive poly-(NIPAAm-co-HMAAm)/polyurethane core-shell nanofiber mat on non-vascular nitinol stents. Nanomedicine: Nanotechnology, Biology and Medicine. 2017;13(2):527–38.

[190] El-Sayed AH, Mekawy M, El-Gendy A, El-Sayed N, Aly A. Magnetization measurements of PdNi, PdCo and CuNi ferromagnetic thermoseeds. International Journal of Pure and Applied Physics. 2007;3:155–62.

[191] Wu J, Wang H, Zhang H, Wei L, Wang X, Wang X, et al. Stainless steel hollow microspheres for arterial embolization hyperthermia. Journal of Medical and Biological Engineering. 2017;37 (6):810–19.

[192] Aguilar LE, Thomas RG, Moon MJ, Jeong YY, Park CH, Kim CS. Implantable chemothermal brachytherapy seeds: A synergistic approach to brachytherapy using polymeric dual drug delivery and hyperthermia for malignant solid tumor ablation. European Journal of Pharmaceutics and Biopharmaceutics. 2018;129:191–203.

[193] Rezk AI, Obiweluozor FO, Choukrani G, Park CH, Kim CS. Drug release and kinetic models of anticancer drug (BTZ) from a pH-responsive alginate polydopamine hydrogel: Towards cancer chemotherapy. International Journal of Biological Macromolecules. 2019;141:388–400.

[194] Aguilar LE, Chalony C, Kumar D, Park CH, Kim CS. Phenol-Boronic surface functionalization of gold nanoparticles; to induce ROS damage while inhibiting the survival mechanisms of cancer cells. International Journal of Pharmaceutics. 2021;596:120267.

[195] Jin Z, Chen Z, Wu K, Shen Y, Guo S. Investigation of migration-preventing tracheal stent with high dose of 5-fluorouracil or paclitaxel for local drug delivery. ACS Applied Bio Materials. 2018;1(5):1328–36.

[196] Liu J, Wang Z, Wu K, Li J, Chen W, Shen Y, et al. Paclitaxel or 5-fluorouracil/esophageal stent combinations as a novel approach for the treatment of esophageal cancer. Biomaterials. 2015;53:592–99.

[197] Fouladian P, Kohlhagen J, Arafat M, Afinjuomo F, Workman N, Abuhelwa AY, et al. Three-dimensional printed 5-fluorouracil eluting polyurethane stents for the treatment of oesophageal cancers. Biomaterials Science. 2020;8(23):6625–36.

[198] Li X, Wu Z, Mei Q, Guo M, Fu X, Han W. Long non-coding RNA HOTAIR, a driver of malignancy, predicts negative prognosis and exhibits oncogenic activity in oesophageal squamous cell carcinoma. British Journal of Cancer. 2013;109(8):2266–78.

[199] Hemmatzadeh M, Mohammadi H, Karimi M, Musavishenas MH, Baradaran B. Differential role of microRNAs in the pathogenesis and treatment of Esophageal cancer. Biomedicine & Pharmacotherapy. 2016;82:509–19.

[200] Aguilar LE, Tumurbaatar B, Ghavaminejad A, Park CH, Kim CS. Functionalized non-vascular nitinol stent via electropolymerized polydopamine thin film coating loaded with bortezomib adjunct to hyperthermia therapy. Scientific Reports. 2017;7(1):1–11.

[201] Bucharskaya AB, Maslyakova GN, Chekhonatskaya ML, Terentyuk GS, Navolokin NA, Khlebtsov BN, et al. Plasmonic photothermal therapy: Approaches to advanced strategy. Lasers in Surgery and Medicine. 2018;50(10):1025–33.

[202] Allison RR, Moghissi K. Photodynamic therapy (PDT): PDT mechanisms. Clinical Endoscopy. 2013;46(1):24.

[203] Yoon I, Li JZ, Shim YK. Advance in photosensitizers and light delivery for photodynamic therapy. Clinical Endoscopy. 2013;46(1):7.

[204] Yang K, Zhao S, Li B, Wang B, Lan M, Song X. Low temperature photothermal therapy: Advances and perspectives. Coordination Chemistry Reviews. 2021:214330.

[205] He S, Song J, Qu J, Cheng Z. Crucial breakthrough of second near-infrared biological window fluorophores: Design and synthesis toward multimodal imaging and theranostics. Chemical Society Reviews. 2018;47(12):4258–78.

[206] Ider M, Abderrafi K, Eddahbi A, Ouaskit S, Kassiba A. Silver metallic nanoparticles with surface plasmon resonance: Synthesis and characterizations. Journal of Cluster Science. 2017;28(3):1051–69.

[207] Chandran N, Bayal M, Pilankatta R, Nair SS. Tuning of Surface Plasmon Resonance (SPR) in Metallic Nanoparticles for Their Applications in SERS. Nanomaterials for Luminescent Devices, Sensors, and Bio-imaging Applications: Springer; 2021. p. 39–66.

[208] Prevo BG, Esakoff SA, Mikhailovsky A, Zasadzinski JA. Scalable routes to gold nanoshells with tunable sizes and response to near-infrared pulsed-laser irradiation. Small. 2008;4(8):1183–95.
[209] Mahmoud MA, El-Sayed MA. Gold nanoframes: Very high surface plasmon fields and excellent near-infrared sensors. Journal of the American Chemical Society. 2010;132(36):12704–10.
[210] Han HS, Choi KY. Advances in nanomaterial-mediated photothermal cancer therapies: Toward clinical applications. Biomedicines. 2021;9(3):305.
[211] Fernandes N, Rodrigues CF, Moreira AF, Correia IJ. Overview of the application of inorganic nanomaterials in cancer photothermal therapy. Biomaterials Science. 2020;8 (11):2990–3020.
[212] Wang Y, Meng H-M, Song G, Li Z, Zhang X-B. Conjugated-polymer-based nanomaterials for photothermal therapy. ACS Applied Polymer Materials. 2020;2(10):4258–72.
[213] Min H, Qi Y, Zhang Y, Han X, Cheng K, Liu Y, et al. A Graphdiyne oxide-based iron sponge with Photothermally enhanced tumor-specific Fenton chemistry. Advanced Materials. 2020;32(31):2000038.
[214] Korupalli C, Kalluru P, Nuthalapati K, Kuthala N, Thangudu S, Vankayala R. Recent advances of polyaniline-based biomaterials for phototherapeutic treatments of tumors and bacterial infections. Bioengineering. 2020;7(3):94.
[215] Zhou Y, Hu Y, Sun W, Lu S, Cai C, Peng C, et al. Radiotherapy-sensitized tumor photothermal ablation using γ-polyglutamic acid nanogels loaded with polypyrrole. Biomacromolecules. 2018;19(6):2034–42.
[216] Guo B, Zhao J, Wu C, Zheng Y, Ye C, Huang M, et al. One-pot synthesis of polypyrrole nanoparticles with tunable photothermal conversion and drug loading capacity. Colloids and Surfaces B: Biointerfaces. 2019;177:346–55.
[217] Yu C, Xu L, Zhang Y, Timashev PS, Huang Y, Liang X-J. Polymer-based nanomaterials for noninvasive cancer photothermal therapy. ACS Applied Polymer Materials. 2020;2 (10):4289–305.
[218] Ding F, Gao X, Huang X, Ge H, Xie M, Qian J, et al. Polydopamine-coated nucleic acid nanogel for siRNA-mediated low-temperature photothermal therapy. Biomaterials. 2020;245:119976.
[219] Cheng W, Zeng X, Chen H, Li Z, Zeng W, Mei L, et al. Versatile polydopamine platforms: Synthesis and promising applications for surface modification and advanced nanomedicine. Acs Nano. 2019;13(8):8537–65.
[220] Adam JK, Odhav B, Bhoola KD. Immune responses in cancer. Pharmacology & Therapeutics. 2003;99(1):113–32.
[221] Lanza R, Russell DW, Nagy A. Engineering universal cells that evade immune detection. Nature Reviews Immunology. 2019;19(12):723–33.
[222] Lopes C, Morgado S, Plácido AI, Roque F, Morgado M. Critical analysis of immune checkpoint inhibitor immunotoxicity management. SN Comprehensive Clinical Medicine. 2021;3 (1):84–94.
[223] Picanço-Castro V, Swiech K, Malmegrim KCR, Covas DT. CAR-T cells for cancer treatment: Current design and next frontiers. In: Swiech K, Malmegrim KCR, Picanço-Castro V, editors. Chimeric Antigen Receptor T Cells: Development and Production. New York, NY: Springer US; 2020. p. 1–10.
[224] Arai K, Nishida J, Hayashida K, Hatake K, Kitamura T, Miyajima A, et al. Coordinate regulation of immune and inflammatory responses by cytokines. Rinsho Byori the Japanese Journal of Clinical Pathology. 1990;38(4):347–53.
[225] Yamada H. Immunomodulators. In: Ōmura S, editor. The Search for Bioactive Compounds from Microorganisms. New York, NY: Springer New York; 1992. p. 171–97.

[226] Morse MA, Clay TM, Lyerly HK. Biology of the cancer vaccine immune response. In: Morse MA, Clay TM, Lyerly HK, editors. Handbook of Cancer Vaccines. Totowa, NJ: Humana Press; 2004. p. 3–10.
[227] Zarnani A-H, Jafari D, Bozorgmehr M, Shabani M, Barzegar-Yarmohammadi L, Ghaemimanesh F, et al. Monoclonal antibodies for cancer immunotherapy. In: Rezaei N, editor. Cancer Immunology: Bench to Bedside Immunotherapy of Cancers. Cham: Springer International Publishing; 2021. p. 273–312.
[228] Engeland CE, Bell JC. Introduction to oncolytic virotherapy. In: Engeland CE, editor. Oncolytic Viruses. New York, NY: Springer New York; 2020. p. 1–6.
[229] Weber JS, Mulé JJ. Cancer immunotherapy meets biomaterials. Nature Biotechnology. 2015;33(1):44–45.
[230] Wang C, Ye Y, Hu Q, Bellotti A, Gu Z. Tailoring biomaterials for cancer immunotherapy: Emerging trends and future outlook. Advanced Materials. 2017;29(29):1606036.
[231] Qiu H, Min Y, Rodgers Z, Zhang L, Wang AZ. Nanomedicine approaches to improve cancer immunotherapy. Wiley Interdisciplinary Reviews: Nanomedicine and Nanobiotechnology. 2017;9(5):e1456.
[232] Ben-Akiva E, Witte SE, Meyer RA, Rhodes KR, Green JJ. Polymeric micro-and nanoparticles for immune modulation. Biomaterials Science. 2019;7(1):14–30.
[233] Figdor CG, de Vries IJM, Lesterhuis WJ, Melief CJ. Dendritic cell immunotherapy: Mapping the way. Nature Medicine. 2004;10(5):475–80.
[234] Shi Y, Liu CH, Roberts AI, Das J, Xu G, Ren G, et al. Granulocyte-macrophage colony-stimulating factor (GM-CSF) and T-cell responses: What we do and don't know. Cell Research. 2006;16(2):126–33.
[235] Jana S, Kumar Sen K, Gandhi A. Alginate based nanocarriers for drug delivery applications. Current Pharmaceutical Design. 2016;22(22):3399–410.
[236] Ali OA, Tayalia P, Shvartsman D, Lewin S, Mooney DJ. Inflammatory cytokines presented from polymer matrices differentially generate and activate DCs in situ. Advanced Functional Materials. 2013;23(36):4621–28.
[237] Singh A, Peppas NA. Hydrogels and scaffolds for immunomodulation. Advanced Materials. 2014;26(38):6530–41.
[238] Fenton OS, Olafson KN, Pillai PS, Mitchell MJ, Langer R. Advances in biomaterials for drug delivery. Advanced Materials. 2018;30(29):1705328.
[239] Dhayani A, Kalita S, Mahato M, Srinath P, Vemula PK. Biomaterials for topical and transdermal drug delivery in reconstructive transplantation. Nanomedicine. 2019;14(20):2713–33.
[240] Cai L, Xu J, Yang Z, Tong R, Dong Z, Wang C, et al. Engineered biomaterials for cancer immunotherapy. MedComm. 2020;1(1):35–46.
[241] Kim JH, Shin JU, Kim SH, Noh JY, Kim HR, Lee J, et al. Successful transdermal allergen delivery and allergen-specific immunotherapy using biodegradable microneedle patches. Biomaterials. 2018;150:38–48.
[242] Siu KS, Chen D, Zheng X, Zhang X, Johnston N, Liu Y, et al. Non-covalently functionalized single-walled carbon nanotube for topical siRNA delivery into melanoma. Biomaterials. 2014;35(10):3435–42.
[243] Han X, Li H, Zhou D, Chen Z, Gu Z. Local and targeted delivery of immune checkpoint blockade therapeutics. Accounts of Chemical Research. 2020;53(11):2521–33.
[244] Riley RS, June CH, Langer R, Mitchell MJ. Delivery technologies for cancer immunotherapy. Nature Reviews Drug Discovery. 2019;18(3):175–96.

[245] Sau S, Alsaab HO, Bhise K, Alzhrani R, Nabil G, Iyer AK. Multifunctional nanoparticles for cancer immunotherapy: A groundbreaking approach for reprogramming malfunctioned tumor environment. Journal of Controlled Release. 2018;274:24–34.
[246] Goldberg MS. Improving cancer immunotherapy through nanotechnology. Nature Reviews Cancer. 2019;19(10):587–602.
[247] Yin H, Kanasty RL, Eltoukhy AA, Vegas AJ, Dorkin JR, Anderson DG. Non-viral vectors for gene-based therapy. Nature Reviews Genetics. 2014;15(8):541–55.
[248] Kranz LM, Diken M, Haas H, Kreiter S, Loquai C, Reuter KC, et al. Systemic RNA delivery to dendritic cells exploits antiviral defence for cancer immunotherapy. Nature. 2016;534(7607):396–401.

Chapter 7
Biomaterials in drug delivery

7.1 Pharmacological concepts

Medicinal compounds (drugs) are used whenever disease or injuries are experienced by the body in order to correct the imbalance caused by these illnesses. Plants and herbs are the first medicinal compounds humans have used to counteract the effects of disease and injuries even without the extensive and systematic knowledge of medicine back then. Medicinal plants were usually eaten, drink, or mashed and put on top of injuries to accelerate healing and removal or alleviation of pain. As modern medicine and pharmaceutical science have progressed, isolation of the active compounds in these medicinal plants has been made and synthetic versions of these organic compounds have been created in the laboratories. In order to maximize their effects, the delivery method to the target organ and tissues must be identified and used. More importantly, pharmacological understanding on the mechanism of action for each drug must be established. Medicinal compounds can be useful or harmful depending on the manner it is taken. It must be considered by researchers exploring the drug delivery field in order to design an effective drug delivery system or platform.

7.1.1 Conventional delivery methods for drug compounds

There are several ways modern medicine/drugs are being delivered (Figure 7.1). Right now, there are established ways of giving drugs which can be classified under two general classes. First is the local drug delivery, which means that the drug is directly administered onto the affected tissue or organ where it is needed. This includes topical administration, local drug release via medical implements (stents, bone implants, and wound dressings) [1–5], and other types of direct injection such as intravitreal injection, intracardiac injection [6, 7], intraperitoneal injection [8], and intracerebroventricular injection [9]. Second is the systemic delivery where the drug enters the systemic blood circulation and affects the whole body [10]. This encompasses oral and parenteral administration of drugs; oral and parenteral administration differs by how the drugs entered the body [11]. As the name suggests, oral administration uses the gastrointestinal tract through the mouth by swallowing, while parenteral administration is any drug entering the body not by swallowing. Examples of this are intravenous, subcutaneous, intramuscular, intradermal injections, implants and patches, any mucosal administration via nasal inhalation of vapor or spray, buccal and sublingual route, rectal suppository, and vaginal route. Different parenteral delivery methods will also give out varied bioavailability. In pharmacology, bioavailability is a subcategory of absorption and is the fraction of an

https://doi.org/10.1515/9783110655377-007

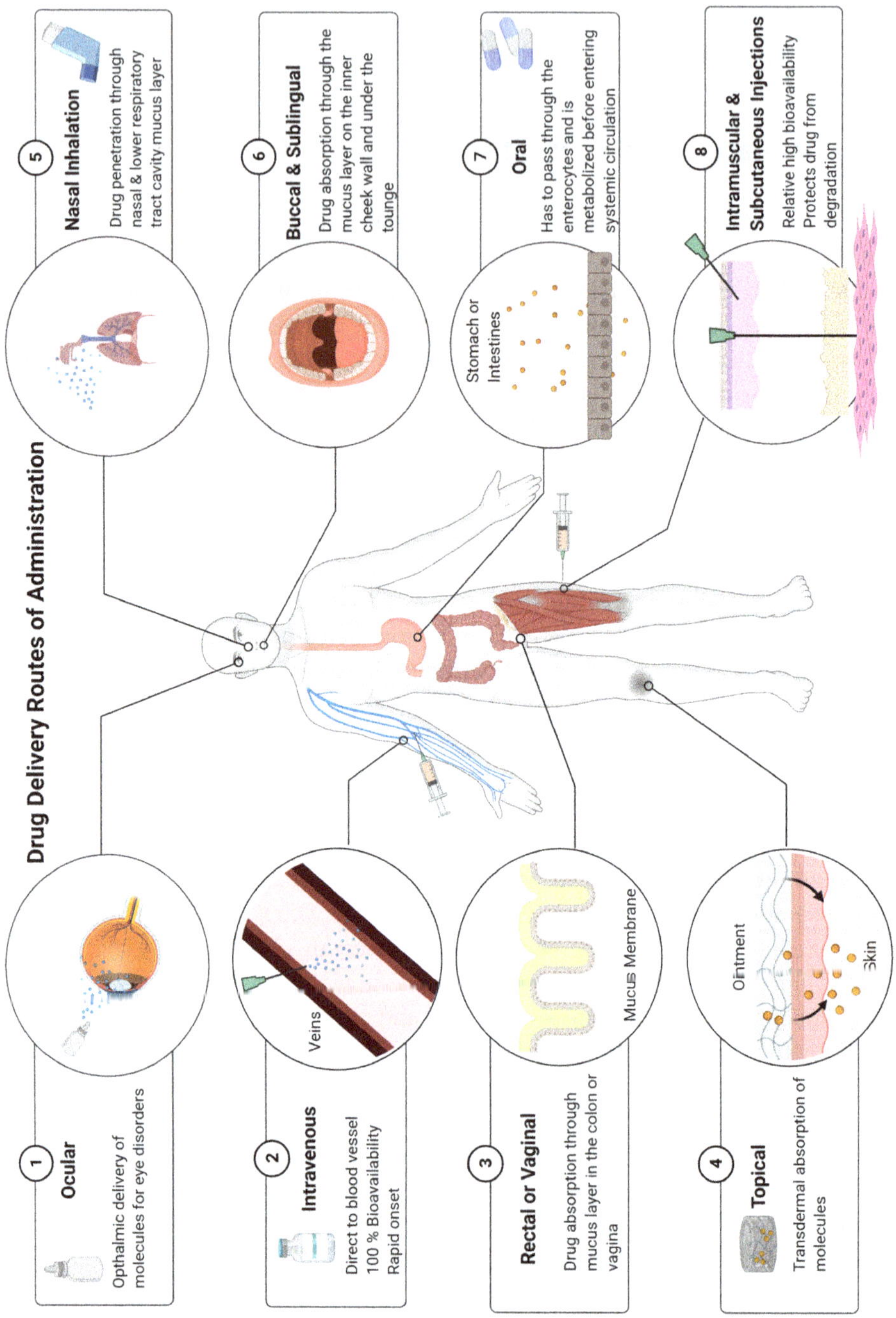

Figure 7.1: Delivery routes used for administering therapeutics inside the body.

administered drug that reaches the systemic circulation. When a medication is administered intravenously, its bioavailability is 100%. However, when a medication is administered via routes other than intravenous, its bioavailability is generally lower than that of intravenous due to intestinal endothelium absorption and first-pass metabolism. Thereby, mathematically, bioavailability equals the ratio of comparing the area under the plasma drug concentration curve versus time (AUC) for the extravascular formulation to the AUC for the intravascular formulation. AUC is used because AUC is proportional to the dose that has entered the systemic circulation.

7.1.2 Drug classifications

Modern drugs can be classified into several classes, the United States Pharmacopeia (USP) currently categorizes a drug or drug component under one of 49 different therapeutic classes:

1. Analgesics, including opioids and non-opioids
2. Anesthetics
3. Antibacterial, including antibiotics
4. Anticonvulsants
5. Antidementia agents
6. Antidepressants
7. Antidotes and antitoxins
8. Antiemetics
9. Antifungals
10. Anti-inflammatory agents, including corticosteroids and nonsteroidal anti-inflammatory drugs (NSAIDs)
11. Antimigraine agents
12. Antimyasthenic agents
13. Antimycobacterial
14. Antineoplastics
15. Antiparasitic
16. Antiparkinsonian agents
17. Antipsychotics
18. Antivirals, including HIV antiretrovirals and direct-acting hep C drugs
19. Anxiolytic (antianxiety) agents
20. Bipolar agents
21. Blood glucose regulators, insulin and other diabetes medications
22. Blood products, including anticoagulants
23. Cardiovascular agents, including beta-blockers and ACE inhibitors
24. Central nervous system agents, including amphetamines
25. Chemotherapeutics
26. Dental and oral agents

27. Dermatological (skin) agents
28. Enzyme replacement agent
29. Gastrointestinal agents, H_2 blockers and proton pump inhibitors
30. Genitourinary (genital and urinary tract) agents
31. Hormonal agents (adrenal)
32. Hormonal agents (pituitary)
33. Hormonal agents (prostaglandins)
34. Hormonal agents, estrogen, testosterone, and anabolic steroids
35. Hormonal agents (thyroid)
36. Hormone suppressant (adrenal)
37. Hormone suppressant (parathyroid)
38. Hormone suppressant (pituitary)
39. Hormone suppressant (sex hormones)
40. Hormone suppressant (thyroid)
41. Immunological agents, including vaccines and disease-modifying anti-rheumatic drugs (DMARDs)
42. Inflammatory bowel disease agents
43. Metabolic bone disease agents
44. Ophthalmic (eye) agents
45. Otic (ear) agents
46. Respiratory tract agents, including antihistamines and bronchodilators
47. Sedatives and hypnotics
48. Skeletal muscle relaxants
49. Therapeutic nutrients, minerals, and electrolytes

With numerous drugs being discovered, the same goes their classification. Thousands of different drug classes and subclasses are being used by different pharmacopeia. In order to create order within this chaos, the World Health Organization created a simplified multidimensional system called the Anatomical Therapeutic Chemicals (ATC) classification system, which reduces drug categories into five levels. The ATC system can classify the active ingredient of a drug under a strict hierarchy so that it is appropriately used and not mistaken for another drug.
- **Level 1:** Describes the organ system the drug treats.
- **Level 2:** Describes the drug's therapeutic effect.
- **Level 3:** Describes the mechanism/mode of action.
- **Level 4:** Describes the general chemical properties of the drug.
- **Level 5:** Describes the chemical components that make up the drug (essentially the chemical name of the drug, such as finasteride or ibuprofen).

7.1.3 Drug-to-drug interactions

The therapeutic effect of one drug can be hindered or reduced by another compound if that is taken at the same treatment regimen. It is important to understand these drug–drug interactions [12]. For example, the chemotherapeutic bortezomib are usually neutralized by phenolics [13] both found in food and nutritional supplements, due to the phenolic competitively binding to the active site of the drug rendering it useless. This is also known as the antagonistic effect drug–drug interactions. On the other hand, some drugs can enhance the other drug therapeutic effect by complimenting their mechanism of action; this is called synergistic or additive effect. The difference between synergistic and additive effects are when the results of two drugs are used together, their effects can be additive and the result is what you expect when you add together the effect of each drug taken independently; synergism happens when combining the drugs leads to a larger effect than expected. This drug combination studies are usually done using the Chou–Talalay method [14–16], based on the median-effect equation, derived from the mass-action law principle, which is the unified theory that provides the common link between single entity and multiple entities, and first-order and higher order dynamics. Drug combination studies are important to determine whether the treatment regimen and at specific dosing can lead to a therapeutic outcome. This is highly applicable in people undergoing chronic treatment such as for HIV, cancer, tuberculosis, and metabolic diseases (diabetes) and others who needs to take a myriad of drugs to maintain.

7.1.4 Adverse and side effects of therapeutic compounds

Many drugs can cause unwanted effects that are not intended during the treatment regimen, and these are called adverse and side effects of drugs [17, 18]. These are unavoidable consequences of taking medicinal compounds and they could vary from slight inconveniences to detrimental. Adverse (high risks to detrimental) and side effects (nuisance effect) are pronounced in systemic delivery where the compound can influence the rest of the body and not just the target tissue or organ. Minimizing adverse and side effects are an important aspect of drug delivery, and there are several ways biomaterials are used in order to do so [19]. Biomaterials can control the release of drugs so that the concentration level of it in the body can be sustained within the therapeutic range. Furthermore, toxicity can be reduced by delivering the compound only at the target tissue. Lastly, understanding more of the mechanisms of action of each active compound inside the body may help counteract the adverse and side effects of a compound.

7.1.5 Therapeutic range

Each therapeutic compound has a set amount of time inside the body when it is at its most effective. This is also set by how much concentration of compound is present in the blood. This is called the therapeutic range, and it varies from compound to compound as each is processed, metabolized, and excreted differently inside the body. A drug compound is usually measured by its half-life, meaning how long for the compound to reduce its concentration in half in the blood serum [20]. Some drugs have longer half-life compared to others due to a longer circulation time giving out a longer effect, while others have shorter half-life [21]. For example in insulin treatment for diabetes, there are types such as rapid-acting [22–24], which starts to work within a few minutes and lasts a couple of hours, while there are long-acting [25, 26], which can work for an entire day, and intermediate between long and rapid called short-acting, which takes about 30 min to work fully and lasts 3–6 h [22, 27]. For a compound to be effective and cause therapeutic effects it must maintain its blood serum concentration at a level that it is not too high or not too low. Each compound has its own set of therapeutic ranges, some can be narrow, and some can be wide. Maintaining this therapeutic range for a longer time is one of the goals for most drug delivery systems in order to reduce the need for frequent dosing [21, 28, 29]. It can reduce the burden on the organs responsible for metabolizing and excreting therapeutic compounds and lead to a better treatment regimen.

7.1.6 Pharmacokinetics and pharmacodynamics

In simple words, pharmacokinetics (PK) is "what the body does to the drug," and pharmacodynamics (PD) describes the intensity of a drug effect in relation to its concentration in a body fluid, usually at the site of drug action [30]. It can be simplified to "what the drug does to the body." PK can be further subdivided into four processes as a drug is taken into the body. These processes or phases are called: absorption, distribution, metabolism, and excretion (ADME) [31]. The four steps are: Absorption, which describes how the drug moves from the site of administration to the site of action, Distribution of the drug happens uniformly inside body through the vascular system, except for tissues that has natural physiological barriers that limits or controls passage of substances (i.e. blood brain barrier). Metabolism of the drug is its conversion into more water-soluble metabolites which prepares it for excretion. The excretion process after this conversion happens through numerous pathways such as sweating, respiration, urination etc. but the most common excretion pathways is through urination as the kidneys is the primary organ that filters the blood for water soluble metabolites to which the drug has been converted into. The effect of the drug inside the body is determined by the drug's PD properties which is characterized its chemical/molecular structure. Properties such as number of active sites and its cell-receptor susceptibility, and reactivity to other substances found inside the body (a.k.a pharmacodynamic interactions). Different classes of drugs have different PK and PD properties inside the body. To give an example, protein based therapeutics such as insulin, has to be administered exclusively via subcutaneous injection as its absorption inside the gut is impossible, whereas smaller peptides such as octreotide can have some sort of absorption in the stomach and intestines given the right conditions. Although they are the same peptide class, but their molecular structure and inherent size differences affect their PK performance. As to their PD, insulin normalizes the blood glucose level in the body by facilitating the transport of glucose molecules from the blood stream towards the cells for metabolism. Octreotide on the other hand relieves flushing in the face and diarrhea associated with GI tumors by binding to somatostatin receptors which in turn leads to smooth muscle contraction in the blood vessels. Lastly, limitations on a drug's PD or PK performance can be solved by the use of biomaterials, for example, insulin is now being explored to be given orally with the use nanoparticle technology.

7.2 Biomaterials' role in drug delivery

Taking therapeutic compounds, no matter which class, has its own limitations. As the body has its own physiological ways to maintain homeostasis, any compound taken in whatever method will be metabolized and excreted in the same manner. There are also sets of physiological and functional barriers that drug compounds encounter limiting their therapeutic effects [32–35]. Drug administration requires multiple repeated dosing to maintain a therapeutic dose while under the treatment regimen. Generally, these methods lead to a rapid drug-level reduction in the blood plasma, due to it being metabolized, degraded, or transported into other tissues which it is not needed. This in turn creates a reduced therapeutic efficacy as well as adverse or side effects in the body. Here, biomaterials are employed in order to improve drug delivery methods and increase their PK and dynamics, reduce adverse and side effects, and ultimately increase therapeutic effects. Biomaterials can serve as a platform to create a sustained or controlled release of therapeutics inside the body. Drugs are incorporated into the physical and chemical structure of the biomaterials and are subsequently released via multiple mechanisms. This could be from degradation of the biomaterial [36], surface erosion [37], phase transition [38], response to stimuli [39, 40], and passive diffusion [41]. All these mechanisms of drug release follow two kinetics. First-order release kinetics releases the unequal concentration of drug from the platform, thus having a lowering amount over time [42]. This is also known as sustained release mechanism and is one of the most common drug release mechanisms for most biomaterial drug delivery platforms. The second type is called zero-order release kinetics [43] that delivers drugs at rates that do not change with time, maintaining the therapeutic dose and avoiding inefficiencies of the drug concentration highs and lows of conventional drug formulations. This is known as controlled drug release systems and is more difficult to achieve and engineer compared to sustained release systems. Controlled release systems give more benefits to drug delivery as it can minimize the need for more frequent dosing, and it maximizes the amount of drug being utilized inside the body. Furthermore, it can reduce the related toxicities of the drug by maintaining only at the therapeutic range. Controlled release systems also have the capability to be able to target specific tissues by exploiting ligand-receptor specificity in cells. This can further reduce unwanted effects on irrelevant tissues in the body and can increase its therapeutic effects with minimal dosing requirements as all the drug is concentrated in the target site. This is highly beneficial on all therapeutic areas but most specifically in chemotherapy as these classes of drugs can harm normal cells as well.

Creating controlled release drug delivery platforms using biomaterials is a daunting task, and it involves multiple disciplines. There are several design considerations for creating such drug delivery platforms. First is the incorporation of sufficient drug concentration within the platform, as having too little will require multiple dosing or insufficient therapeutic range. Second is protecting the drug from degradation from the environment and within the body, ensuring that the bioactivity or effectiveness of the drug is not hindered by the biomaterial. Third is controlling the release rate over time. This will depend on the range of the treatment and can be as short from several days to months and years. Lastly is to ensure that the biomaterial used for the platform is not causing degradation products that can harm the body. These degradation products should also not interfere with the bioactivity of the drug payload and must be able to be excreted from the body without causing harm.

There are several classes of biomaterials that are used in drug delivery, but primarily, polymers have been the main material of choice for researchers to serve as the drug delivery platform that can enact multiple types of release kinetics depending on the desired application [36, 44, 45]. Polymers have the special properties of being mostly biocompatible but also carry multiple types of therapeutics, whether chemically or physically. Drugs can be encapsulated [46, 47] or entrapped [48, 49] in a polymeric network via two means. Physical entrapment is when the drug does not interact with the polymer, except for a few hydrogen bonding between the drug and the polymeric drug delivery platform, while chemically carrying polymers use reversible covalent or ionic bonds and stimuli-responsive grafting or conjugation using their respective functional groups [50, 51]. These two types of drug carrying capabilities have their own advantages. While physical entrapment can carry more drug in weight percentage ratio compared to chemically grafted drugs, it may lack the capability to release the drug using specific physiological conditions and may only rely only with diffusion or polymeric erosion to release its drug payload. This can cause burst release and possible related toxicity due to the uncontrolled manner of delivery [52]. For drugs that are chemically grafted towards a polymeric carrier, specific physiological conditions must be met for the grafting to be reversed. This grafting or conjugation reversal means the liberation of the drug from the polymeric carrier, thus releasing the drug in a controlled manner.

Other classes of biomaterials that are nonpolymeric are also being used for drug delivery, albeit their utility is not the same as with polymers, as some of them are deemed less versatile than polymeric materials in terms of providing a controlled delivery of drugs. Inorganic materials such as bioceramics [53, 54], and metals [55, 56] can also serve as drug delivery platforms and can carry drugs using physical entrapment. Examples of this are mesoporous silica that can hold a high amount of drug due to its high surface to volume ratio [54, 57], bioceramic implants such as hydroxyapatite that can entrap growth factors and antibacterial agents onto their porous network [54, 58], and gold nanoparticles that can carry therapeutics using chemosorption due to its special surface properties [59, 60].

7.2.1 Biomaterials in systemic drug delivery

There are several drug delivery routes where biomaterials are used. From local to parenteral drug delivery, researchers have found applications for biomaterials. This is due to the inherent advantages and disadvantages of conventional drug delivery methods, where biomaterials can both enhance and reduce, respectively. In systemic drug delivery, for example, injection rout (e.g., IV, IM, and SQ) can be applied to most of the drug classes, have rapid onset or effect, and have 50–100% bioavailability. However, injectable drugs are rapidly cleared from the body, thus requiring frequent dosing. It is also known that frequent injection of drugs has its risks such as septicemia [61–63] (infection caused by tainted needles), and phlebitis [64, 65] (inflammation of the vessels caused by frequent injection). Furthermore, it can cause liver toxicity [66] and immunogenicity [67, 68] as the drugs are directly routed into the systemic circulation at high concentration. Another example is with oral delivery. Drugs taken orally typically have high adherence due to their simplicity of administration, but the harsh environment in the gastrointestinal tract can limit which classes of drugs can be taken through this route. Only drugs with high stability under variations of pH and enzymatic degradation can be taken orally. Even if the drugs can survive the GI environment, the bioavailability is typically low due to the natural physiological barrier in the gut and due to the hepatic first-pass effect [69, 70]. The hepatic first pass metabolism affect drugs that are administered orally in contrast to other routes (Figure 7.2). Drugs that are absorbed from the intestine get metabolized first into the liver before reaching the systemic circulation. Also, for many drugs, much of the dose is reduced by xenobiotic metabolism (bacteria in the gut) before reaching the tissues [71]. Since some drugs are metabolized by gut flora or digestive enzymes, the first-pass effect refers to the combined effect of metabolism by the liver and in the gut further reducing their bioavailability.

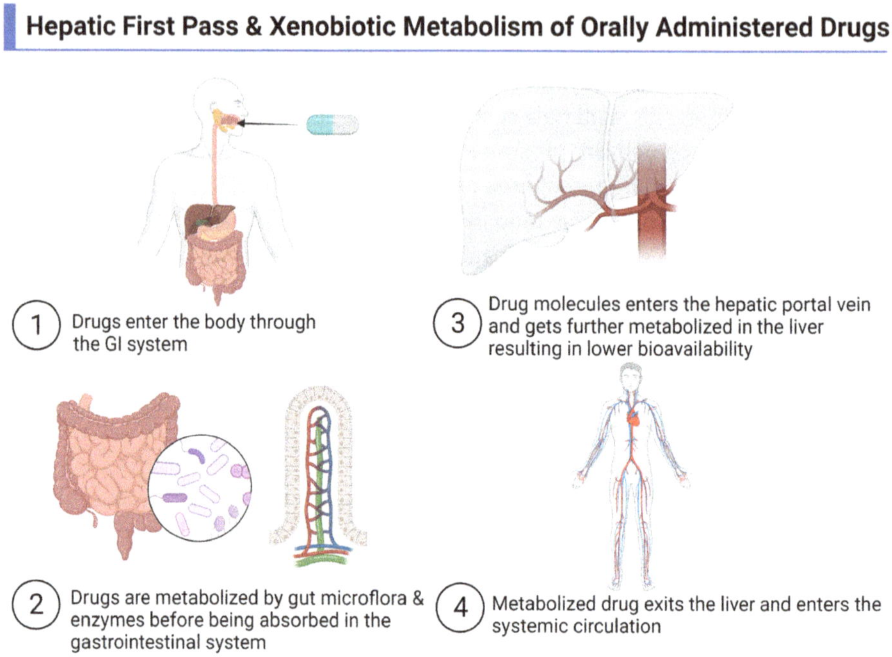

Figure 7.2: Hepatic first pass and xenobiotic metabolism reduces the bioavailability of orally delivered drugs.

7.2.1.1 Biomaterials used for intravenous drug delivery: strategies in increasing pharmacokinetics

Here, several biomaterials' classes and subclasses have been explored to mitigate these disadvantages. In injectable drugs, nanoparticles have been explored to increase PK properties of drugs that are otherwise easily excreted in the body. Nanoparticles when introduced directly to the systemic circulation via IV injection increase the PK properties of drugs by allowing them to circulate for a longer time. They do so by protecting the drug molecule from the body's natural excretion and metabolic pathways. Depending on the type and design of nanoparticle, it can protect the drug molecule by physical encapsulation or chemical engraftment. Drugs and their metabolites are easily excreted inside the body, primarily by the kidneys through the urine, and other methods such as sweating. This drug elimination from the body is determined by its half-life, and it depends on how quickly the drug is eliminated from the plasma depending on its mechanism of action. It can be eliminated from the body, or it can be translocated to another body fluid compartment such as the intracellular fluid or it can be destroyed in the blood. The removal of a drug from the plasma is known as clearance, the distribution of the drug in the various body tissues is known as the volume of distribution, and both PK parameters are important in determining the half-life of a drug. The advent of nanoparticle drug delivery systems allows for the delay of these processes inside the body (Figure 7.3). The biomaterial used in creating these nanoparticles, whether it is based on organic materials such as polymers, peptides, lipids, and inorganic metals and ceramics or a combination of them, can avoid the metabolism and excretion of its drug payload until it is delivered onto the target site. Physical entrapment of drugs is usually done in nanoparticles by having cavities of physical spaces where the drug molecule can be housed. Mesoporous silica and ceramic nanoparticles are examples of this. Usually these are combined with a polymeric release component that will allow for a targeting ligand to be attached and/or a stimulus triggering mechanism to be incorporated. Micelles and liposomes are closed bilayer structures spontaneously formed by hydrated phospholipids also follows physical entrapment of drug molecules inside their cavity, as they form nanoscaled membranes that can house and protect the drug payload [72, 73]. The surfaces of these liposomes can also be functionalized with various targeting ligands to selectively bind to cells of interests. Chemical engraftment follows multiple chemical modifications of the polymer-drug, protein-drug, or lipid-drug conjugates and examples of this are dendrimers, supramolecular surface modified polymeric and inorganic metal nanoparticles. However, it must be taken into consideration that the design of the nanoparticle can lead to opsonization [74–76] where the whole nanoparticle can be rendered less effective by the body's immune system, because even though nanoparticles can protect the drug payload from the body's rapid metabolism and excretion, nanoparticles are also targeted by the immune system due to the formation of the protein corona around nanoparticles [77, 78]. This process is dependent on several factors, including nanoparticle size, surface charge, hydrophobicity, and surface chemistry. Thus, further modifications on the nanoparticle must be done in order to evade this process. One effective strategy researchers employ is to modify the nanoparticles' surface charge that can avoid the adsorption of plasma proteins [79, 80], including serum albumin, apolipoproteins, complement components, and immunoglobulins. Positively charged nanoparticles tend to adsorb more serum proteins than negatively charged nanoparticles [81]; this is due to the electrostatic interaction of proteins toward the nanoparticle's surface [82]. Proteins in general are negatively charged, thus making them bind with affinity toward positively charged nanoparticles. Adding negatively charged surfactants and functionalizing with poly(ethylene glycol) (PEG), or PEGylation can mitigate this effect, thus increasing the PK of the nanoparticle and subsequently its drug payload. However, it must be understood that nanoparticles can enter the cell membrane with higher efficiency if they are positively charged. So even if the PK of the nanoparticle are increased, its PD or its effects toward the target site might be lessened. But all these depend on the manner of drug delivery; if the payload is needed to be released intracellularly, then having a better PD property on the nanoparticle might be desired, but if the drug is needed to be released extracellularly, then having an improved PK is a better strategy.

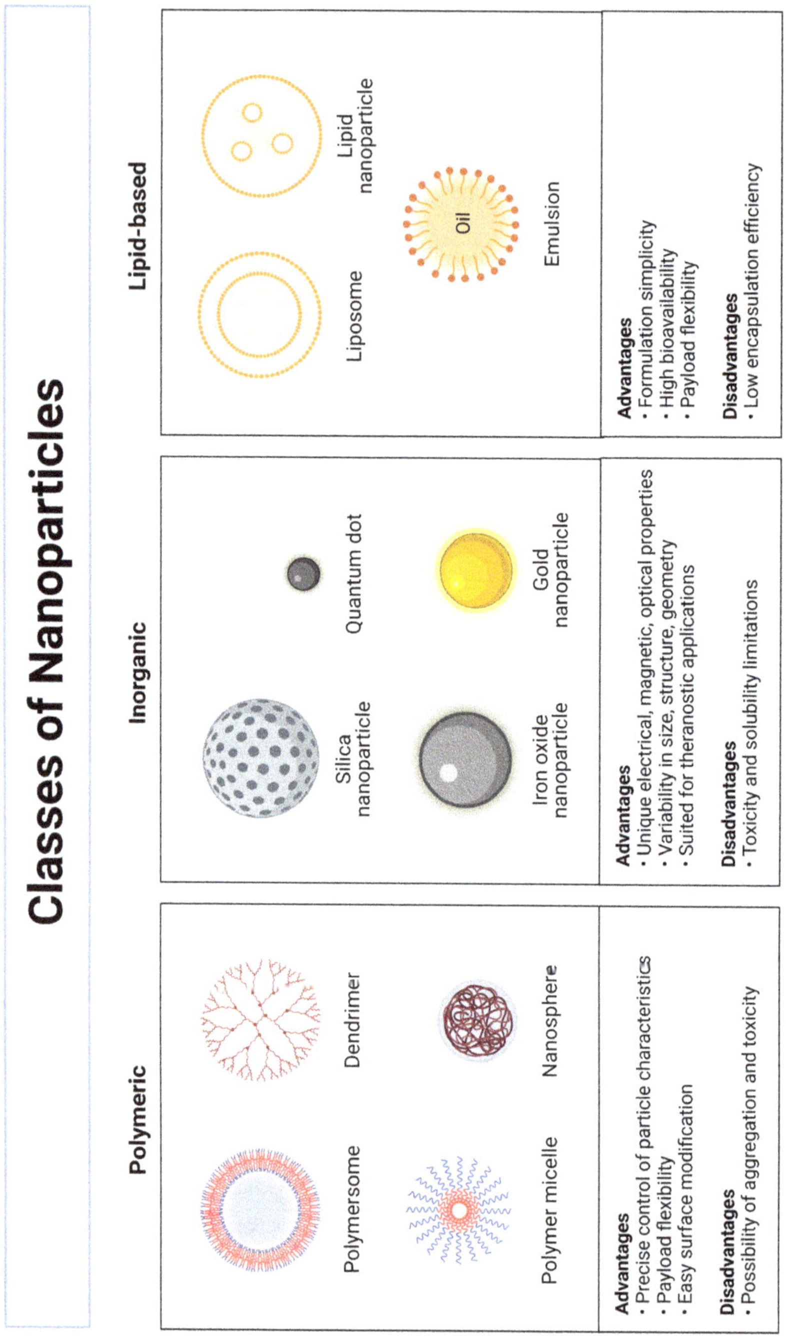

Figure 7.3: Classes of nanoparticles used as drug delivery systems.

Other more "active stealth" strategy can enhance the PK of a nanoparticle by attaching ligands that instruct or fool immune cells not to uptake and eliminate them [83–85]. This is done by attaching CD47 "self" peptides on the surface of nanoparticles to avoid phagocytic clearance [86–88]. CD47 are in-silico designed and synthesized and subsequently attached on the surface of 160 nm nanobeads. The surface modified nanobeads were administered to NSG (nonobese diabetic (NOD) severe combined immunodeficient IL2rynull) mice; the peptides significantly increased the nanoparticle circulation by delaying phagocytic clearance by the liver and spleen. Furthermore, nanoparticles functionalized with the self-peptide showed greater accumulation in A549 tumors within 10 min of administration, with incorporation of paclitaxel within the nanobeads resulting in substantial tumor shrinkage compared with the conventional formulation of the drug [89, 90].

Another variation of creating stealth nanoparticles for systemic evasion of opsonization is with the use of cell membranes isolated from immune cells (leukocytes) [91, 92]. Upon surface functionalization with leukocytic membranes, particles showed about a tenfold decrease in protein (IgG and albumin) adsorption to the surface. Consequently, the biomimetic coating resulted in substantially less particle uptake in murine J774 macrophages (~75% decrease) and human THP-1 phagocytic cells (~50% decrease), especially when the coating was from the same donor species. When the platform was examined in murine models, functionalized particles accumulated to a lesser degree in the liver after systemic administration. Time-dependent accumulation of coated particles was shown to be independent of the phagocytic effects of Kupffer cells, with particles primarily associated with liver endothelium. This reduction in MPS uptake led to an enhanced accumulation of particles (about twofold) in a murine model of melanoma.

A unique way of increasing PK properties of nanoparticles is taking advantage of ubiquitous protein corona that forms in their surfaces to target specific cells. As discussed, protein adsorption may increase the phagocytic affinity of nanoparticles toward immune cells, but when it comes to increasing the translocation of nanoparticles in the blood–brain barrier (BBB) (Figure 7.4), this protein corona formation may be taken advantage of as they can mimic lipoproteins that are usually easily translocated in the BBB. Using poly (butyl cyanoacrylate) nanoparticles coated with polysorbate 80, researchers have shown the effective adsorption of apolipoprotiens to the polysorbate 80 surface and enable it to traverse the BBB in vivo, through low-density lipoprotein–mediated endocytosis pathway [93]. Furthermore, this strategy can also target immune cells such as macrophages that are involved in inflammatory diseases, such as rheumatoid arthritis, cancer, and atherosclerosis. It is a way of using nanoparticle limitations to its advantage, demonstrating that understanding physiology can lead to better utilization of biomaterials.

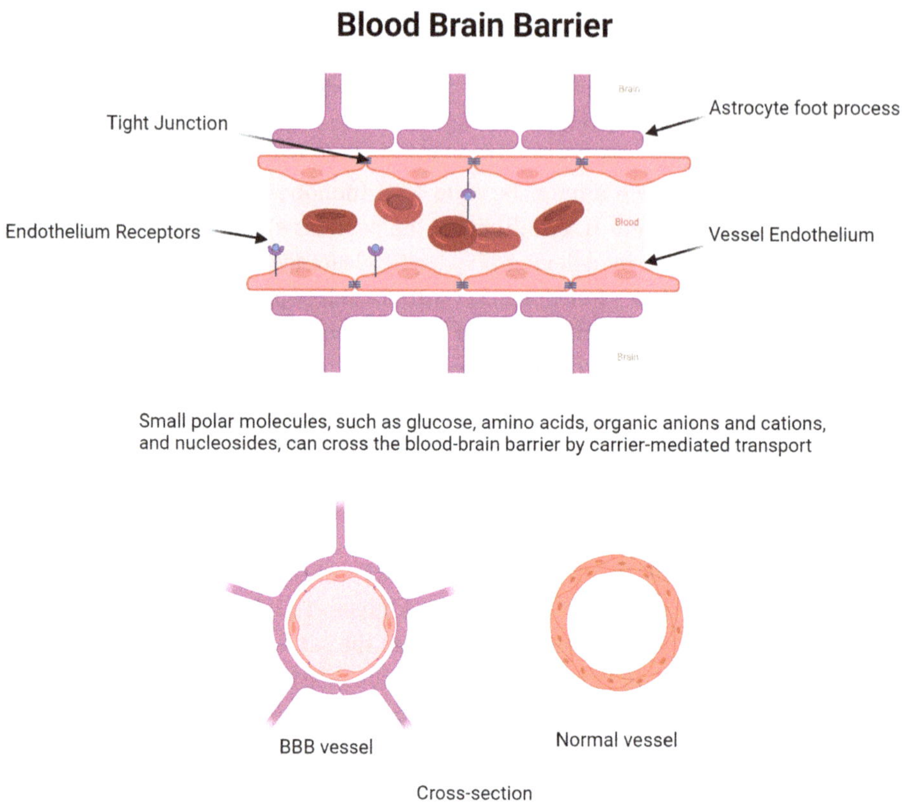

Figure 7.4: The blood–brain barrier protects the brain from substances that can harm the brain. Compared to normal blood vessels, the blood vessels in the brain are tightly packed and surrounded with astrocyte foot process that protects the leakage of any harmful substances or pathogens.

7.2.1.2 Biomaterials used for oral drug delivery: traditional formulation strategies (e.g., tablets and capsules)

In almost all drug delivery methods, oral delivery has been proven to be the simplest and pose the highest adherence toward patients. This has been the preferred way of delivery for most drugs that can survive and be absorbed in the gut. However, not all drugs can be administered orally due to the harsh environment in the gastrointestinal system. Chemically stable drugs can survive this condition, but they can also suffer from lowered pk and pd effects due to the natural physiological barriers present in the GI. First, there are the enzymes that will try to digest all large molecules that are introduced in the GI; then the acidic conditions in the stomach to further reduce the chemical structure of the materials into its smaller components. The GI also has a very selective cell epithelial barrier (Figure 7.5) [94]. Small portions of drugs usually start being absorbed in the gastric epithelium. The anatomy of the gastric epithelial cells is

Figure 7.5: The human gut environment made up of various enterocytes, xenobacteria, and immune cells. This helps the gut absorb nutrients from digested food and protects the body from entry of pathogens. The xenobacteria helps in maintaining balance in the gut by helping digesting food and secreting metabolites and neurotransmitters that can help the brain function better. The enterocyte is also a potent physiological barrier that only allows small molecules to pass through, making large molecule therapeutics such as peptides unsuitable for oral delivery.

unique as they consist of folded branching, tubular gastric glands that reach deep into the muscularis mucosa [95–97]. In the acid-secreting mucosa, surface mucous cells, mucous neck cells, acid-secreting parietal cells, and pepsinogen-secreting chief cells all reside in characteristic locations. These cells start chemically digesting macerated food from the mouth and the esophagus, and the same goes with most drugs that are introduced orally. However, the bulk of absorption happens in the intestines as the rate for gastric emptying limits the absorption rate of certain drugs in the stomach. The intestinal epithelium constitutes barrier surfaces that separate mammalian hosts from the external environment. Furthermore, gastrointestinal tract is the largest of these barriers and is specially adapted to colonization by commensal bacteria that aid in digestion and markedly influence the development and function of the mucosal immune system.

In these regards, targeting certain areas in the GI for absorption can be done using enteric coatings. Enteric coatings can enable drug dissolution only at specific sites of the GI tract [98, 99]. pH-responsive polymers are used as enteric coatings as they utilize the pH gradient present in the alimentary tract [100, 101]. The stomach has a mean pH of 1–3, while the duodenum and the rest of the intestines may vary from 5 to 7. This pH gradient has been utilized to allow where the drug can be absorbed. This strategy can also protect pH-sensitive drugs from being destabilized by the enzymes and acidic condition in the stomach, prevent gastric distress/ulceration or nausea due to irritation caused by certain drugs, deliver drugs that are absorbed more efficiently in the small intestine to their primary absorption site in their most concentrated form, and provide a delayed release component to repeat action tablets.

Several classes of polymeric materials have been used as enteric coatings. Polymers with isoelectric points corresponding to the pH of its surrounding can be used for this pH-responsive enteric coating applications [102]. The isoelectric point (pI) of the polymer will determine its site of dissolution in the GI. pI is the pH at which a particular molecule carries no net electrical charge, thus limiting dissolution of the polymer. At a pH below a polymer's pI, that polymer will carry a net positive charge; at a pH above its pI, the molecule will carry a net negative charge. Key properties of enteric coating materials must include: resistance to gastric fluids, ready susceptibility or permeability to intestinal fluids, high compatibility with other film-coating additives and the tablet being coated, capacity to form continuous film with adequate mechanical properties, ease of application without special equipment, ability to be printed or to allow film to be applied to debossed tablets, nontoxic with no pharmacologic activity, stable alone and in coating solution.

Enteric polymers are added to the tablet formulation coating or be made as a capsule to house the drug with a suitable plasticizer to allow for uniform coating and film formation [107, 108]. A plasticizer is a substance that is added to a material to make it softer and more flexible, to increase its plasticity, to decrease its viscosity, or to decrease friction during its handling in manufacture [107, 109]. There should be chemical similarities between the polymers and its plasticizers. It is supposed that the function of the plasticizer is that its molecules are embedded between the individual polymer chains and disintegrate the polymer–polymer interactions. One of the most important properties of the polymers is known as the glass transition temperature (Tg), which measures chain mobility. This is a temperature at which a polymer transforms from a hard, glassy material to a soft, rubbery material. Plasticizers lower the glass transition temperature of the polymers, making it easier to handle and allow for film formation during the coating process.

7.2.1.3 Biomaterials used for novel oral drug delivery technologies (nanoparticles and oral devices)

Novel drug delivery methods applied in oral route are also being researched upon as there is the emerging need for drugs that are conventionally administered via injection route to be shifted to oral administration. It has been mentioned that oral delivery is the most preferred way of delivering drugs in the body due to its simplicity and high adherence. However, the harsh environment inside the gut limits what type of drugs can be administered via this route. For example, protein-based drugs such as insulin, octreotide, and other long- and medium-chained therapeutic peptides are unable to be absorbed in the body via oral route as they are denatured and digested into smaller amino acids before they can traverse the gastrointestinal epithelium. Thus, certain techniques using biomaterials are employed to increase bioavailability

Table 7.1: Example of polymers used as an enteric coating.

Polymer	Description	Ref.
Cellulose acetate phthalate (CAP)	CAP is insoluble in water, alcohols, hydrocarbons, and chlorinated hydrocarbons but soluble in ketones, ethers, alcohols, esters, and in certain solvent mixtures. It exhibits rapid dissolution at a pH > 6 and is relatively permeable to moisture and gastric juices. Due to its high moisture permeability, CAP is susceptible to hydrolytic decomposition.	[103]
Cellulose acetate trimellitate (CAT)	Trimellitic anhydride contains an additional free carboxyl group over that of phthalic anhydride, and hence CAT contains a greater concentration of acidic groups for a given degree of substitution than CAP rendering it more soluble in aqueous media. Also, the pK_a of CAT is between 4.1 and 4.3 which is slightly lower than CAP. With a relatively low pK_a value and greater functional group concentration, CAT is the most soluble enteric cellulose derivative with the onset of dissolution occurring at pH 4.7–5.0. This useful property makes CAT ideal for targeted drug release to the proximal regions of the small intestine.	[103]

Table 7.1 (continued)

Polymer	Description	Ref.
Hydroxyl propyl methyl cellulose phthalate (HPMCP)	HPMCP is characteristically insoluble in gastric fluids but swellable and rapidly soluble in the upper intestine. It may be plasticized with diethylphthalate, acetylated monoglyceride, or triacetin. Mechanically HPMCP is a more flexible polymer and on a weight basis will not require as much plasticizer as CAP or CAT.	[104]
Hydroxyl propyl methyl cellulose acetate succinate (HPMCAS)	HPMCAS is insoluble in acidic media, yet soluble in neutral pH according to ionization of free carboxyl groups on the polymer backbone. It has been demonstrated to be more chemically stable than CAP and HPMCP as indicated by a substantial reduction in the evolution of free acid when stored at 60 °C and 100% relative humidity (RH).	[105]
Polyvinyl acetate phthalate (PVAP)	The onset of aqueous dissolution of PVAP begins at a pH of about 5.0 allowing for enteric release as well as the potential for targeted drug release to the proximal small intestine. Although structurally similar to CAP (containing the dicarboxylic phthalic acid in a partially esterified form), PVAP is relatively more stable to hydrolysis than CAP due to its lower moisture permeability.	[106]
Methacrylic acid copolymers	This group of polymers is widely used for enteric coating applications as they contain free carboxylic acid groups that are ionized whenever the pH of the environment exceeds 5.5. Several different types of Eudragit polymers with enteric release capabilities are commercially available in a wide range of different physical forms (aqueous dispersion, organic solution, granules, and powders). The pH at which these polymers dissolve is dependent on the content of the carboxylic acid in the copolymer. Methacrylic acid methylmethacrylate copolymers (Eudragit L and S), and methacrylic acid ethyl acrylate copolymer (Eudragit L30D) are the preferred choice of coating polymers for enteric formulations. They allow targeting of specific areas of the intestine.	[107]

Table 7.2: Classification of plasticizers.

Types	Plasticizers	Ref.
Polyos	Glycerol (glycerin), propylene glycol, polyethylene glycols (PEG)	[110]
Organic esters	Phthalate esters (diethyl, dibutyl), dibutyl sebacate, citrate esters (triethyl, acetyl triethyl, acetyl tributyl), triacetin	[110]
Oils/glycerides	Castor oil, acetylated monoglycerides, fractionated coconut oil	[110]

of orally administered peptide-based therapeutics. Take insulin for example; when it was first discovered by Sir Frederick G Banting, Charles H Best and JJR Macleod at the University of Toronto in 1921 and subsequently purified by James B Collip, it was first administered orally to type I diabetic patients but to no therapeutic success, before they found a better way to administer it via subcutaneous injection. This is due to the insulin being denatured in the GI and subsequently broken down into smaller amino acid components. Until now, the only way insulin can be administered is via subcutaneous injection which poses low adherence due to daily needle handling.

Oral peptides have seen a recent advancement in the clinic as the very first therapeutic peptide GLP-1 (semaglutide) from Novo Nordisk has been successfully marketed [111]. GLP-1 has been formulated with a permeation enhancer salcaprozate sodium (SNAC) that allows for a transient permeability of the gastric epithelium and increasing its pH to neutral conditions to both allow for GLP-1 to be absorbed and be protected from the stomach's acidic condition [111, 112]. However, this technique only poses less than 1% bioavailability and strict dosing regimen. Also, this permeation enhancer does not work consistently with other therapeutic peptides. Thus, further research is being conducted to find ways to increase bioavailability of orally delivered peptides.

Nanotechnology has also been employed in oral drug administration of therapeutic peptide payloads. Nanoparticles can be designed to house therapeutic peptides to protect them from enzymatic degradation and acidic environment in the gut. This is with the intent to allow an efficient translocation of the nanoparticle from the gastrointestinal epithelium toward the systemic circulation.

There are four mechanisms by which nanoparticles can traverse the epithelial cell lining in the gut (Figure 7.6) [113]. The transport mechanisms are receptor-mediated transport [114, 115]; paracellular transport [117]; transcellular transport [118]; and M-cell-mediated transport (i.e., phagocytosis by M cells) [119]. Receptor-mediated transport utilizes different surface receptors present on the enterocytes that facilitates endocytosis of the nanoparticle. Paracellular transport is the route where nanoparticles open the tight junctions in between the enterocytes to allow translocation toward the systemic circulation. Transcellular transport is the route of nanoparticle translocation when they are directly uptaken by the enterocytes using endosomal trafficking and sending them to the basal side of the cell and eventually the systemic circulation. Lastly, the M-cell-mediated transport utilizes the macrophages present along the gastrointestinal tract (aka Peyer's patch cells) to uptake nanoparticles and translocate them toward the lymphatic system and consequently the systemic circulation.

There are various ways by which nanoparticles can achieve these inherent transport mechanisms in the GI.

In the human digestive system, mucus is used as a lubricant for food materials that must pass over membranes. Mucus is extremely important in the gastrointestinal tract as it forms a protective layer [127] all throughout the alimentary canal that helps reduce inflammation by mitigating bacterial interaction that may be present along the membrane layer of the GI. Whereas the mucus layer found in the gastric

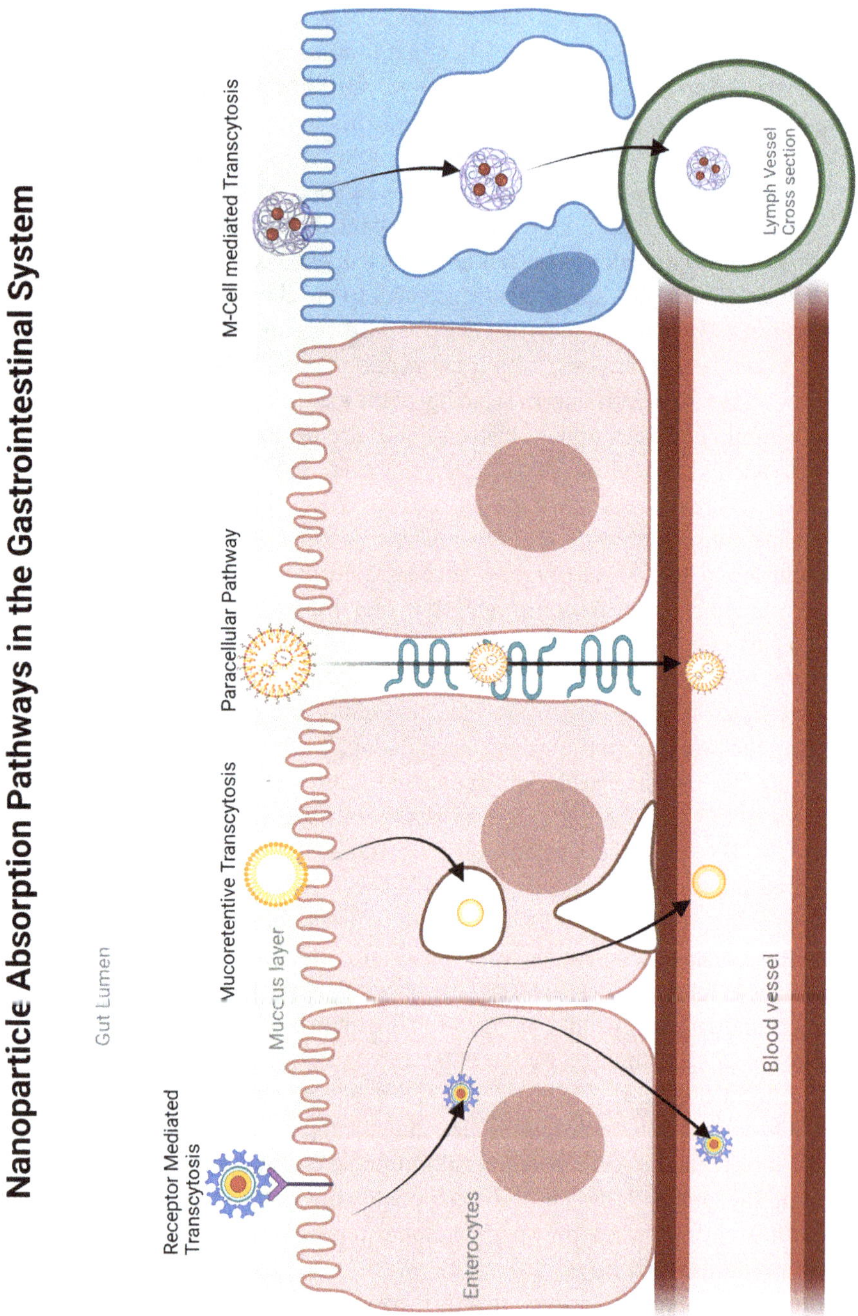

Figure 7.6: Nanoparticle drug delivery systems can pass through the GI lining via four pathways: receptor-mediated transcytosis, mucoretentitive transcytosis, paracellular pathway, and M-cell-mediated transcytosis.

Table 7.3: Strategies used in nanoparticle formulation for oral delivery of peptides.

Strategy	Description	Example of materials used	Ref.
Mucoadhesion	Increasing retention time in the surface of enterocytes therefore increasing the likelihood of absorption thru transcellular uptake	Chitosan, alginate, polyacrylic acid, PLGA, thiomers	[120]
Cationic or charge modification	Increases enterocyte uptake Via amphiphilic interaction of the nanoparticles to the cell membrane (lipidation, or PEGylation)	Lipid suspension, liposomes, SEDDS Submicroemulsion, HDV-I, SEDDS	[121]
Cell-penetrating peptides	Increases enterocyte uptake via amphiphilic interaction of the nanoparticles to the cell membrane (lipidation, or PEGylation)	Penetratin, oligoarginine, transportan, TAT	[122–124]
Receptor-ligand	Attaching targeting ligands to increase absorption in target cells in the GIT.	Transferrin, immunoglobulins Biotin, lectins, folate, bile salts Vitamin B_{12}	[125, 126]

mucosa protects the stomach lining from the highly acidic environment within it [128]. The mucus that is secreted inside the GI mucous membrane is made from mostly water (95%), the mucin secretions from the goblet cells, and the submucosal glands (2–3% glycoproteins), proteoglycans (0.1–0.5%), lipids (0.3–0.5%), proteins, and DNA. The major mucins secreted are called MUC5AC and MUC5B; these are large polymers that give the mucus its high viscoelastic properties [129]. MUC5AC is the main gel-forming mucin secreted by goblet cells, in the form of threads, and thin sheets [130, 131]. MUC5B is a polymeric protein secreted from submucosal glands and some goblet cells, and this is in the form of strands [132]. Micro- and nano-sized particles capable of mucoadhesion have been one of the strategies nanotechnologists are exploring to allow for increased bioavailability of therapeutic peptide payloads that are delivered orally. Mucoadhesion is commonly defined as the adhesion between two materials, at least one of which is a mucosal surface. This is achieved by using a polymeric material that has a unique physicochemical property that allows for high adhesion interaction specially on the mucosal tissue inside the body. Since the GI is lined with this mucous membrane, this strategy can be a way to increase the retention time of nanoparticles loaded with drugs onto the surface of the enterocytes and subsequently increase their chances to be uptaken. Nanoparticles made up of chitosan, alginate, polyacrylic acid (PLA), polylactic glycolic acid (PLGA), and thiomers exhibit mucoadhesive properties due to the presence of functional groups or moieties that can interact with the mucus layer of the GI. These functional groups can be in the form of carboxyl, hydroxyl, or catechol moieties. The adhesion process of the nanoparticles can be described by various mucoadhesion theories [133, 134]. First is the diffusion theory where the polymeric chains from the mucoadhesive particles interpenetrate into glycoprotein mucin chains and reach a sufficient depth within the opposite matrix to allow formation of a semipermanent bond. Second is the electrostatic theory where transfer of electrons occurs across the adhesive interface and adhering surface. This results in the establishment of the electrical double layer at the interface and a series of attractive forces responsible for maintaining contact between the two layers. Third is the fracture theory that describes after an initial contact between two surfaces, the materials adhere because of surface forces acting between the chemical structures at the two surfaces. When polar molecules or groups are present, they reorientate at the interface. Chemisorption can occur when adhesion is particularly strong. The theory maintains that adherence to tissue is due to the net result of one or more secondary forces (van der Waal's forces, hydrogen bonding, and hydrophobic bonding).

However, just pure localization or retention of the nanoparticles in the mucosal layer does not constitute a successful uptake and translocation from the enterocytes. There are certain modification considerations that need to be addressed if nanoparticles can pass through the gastrointestinal epithelium. One strategy is through the modification of the surface charge of the nanoparticle to increase enterocyte uptake via amphiphilic interaction of the nanoparticles to the cell membrane [135]. Cationic liposomes are great candidates for this application as they can create favorable interactions with the negatively charged surface of the cell membrane. Furthermore, adding PEG, otherwise known as PEGylation, can also modify the nanoparticle's surface charge to be more cationic [136]. It also creates a surface that degradation enzymes may find difficult to interact with. The initial interaction between the nanoparticles and the cell membrane may dictate whether the nanoparticles can be uptaken via phagocytosis or endocytosis [137]. But it must be noted that there is a contradictory property requirement for having mucoadhesion and cationic surface charge [120, 138]. Thus, strategies of having a charge reversal property depending on the presence of specific enzymes, changes in pH, or redox agents have also been investigated [139–142]. This charge reversal strategy allows for bi-functional properties to be incorporated with a single nanoparticle platform. Mucoadhesion, for example, requires a negatively charged functional moieties to happen, but this limits the interaction of the nanoparticle toward the cell membrane of the enterocytes. Eventually, when the proper environmental stimuli condition is met, then the surface charge property can be reversed into a cationic surface leading to a much efficient interaction between the particles and the cell membrane of the enterocytes.

Strategies for surface charge modification such as introducing cationic cell-penetrating peptides (CPP) in nanoparticles have also emerged as versatile tools to facilitate efficient oral drug delivery [122, 143]. These small peptide sequences are coupled to a biologically active cargo and can transport it across the enterocyte's plasma membrane. CPPs are divided into three classes depending on their amino acid sequences [124, 144]: cationic, amphipathic, and hydrophobic, with cationic peptides being the largest group. Cationic CPPs are characterized by large contents of lysine, ornithine, arginine, and histidine. The most widely known sequence is the one derived from the transactivator protein of human deficiency virus type 1 (HIV), aka Tat. CPPs usually enter cells via energy-independent direct penetration or energy-dependent endocytotic processes [145, 146]. For both major processes, it is theorized that positively charged CPPs, e.g., penetratin, Tat, sC18, VP22, and the poly-arginine, interact with the negatively charged groups that are present at the surface of enterocytes, which allows for binding and aggregation, and subsequent induction of the specific entry mechanisms (direct or endocytotic). CPPs can be attached in various nanoparticle iterations (organic, inorganic, or composite NPs) using electrostatic or covalent attachment. CPPs can be immobilized on the surfaces of these NP's. There are certain conjugation strategies based on the template nanoparticle used and how it is intended to be delivered. Organic nanoparticles may utilize electrostatic and thiol-maleimide coupling [147], while inorganic nanoparticles are using covalent metallic-sulfide covalent bonding (e.g., Pd-S coupling, Au-S coupling, or Ag-S coupling) [148].

Certain receptors are also targeted for oral delivery to enhance translocation of nanoparticles toward the systemic circulation. There are wide arrays of receptors that researchers have been investigating not only for increased uptake but also for an efficient intracellular trafficking inside the enterocytes. One major hurdle for nanoparticles once inside the cells is that they are usually relocated back to the apical side of the enterocytes or placed in intracellular compartments for elimination. This makes nanoparticles unable to enter the systemic circulation, even though they have successfully entered the GI epithelium. Hence, choosing the right receptor pathway for an efficient intracellular trafficking must be ensured. In one example, researchers had used albumin to target a different receptor pathway (neonatal Fc receptor, FcRn) in the intestine to translocate NP to basal epithelium and eventual entry to systemic circulation [149, 150]. FcRn works as a dual-binding receptor transporting immunoglobulin G (IgG) and albumin across the polarized cell barrier, protecting them from intracellular catabolic degradation [151]. The nanoparticle carrier they used is a mesoporous silica NP used as a template to conjugate albumin and insulin, using carbodiimide crosslinker chemistry and a pH-responsive (HPMC Hypromellose acetate succinate) that allows for targeted release and uptake at the intestinal region [152, 153]. Another interesting strategy that targets intestinal enterocytes had utilized the glucose receptor (GLUT-2) upregulation when the subject's GI is conditioned before NP introduction [154]. The increase of this receptor is observed when animal subjects are fed with a high-glucose solution beforehand, and the subsequent oral gavage of the nanoparticles decorated with a fructose ligand (Fru) that can selectively bind to it. Here, Fru-PEG NPs underwent internalization and basolateral exocytosis via a glucose transporter type 2 (GLUT2)-dependent process. Their results showed pharmacological bioavailability of 18.56% in mouse model (blood glucose lowering at 65%) with longer-acting results with a combination of AceK and an 8.8-fold increase of trans-epithelial transport in vitro (Caco-2) compared to PEG NP. Although these are very promising results, with new GI receptor-ligand mechanisms being explored and discovered to increase translocation of NP with a significant proof of concept results on glucose lowering in mouse models, this method requires multiple nanoparticle synthesis steps limiting the clinical translation.

7.2.1.4 Oral drug delivery devices

Oral devices are a new class of drug delivery systems that utilize the ease of administration of orally formulated drugs. Oral drug delivery devices are now an emerging platform for delivering therapeutics that are conventionally delivered via subcutaneous injection devices. The rationale for such devices arises from the need to find an alternative way of administering drugs without the burden of handling of needle injectors while benefiting from the ease and convenience of oral administration. Examples of these oral devices are the SOMA®being developed by Novo Nordisk A/S together with Massachusetts Institute of Technology, the RANI® pill from RANI therapeutics, and the Progenity® device [155, 156]. Other academics also reported the use of the LUMI device [156–158], an intestinal targeting device that unfolds a microneedle array injector for delivering therapeutic macromolecules.

The very first commercialized oral device is called OROS® (osmotic-controlled release oral delivery system) [159, 160]. It is an advanced controlled release oral drug delivery system in the form of a rigid tablet with a semi-permeable outer membrane and one or more small laser drilled holes in it. As the tablet passes through the body, water is absorbed through the semipermeable membrane via osmosis, and the resulting osmotic pressure is used to push the active drug through the laser drilled opening(s) in the tablet and into the gastrointestinal tract. OROS is a trademarked name owned by ALZA Corporation, which pioneered the use of osmotic pumps for oral drug delivery. Osmotic release systems have several major advantages over other controlled-release mechanisms. They are significantly less affected by factors such as pH, food intake, GI motility, and differing intestinal environments. Using an osmotic pump to deliver drugs has additional inherent advantages regarding control over drug delivery rates. This allows for much more precise drug delivery over an extended period, which results in much more predictable PK. However, osmotic release systems are relatively complicated, are somewhat difficult to manufacture, and may cause irritation or even blockage of the GI tract due to prolonged release of irritating drugs from the nondeformable tablet.

Another oral device under clinical development is the SOMA® (Figure 7.7) [161, 162]. The SOMA® (self-orienting milliactuator) is an oral injection device that delivers either a millimeter-sized solid peptide or a traditional liquid formulation toward the gastric lining. The ingenuous part of this device is its self-righting capabilities. For it

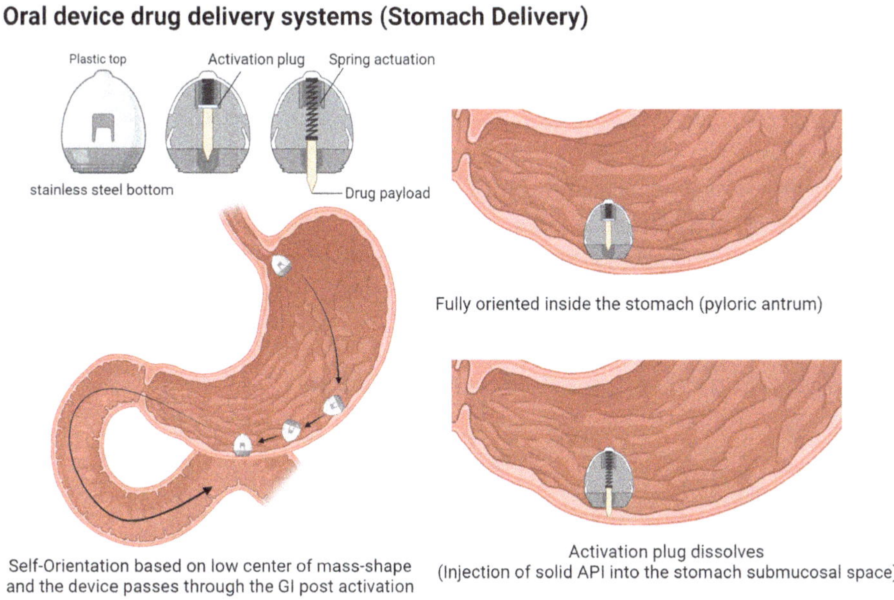

Figure 7.7: Self-orientating milliactuator (SOMA®) device, an oral device under research and development toward potential commercialization. Oral devices allow for an oral route of administration of molecules that has low stability in the presence of low pH and enzymatic degradation of the gastrointestinal system. These include therapeutic peptides (insulin, octreotide, GLP-1) and nucleotides (mRNA) which are exclusively delivered via subcutaneous injection.

to deliver its drug payload, the device must self-right itself so that the actuation mechanism of the device is perpendicular to the gastric tissue. It can do so because the shape of the SOMA® has been inspired by the shell of the leopard tortoise and the distribution of its mass it calculated to have a lower center of gravity, allowing it to always stand upright even after ingestion in the stomach. Researchers must use stainless steel and a stable rigid light weight polymer to achieve this low center of gravity in the device. The activation of the SOMA® comes from a dissolvable sugar tablet; once this tablet is exposed to gastric liquid, it will release a compressed spring that will then accelerate the solid peptide formulation of insulin and implant the drug toward the gastric submucosa. This in turn mimics the route of a subcutaneous injection but aimed toward the highly vascularized submucosal gastric tissue. The PK results are comparable to a subcutaneous injection in pig models, and they have shown glucose-lowering effects as well.

A similar concept under clinical development is with the RANI pill (see Figure 7.8). This pill is envisioned to deliver therapeutic peptides toward the intestinal submucosa using the same solid formulation injection strategy [163, 164]. The difference of targeting the intestinal submucosa is that the device does not require self-orienting principles but more of a tissue proximity capability, as the intestines are tubular in form in contrast to the stomach which has a larger lumen. The RANI pill also requires an enteric coating for it to be protected from the acidic environment of the stomach [165]. The size of the RANI pill is approximately the size of a standard fish oil or calcium pill, which constitutes to a double zero capsule size. With its new design, the RANI pill incorporates an inflating balloon that pushes a solid octreotide formulation toward the intestinal submucosa [166]. The balloon then deflates and gets eliminated from the GI through defecation. Their recent clinical findings have shown that the inflation of the device does not create any discomfort for the subjects and that the drug payload can be detected on the bloodstream.

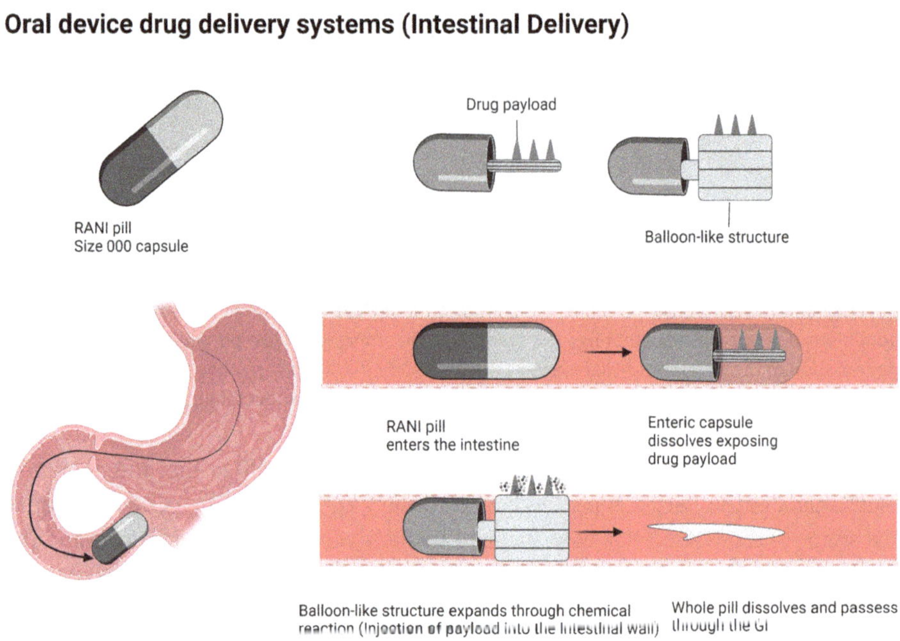

Figure 7.8: RANI® pill oral device targets the intestinal wall and delivers therapeutic peptides via submucosal injection. The device utilizes an enteric capsule that is stable under acidic pH and dissolves under neutral pH. The device also uses a chemical reaction to inflate a balloon-like structure to drive the solid drug payload toward the intestinal submucosa. The device is completely biodegradable inside the body as it passes through the GI tract.

Progenity is also developing an intestinal drug delivery device but in contrast to the RANI pill. The device uses liquid formulation that is jetted toward the intestinal mucosa, creating a needleless injection modality. Currently, they have two devices under clinical development. The first one is for oral biotherapeutics that can deliver antibodies (adalimumab) and GLP-1 agonist. The second one is a targeted drug delivery device for GI disorders. This device can locate GI pathologies such as inflammatory bowel disease (IBD) and deliver its therapeutic payload (Adalimumab, Tofacitinib) locally where it is needed. The advantage of using a liquid formulation compared to solid formulations is the stability of liquid formulations is higher compared to solid formulated peptides. Its bioavailability can also be increased as liquid formulations can dissipate and be readily uptaken by the microvasculature in the intestinal submucosa.

Lastly, the LUMI device [156–158] created by researchers in MIT, USA, utilizes an unfolding array of microneedles to deliver macromolecule therapeutics in the intestinal lumen [157]. The LUMI uses a solid peptide formulation entrapped in these microneedles, which are folded and kept in its cylindrical casing. The casing also incorporates an enteric coating which dissolves once the device enters the intestines. Intestinal fluid activates the device by dissolving a sugar pellet that holds a compressed spring. Once the spring is actuated it deploys the microneedle array into lumen of the intestines. The array has three unfolding arms which can latch on to the intestinal wall delivering the therapeutic payload directly toward the microvasculature of the intestines. Afterward, the arms that holds the microneedles dissolves as the polymer used to make these arms are pH dissolvable and the casing then gets eliminated from the body via defecation. Their animal model results have shown successful attachment of the microneedles in the intestines and also delivered the drug payload toward the systemic circulation. They have also shown a similar human insulin level delivered via both intestinal microneedle patch and subcutaneous injection of dissolved microneedles and the insulin had maintained its potency by lowering the blood glucose level of the animals like a subcutaneous injection.

All these oral devices had functionalities that rely on the development of biomaterials, especially pH dependent, thermoresponsive, and biodegradable/compatible polymers. The devices intended to deliver their therapeutic payload in the intestines must first survive the acidic environment of the stomach. Hence, incorporating an enteric coating is a must on these devices. An underlying challenge for all oral devices is ensuring sufficient tissue proximity during dosing. So, ideas like creating an inflatable balloon, self-righting shape, and unfolding arms are incorporated in these oral devices. Furthermore, since this is a new modality and new class of drug delivery devices, there are still uncertainties that must be answered before it can be successfully and routinely administered to humans. These devices can be lodged inside the GI tract and cause perforation, alongside with the risk of faulty dosing mechanism where the patient will have no idea if the drug has been successfully dosed unlike in other methods such as subcutaneous injection. Furthermore, there is the environmental impact of the oral device space since they will all end up in the sewer systems and eventually reach the oceans. And since they are all classified as medical devices, thinking that they might end up polluting our waters might not be the best trade-off for dosing convenience. However, ensuring that all or at least some components can be degraded by the environment might solve this environmental issue.

7.2.2 Biomaterials in alternative drug delivery

There are alternative ways of delivering drugs toward the body besides from direct injection and oral delivery. These methods are alternative ways for drugs to gain access to the systemic circulation. Examples of these are implants, nasal, transdermal, mucosal, and ocular deliveries. All these routes are made possible by developing the right biomaterial vehicles for the drug (Figure 7.9).

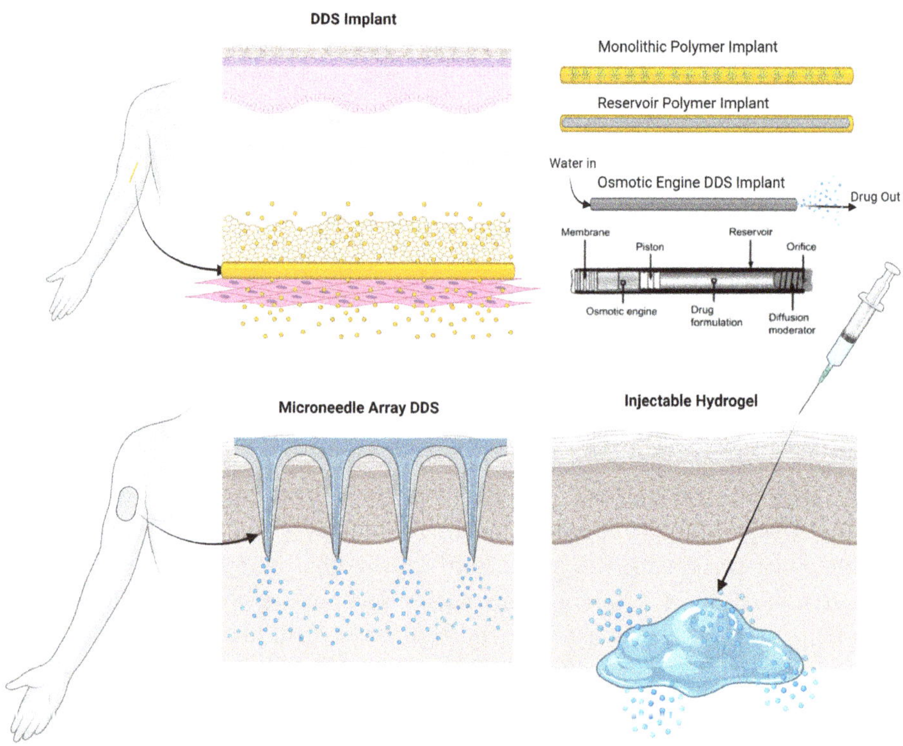

Figure 7.9: Drug delivery systems using subcutaneous implants, transdermal microneedle patches, and injectable hydrogels. These drug delivery systems can provide first- and zero-order release kinetics using biomaterials.

7.2.2.1 Biomaterials in drug delivery implant technologies

Drug delivery system (DDS) implants are devices that are surgically placed inside the body intended to deliver drugs toward the systemic circulation via a controlled manner. The drugs being delivered are usually needed by the body to control or maintain physiological functions. DDS implants became popular due to its convenience as the patient receiving it does not have to keep in mind taking a daily dose of the medicine. It also bypasses the first-pass metabolism delivering more drugs into the body. This in turn generates multiple benefits; they may minimize potential side effects of therapy while offering the opportunity for increased patient compliance, opportunity to deliver drugs that are not suited for daily oral administration, and early removal if adverse effects require termination of treatment. Implantable devices require a clinician for insertion, and the insertion itself is usually a minimally to noninvasive process. Furthermore, this insertion will only be a onetime process, up until the drug in the implant is depleted. This may be as seen as a disadvantage for DDS implants, but the prolonged drug delivery that it can achieve can remove the need for patient regimental dosing, overcoming this disadvantage.

Classification of DDS implants is not as direct as other drug delivery systems as multiple device variants fall into hybrid categories. Thus, a more general and broad classification is used: passive and active implants [167, 168]. As the name suggests, passive implants use passive drug release mechanisms (diffusion, erosion, etc.), and the active implants utilize energy sources to drive the release of drugs (osmotic pumps, microelectromechanical drives, and pumps MEMS). The passive implants are usually made from polymeric materials, while the active implants are from biomedical metallic alloys.

Passive implants can be further subdivided into two classifications: degradable and nondegradable polymeric implants. The drug payload is usually entrapped in the polymeric network of the implant. Several parameters can be adjusted to have varied drug release profile, and these are: drug type/concentration, polymer type, implant design, and surface properties of the implant.

Nonbiodegradable implants are usually made from polymers such as silicones, poly(urethanes), poly(acrylates), or copolymers such as poly (ethylene vinyl acetate). These polymers are elastic and soft and usually the size of a matchstick when fully manufactured. This type of implant can be further classified as monolithic or reservoir type [169, 170]. Monolithic implants are when the drug payload is homogenously dispersed all throughout the polymer network while the reservoir type, as the name suggests, contains the drug in a cavity or space (reservoir). Both implant types use the permeability polymer used to control the release of the drug payload. The drug must diffuse toward the tissue and enter the capillaries to which they are implanted to enter the systemic circulation and enact their therapeutic value. These types of implants are very common in birth control to deliver hormones that can disrupt the maturation of ovum in women. This is very convenient as compared to oral contraceptives which, if a user misses a dose, can lead to unwanted pregnancy. However, in implantable contraceptives, the user can implant and forget about the treatment that usually last for 3–6 months. One setback for nondegradable implant is the need to remove it once the drug payload is depleted. This may leave unwanted scarring and, if left inside the body, may cause long-term foreign body response.

Thus, the need for biodegradable implants arose as in contrast with nondegradable DDS implants, they remove the need for surgical removal of the implant once the drug has been all used. The polymers or block copolymers used in making biodegradable implants can be broken down into monomeric components that are excreted or absorbed by the body. Examples of these polymers are poly(caprolactone) (PCL), poly (lactic acid) (PLA) or poly (lactic-co-glycolic acid) (PLGA) and natural materials made from polysaccharides (Cellulose), or composites of natural and synthetic polymers. Fine-tuning the molecular properties (Mn, Mz, Mw), hydrophobicity and hydrophilicity, cross-linking degree, and surface morphology (roughness and porosity) can influence how fast the polymer can be degraded.

The drug release mechanism also varies from both degradable and nonbiodegradable DDS implants. In nondegradable implants, reservoir types separate the drug compartment by enclosing it inside a cavity and lined it with the polymer membrane that serves as the mediator for the diffusion of the drug towards the body [170]. This type of implant can create a more constant release rate as it is not affected by the concentration gradient of the drug present in the polymer matrix, but mostly governed by the thickness and permeability of polymeric membrane (zero-order release). Whereas in matrix-type systems [171], which consist of a rate-controlling medium such as a polymer with drug homogenously dispersed in its matrix, the drug release mechanism is predominantly governed by Fickian diffusion [172], which is dependent on the drug concentration gradient, diffusion distance, and the degree of swelling of the polymer. These systems allow for a slow diffusion of the drug through the polymeric material, helping to sustain the drug release, and are dependent on the concentration of the drug in the polymeric matrix. This is also known as first-order release kinetics.

For biodegradable DDS implants, the release mechanism is mostly governed by the degradation process of the polymer. It must be said that biodegradable DDS implants can also be made into reservoir or matrix type systems. Drug release from a reservoir system is controlled by the rate of polymer degradation or drug dissolution into, and then diffusion through, the polymer wall, or a combination of both. Drug release from a drug polymer mixture is controlled by diffusion, swelling, or erosion. The release of drug from the system will be dependent on: the solubility and permeability of the drug in the polymer, the drug load, and the in vivo degradation rate of the polymer.

Active DDS implants have a different way to deliver the drug payload as they use energy to drive the release of drugs in a much consistent manner. In contrast to passive implants, active DDS implants present a higher degree of control in drug release but involve a much complex manufacturing and development cost. Active DDS implants can be made using both polymeric and metallic components, and these two mainly use osmotic pumps as the main driving engine to deliver the drug payload. Polymeric osmotic implants consist mainly of a semipermeable membrane that surrounds a drug reservoir and an orifice that allows for the drug release [173]. This design within the implant creates high pressures that will force drug release through the orifice. The release kinetics follow a zero-order release [174] due to the consistent osmotic pressure build-up within the implant. The metallic implants that use osmotic pumps are more complex in structure with multiple components. The metal casing is made from biomedical-grade alloys (titanium or stainless steel) [175, 176] to prevent corrosion and leaching of metal ions in the body [177]. This implant has two openings with one side dedicated for the polymeric membrane that allows only for water to rush toward the osmotic engine. This osmotic engine usually contains sodium chloride, drives a rubber piston that then pushes on the drug formulation located inside the reservoir of the implant [178]. At the release end of the implant is where a diffusion moderator is located that regulates the amount of drug being released at the orifice of the implant. This design is commonly known as the DUROS® osmotic pump [173] and was successfully put into clinic as Viadur® (leuprolide acetate) [179]. This implant is meant for treating prostate cancer, with the delivery of leuprolide acetate continuously at a rate of 125 µg/day for 1 year.

One challenge for DDS implants is that their application is currently limited to very potent drugs. For an implant to be commercially viable, its size should not be larger than 5 cm length × 0.5 cm diameter, if referenced with the DUROS® osmotic pump. Knowing this size limitation, an implant should be able to compact as much drug as possible in this limited space, thus severely limiting what types of drugs can be used for implant applications. Furthermore, the implant should be able to last at least three months to substantiate the risk-benefit costs of the treatment. If its drug delivery lifespan is shorter than three months, then doing daily to weekly subcutaneous injections will be a better method to deliver these therapeutics. Implants would have been a highly desirable route for people living with chronic metabolic disease like diabetes or heart diseases, as patients who suffer from these diseases require daily administration of drugs to maintain a healthy lifestyle. However, the current anachronism for this certain drug delivery application within these therapy areas is that the drugs are usually not potent enough to be fitted inside a small implant. Further research on drug discovery and formulations must go hand in hand together with implant development to realize DDS implants for chronic diseases.

7.2.2.2 Injectable hydrogel drug delivery implants

A new class of implantable or injectable biomaterials that can deliver therapeutics are now being developed to expand the therapeutic areas to which biomaterials can be used for long-term drug delivery. Hydrogels are one of the promising types of biomaterials that can provide sustained or controlled release of drugs while also having the convenience of ease in implantation due to their injectability using a normal hypodermic syringe and biodegradability over time. In contrast to traditional DDS implants which require a highly trained clinician to administer the implant device, injectable hydrogels require less training with subsequently less complications for the patients.

A hydrogel is a cross-linked hydrophilic polymer that is stable in water [180, 181]. They are highly absorbent, yet maintain well-defined solid structures. These properties are valuable within biomedical applications. For example, hydrogels are a very suitable drug delivery platform as they can carry high concentration of drugs through their polymeric network, exhibit high biocompatibility, easy injectability, and tunable biodegradation.

A clear example comes from the Wyss Institute, where they had developed an injectable hydrogel for delivering vascular endothelial growth factor (VEGF) [182, 183]. The hydrogel is made from alginate, a polysaccharide commonly found in seaweed. Alginate can be cross-linked easily with any divalent or trivalent ions; it can also be tailored to have shear thinning properties for it to be extruded from a syringe. In their research VEGF (vascular endothelial growth factor)-alginate hydrogel is compared to pure VEGF injection, and since VEGF has a short half-life, it creates less therapeutic effects in comparison to VEGF-alginate hydrogel. The hydrogel allows for a sustained release of the VEGF in the affected area and protecting it from degradation and rapid elimination from the body. Once the VEGF has been released, the alginate hydrogel is degraded inside the body via hydrolysis and can be easily eliminated from the kidneys. This is applied in vascular diseases in the lower extremities where blood flow has been restricted due to plaque build-ups along the major vessels leading to tissue ischemia. The VEGF-alginate hydrogel prevents further damage in the lower extremities and resumes blood perfusion in the affected muscles by growing new vessels.

Choosing the right hydrogel is crucial in creating an implantable drug delivery hydrogel system. The hydrogel is placed near to where it is needed giving a more localized drug delivery, increasing the PK and PD of the drug payload. Natural biopolymers are highly suited in this application as they won't increase immune response and can exhibit rapid degradation while allowing a sustained release of its drug payload. Hydrogels can also be fine-tuned to either graft the drug in its molecular structure or just physically entrap the drug in its polymeric network. The former can create a stimuli-responsive release which can pave for a more controlled release while the latter can release the drug via diffusion or through hydrogel erosion.

7.2.2.3 Biomaterials in transdermal and mucosal drug delivery

Delivering drugs through the skin and the mucous membrane has sparked great interest from researchers as this route can enhance the bioavailability of a drug as it can bypass most of the physiological barriers from conventional routes such as oral delivery [184–186]. It is considered as a noninvasive drug delivery route into the body through the skin and. unlike direct administration routes that use needles which can create several complications associated with needle handling, transdermal and mucosal delivery can create a safer and painless delivery route for patients.

One example of these novel oral drug delivery technologies is the use of oral or buccal patches that deliver the drug payload using microneedles [187]. The buccal cavity is surrounded by highly vascularized mucosal tissue, thus making it attractive as a site to deliver drugs bypassing the first pass effect and the harsh environment of the GI. Microneedles (MN) made up of polyvinylpyrrolidone (PVP) were fabricated using a simple polymer casting method to introduce human insulin (HI) [188–190] and human growth hormone (hGH) [191, 192] in the vasculature inside the buccal mucosa of swine animal models. The simple fabrication technique also ensures that the

HI and hGH chemical structure remains the same, keeping its bioactivity. Their results have shown that the MNs successfully delivered HI and hGH when applied to the palate and buccal mucosa of swine with MN application to the palate and buccal mucosa delivered a similar amount of HI as subcutaneous injection consisting of MN tips dissolved in saline. Researchers have also investigated the overall comfort of wearing this microneedle patch inside the mouth [193], and they have done so by doing initial human acceptability study with bare microneedle patch without the HI and hGH. Varied preferences for the volunteers have been shown in this study, with the palate being the preferred site of patch application, followed by the cheek, lip, tongue, and sublingual areas. This strategy demonstrates that there are also other oral routes that can be used to administer sensitive therapeutic payloads which can be at par with traditional administration route (subcutaneous). Although it may be not as convenient as with traditional oral route, as it is still very difficult to overcome the physiological barriers of the GI, this strategy using a pragmatic solution can enhance adherence of patients with needle fear to the treatment regimen.

Transdermal drug delivery also has seen recent advances within the previous years. Since the skin can be used as an entry point for administering drugs, researchers are finding ways how it can be accessed without compromising the number one function of the skin, which is being the primary barrier from pathogens [194–197]. The skin as a barrier comes from the epidermis, which is its outermost layer, and has properties of blocking external substances. The barrier effect is very significant in the transport of substances. In transdermal delivery, smaller molecular weight substances have easier absorption in the body as it can utilize an intracellular pathway. However, large molecular weight drugs are blockaded due to the physical barriers the skin possesses. Furthermore, the vascular system in the dermal layer can inhibit transdermal drug delivery [198, 199]. Here, a single layer of endothelial cells terminating in the papillary loops of the superficial arteriovenous plexus near the dermal-epidermal junction in the upper dermis represents the interface between the tissues surrounding the skin and the human vasculature. This layer actively responds to stimuli such as pressure, shear, osmotic pressure, heat, chemokines, and cytokines inducing vasodilation or constriction, thus modulating the permeability of the vasculature of the skin. This is the most challenging hurdle for transdermal drug delivery as it seeks to resolve the barrier effect of the stratum corneum, deliver the drug to the skin, and enter the vasculature to reach the systemic circulation and eventually the target organ or tissue. This must be done in a way that it is transient and will not impede the primary function of the skin.

There are new strategies how this can be achieved using active delivery. This means that an external device or equipment is used to enhance the skin's permeability to drug compounds. Using external stimuli, such as electrical, mechanical, or physical stimuli, the skin's permeability to drugs and other therapeutic biomolecules can be achieved, as compared to the delivery of drugs by topical application on the skin.

7.2.2.4 Iontophoresis

Iontophoresis enhances the transport of ions across the membrane using a small externally applied potential difference (<0.5 mA/cm^2) [200, 201]. This technique has been proven to enhance skin permeation and increase release rate of several drugs with low absorption properties. Some proof-of-concept experiments in vivo have shown transport of ionic or nonionic drugs by the application of an electrochemical potential gradient. However, the efficacy of iontophoresis depends on the polarity, valency, and mobility of the drug molecule, the nature of the applied electrical cycle, and the formulation containing the drug. In particular, the dependence on current makes drug absorption through iontophoresis less dependent on biological parameters, unlike most other drug delivery systems.

7.2.2.5 Sonophoresis

Ultrasound waves can also be used to improve transdermal delivery of large molecular compounds [202, 203]. Some experiments have shown that low-frequency ultrasound is more effective because it facilitates drug movement by creating an aqueous path in the perturbed bilayer through cavitation. The drug must be mixed with a specific coupler, such as a gel or a cream [198, 204], which transmits ultrasonic waves to the skin and disturbs the skin layers, thereby creating an aqueous path through which the drug can be injected. Drugs typically pass through passages created by the application of ultrasonic waves with energy values between 20 kHz and 16 MHz. Ultrasound waves can also increase the local temperature of the skin promoting vasodilation, which further promotes drug penetration.

7.2.2.6 Electroporation

High-voltage electric pulses ranging from 5 to 500 V are used for short exposure times (~ms) to the skin, which leads to the formation of small pores in the stratum corneum that improves permeability and aid drug diffusion [205, 206]. However, this method has disadvantages as it can only deliver smaller drug loads and the massive cellular perturbation can sometimes lead to cell death and denaturation of heat sensitive drugs. This is due to the high voltage being used during the perturbation period.

7.2.2.7 Photomechanical waves

Photodynamic waves transmitted to the skin can penetrate the stratum corneum, creating a transient permeation window for drugs [207–209]. The light waves produce heat and slightly ablates the skin under low radiation exposure at 5–7 J/cm^2. In comparison to direct heating or ablation methods, light waves are less direct and harmful which can have a longer time span of permeation for the drugs and other biomolecules. The wave generated by a single laser pulse also showed increased skin permeability within minutes, allowing macromolecules to diffuse into the skin.

7.2.2.8 Microneedles

A more direct transdermal drug delivery using millimeter- to micron-sized needles is one of the most popular methods for transdermal drug delivery and is an active area of current research with studies ranging from creating optimal needle geometry to combining it with other novel drug delivery systems such as nanoparticles and being able to react to external stimuli to achieve controlled release [210, 211]. As the name suggests, micron-sized needles pierce the superficial layer of the skin, resulting in drug diffusion across the epidermal layer. Because these microneedles are short and thin, these deliver drugs directly to the blood capillary area for active absorption, which helps in avoiding pain. Microneedles are usually combined with adhesives to maintain its anchorage on the epidermis [212, 213]. Microneedles are usually made from polymers but more active microneedle-based device use metals in applications such as biosensing to create a feedback loop on a drug delivery patch [214, 215]. This type of metallic microneedles is made using laser-mediated fabrication techniques. The 3D structure of a metallic microneedle is generated through cutting or ablating on a flat metal surface using a high-precision laser. Polymeric microneedles, on the other hand, are conventionally fabricated using photolithography, as this method can manufacture dissolving/hydrogel microneedles or silicon microneedles via making an inverse mold based on the microneedle structure through etching of photoresist.

7.2.3 Biomaterials in gene therapy agent delivery

Genes are the biomolecular instructions of life. It is a basic unit of heredity made from nucleotide sequences called the DNA. The DNA holds the code that allows for the synthesis of a gene product, either RNA or protein. During gene expression, the DNA is first copied into RNA and RNA into various proteins. This is also known as the transcription-translation-synthesis process which is the central dogma of biochemistry and biology. Genes allows for cell replication and creation of proteins needed for overall homeostasis in the body. However, it may also mean that a fault in the genetic sequence in a person may also be transmitted to its offspring if it continues to be selected. Almost all diseases have some degree of genetic factor in it, and there are scientifically identified diseases that the sole cause is the faulty gene creating defective proteins and/or anatomic malformations. There are different ways a defective gene may form. Exposure to mutagenic compounds and/or ionizing radiation can create random alterations on the genes, and if these genes are not sequestered by the body through its housekeeping mechanisms, it may lead to diseases. Hence, clinicians, geneticists, and biomedical engineers have thought of targeting this main culprit of wide arrays of diseases.

Gene therapy is an emerging therapeutic modality which has the potential to be applied to rare and common diseases [216]. This is done by the introduction of new

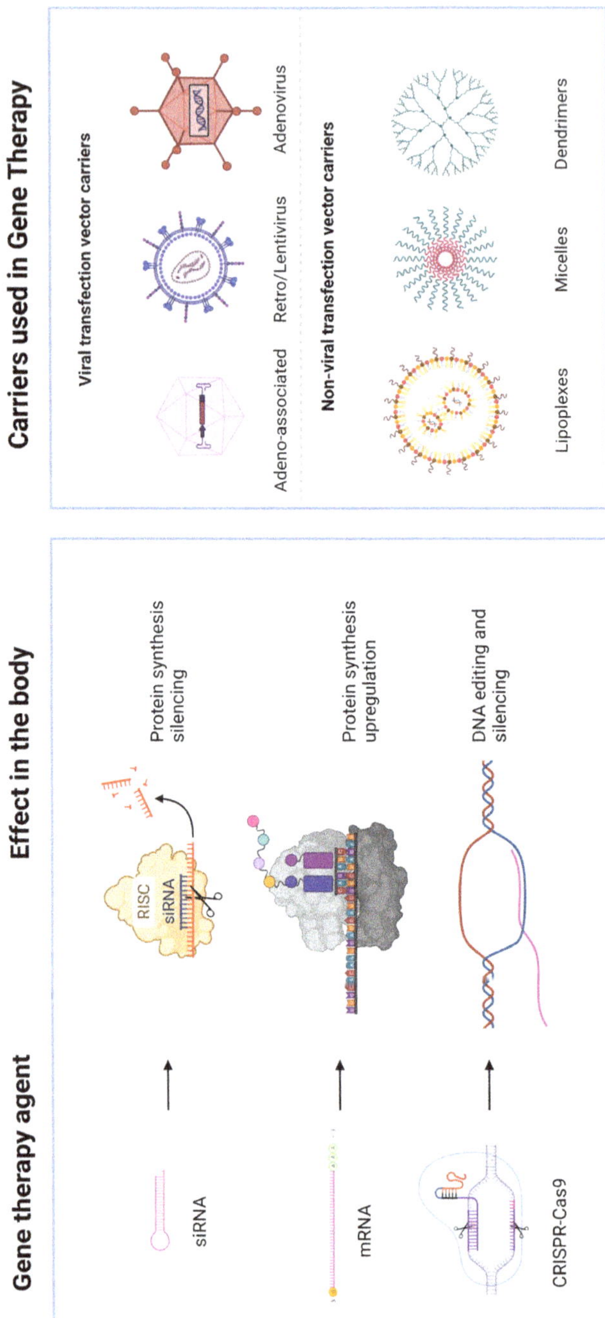

Figure 7.10: Gene therapy agents and their effect toward the cells, and the current in vivo carriers under research and development.

genetic code (e.g., mRNA) [217], gene suppressing agents (e.g., RNAi) [218], or altering the genetic sequence itself (e.g. CRISPR-Cas9) [219]. All these techniques require delivering the therapeutic gene therapy agent into the target cells which are expressing the faulty genes (see Figure 7.10). As part of a new therapeutic modality gene therapy had seen significant clinical translation for the past four years. The very first clinically approved gene therapy product is Patisiran with a commercial name of ONPATTRO® [220]. This therapeutic agent is an siRNA (small interfering Ribonucleic acid) encapsulated inside a liposome. Its main indication is for the treatment of polyneuropathy in people with hereditary transthyretin-mediated amyloidosis, a fatal rare disease that is estimated to affect 50,000 people worldwide. This type of amyloidosis is characterized by the overproduction of a protein called transthyretin that affects the kidneys and the heart in its late stage. It is also characterized by the systemic deposition of amyloidogenic variants of the transthyretin protein, especially in the peripheral nervous system, causing a progressive sensory and motor polyneuropathy. Before the advent of gene therapy, the conventional treatment of this condition is through liver transplantation to replace the tissue that creates the defective protein. The main cause of this disease is the mutation on the TTR gene, located on the 18th chromosome; this mutation causes the transthyretin protein to misfold and aggregate into a variety of structures which is called amyloid fibrils. Here, interfering RNA can be of use to silence the TTR gene faulty code instructions. Interfering RNA is part of the housekeeping mechanism of the cell to silence the translation of mRNA (messenger RNA) into proteins in the ribosomes. Generally, mRNA is made from the nucleus during the transcription phase and is exported into the ribosomes to create various kinds of proteins. If the gene that allows for the creation of this mRNA is kept and not silenced inside the nucleus, they may initiate various kinds of pathologies inside the body. There are two types of RNAi. The first one is called the microRNA, and the second is the small interfering RNA. They have basically the same function meant to stop the synthesis of a faulty protein, but they may differ in their mechanism in doing so. The microRNA is responsible for competitively binding into mRNA for it to be inhibited inside the cell. siRNA, on the other hand, cleaves the mRNA into fragments to silence it.

Delivering gene therapeutic agent poses numerous challenges as these nucleotide sequences are notorious for having short half-life, leading to low therapeutic efficacy. The causes for short half-life ranges from 5 to 30 min and further elaborated in (Table 7.4) [221].

Hence, chemical modifications or encapsulation vehicles are needed to ensure that the transfection of the RNAi payload happens into the target cells. In this section, chemical modifications of RNAi will not be further discussed as they are out of scope for biomaterial science. In short, chemical modification of naked RNAi utilizes modification on the phosphate, ribose, and base structures of the RNAi and the addition of a conjugate (e.g., GalNac) [225]. The effect of these modifications allows for increased storage stability, avoid nuclease degradation, and increase

Table 7.4: Challenges in gene therapy delivery in vivo.

Factors	Effect	Ref.
Nucleases	Nucleases degrade extracellular nucleotides as part of maintaining homeostasis and body's defense from pathogens	[222]
Immune system	Immune cells roam for nonnative materials in the body and promptly eliminates it via phago- or pinocytosis	[223]
Serum stability	Nucleotides are bound by weak hydrogen bonds and can be dissociated easily by blood components especially serum proteins	[224]
Specific targeting	The gene therapy agent ends up in the tissue where it is not needed	[225]
Endocytosis and lack of endosomal escape	Gene therapy agent is kept in the endosomes and is degraded inside it due to its acidic environment	[226]

transfection to target cells. The other paradigm for RNAi therapy is the utilization of liposomes. Liposome encapsulation of RNAi is an emerging and clinically approved method as evidenced using Patisiran in the clinic. Liposomes are closed bilayer structures spontaneously formed by hydrated phospholipids that are widely used as efficient delivery systems of nucleotides as they provide a suitable environment for the nucleotide while avoiding nuclease degradation and increase transfection to target cells by allowing for ligand modification as well [227]. However, this would also mean that the current route of delivery for Patisiran is only through intravenous infusion. Encapsulation inside a liposome reduces the concentration of the drug per volume available, but it can increase the circulation time inside the body [220].

Another success story for gene therapy is the development of the first vaccines against COVID-19 virus (sars-cov-2) which become a pandemic in the year 2020. The scramble to sequence the viral genome paved the way for both Moderna and Pfizer BioNtech to develop an mRNA-based vaccine that is delivered via intramuscular injection [228]. The mRNA they developed instructs the body to create the spike proteins present on the surfaces of the sars-cov-19 virus so whenever the body is truly exposed to the virus, the immune cells can readily create antibodies to fight off the infection and avoid severe complications and hospitalizations. Both Moderna and Pfizer BioNtech utilize lipoplex encapsulation technology to protect the mRNA payload [229]. As discussed earlier, messenger RNA is responsible for the translation of various proteins in the body. Hence mRNA can be used to treat various diseases not just viral infection but also cancer. mRNA-based cancer vaccines are somewhat like what a viral vaccine may do. But instead of using viral protein sequences, surface proteins of cancer cells are used to create mRNA vaccines which are then inoculated to the immune cells to attack any malignant cells if they ever grow inside the body [230].

The ability of editing genes or completely silencing them from the genetic sequence in the body opens a wide range of possibilities. From having designer babies to treating ultra-rare uncurable diseases, gene editing comes with a landscape of promise and ethical dilemmas. The discovery of CRISPR-Cas9 has demonstrated that we can remove and edit any gene on-demand. This discovery came about with researchers observing some microorganisms using CRSPR-cas9 to remove viral genome in their genetic sequences [219]. This is what bacteria use as a pseudo acquired immune response or defense mechanism to viral infections. CRISPR is an acronym for clustered regularly interspaced short palindromic repeats and is a family of DNA sequences found in the genomes of prokaryotic bacteria and archaea while cas9 is an enzyme that uses CRISPR sequences as a guide to recognize and cleave specific strands of DNA that are complementary to the CRISPR sequence. Together, they are the basis of CRISPR-Cas9 that can be used to edit genes within an organism. Delivery methods of CRISPR-Cas9 follow the same hurdles stated in Table 7.4.

Introducing gene therapy agents into the body follows two dogmas: viral gene vector [231] and nonviral gene vectors [232]. Viral gene vectors, as the name suggests, is the use of attenuated viruses (nonharmful) as carriers for the gene therapy agent (RNAi, mRNA, or CRISPR-cas9). The salient point of using viruses comes about with their ability to transfect cell targets with ultra-high specificity. The current viral vectors under development are adeno-associated virus (AAV), retro/lenti virus, and adenovirus. Their surface spike proteins allow them to specifically transfect target cells while providing a safe environment for the gene therapy agent payload. Viruses in a way are nature's number one gene transfecting mechanism and using them to therapeutic advantage is only logical. However, since they are virus derived, they can trigger immune response in the body and there are several safety issues concerning using viral vectors. Furthermore, there are five significant challenges viral vectors are facing:

1) **Vector concentration**
   Standard upstream bioprocesses can deliver a concentration of up to 200 trillion vector genomes per cell, but gene therapies require a concentration of up to 100 quadrillion per cell. As a result, the virus needs to be concentrated between 100 and 10,000 times [233].
2) **Blank capsids**
   Blank or empty vectors can account for up to 90% of the batch and reduce the effectiveness of the therapy, requiring that a patient receive a higher dose [233].
3) **Oncogenic host cell DNA**
   When malignant cell lines are used to grow viral vectors, residual DNA during purification must be ensured as the presence of the oncogenic genes may cause tumorigenicity into the patient. It is nearly impossible to neutralize these DNA fragments in vectors that contain host cell DNA [234].

4) **Immunogenic protein impurities**
   Capsid protein can elicit a response that could ultimately reduce the effectiveness of the therapy [223].
5) **Harvesting of viral vectors from cells**
   Viral vectors can be difficult to purify from its host cells, leading to a slower and more expensive production process [233].

To solve this problem, nonviral vectors are being developed such as with the case of nanocarriers (liposomes, dendrimers, micelles, and other inorganic and organic nanoparticles). Nonviral vectors can be designed like a virus to have targeting ligands to transfect specific cell lines in the body. This, however, requires a lot of basic science research to identify which receptor is expressed in given populations consistently, given different disease processes. However, once this is identified, loading targeting ligands on nanocarriers is not a complicated task. Nonviral vectors are low in their cytotoxicity, immunogenicity, and mutagenesis, in contrast to viral vectors, hence attracting more and more researchers to explore this delivery system, but they do not carry ideal characteristics and have faced critical challenges, including gene transfer efficiency, specificity, and gene expression duration [226, 232].

7.2.4 Biomaterials allow drug delivery capabilities in medical devices

Other medical devices such as stents (vascular and nonvascular), bone implants, contact lenses, and wound dressings are now being endowed with multifunctionality by incorporating drug delivery capabilities on them [1–5]. Conventionally, these devices are meant only as intended as they are designed to be. But in biomedical devices, it is always an advantage if the device can also perform other things that can increase their intended performance.

For example, stents (vascular or nonvascular) are created to create patency inside a bodily orifice by putting a radial support. But over time, this mechanical support is not enough to maintain the patency it creates. Restenosis due to re-accumulation of plaque or ingrowth of cancer tumor can occur [235, 236]. Thus, being able to release drugs that can prevent these events is useful to prolonging or increasing the device's performance. This is done using multiple strategies. Addition of drug-eluting capabilities from polymeric membranes and thin film coatings can load anti-restenotic drugs such as paclitaxel, heparin, or warfarin [237, 238].

Bone implants and prosthesis can now increase their osteogenic properties to improve anchorage of local bone onto their surfaces. Coating their surfaces with osteogenic growth factors can increase the bone activity [239] in contrast to just simply having a nonreactive surface. Furthermore, antibacterial drugs can now be loaded onto the surfaces of metallic and ceramic bone implants and prosthesis to avoid biofilm formation and reduce the risk of nosocomial infection post-implantation of the device.

Lastly, wound dressings are not just made to prevent pathogens to enter the wound, but they can also now release drugs that can actively prevent infection, reduce inflammation, and promote healing by incorporation of growth factors [240, 241].

All of these are now possible with the use of biomaterials as drug delivery platforms.

7.3 Conclusion

Biomaterials have shown their value in the drug delivery, as they can be used to protect, carry, and even enhance the therapeutic effects of various drug products. Drugs and various therapeutics have several limitations such as reduced bioavailability, rapid half-life and clearance, and toxicity. But incorporating biomaterials as a drug delivery platform can limit and even mitigate these limitations. The advent of nanotechnology using nanoparticles has also significantly improved the capabilities of various drug products by allowing to circumvent the physiological barriers of the body. Novel devices that utilize various kinds of biomaterials are also being researched upon to improve the field of drug delivery. It has been shown that there is still room for improvement for drug delivery technologies as creating a robust drug delivery platform for most drugs is still very elusive. Tailored biomaterials depending on application are still the current trend of research for this field. Advancement in biomaterials science will also carry improvements in the field of drug delivery.

References

[1] Zilberman M, Elsner JJ. Antibiotic-eluting medical devices for various applications. Journal of Controlled Release. 2008;130(3):202–15.

[2] Raval A, Parikh J, Engineer C. Mechanism of controlled release kinetics from medical devices. Brazilian Journal of Chemical Engineering. 2010;27(2):211–25.

[3] Livingston M, Tan A. Coating techniques and release kinetics of drug-eluting stents. Journal of Medical Devices. 2016;10:1.

[4] Sarigol-Calamak E, Hascicek C. Tissue scaffolds as a local drug delivery system for bone regeneration. Cutting-Edge Enabling Technologies for Regenerative Medicine. 2018:475–93.

[5] Kiaee G, Mostafalu P, Samandari M, Sonkusale S. A pH-mediated electronic wound dressing for controlled drug delivery. Advanced Healthcare Materials. 2018;7(18):1800396.

[6] Carlsson L, Clarke JC, Yen C, Gregoire F, Albery T, Billger M, et al. Biocompatible, purified VEGF-A mRNA improves cardiac function after intracardiac injection 1 week post-myocardial infarction in swine. Molecular Therapy-Methods & Clinical Development. 2018;9:330–46.

[7] Altman PA, Sievers R, Lee R. Exploring heart lymphatics in local drug delivery. Lymphatic Research and Biology. 2003;1(1):47–54.

[8] Cho EJ, Sun B, Doh K-O, Wilson EM, Torregrosa-Allen S, Elzey BD, et al. Intraperitoneal delivery of platinum with in-situ crosslinkable hyaluronic acid gel for local therapy of ovarian cancer. Biomaterials. 2015;37:312–19.

[9] P Blumling III J, Silva GA. Targeting the brain: Advances in drug delivery. Current Pharmaceutical Biotechnology. 2012;13(12):2417–26.
[10] Loira-Pastoriza C, Todoroff J, Vanbever R. Delivery strategies for sustained drug release in the lungs. Advanced Drug Delivery Reviews. 2014;75:81–91.
[11] Qureshi OS, Kim H-S, Zeb A, Choi J-S, Kim H-S, Kwon J-E, et al. Sustained release docetaxel-incorporated lipid nanoparticles with improved pharmacokinetics for oral and parenteral administration. Journal of Microencapsulation. 2017;34(3):250–61.
[12] Ryu JY, Kim HU, Lee SY. Deep learning improves prediction of drug–drug and drug–food interactions. Proceedings of the National Academy of Sciences. 2018;115(18):E4304–E11.
[13] Cao J, Han J, Xiao H, Qiao J, Han M. Effect of tea polyphenol compounds on anticancer drugs in terms of anti-tumor activity, toxicology, and pharmacokinetics. Nutrients. 2016;8(12):762.
[14] Elwakeel A, Soudan H, Eldoksh A, Shalaby M, Eldemellawy M, Ghareeb D, et al. Implementation of the Chou-Talalay method for studying the in vitro pharmacodynamic interactions of binary and ternary drug combinations on MDA-MB-231 triple negative breast cancer cells. Synergy. 2019;8:100047.
[15] Banerjee V, Sharda N, Huse J, Singh D, Sokolov D, Czinn SJ, et al. Synergistic potential of dual andrographolide and melatonin targeting of metastatic colon cancer cells: Using the Chou-Talalay combination index method. European Journal of Pharmacology. 2021;897:173919.
[16] Ashton JC. Drug combination studies and their synergy quantification using the Chou–Talalay method. Cancer Research. 2015;75(11):2400.
[17] Pichler WJ, Campi P. Adverse side effects to biological agents. Drug Hypersensitivity. 2007:151–65.
[18] Berger SI, Iyengar R. Role of systems pharmacology in understanding drug adverse events. Wiley Interdisciplinary Reviews. Systems Biology and Medicine. 2011;3(2):129–35.
[19] Smolensky MH, Peppas NA. Chronobiology, drug delivery, and chronotherapeutics. Advanced Drug Delivery Reviews. 2007;59(9–10):828–51.
[20] Smith DA, Beaumont K, Maurer TS, Di L. Relevance of half-life in drug design: Miniperspective. Journal of Medicinal Chemistry. 2017;61(10):4273–82.
[21] Kontermann RE. Half-life extended biotherapeutics. Expert Opinion on Biological Therapy. 2016;16(7):903–15.
[22] Rys P, Pankiewicz O, Łach K, Kwaskowski A, Skrzekowska-Baran I, Malecki M. Efficacy and safety comparison of rapid-acting insulin aspart and regular human insulin in the treatment of type 1 and type 2 diabetes mellitus: A systematic review. Diabetes & Metabolism. 2011;37(3):190–200.
[23] Kerr D, Wizemann E, Senstius J, Zacho M, Ampudia-Blasco FJ. Stability and performance of rapid-acting insulin analogs used for continuous subcutaneous insulin infusion: A systematic review. Journal of Diabetes Science and Technology. 2013;7(6):1595–606.
[24] Avgerinos I, Papanastasiou G, Karagiannis T, Michailidis T, Liakos A, Mainou M, et al. Ultra-rapid-acting insulins for adults with diabetes mellitus: A systematic review and meta-analysis. Diabetes, Obesity & Metabolism. 2021.
[25] Peterson GE. Intermediate and long-acting insulins: A review of NPH insulin, insulin glargine and insulin detemir. Current Medical Research and Opinion. 2006;22(12):2613–19.
[26] Tricco AC, Ashoor HM, Antony J, Beyene J, Veroniki AA, Isaranuwatchai W, et al. Safety, effectiveness, and cost effectiveness of long acting versus intermediate acting insulin for patients with type 1 diabetes: Systematic review and network meta-analysis. Bmj. 2014:349.
[27] Pouwer F, Hermanns N. Insulin therapy and quality of life. A Review. Diabetes/metabolism Research and Reviews. 2009;25(S1):S4–S10.

[28] Bech EM, Pedersen SL, Jensen KJ. Chemical Strategies for Half-Life Extension of Biopharmaceuticals: Lipidation and its Alternatives: ACS Publications; 2018. p. 577–80.
[29] Sleep D, Cameron J, Evans LR. Albumin as a versatile platform for drug half-life extension. Biochimica Et Biophysica Acta (Bba)-general Subjects. 2013;1830(12):5526–34.
[30] Andes D, Craig WA. Animal model pharmacokinetics and pharmacodynamics: A critical review. International Journal of Antimicrobial Agents. 2002;19(4):261–68.
[31] Jambhekar SS, Breen PJ. Basic Pharmacokinetics: Pharmaceutical Press London; 2009.
[32] Bertrand N, Leroux J-C. The journey of a drug-carrier in the body: An anatomo-physiological perspective. Journal of Controlled Release. 2012;161(2):152–63.
[33] Bicker J, Alves G, Fortuna A, Falcão A. Blood–brain barrier models and their relevance for a successful development of CNS drug delivery systems: A review. European Journal of Pharmaceutics and Biopharmaceutics. 2014;87(3):409–32.
[34] Smith QR. A review of blood-brain barrier transport techniques. The Blood-Brain Barrier. 2003:193–208.
[35] Davson H. Review lecture. The blood-brain barrier. The Journal of Physiology. 1976;255 (1):1–28.
[36] Kamaly N, Yameen B, Wu J, Farokhzad OC. Degradable controlled-release polymers and polymeric nanoparticles: Mechanisms of controlling drug release. Chemical Reviews. 2016;116(4):2602–63.
[37] Rothstein SN, Federspiel WJ, Little SR. A unified mathematical model for the prediction of controlled release from surface and bulk eroding polymer matrices. Biomaterials. 2009;30 (8):1657–64.
[38] Qiu J, Xu J, Xia Y. Nanobottles for controlled release and drug delivery. Advanced Healthcare Materials. 2021;10(4):2000587.
[39] Gu H, Mu S, Qiu G, Liu X, Zhang L, Yuan Y, et al.Redox-stimuli-responsive drug delivery systems with supramolecular ferrocenyl-containing polymers for controlled release. Coordination Chemistry Reviews. 2018;364:51–85.
[40] Chen X, Cheng X, Soeriyadi AH, Sagnella SM, Lu X, Scott JA, et al. Stimuli-responsive functionalized mesoporous silica nanoparticles for drug release in response to various biological stimuli. Biomaterials Science. 2014;2(1):121–30.
[41] Bhowmik D, Gopinath H, Kumar BP, Duraivel S, Kumar KS. Controlled release drug delivery systems. The Pharma Innovation. 2012;1:10.
[42] Dash S, Murthy PN, Nath L, Chowdhury P. Kinetic modeling on drug release from controlled drug delivery systems. Acta Poloniae Pharmaceutica. 2010;67(3):217–23.
[43] Tang L, Zhao C-Y, Wang X-H, Li R-S, Yang J-R, Huang Y-P, et al. Macromolecular crowding of molecular imprinting: A facile pathway to produce drug delivery devices for zero-order sustained release. International Journal of Pharmaceutics. 2015;496(2):822–33.
[44] Rambhia KJ, Ma PX. Controlled drug release for tissue engineering. Journal of Controlled Release. 2015;219:119–28.
[45] Kanjickal DG, Lopina ST. Modeling of drug release from polymeric delivery systems – a review. Critical Reviews™ in Therapeutic Drug Carrier Systems. 2004;21:5.
[46] Zhang L, Chan JM, Gu FX, Rhee J-W, Wang AZ, Radovic-Moreno AF, et al. Self-assembled lipid– polymer hybrid nanoparticles: A robust drug delivery platform. ACS Nano. 2008;2(8):1696–702.
[47] Zhang H, Liu D, Shahbazi MA, Mäkilä E, Herranz-Blanco B, Salonen J, et al. Fabrication of a multifunctional Nano-in-micro drug delivery platform by microfluidic templated encapsulation of porous silicon in polymer matrix. Advanced Materials. 2014;26 (26):4497–503.

[48] Khalid M, El-Sawy HS. Polymeric nanoparticles: Promising platform for drug delivery. International Journal of Pharmaceutics. 2017;528(1–2):675–91.
[49] Das M, Nariya P, Joshi A, Vohra A, Devkar R, Seshadri S, et al.Carbon nanotube embedded cyclodextrin polymer derived injectable nanocarrier: A multiple faceted platform for stimulation of multi-drug resistance reversal. Carbohydrate Polymers. 2020;247:116751.
[50] Fleige E, Quadir MA, Haag R. Stimuli-responsive polymeric nanocarriers for the controlled transport of active compounds: Concepts and applications. Advanced Drug Delivery Reviews. 2012;64(9):866–84.
[51] Xin Y, Yuan J. Schiff's base as a stimuli-responsive linker in polymer chemistry. Polymer Chemistry. 2012;3(11):3045–55.
[52] Ud Din F, Aman W, Ullah I, Qureshi OS, Mustapha O, Shafique S, et al. Effective use of nanocarriers as drug delivery systems for the treatment of selected tumors. International Journal of Nanomedicine. 2017;12:7291.
[53] Dang W, Yi K, Ju E, Jin Y, Xu Y, Wang H, et al. 3D printed bioceramic Scaffolds as a universal therapeutic platform for synergistic therapy of osteosarcoma. ACS Applied Materials & Interfaces. 2021;13(16):18488–99.
[54] Arcos D, Vallet-Regi M. Bioceramics for drug delivery. Acta Materialia. 2013;61(3):890–911.
[55] Li Y, Jin J, Wang D, Lv J, Hou K, Liu Y, et al.Coordination-responsive drug release inside gold nanorod@ metal-organic framework core–shell nanostructures for near-infrared-induced synergistic chemo-photothermal therapy. Nano Research. 2018;11(6):3294–305.
[56] He L, Liu Y, Lau J, Fan W, Li Q, Zhang C, et al. Recent progress in nanoscale metal-organic frameworks for drug release and cancer therapy. Nanomedicine. 2019;14(10):1343–65.
[57] Zhu M, Wang H, Liu J, He H, Hua X, He Q, et al. A mesoporous silica nanoparticulate/ β-TCP/BG composite drug delivery system for osteoarticular tuberculosis therapy. Biomaterials. 2011;32(7):1986–95.
[58] Mondal S, Pal U 3D hydroxyapatite scaffold for bone regeneration and local drug delivery applications. Journal of Drug Delivery Science and Technology. 2019;53:101131.
[59] Bolaños K, Kogan MJ, Araya E. Capping gold nanoparticles with albumin to improve their biomedical properties. International Journal of Nanomedicine. 2019;14:6387.
[60] Sierpe R, Lang E, Jara P, Guerrero AR, Chornik B, Kogan MJ, et al. Gold nanoparticles interacting with β-cyclodextrin–phenylethylamine inclusion complex: A ternary system for photothermal drug release. ACS Applied Materials & Interfaces. 2015;7(28):15177–88.
[61] Salmon AM, Dwyer R, Jauncey M, van Beek I, Topp L, Maher L. Injecting-related injury and disease among clients of a supervised injecting facility. Drug and Alcohol Dependence. 2009;101(1–2):132–36.
[62] Wright T, Hope V, Ciccarone D, Lewer D, Scott J, Harris M. Prevalence and severity of abscesses and cellulitis, and their associations with other health outcomes, in a community-based study of people who inject drugs in London, UK. Plos One. 2020;15(7):e0235350.
[63] Larney S, Peacock A, Mathers BM, Hickman M, Degenhardt L. A systematic review of injecting-related injury and disease among people who inject drugs. Drug and Alcohol Dependence. 2017;171:39–49.
[64] Simin D, Milutinović D, Turkulov V, Brkić S. Incidence, severity and risk factors of peripheral intravenous cannula-induced complications: An observational prospective study. Journal of Clinical Nursing. 2019;28(9–10):1585–99.
[65] Kawada K, Ohta T, Tanaka K, Kadoguchi N, Yamamoto S, Morimoto M. Risk factors of nicardipine-related phlebitis in acute stroke patients. Journal of Stroke and Cerebrovascular Diseases. 2016;25(10):2513–18.
[66] Reust CE. Common adverse effects of antiretroviral therapy for HIV disease. American Family Physician. 2011;83(12):1443–51.

[67] Tovey MG, Lallemand C. Immunogenicity and other problems associated with the use of biopharmaceuticals. Therapeutic Advances in Drug Safety. 2011;2(3):113–28.

[68] Hoang Thi TT, Pilkington EH, Nguyen DH, Lee JS, Park KD, Truong NP. The importance of poly (ethylene glycol) alternatives for overcoming PEG immunogenicity in drug delivery and bioconjugation. Polymers. 2020;12(2):298.

[69] Sharpe LA, Daily AM, Horava SD, Peppas NA. Therapeutic applications of hydrogels in oral drug delivery. Expert Opinion on Drug Delivery. 2014;11(6):901–15.

[70] Zhang Y, Benet LZ. The gut as a barrier to drug absorption. Clinical Pharmacokinetics. 2001;40(3):159–68.

[71] Spanogiannopoulos P, Bess EN, Carmody RN, Turnbaugh PJ. The microbial pharmacists within us: A metagenomic view of xenobiotic metabolism. Nature Reviews. Microbiology. 2016;14(5):273–87.

[72] Quan L, Zhang Y, Crielaard BJ, Dusad A, Lele SM, Rijcken CJ, et al. Nanomedicines for inflammatory arthritis: Head-to-head comparison of glucocorticoid-containing polymers, micelles, and liposomes. ACS Nano. 2014;8(1):458–66.

[73] Al Sawaftah N, Paul V, Awad N, Husseini GA. Modeling of Anti-cancer Drug Release Kinetics from Liposomes and Micelles: A review. IEEE Transactions on NanoBioscience. 2021.

[74] Wani TU, Raza SN, Khan NA. Nanoparticle opsonization: Forces involved and protection by long chain polymers. Polymer Bulletin. 2020;77(7):3865–89.

[75] Owens III DE, Peppas NA. Opsonization, biodistribution, and pharmacokinetics of polymeric nanoparticles. International Journal of Pharmaceutics. 2006;307(1):93–102.

[76] Meng H, Xue M, Xia T, Ji Z, Tarn DY, Zink JI, et al. Use of size and a copolymer design feature to improve the biodistribution and the enhanced permeability and retention effect of doxorubicin-loaded mesoporous silica nanoparticles in a murine xenograft tumor model. ACS Nano. 2011;5(5):4131–44.

[77] Oh JY, Kim HS, Palanikumar L, Go EM, Jana B, Park SA, et al. Cloaking nanoparticles with protein corona shield for targeted drug delivery. Nature Communications. 2018;9(1):1–9.

[78] Barbero F, Russo L, Vitali M, Piella J, Salvo I, Borrajo ML, et al. Formation of the Protein Corona: The Interface between Nanoparticles and the Immune System. Seminars in Immunology: Elsevier; 2017. p. 52–60.

[79] Gao H, He Q. The interaction of nanoparticles with plasma proteins and the consequent influence on nanoparticles behavior. Expert Opinion on Drug Delivery. 2014;11(3):409–20.

[80] Gref R, Lück M, Quellec P, Marchand M, Dellacherie E, Harnisch S, et al. 'Stealth'corona-core nanoparticles surface modified by polyethylene glycol (PEG): Influences of the corona (PEG chain length and surface density) and of the core composition on phagocytic uptake and plasma protein adsorption. Colloids and Surfaces. B, Biointerfaces. 2000;18(3–4):301–13.

[81] Zhang T-X, Zhu G-Y, Lu B-Y, Zhang C-L, Peng Q. Concentration-dependent protein adsorption at the nano–bio interfaces of polymeric nanoparticles and serum proteins. Nanomedicine. 2017;12(22):2757–69.

[82] Murthy AK, Stover RJ, Hardin WG, Schramm R, Nie GD, Gourisankar S, et al. Charged gold nanoparticles with essentially zero serum protein adsorption in undiluted fetal bovine serum. Journal of the American Chemical Society. 2013;135(21):7799–802.

[83] Shetab Boushehri MA, Lamprecht A. Challenges of using lipopolysaccharides for cancer immunotherapy and potential delivery-based solutions thereto. Therapeutic Delivery. 2019;10(3):165–87.

[84] Chauhan G, Madou MJ, Kalra S, Chopra V, Ghosh D, Martinez-Chapa SO. Nanotechnology for COVID-19: Therapeutics and vaccine research. ACS Nano. 2020;14(7):7760–82.

[85] Aminin D, Wang YM. Macrophages as a "weapon" in anticancer cellular immunotherapy. The Kaohsiung Journal of Medical Sciences. 2021.

[86] Hayat SMG, Bianconi V, Pirro M, Sahebkar A. Stealth functionalization of biomaterials and nanoparticles by CD47 mimicry. International Journal of Pharmaceutics. 2019;569:118628.
[87] Gheibi Hayat SM, Jaafari MR, Hatamipour M, Jamialahmadi T, Sahebkar A. Harnessing CD47 mimicry to inhibit phagocytic clearance and enhance anti-tumor efficacy of nanoliposomal doxorubicin. Expert Opinion on Drug Delivery. 2020;17(7):1049–58.
[88] Vandchali NR, Moadab F, Taghizadeh E, Tajbakhsh A, Gheibihayat SM. CD47 functionalization of nanoparticles as a poly (ethylene glycol) alternative: A novel approach to improve drug delivery. Current Drug Targets. 2021;22(15):1750–59.
[89] Sun H, Dong Y, Feijen J, Zhong Z. Peptide-decorated polymeric nanomedicines for precision cancer therapy. Journal of Controlled Release. 2018;290:11–27.
[90] Barui AK, Oh JY, Jana B, Kim C, Ryu JH. Cancer-targeted nanomedicine: Overcoming the barrier of the protein corona. Advanced Therapeutics. 2020;3(1):1900124.
[91] Parodi A, Quattrocchi N, Van De Ven AL, Chiappini C, Evangelopoulos M, Martinez JO, et al. Synthetic nanoparticles functionalized with biomimetic leukocyte membranes possess cell-like functions. Nature Nanotechnology. 2013;8(1):61–68.
[92] Parodi A, Molinaro R, Sushnitha M, Evangelopoulos M, Martinez JO, Arrighetti N, et al. Bio-inspired engineering of cell-and virus-like nanoparticles for drug delivery. Biomaterials. 2017;147:155–68.
[93] Wilson B, Samanta MK, Santhi K, Kumar KPS, Paramakrishnan N, Suresh B. Poly (n-butylcyanoacrylate) nanoparticles coated with polysorbate 80 for the targeted delivery of rivastigmine into the brain to treat Alzheimer's disease. Brain Research. 2008;1200:159–68.
[94] Odenwald MA, Turner JR. The intestinal epithelial barrier: A therapeutic target?. Nature Reviews. Gastroenterology & Hepatology. 2017;14(1):9–21.
[95] Chimmalgi M, Sant S. Study of foetal stomach under light microscope. Journal of the Anatomical Society of India. 2005;54(2):22–27.
[96] Zaher M, El-Ghareeb A-W, Hamdi H, AbuAmod F. Anatomical, histological and histochemical adaptations of the avian alimentary canal to their food habits: I-Coturnix coturnix. Life Science Journal. 2012;9(3):253–75.
[97] Dimaline R, Varro A. Attack and defence in the gastric epithelium–a delicate balance. Experimental Physiology. 2007;92(4):591–601.
[98] Pinto JF. Site-specific drug delivery systems within the gastro-intestinal tract: From the mouth to the colon. International Journal of Pharmaceutics. 2010;395(1–2):44–52.
[99] Sinha V, Kumria R. Binders for colon specific drug delivery: An in vitro evaluation. International Journal of Pharmaceutics. 2002;249(1–2):23–31.
[100] Liu L, Yao W, Rao Y, Lu X, Gao J. pH-Responsive carriers for oral drug delivery: Challenges and opportunities of current platforms. Drug Delivery. 2017;24(1):569–81.
[101] Fadda H, Basit A. Dissolution of pH responsive formulations in media resembling intestinal fluids: Bicarbonate versus phosphate buffers. Journal of Drug Delivery Science and Technology. 2005;15(4):273–79.
[102] Li K, Zhang L, Li J, Xue Y, Zhou J, Han Y. pH-Responsive ECM Coating on Ti Implants for Antibiosis in Reinfected Models. ACS Applied Bio Materials. 2021.
[103] Thoma K, Bechtold K. Influence of aqueous coatings on the stability of enteric coated pellets and tablets. European Journal of Pharmaceutics and Biopharmaceutics. 1999;47(1):39–50.
[104] Park HJ, Lee GH, Jun JH, Son M, Choi YS, Choi MK, et al. Formulation and in vivo evaluation of probiotics-encapsulated pellets with hydroxypropyl methylcellulose acetate succinate (HPMCAS). Carbohydrate Polymers. 2016;136:692–99.
[105] Fang Y, Wang G, Zhang R, Liu Z, Liu Z, Wu X, et al. Eudragit L/HPMCAS blend enteric-coated lansoprazole pellets: Enhanced drug stability and oral bioavailability. AAPS PharmSciTech. 2014;15(3):513–21.

[106] Zaid AN. Development and stability evaluation of enteric coated Diclofenac sodium tablets using Sureteric. Pakistan Journal of Pharmaceutical Sciences. 2012;25(1):59–64.
[107] Thakral S, Thakral NK, Majumdar DK. Eudragit®: A technology evaluation. Expert Opinion on Drug Delivery. 2013;10(1):131–49.
[108] Yang Q, Yuan F, Xu L, Yan Q, Yang Y, Wu D, et al. An update of moisture barrier coating for drug delivery. Pharmaceutics. 2019;11(9):436.
[109] Khatri P, Desai D, Shelke N, Minko T. Role of plasticizer in membrane coated extended release oral drug delivery system. Journal of Drug Delivery Science and Technology. 2018;44:231–43.
[110] Lin WJ, Lee HK, Wang DM. The influence of plasticizers on the release of theophylline from microporous-controlled tablets. Journal of Controlled Release. 2004;99(3):415–21.
[111] Twarog C, Fattah S, Heade J, Maher S, Fattal E, Brayden DJ. Intestinal permeation enhancers for oral delivery of macromolecules: A comparison between salcaprozate sodium (SNAC) and sodium caprate (C10). Pharmaceutics. 2019;11(2):78.
[112] Fattah S, Ismaiel M, Murphy B, Rulikowska A, Frias JM, Winter DC, et al. Salcaprozate sodium (SNAC) enhances permeability of octreotide across isolated rat and human intestinal epithelial mucosae in Ussing chambers. European Journal of Pharmaceutical Sciences. 2020;154:105509.
[113] Bannunah AM, Vllasaliu D, Lord J, Stolnik S. Mechanisms of nanoparticle internalization and transport across an intestinal epithelial cell model: Effect of size and surface charge. Molecular Pharmaceutics. 2014;11(12):4363–73.
[114] Almalik A, Karimi S, Ouasti S, Donno R, Wandrey C, Day PJ, et al. Hyaluronic acid (HA) presentation as a tool to modulate and control the receptor-mediated uptake of HA-coated nanoparticles. Biomaterials. 2013;34(21):5369–80.
[115] Xu S, Olenyuk BZ, Okamoto CT, Hamm-Alvarez SF. Targeting receptor-mediated endocytotic pathways with nanoparticles: Rationale and advances. Advanced Drug Delivery Reviews. 2013;65(1):121–38.
[116] Bourganis V, Kammona O, Alexopoulos A, Kiparissides C. Recent advances in carrier mediated nose-to-brain delivery of pharmaceutics. European Journal of Pharmaceutics and Biopharmaceutics. 2018;128:337–62.
[117] Reinholz J, Landfester K, Mailänder V. The challenges of oral drug delivery via nanocarriers. Drug Delivery. 2018;25(1):1694–705.
[118] Roger E, Kalscheuer S, Kirtane A, Guru BR, Grill AE, Whittum-Hudson J, et al. Folic acid functionalized nanoparticles for enhanced oral drug delivery. Molecular Pharmaceutics. 2012;9(7):2103–10.
[119] Des Rieux A, Fievez V, Théate I, Mast J, Préat V, Schneider Y-J. An improved in vitro model of human intestinal follicle-associated epithelium to study nanoparticle transport by M cells. European Journal of Pharmaceutical Sciences. 2007;30(5):380–91.
[120] Schattling P, Taipaleenmäki E, Zhang Y, Städler B. A polymer chemistry point of view on mucoadhesion and mucopenetration. Macromolecular Bioscience. 2017;17(9):1700060.
[121] Araujo GP, Martins FT, Taveira SF, Cunha-Filho M, Marreto RN. Effects of formulation and manufacturing process on drug release from solid self-emulsifying drug delivery systems prepared by high shear mixing. AAPS PharmSciTech. 2021;22(8):254.
[122] Derakhshankhah H, Jafari S. Cell penetrating peptides: A concise review with emphasis on biomedical applications. Biomedicine & Pharmacotherapy. 2018;108:1090–96.
[123] Kerkis A, Hayashi MA, Yamane T, Kerkis I. Properties of cell penetrating peptides (CPPs). IUBMB Life. 2006;58(1):7–13.
[124] Milletti F. Cell-penetrating peptides: Classes, origin, and current landscape. Drug Discovery Today. 2012;17(15–16):850–60.

[125] Ding H-M, Ma Y-Q. Role of physicochemical properties of coating ligands in receptor-mediated endocytosis of nanoparticles. Biomaterials. 2012;33(23):5798–802.

[126] Uematsu S, Akira S. Toll-Like Receptors (TLRs) and their Ligands. Toll-like Receptors (TLRs) and Innate Immunity. 2008. p. 1–20.

[127] Johansson ME, Sjövall H, Hansson GC. The gastrointestinal mucus system in health and disease. Nature Reviews. Gastroenterology & Hepatology. 2013;10(6):352–61.

[128] Li L, Lieleg O, Jang S, Ribbeck K, Han J. A microfluidic in vitro system for the quantitative study of the stomach mucus barrier function. Lab on a Chip. 2012;12(20):4071–79.

[129] Rubin BK. Mucus, phlegm, and sputum in cystic fibrosis. Respiratory Care. 2009;54 (6):726–32.

[130] Ma J, Rubin BK, Voynow JA. Mucins, mucus, and goblet cells. Chest. 2018;154(1):169–76.

[131] Ostedgaard LS, Moninger TO, McMenimen JD, Sawin NM, Parker CP, Thornell IM, et al. Gel-forming mucins form distinct morphologic structures in airways. Proceedings of the National Academy of Sciences. 2017;114(26):6842–47.

[132] Ermund A, Meiss LN, Rodriguez-Pineiro AM, Bähr A, Nilsson HE, Trillo-Muyo S, et al. The normal trachea is cleaned by MUC5B mucin bundles from the submucosal glands coated with the MUC5AC mucin. Biochemical and Biophysical Research Communications. 2017;492(3):331–37.

[133] Gómez-Guillén MC, Montero MP. Enhancement of oral bioavailability of natural compounds and probiotics by mucoadhesive tailored biopolymer-based nanoparticles: A review. Food Hydrocolloids. 2021:106772.

[134] Sarparanta MP, Bimbo LM, Mäkilä EM, Salonen JJ, Laaksonen PH, Helariutta AK, et al. The mucoadhesive and gastroretentive properties of hydrophobin-coated porous silicon nanoparticle oral drug delivery systems. Biomaterials. 2012;33(11):3353–62.

[135] Jung T, Kamm W, Breitenbach A, Kaiserling E, Xiao J, Kissel T. Biodegradable nanoparticles for oral delivery of peptides: Is there a role for polymers to affect mucosal uptake?. European Journal of Pharmaceutics and Biopharmaceutics. 2000;50(1):147–60.

[136] Saqafi B, Rahbarizadeh F. Effect of PEI surface modification with PEG on cytotoxicity and transfection efficiency. Micro & Nano Letters. 2018;13(8):1090–95.

[137] Oh N, Park J-H. Endocytosis and exocytosis of nanoparticles in mammalian cells. International Journal of Nanomedicine. 2014;9(Suppl 1):51.

[138] Wu L, Shan W, Zhang Z, Huang Y. Engineering nanomaterials to overcome the mucosal barrier by modulating surface properties. Advanced Drug Delivery Reviews. 2018;124:150–63.

[139] Chen X, Liu L, Jiang C. Charge-reversal nanoparticles: Novel targeted drug delivery carriers. Acta Pharmaceutica Sinica B. 2016;6(4):261–67.

[140] Thomas RG, Surendran SP, Jeong YY. Tumor Microenvironment-Stimuli Responsive Nanoparticles for Anticancer Therapy. Frontiers in Molecular Biosciences. 2020;7:414.

[141] Yu W, Liu R, Zhou Y, Gao H. Size-tunable strategies for a tumor targeted drug delivery system. ACS Central Science. 2020;6(2):100–16.

[142] Bernkop-Schnürch A. Strategies to overcome the polycation dilemma in drug delivery. Advanced Drug Delivery Reviews. 2018;136:62–72.

[143] Bartlett RL, Medow MR, Panitch A, Seal B. Hemocompatible poly (NIPAm-MBA-AMPS) colloidal nanoparticles as carriers of anti-inflammatory cell penetrating peptides. Biomacromolecules. 2012;13(4):1204–11.

[144] Manavalan B, Subramaniyam S, Shin TH, Kim MO, Lee G. Machine-learning-based prediction of cell-penetrating peptides and their uptake efficiency with improved accuracy. Journal of Proteome Research. 2018;17(8):2715–26.

[145] Lin J, Alexander-Katz A. Cell membranes open “doors” for cationic nanoparticles/ biomolecules: Insights into uptake kinetics. Acs Nano. 2013;7(12):10799–808.

[146] Di Pisa M, Chassaing G, Swiecicki J-M. Translocation mechanism (s) of cell-penetrating peptides: Biophysical studies using artificial membrane bilayers. Biochemistry. 2015;54 (2):194–207.
[147] Wang XD, Rabe KS, Ahmed I, Niemeyer CM. Multifunctional silica nanoparticles for covalent immobilization of highly sensitive proteins. Advanced Materials. 2015;27(48):7945–50.
[148] Qi Y, Wang J, Wang X, Cheng JJ, Wen Z. Selective adsorption of Pb (II) from aqueous solution using porous biosilica extracted from marine diatom biomass: Properties and mechanism. Applied Surface Science. 2017;396:965–77.
[149] Xu Y, Shrestha N, Préat V, Beloqui A. Overcoming the intestinal barrier: A look into targeting approaches for improved oral drug delivery systems. Journal of Controlled Release. 2020;322:486–508.
[150] Hashem L, Swedrowska M, Vllasaliu D. Intestinal uptake and transport of albumin nanoparticles: Potential for oral delivery. Nanomedicine. 2018;13(11):1255–65.
[151] Martins JP, Kennedy PJ, Santos HA, Barrias C, Sarmento B. A comprehensive review of the neonatal Fc receptor and its application in drug delivery. Pharmacology & Therapeutics. 2016;161:22–39.
[152] Zhang D-X, Esser L, Vasani RB, Thissen H, Voelcker NH. Porous silicon nanomaterials: Recent advances in surface engineering for controlled drug-delivery applications. Nanomedicine. 2019;14(24):3213–30.
[153] Martins JP, D'Auria R, Liu D, Fontana F, Ferreira MP, Correia A, et al. Engineered multifunctional albumin-decorated porous silicon nanoparticles for FcRn translocation of insulin. Small. 2018;14(27):1800462.
[154] Primec M, Škorjanc D, Langerholc T, Mičetić-Turk D, Gorenjak M. Specific Lactobacillus probiotic strains decrease transepithelial glucose transport through GLUT2 downregulation in intestinal epithelial cell models. Nutrition Research. 2021;86:10–22.
[155] Abramson A, Frederiksen MR, Vegge A, Jensen B, Poulsen M, Mouridsen B, et al. Oral delivery of systemic monoclonal antibodies, peptides and small molecules using gastric auto-injectors. Nature Biotechnology. 2021:1–7.
[156] Zizzari AT, Pliatsika D, Gall FM, Fischer T, Riedl R. New perspectives in oral peptide delivery. Drug Discovery Today. 2021.
[157] Abramson A, Caffarel-Salvador E, Soares V, Minahan D, Tian RY, Lu X, et al. A luminal unfolding microneedle injector for oral delivery of macromolecules. Nature Medicine. 2019;25(10):1512–18.
[158] Wilson LM, Castle JR. Recent advances in insulin therapy. Diabetes Technology & Therapeutics. 2020;22(12):929–36.
[159] Modi NB, Wang B, Noveck RJ, Gupta SK. Dose-proportional and stereospecific pharmacokinetics of methylphenidate delivered using an osmotic, controlled-release oral delivery system. The Journal of Clinical Pharmacology. 2000;40(10):1141–49.
[160] Conley R, Gupta SK, Sathyan G. Clinical spectrum of the osmotic-controlled release oral delivery system (OROS), an advanced oral delivery form. Current Medical Research and Opinion. 2006;22(10):1879–92.
[161] Abramson A, Caffarel-Salvador E, Khang M, Dellal D, Silverstein D, Gao Y, et al. An ingestible self-orienting system for oral delivery of macromolecules. Science. 2019;363(6427):611–15.
[162] Ko MJ, Kang MJ, Ko KJ, Ki YO, Chang HJ, Kwon J-Y. Clinical usefulness of Schedule for Oral-Motor Assessment (SOMA) in children with dysphagia. Annals of Rehabilitation Medicine. 2011;35(4):477.
[163] Service RF. Pills give patients a shot inside the stomach. American Association for the Advancement of Science; 2019.

[164] Vllasaliu D, Thanou M, Stolnik S, Fowler R. Recent advances in oral delivery of biologics: Nanomedicine and physical modes of delivery. Expert Opinion on Drug Delivery. 2018;15(8):759–70.

[165] Søgaard PP, Lind M, Christiansen CR, Petersson K, Clauss A, Caffarel-Salvador E. Future perspectives of oral delivery of next generation therapies for treatment of skin diseases. Pharmaceutics. 2021;13(10):1722.

[166] Luo Z, Paunović N, Leroux J-C. Physical methods for enhancing drug absorption from the gastrointestinal tract. Advanced Drug Delivery Reviews. 2021.

[167] Qin S, Xu K, Nie B, Ji F, Zhang H. Approaches based on passive and active antibacterial coating on titanium to achieve antibacterial activity. Journal of Biomedical Materials Research. Part A. 2018;106(9):2531–39.

[168] Cha GD, Kang D, Lee J, Kim DH. Bioresorbable electronic implants: History, materials, fabrication, devices, and clinical applications. Advanced Healthcare Materials. 2019;8(11):1801660.

[169] Quarterman JC, Geary SM, Salem AK. Evolution of drug-eluting biomedical implants for sustained drug delivery. European Journal of Pharmaceutics and Biopharmaceutics. 2021;159:21–35.

[170] Stewart SA, Domínguez-Robles J, Donnelly RF, Larrañeta E. Implantable polymeric drug delivery devices: Classification, manufacture, materials, and clinical applications. Polymers. 2018;10(12):1379.

[171] Jain KK. An overview of drug delivery systems. Drug Delivery Systems. 2020:1–54.

[172] Fu Y, Kao WJ. Drug release kinetics and transport mechanisms of non-degradable and degradable polymeric delivery systems. Expert Opinion on Drug Delivery. 2010;7(4):429–44.

[173] Patil P, Uphade K, Saudagar R. A Review: Osmotic Drug Delivery System. Pharma Science Monitor. 2018;9:2.

[174] Tobias IS, Lee H, Engelmayr GC Jr, Macaya D, Bettinger CJ, Cima MJ. Zero-order controlled release of ciprofloxacin-HCl from a reservoir-based, bioresorbable and elastomeric device. Journal of Controlled Release. 2010;146(3):356–62.

[175] Khan W, Muntimadugu E, Jaffe M, Domb AJ. Implantable Medical Devices. Focal controlled drug delivery: Springer; 2014. p. 33–59.

[176] Kleiner LW, Wright JC, Wang Y. Evolution of implantable and insertable drug delivery systems. Journal of Controlled Release. 2014;181:1–10.

[177] Talha M, Ma Y, Lin Y, Pan Y, Kong X, Sinha O, et al. Corrosion performance of cold deformed austenitic stainless steels for biomedical applications. Corrosion Reviews. 2019;37(4):283–306.

[178] Herrlich S, Spieth S, Messner S, Zengerle R. Osmotic micropumps for drug delivery. Advanced Drug Delivery Reviews. 2012;64(14):1617–27.

[179] Pagar SP, Kale HB, Surawase RK, Pawar D, Bhamare S. A Review: Controlled Porosity Osmotic Pump and its Historical Framework. 2020.

[180] Liu R, Zhang S, Chen X. Injectable hydrogels for tendon and ligament tissue engineering. Journal of Tissue Engineering and Regenerative Medicine. 2020;14(9):1333–48.

[181] Liu J, Lin S, Liu X, Qin Z, Yang Y, Zang J, et al. Fatigue-resistant adhesion of hydrogels. Nature Communications. 2020;11(1):1–9.

[182] Wang T, Guo S, Zhang H, Chen Y, Cai Y. Injectable hydrogel delivering bone morphogenetic protein-2, vascular endothelial growth factor, and adipose-derived stem cells for vascularized bone tissue engineering. Journal of Drug Delivery Science and Technology. 2020;57:101637.

[183] Des Rieux A, De Berdt P, Ansorena E, Ucakar B, Damien J, Schakman O, et al. Vascular endothelial growth factor-loaded injectable hydrogel enhances plasticity in the injured spinal cord. Journal of Biomedical Materials Research. Part A. 2014;102(7):2345–55.
[184] Dhayani A, Kalita S, Mahato M, Srinath P, Vemula PK. Biomaterials for topical and transdermal drug delivery in reconstructive transplantation. Nanomedicine. 2019;14(20):2713–33.
[185] Cai L, Xu J, Yang Z, Tong R, Dong Z, Wang C, et al. Engineered biomaterials for cancer immunotherapy. MedComm. 2020;1(1):35–46.
[186] Kim JH, Shin JU, Kim SH, Noh JY, Kim HR, Lee J, et al. Successful transdermal allergen delivery and allergen-specific immunotherapy using biodegradable microneedle patches. Biomaterials. 2018;150:38–48.
[187] Han X, Li H, Zhou D, Chen Z, Gu Z. Local and targeted delivery of immune checkpoint blockade therapeutics. Accounts of Chemical Research. 2020;53(11):2521–33.
[188] Lee I-C, Wu Y-C, Tsai S-W, Chen C-H, Wu M-H. Fabrication of two-layer dissolving polyvinylpyrrolidone microneedles with different molecular weights for in vivo insulin transdermal delivery. Rsc Advances. 2017;7(9):5067–75.
[189] Al-Japairai KAS, Mahmood S, Almurisi SH, Venugopal JR, Hilles AR, Azmana M, et al. Current trends in polymer microneedle for transdermal drug delivery. International Journal of Pharmaceutics. 2020;587:119673.
[190] Jin X, Zhu DD, Chen BZ, Ashfaq M, Guo XD. Insulin delivery systems combined with microneedle technology. Advanced Drug Delivery Reviews. 2018;127:119–37.
[191] Sun W, Araci Z, Inayathullah M, Manickam S, Zhang X, Bruce MA, et al. Polyvinylpyrrolidone microneedles enable delivery of intact proteins for diagnostic and therapeutic applications. Acta Biomaterialia. 2013;9(8):7767–74.
[192] Jamaledin R, Di Natale C, Onesto V, Taraghdari ZB, Zare EN, Makvandi P, et al. Progress in microneedle-mediated protein delivery. Journal of Clinical Medicine. 2020;9(2):542.
[193] Prausnitz MR, Gomaa Y, Li W. Microneedle patch drug delivery in the gut. Nature Medicine. 2019;25(10):1471–72.
[194] Kolarsick PA, Kolarsick MA, Goodwin C. Anatomy and physiology of the skin. Journal of the Dermatology Nurses' Association. 2011;3(4):203–13.
[195] Schommer NN, Gallo RL. Structure and function of the human skin microbiome. Trends in Microbiology. 2013;21(12):660–68.
[196] Proksch E, Brandner JM, Jensen JM. The skin: An indispensable barrier. Experimental Dermatology. 2008;17(12):1063–72.
[197] Kalafa EG, Hixona K, Kadakia P, Dunn A, Sell S. 9.1. 1 Skin Anatomy and Physiology. Electrospun Materials for Tissue Engineering and Biomedical Applications: Research, Design and Commercialization. 2017. p. 179.
[198] Jeong WY, Kwon M, Choi HE, Kim KS. Recent advances in transdermal drug delivery systems: A review. Biomaterials Research. 2021;25(1):1–15.
[199] Carter P, Narasimhan B, Wang Q. Biocompatible nanoparticles and vesicular systems in transdermal drug delivery for various skin diseases. International Journal of Pharmaceutics. 2019;555:49–62.
[200] Azad Khan MY, Asif M, Chauhan I, Singh AP, Sharma R, Singh P, et al. Iontophoretic drug delivery: History and applications. Journal of Applied Pharmaceutical Science. 2011;1(03):11–24.
[201] Dixit N, Bali V, Baboota S, Ahuja A, Ali J. Iontophoresis-an approach for controlled drug delivery: A review. Current Drug Delivery. 2007;4(1):1–10.
[202] Seah BC-Q, Teo BM. Recent advances in ultrasound-based transdermal drug delivery. International Journal of Nanomedicine. 2018;13:7749.

[203] Chen B, Wei J, Iliescu C. Sonophoretic enhanced microneedles array (SEMA) – Improving the efficiency of transdermal drug delivery. Sensors and Actuators. B, Chemical. 2010;145(1):54–60.

[204] Meshali M, Abdel-Aleem H, Sakr F, Nazzal S, El-Malah Y. Effect of gel composition and phonophoresis on the transdermal delivery of ibuprofen: In vitro and in vivo evaluation. Pharmaceutical Development and Technology. 2011;16(2):93–101.

[205] Gehl J. Electroporation: Theory and methods, perspectives for drug delivery, gene therapy and research. Acta Physiologica Scandinavica. 2003;177(4):437–47.

[206] Liu Z, Liang X, Liu H, Wang Z, Jiang T, Cheng Y, et al. High-throughput and self-powered electroporation system for drug delivery assisted by microfoam electrode. ACS Nano. 2020;14(11):15458–67.

[207] McGraw E, Dissanayaka RH, Vaughan JC, Kunte N, Mills G, Laurent GM, et al. Laser-assisted delivery of molecules in fungal cells. ACS Applied Bio Materials. 2020;3(9):6167–76.

[208] Wong TW. Electrical, magnetic, photomechanical and cavitational waves to overcome skin barrier for transdermal drug delivery. Journal of Controlled Release. 2014;193:257–69.

[209] Soukos NS, Socransky SS, Mulholland SE, Lee S, Doukas AG. Photomechanical drug delivery into bacterial biofilms. Pharmaceutical Research. 2000;17(4):405–09.

[210] Nagarkar R, Singh M, Nguyen HX, Jonnalagadda S. A review of recent advances in microneedle technology for transdermal drug delivery. Journal of Drug Delivery Science and Technology. 2020:101923.

[211] Prausnitz MR. Microneedles for transdermal drug delivery. Advanced Drug Delivery Reviews. 2004;56(5):581–87.

[212] Yang SY, O'Cearbhaill ED, Sisk GC, Park KM, Cho WK, Villiger M, et al. A bio-inspired swellable microneedle adhesive for mechanical interlocking with tissue. Nature Communications. 2013;4(1):1–10.

[213] Jang S-J, Doshi T, Nerayo J, Caprio A, Alaie S, Auge J, et al. Microneedle patterning of 3D nonplanar surfaces on implantable medical devices using soft lithography. Micromachines. 2019;10(10):705.

[214] Cahill EM, Keaveney S, Stuettgen V, Eberts P, Ramos-Luna P, Zhang N, et al. Metallic microneedles with interconnected porosity: A scalable platform for biosensing and drug delivery. Acta Biomaterialia. 2018;80:401–11.

[215] Teymourian H, Tehrani F, Mahato K, Wang J. Lab under the skin: Microneedle based wearable devices. Advanced Healthcare Materials. 2021:2002255.

[216] Bulaklak K, Gersbach CA. The once and future gene therapy. Nature Communications. 2020;11(1):5820.

[217] Weissman D, Karikó K. mRNA: Fulfilling the promise of gene therapy. Molecular Therapy. 2015;23(9):1416–17.

[218] Caplen NJ. RNAi as a gene therapy approach. Expert Opinion on Biological Therapy. 2003;3(4):575–86.

[219] Ran FA, Hsu PD, Wright J, Agarwala V, Scott DA, Zhang F. Genome engineering using the CRISPR-Cas9 system. Nature Protocols. 2013;8(11):2281–308.

[220] Hoy SM. Patisiran: First global approval. Drugs. 2018;78(15):1625–31.

[221] Gonçalves GAR, Paiva RMA. Gene therapy: Advances, challenges and perspectives. Einstein (Sao Paulo). 2017;15(3):369–75.

[222] Silva G, Poirot L, Galetto R, Smith J, Montoya G, Duchateau P, et al. Meganucleases and other tools for targeted genome engineering: Perspectives and challenges for gene therapy. Current Gene Therapy. 2011;11(1):11–27.

[223] Hamilton BA, Wright JF. Challenges posed by immune responses to AAV Vectors: Addressing root causes. Frontiers in Immunology. 2021:12.

[224] Jafari M, Xu W, Pan R, Sweeting CM, Karunaratne DN, Chen P. Serum stability and physicochemical characterization of a novel amphipathic peptide C6M1 for SiRNA delivery. Plos One. 2014;9(5):e97797.

[225] Hu B, Zhong L, Weng Y, Peng L, Huang Y, Zhao Y, et al. Therapeutic siRNA: State of the art. Signal Transduction and Targeted Therapy. 2020;5(1):101.

[226] Zu H, Gao D. Non-viral vectors in gene therapy: Recent Development, challenges, and prospects. The AAPS Journal. 2021;23(4):78.

[227] Tenchov R, Bird R, Curtze AE, Zhou Q. Lipid nanoparticles – from liposomes to mRNA vaccine delivery, a landscape of research diversity and advancement. ACS Nano. 2021.

[228] Yurkovetskiy L, Wang X, Pascal KE, Tomkins-Tinch C, Nyalile TP, Wang Y, et al. Structural and functional analysis of the D614G SARS-CoV-2 spike protein variant. Cell. 2020;183 (3):739–51.e8.

[229] Corbett KS, Edwards DK, Leist SR, Abiona OM, Boyoglu-Barnum S, Gillespie RA, et al. SARS-CoV-2 mRNA vaccine design enabled by prototype pathogen preparedness. Nature. 2020;586 (7830):567–71.

[230] Fiedler K, Lazzaro S, Lutz J, Rauch S, Heidenreich R. mRNA cancer vaccines. Recent Results in Cancer Research. 2016;209:61–85.

[231] Bulcha JT, Wang Y, Ma H, Tai PWL, Gao G. Viral vector platforms within the gene therapy landscape. Signal Transduction and Targeted Therapy. 2021;6(1):53.

[232] Yin H, Kanasty RL, Eltoukhy AA, Vegas AJ, Dorkin JR, Anderson DG. Non-viral vectors for gene-based therapy. Nature Reviews. Genetics. 2014;15(8):541–55.

[233] Grieger JC, Soltys SM, Samulski RJ. Production of recombinant adeno-associated virus vectors using suspension HEK293 CELLS and continuous harvest of vector from the culture media for GMP FIX and FLT1 clinical vector. Molecular Therapy. 2016;24(2):287–97.

[234] von Kalle C, Deichmann A, Schmidt M. Vector integration and tumorigenesis. Human Gene Therapy. 2014;25(6):475–81.

[235] Dibra A, Mehilli J, Braun S, Hadamitzky M, Baum H, Dirschinger J, et al. Inflammatory response after intervention assessed by serial C-reactive protein measurements correlates with restenosis in patients treated with coronary stenting. American Heart Journal. 2005;150 (2):344–50.

[236] Drachman DE, Simon DI. Inflammation as a mechanism and therapeutic target for in-stent restenosis. Current Atherosclerosis Reports. 2005;7(1):44–49.

[237] Borhani S, Hassanajili S, Tafti SHA, Rabbani S. Cardiovascular stents: Overview, evolution, and next generation. Progress in Biomaterials. 2018;7(3):175–205.

[238] Aguilar LE, Unnithan AR, Amarjargal A, Tiwari AP, Hong ST, Park CH, et al. Electrospun polyurethane/Eudragit® L100-55 composite mats for the pH dependent release of paclitaxel on duodenal stent cover application. International Journal of Pharmaceutics. 2015;478 (1):1–8.

[239] Wu J, Liu Y, Cao Q, Yu T, Zhang J, Liu Q, et al. Growth factors enhanced angiogenesis and osteogenesis on polydopamine coated titanium surface for bone regeneration. Materials & Design. 2020;196:109162.

[240] Maver T, Hribernik S, Mohan T, Smrke DM, Maver U, Stana-Kleinschek K. Functional wound dressing materials with highly tunable drug release properties. RSC Advances. 2015;5(95):77873–84.

[241] Ambekar RS, Kandasubramanian B. Advancements in nanofibers for wound dressing: A review. European Polymer Journal. 2019;117:304–36.

Chapter 8
Biomaterials in tissue engineering

8.1 Tissue diseases and repair process

Regenerating or growing damaged tissue has always been a goal for us, humans, as it can help prolong the lives of people who suffer from various diseases. However, the human body has very limited capability to regenerate damaged organs and tissues, and even if it heals, scarring creates an imperfect replica. There are some organs in the human body that can regenerate to a certain degree, like the liver and the skin, but even these organs cannot create an exact regenerated copy of themselves. Human tissue in a healthy state naturally regenerates over time; and these tissues create new cells to replace old senescent cells. Some parts of the human body can fully replace a living tissue, given enough time. This regeneration and constant replacement of cells are profoundly expressed in the gastric mucosa with a cell turnout of 2 h, the human skin (epidermis) that can be replaced every 2 weeks, and the human bone within 10 years. But there are some tissues and organs that have a fixed number of cells or have a replication time that is so slow that if they are damaged, regeneration is almost close to impossible, and permanent impairment can result.

To classify this, these types of tissues are divided based on their capability to replicate and proliferate into: continuously dividing tissues, quiescent tissues, and nondividing tissues. Continuously dividing tissues (also known as labile tissues) are comprised of cells that are constantly proliferating to replace dead or sloughed-off cells; examples are epithelium cells found in the skin and gastrointestinal tract. These tissues are rich in stem cells, which helps in their self-renewing ability. Furthermore, stem cells can give rise to more than one type of cell, making them valuable in tissue regeneration. Each stem cell is known to have asymmetrical replication, that is, they can give rise to one daughter cell that differentiates and matures and another daughter cell that remains undifferentiated and capable of perpetuating the renewal cycle. Quiescent tissues [1], on the other hand, normally exist in a nondividing state but may enter the cell cycle in response to certain stimuli or pathology, such as cell injury or cancer. These quiescent tissues are found in the parenchymal cells of the liver, kidney, and pancreas, mesenchymal cells such as fibroblasts and smooth muscle cells, endothelial cells, and lymphocytes. Interestingly, the liver tissues, unlike other organs, have a relatively robust proliferative capacity. When a lobe of the liver is resected for donation, for example, the remaining liver cells proliferate at such a rate that the liver reaches the size prior to resection [2–4]. While this process is commonly described as regeneration, it is more accurately viewed as compensatory growth, since the original lobe itself does not regrow. Last, are tissues that are composed of cells that have left the cell cycle permanently and are therefore unable to proliferate. These nondividing tissues (or permanent tissues) include cardiac and neuronal tissues.

https://doi.org/10.1515/9783110655377-008

Tissue repair in these tissues always leaves permanent evidence of injury, also known as scarring.

There are multiple pathologies that can damage or alter these irreparable or difficult-to-repair tissues. These are physical injuries [5], degenerative and autoimmune diseases [6, 7], perfusion and ischemic accidents, and congenital defects that can all damage or alter organs with limited to detrimental effects. So, it is important to distinguish between these pathologies to create suitable strategies for regenerative materials.

8.1.1 Physical injuries and its healing process

A physical injury occurs when the body has experienced trauma from external sources; examples are blunt force that leads to bruising, fractures, and internal bleeding, extreme sprain and strains that dislocates limbs, and burns from electrical, chemical, and fire sources. All these physical injuries happen to a relatively healthy tissue with no underlying health condition. Therefore, the regular healing process will occur in the affected organ.

Healing is a multistage process that involves several key factors Figure 8.1 [8], and to put it simply, these are the cells, growth factors, and extracellular matrix. The cells both involve the immune and native cells around the injured site. The first step in the healing process is inflammation [9–11]. Inflammation is the immediate response of the body to help contain the damage around the injured site. Its main effect is to reduce the blood flow from the wound to avoid exacerbation of bleeding [12]. Immediately after the injury, the exposed collagen activates coagulation in the injured site, while platelets form an initial plug [13, 14], coagulation cascade factors interact to form fibrin and initiate blood clotting to prevent further bleeding out and entry of pathogens [15]. Furthermore, this clot also serves as a scaffold for incoming cells and as a repository of cytokines and growth factors [11]. Later, immune cells are recruited in the area by cytokines, mainly IL-1, TNF, TGF-B, etc. The immune cells will then clean up the area for dead or damaged cells and whatever bacterial organisms had entered the injury site [16]. At the same time, epithelial cells form basal layers of the epidermis into the injured site, depositing basement membrane components as they migrate. This cell action is modulated by growth factors, mainly EGF and TGF-a that come from activated platelets and macrophages. In addition to this cascade, fibroblast also secretes KGF and Il-6 to enhance keratinocyte migration and proliferation. This whole process happens during the inflammation stage, and while this is happening, exudates, increased localized temperature, and vasodilation are also exhibited by the injured site. Tenderness around the area and increased pain perception are also experienced by the injured person, as pain receptors are also highly affected by the inflammation process.

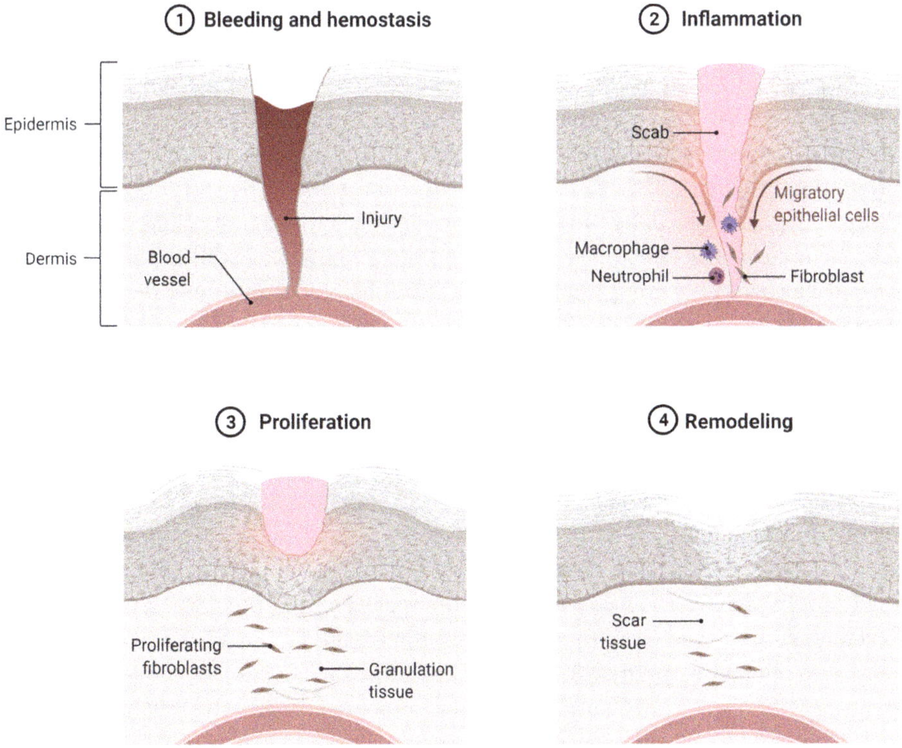

Figure 8.1: Wound healing process: (1) Injury occurs that ruptures the skin and breaks the superficial or deep vessels. Blood clotting factors can subside the bleeding, if the injury is small (hemostasis). (2) A scab will form on top of the wound followed by the migration of macrophages and neutrophils that will target pathogens that were able to enter the skin barrier. Epithelial cells will also migrate to the injury site to start the healing process. (3) Proliferation of fibroblast cells and formation of granulation tissue occurs. (4) The tissue remodels itself with the formation of a scar tissue.

Once the inflammation stage subsides, the proliferation stage begins. The hallmark of this stage is the formation of a substance called granulation tissue [17, 18]. This is characterized by the formation of new blood vessels and extracellular matrix laid by nascent fibroblast cells. These blood vessels are leaky at the beginning, leading to a loose appearance of the healing injured site and appear to be softer in comparison to the surrounding healthy tissue. But, as the vessels mature and fibroblasts create more collagen, it starts to conform to the surrounding tissue in physical properties. All the while, macrophages are functioning by ingesting unwanted cellular debris, bacteria, and foreign materials. In addition to this housekeeping role, macrophages also secrete PDGF, TGF-β, tumor necrosis factor (TNF), IL-1, and KGF to recruit more fibroblasts and keratinocytes; VEGF, FGF, and PDGF to increase angiogenesis [19, 20]; and TGF-β, PDGF, TNF [21], osteopontin (OPN) [22], IL-1, collagenase, and matrix metalloproteinases (MMPs) to facilitate tissue remodeling by production of extracellular matrix [23, 24].

During the latter part of the healing process, the maturation stage starts due to increased type I collagen synthesis in production by fibroblasts cells [25]. The granulation tissue is now being replaced by scar tissue composed of dense collagen, fibroblast, and elastic tissue [26, 27]. As time progresses, the fibroblasts begin to disappear, and, eventually, an acellular scar bridges the region of the wound, covered by intact, fully developed epidermis. Dermal appendages in the region of the wound, however, are not reformed. Given that the body cannot fully regain the loss tissue after an injury and scar tissue replacements are imperfect void fillers, the function of the injured site is also hampered even after a full recovery. This is true for connective tissue in the musculoskeletal system, where injured ligaments and tendons are usually impaired for life even after full rehabilitation. Range of motion of the limb and joints are severely reduced, and hence athletes may take several months to even years to be able to regain their athleticism, and some may have career-ending injuries.

There are also several underlying risk factors for impaired healing process in the body. In people who have diabetes, both types I and II, the granulation phase is severely reduced, and if an injury happens, the wound tends to linger and even fester due to microangiopathy [28, 29]. Poor nutrition may also inhibit the healing process, as it is a calorie and nutrient intensive process [30, 31]. The raw materials for cell, growth factors, and extracellular matrix formation come from food, and if the person suffers from inadequate nutrition, the healing process may also be impeded. Steroidal drugs can also slow down the wound healing process, as it is known to inhibit inflammation and collagen formation. The size and site of injury may also play a role on how fast the healing process is [32]; if the injury site is located where motion or friction frequently occurs, it may create pressure at the site and prolong the healing process. And, probably, the most notorious and detrimental factor leading to delayed wound-healing is bacterial colonization [33]. Bacterial organisms are always present and are seeking an opportunity to colonize the body for habitat and resources. Some pathogenic bacteria will secrete toxins and their mere presence prolongs the inflammation phase of the healing process [33]. The immune cells will be overburdened by the pathogens creating more exudates and pus in the injury site. If bacteria are present in the injury site, the proliferation and maturation stages of the healing process will never occur. In worst cases, the pathogens may enter the blood stream and lead to what is called septicemia, or infection of the blood, and spread around the body and kill the person.

Table 8.1: Wound healing steps and growth factors involved.

Process	Involved growth factors	Ref.
Fibroblast migration	FGF, PDGF	[34]
Fibroblast proliferation	EGF, FGF, PDGF, TGF-a, TNF	[34]
Monocyte migration	FGF, PDGF, TGF-b, TNF	[35]
Macrophage activation	TNF	[35]
Epithelial migration	EGF, FGF, KGF, TFG-a	[36]
Epithelial proliferation	EGF, FGF, KGF, TFG-a	[36]
Angiogenesis	FGF, PDGF, TGF-a, TNF, VEGF	[37]
Collagen synthesis	PDGF, TGF-b	[38]
Collagenase synthesis	EGF, FGF, PDGF,TNF,VEGF	[38]
Wound contraction	FGF, PDGF	[39]

EGF, epidermal growth factor; **FGF**, fibroblast growth factor; **KGF**, keratinocyte growth factor; **PDGF**, platelet-derived growth factor; **TGF**, transforming growth factor; **TNF**, tumor necrosis factor; **VEGF**, vascular endothelial growth factor

8.1.2 Degenerative and autoimmune diseases

A degenerative disease is defined as a disease characterized by worsening condition due to the deterioration of the function and structure of the affected body part, thus causing disability, mortality, and morbidity, which could be premature (Figure 8.2). As with all things in the universe, the human body experiences degeneration over time. Exposure to certain external factors such as toxins, radiation, stress, etc., can affect how fast this process will occur. Genetics also plays a role, as some people are genetically predisposed to developing diseases that deteriorate the healthy state of the body. Different parts of the body experience degeneration to the extent that it affects the body to maintain homeostasis. Aging is one of the common risk factors for many degenerative diseases. It entails diverse biological changes. At the cellular level, a cell is aging when it loses its ability to divide and yet remain metabolically active [40]. At the tissue and organ level, an aged biological structure is one that eventually wears off. Their structure and function are no longer as potent as they were when they were young. This is often attributed to the genetic changes inside the cell, particularly the shortening of the telomeres [41, 42]. Deterioration over time can occur because of telomere degeneration. Consequently, the cell loses its ability to function, upon reaching a certain point. When several cells lose their capacity to divide and to function effectively, the body would, therefore, be impaired. Disease manifests, as symptoms appear. Apart from natural degenerative changes

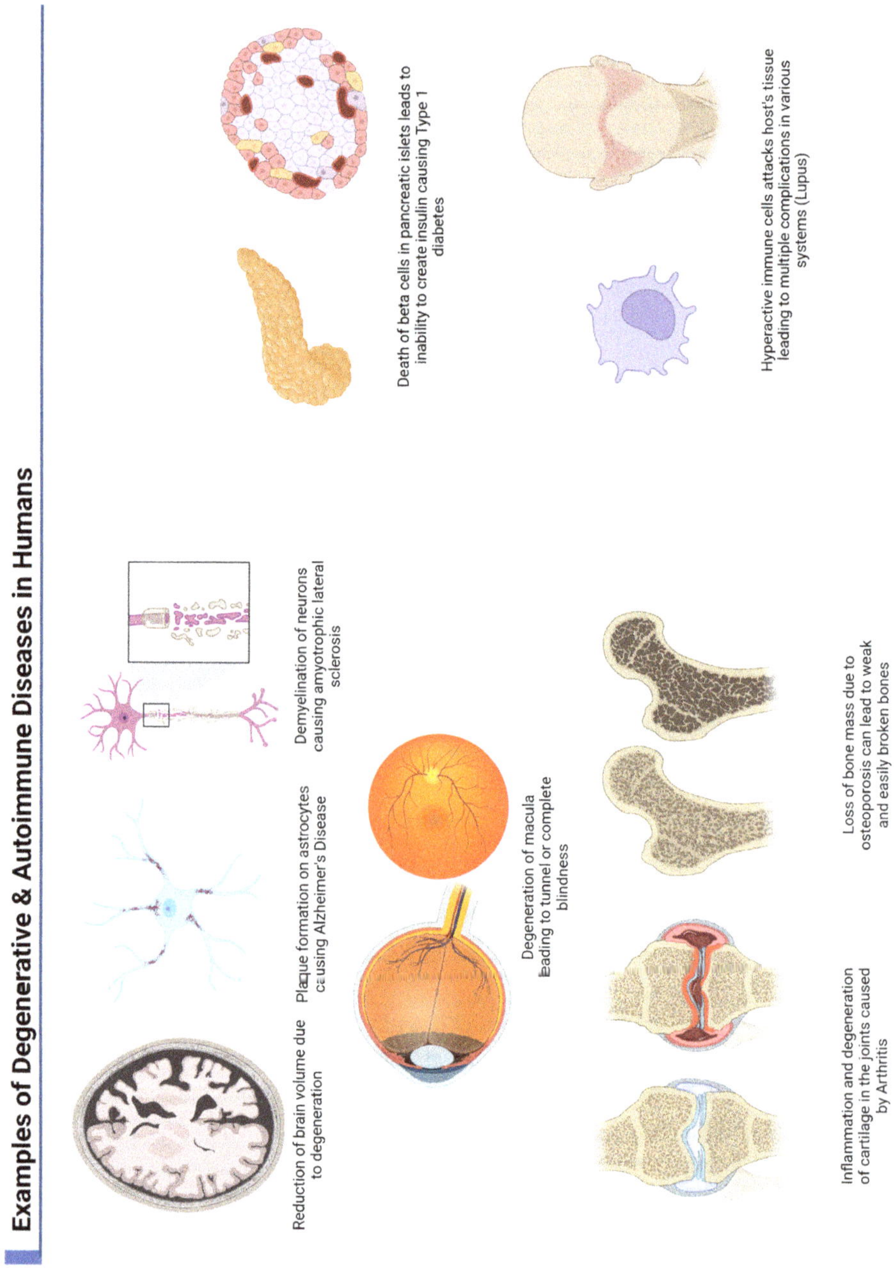

Figure 8.2: Degenerative and autoimmune diseases in humans. Nervous system: Alzheimer's and Parkinson's, and amyotrophic lateral sclerosis. Eyes: macular degeneration. Skeletal system: osteoarthritis and osteoporosis. Pancreas: Type 1 diabetes. Lupus: the immune system attacks healthy cells, tissues, and organs.

caused by DNA damage accumulation from aging, the body may also be susceptible to many stressors, e.g., oxidative stress. Mitochondrial DNA mutations are also an important risk factor.

In particular, the nervous system may experience neurodegenerative diseases such as Alzheimer's disease, Parkinson's disease, amyotrophic lateral sclerosis, Huntington's disease, and multiple sclerosis [43], all of which are characterized by excessive neuronal death (apoptosis) and impairment of their bioactivity via abnormal protein depositions (amyloid-β and tau proteins – Alzheimer's), loss of dopaminergic neurons (Parkinson's), and caspase alterations, as well as certain gene mutations leading to cell death (Amyotrophic lateral sclerosis, Huntington's disease, and multiple sclerosis). Alzheimer's and Parkinson's diseases are the most common neuronal degenerative disease with grave consequences. Alzheimer's is characterized by the overproduction of plaque forming proteins (Amyloid-β and Tau proteins) which condense and cause impairment in neuronal function in the brain. A person with Alzheimer's disease experiences progressive loss of memory and cognitive function due to reduced neuronal connection density and shrinkage in overall brain volume. A person with Alzheimer's would need constant assistance to ensure that they will not endanger themselves. Parkinson's disease, on the other hand, is characterized by death or loss of dopaminergic neurons, which causes dyskinesia (involuntary movements) and stiffening of the muscles in the body. In comparison to Alzheimer's disease, people who have Parkinson's do not exhibit cognitive or memory loss. Huntington's disease is a disease characterized by mental and physical deterioration leading to death. The individual with this disease initially lacks coordination and then, eventually, loses a steady gait, coordinated movement, and mental abilities. Amyotrophic lateral sclerosis (ALS) is a disease caused by the death of neurons that control the voluntary muscles. Initially, there is a weakening of the arms or legs. Eventually, the person with ALS will lose the ability to use hands and feet, speak, and swallow. Multiple sclerosis is a neurodegenerative disease characterized by the gradual accumulation of focal plaques of demyelination particularly in the periventricular areas of the brain. The progression of all these diseases can be slowed down with cocktails of drugs, but a complete cure is still elusive.

There are other examples of degenerative diseases in the body, one of which is macular degeneration [44]. The macula is part of the retina at the back of the eye. It is only about 5 mm across but is responsible for our central vision, most of our color vision, and the fine details of what we see. The macula has a very high concentration of photoreceptor cells – the cells that detect light. They send signals to the brain, which interprets them as images. The rest of the retina processes our peripheral or side vision. Macular degeneration can cause tunnel blindness due to the slow atrophy of the pigmented retinal epithelium that supports the photoreceptors in the eye. If left untreated, this will cause permanent total blindness. The RPE cells cannot regenerate and slowing down the progression of macular degeneration is the only means of treatment for this disease. There are currently two classifications of macular degeneration: the first is wet AMD, which is characterized by overgrowth of new leaky microvasculature on the subretinal space, leading to excessive bleeding that can cause the RPE to die; the other is called dry AMD, which is idiopathic, but also has the hallmarks of sudden atrophy of the supporting RPE cells that leads to the death of photoreceptors. Both photoreceptors and RPE cells are permanent and nondividing tissues, making this degenerative disease very hard to cure.

The skeletal system is also one of the most common body parts that degenerate. Osteoarthritis and osteoporosis are both classified as degenerative diseases that can limit the body's range of motion and locomotion [45, 46]. First, osteoarthritis occurs when the cartilage, which protects the joints and cushions each bone from directly having contact and creating friction, deteriorates over time. It is common to damage the joints in the hands, knee, and hips, and it is commonly observed in individuals of advanced age. The cartilage on these joints wears down due to the reduction of collagen and is difficult to replenish as cartilage, like bone, is surrounded by a perichondrium-like fibrous membrane. This layer is not efficient at regenerating cartilage. Hence, recovery is slow after injury. Furthermore, the lack of active blood flow is the major reason any injury to cartilage takes a long time to heal. Osteoarthritis is a severely debilitating condition and the pain that the person experiences usually leaves them immobile. Osteoporosis on the other hand, is the reduction of bone density making them fragile and more likely to break. Normal bone has porous microscopic morphology, but in an osteoporotic bone, these porosities are enlarged, leaving more voids inside the bone. The main cause of osteoporosis is a drop in the body's estrogen levels. Estrogen is a hormone that helps build and maintain your bones, but in women, estrogen levels significantly decrease due to menopause, and without supplemental estrogen, bone regeneration is reduced.

The immune system can also cause some organs and tissues to deteriorate; this is known as autoimmune disease. The body's immune system is trained to attack non-self-tissues and pathogens. However, in some individuals with overacting immunity, their own organs and tissues seem to be non-self and are attacked by their own immune cells. Some autoimmune diseases [47, 48] target only one organ, such as in Type 1 diabetes, which damages the pancreas, while other diseases, like systemic lupus erythematosus (SLE) [49, 50] affect the whole body.

Type I diabetes, also known as insulin dependent diabetes, is a condition in which the body is unable to produce insulin, a hormone that facilitates the transport of glucose across the membrane of the body's cells. This leads to increase in blood glucose levels, which, if left untreated, will lead to multiple complications. The primary cause of Type I diabetes (T1D), as mentioned, is the destruction of the pancreatic cells by the immune system (most specifically the beta cells), which produces the insulin needed by the body. Most people, who were diagnosed with T1D before the advent of synthetic insulin injections, died within one to two years of diagnosis. The discovery of insulin and its successful isolation back in the 1920s gave T1D patients better quality of life and a chance to survive the disease. However, insulin must be administered daily, sometimes multiple times a day, to ensure that the body can utilize and regulate the glucose in the blood, and if left untreated, T1D can lead to a myriad of complications such as damage in the microvasculature as well as in vital organs, such as the heart and kidneys.

The most common autoimmune condition that affects the whole body is lupus [51, 52]. It is a disease that occurs when your body's immune system attacks your own tissues and organs in a systemic manner. Inflammation caused by lupus can affect many different body systems including your joints, skin, kidneys, blood cells, brain, heart, and lungs. Lupus can be difficult to diagnose because its signs and symptoms often mimic those of other ailments. The most distinctive sign of lupus, a facial rash that resembles the wings of a butterfly unfolding across both cheeks, occurs in many, but not all, cases of lupus. Some people are born with a tendency toward developing lupus, which may be triggered by infections, certain drugs, or even sunlight. While there is no cure for lupus, treatments can help control symptoms.

Inflammation caused by lupus can affect many areas of your body, including the following:

- Kidneys: Lupus can cause serious kidney damage, and kidney failure is one of the leading causes of death among people with lupus.
- Brain and central nervous system: If your brain is affected by lupus, you may experience headaches, dizziness, behavior changes, vision problems, and even strokes or seizures. Many people with lupus experience memory problems and may have difficulty expressing their thoughts.
- Blood and blood vessels: Lupus may lead to blood problems, including a reduced number of healthy red blood cells (anemia) and an increased risk of bleeding or blood clotting. It can also cause inflammation of the blood vessels.
- Lungs: Having lupus increases your chances of developing an inflammation of the chest cavity lining, which can make breathing painful. Bleeding into lungs and pneumonia are also possible.
- Heart: Lupus can cause inflammation of your heart muscle, arteries, or heart membrane. The risk of cardiovascular disease and heart attacks increases greatly, as well.

All these diseases seem to be incurable with the current medical interventions available in the clinic. In these regards, replacing a degenerated organ seems to be a viable way to cure the said diseases.

8.1.3 Perfusion and ischemic accidents

All parts of the human body require constant supply of blood to deliver nutrients and facilitate waste and gas exchange. If a part of the body has limited or restricted blood supply, it will experience ischemia [53]. Ischemia is a vascular disease with resultant damage to, or dysfunction of, tissue including hypoxia and microvascular dysfunction [54]. Hypoxia or low oxygenation in each part of a body, sometimes, results from constriction, embolism, or plaque build-up (Figure 8.3). Ischemia comprises not only insufficiency of oxygen but also reduced availability of nutrients and inadequate removal of metabolic wastes. Ischemia can be classified as partial (poor perfusion) or total. The prolonged reduced perfusion may lead to infarction or tissue death. The effect of this infarction may vary, depending on the affected organ. If it happens in the heart muscle, it will lead to what is called a myocardial infarction, also known as "heart attack," and if it happens in the brain, then it is called a "stroke." Other parts of the body, such as the lungs, kidneys, limbs (legs and arms), and the small and large intestines may also be affected by this condition.

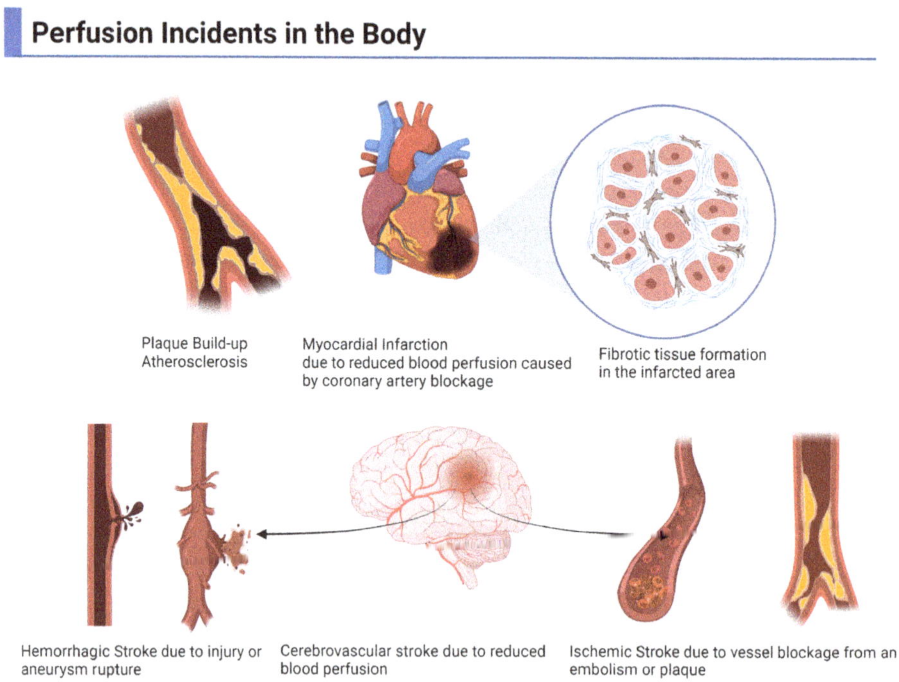

Figure 8.3: Perfusion incidents in the body due to atherosclerosis and bleeding from injury or aneurysm ruptures, and blood embolism. The perfusion of blood is disrupted, which causes the tissue in the heart or brain to die off. This will leave irreversible damage and would require emergency surgical intervention for correction.

Myocardial infarction, as mentioned, is due to prolonged hypoxic condition in the myocardium due to the coronaries being blocked or restricted [55]. The coronaries are the blood vessels that supply oxygenated blood to the myocardium or heart muscles. When this blood supply is disrupted, the patient may experience chest pain, fast and irregular heart beats, and, if not rectified, it may lead to infarction.

The main condition leading to infarction is called atherosclerosis or the build-up of fats, cholesterol, and other substances in and on the arterial walls. This build-up is called plaque that can also lead to thickening or hardening of the arteries. The plaque can cause the arteries to narrow, blocking blood flow and they may also burst, leading to a blood clot.

Cerebrovascular accident or stroke is classified into hemorrhagic and ischemic. The difference between the two is as follows: hemorrhagic stroke is caused by the rupture of the vessels that supply blood to the brain [56]; this bleeding causes increased intracranial pressure that can severely damage the brain and, oftentimes, cause sudden death. Ischemic stroke is the reduction of blood perfusion caused by an embolism or plaque inside the vasculature in the brain tissue, causing the neurons to die off [57].

The inadequate delivery of oxygenated blood to the organs must be resolved either by treating the cause of the inadequate delivery or reducing the oxygen demand of the system that needs it. For example, patients with myocardial ischemia have a decreased blood flow to the heart muscle and are prescribed medications that reduce chronotrophy and ionotrophy to meet the new level of blood delivery supplied by the stenosed vessels. If left untreated, tissues that suffered from an infarction undergo necrosis. The body's response to necrotic tissue is usually very severe, as immune cells are recruited into the infarcted area due to the sudden release of immune cytokines. In detail, the infarcted tissue site is rapidly infiltrated with neutrophils, followed by accumulations of monocytes to remove dead cells and other cellular debris. Furthermore, the body responds to necrotic tissue caused by an infarct by creating an inflammation. Inflammation is the standard response by the body to any kind of damage that limits its effects on the body. In an ischemic infarct, it is possible that inflammation-induced vasodilatation helps increase perfusion in the adjacent areas where the initial ischemia happened. In other forms of injury, increased blood flow and fluid leakage may help dilute and drain away soluble injurious agents, or the cellular response may encase and wall off offending particulate material (e.g., in a granuloma). However, in sterile infarct situations, the inflammatory response may still play a defensive role in containing the damage.

In addition to containing or neutralizing an injurious process, the inflammatory response is also thought to promote repair at sites of tissue damage. To heal an injured site, the dead cells and debris must be cleared, and this is accomplished through the action of the recruited phagocytes. To fill in gaps and restore tissue integrity, epithelial cells or fibroblasts must divide, and a new blood supply must be established; these processes are stimulated by mediators made by inflammatory cells. In addition, the fibrinogen that leaks from inflamed vessels is converted into fibrin fibrils that provide scaffolding for the growth of many of these cells. Thus, the generation of a rapid inflammatory response subserves several different roles in both host defense and tissue repair.

8.1.4 Congenital defects

Congenital defects are structural or functional anomalies that occur during intrauterine life. Also known as birth defects, congenital disorders, or congenital malformations, conditions develop prenatally and may be identified before or at birth, or later in life. Structural defects also lead to physiological disorders, as the body part cannot function. However, some congenital anomalies can be treated with surgical and nonsurgical options, such as cleft lip and palate [58, 59], clubfoot [60], and hernias [61]. Others, including heart defects, neural tube defects [62, 63], and Down syndrome [64], can cause lifelong impacts and are much more difficult to rectify.

There are two classifications of congenital defects: primary structural and primary metabolic. Primary structural congenital defects refer to anomalies that affect the structure of a body part that disrupts normal physiological functions [65]. Primary metabolic defect is also referred to as an inborn error of metabolism [66]. Most of these are single-gene defects, usually heritable. Many affect the structure of body parts, but some simply affect the function. Several causes contribute to congenital defects; these are called teratogenic substances [67]. A teratogen is an agent that can disturb the development of the embryo or fetus. Teratogens halt the pregnancy or produce a congenital malformation (a birth defect). Classes of teratogens include radiation, maternal infections, chemicals, and drugs. Genetics and the age of the parents also play a role in the risk of developing congenital defects. Women who are over the age of 35 have a higher chance of having prenatal malformation during their pregnancy. Fathers also contribute proportionally greater DNA mutations to their offspring via their germ cells than the mother, with the paternal age governing how many mutations are passed on. This is because, as humans age, male germ cells acquire mutations at a much faster rate than female germ cells. Lifestyle choices such as drug addiction and smoking can also cause DNA mutations in the gametes of the parents.

In tissue engineering, primary structural congenital defects can be rectified by regrowing the affected body part using biomaterials. There are several congenital defects that are now being investigated to know if biomaterial-based tissue engineering can help. Examples are spina bifida, urogenital defects, gastroschisis, diaphragmatic hernia, and esophageal atresia.

8.2 Tissue engineering methods and applications

Tissue Engineering is an emerging field in biomedicine to aid in reconstructing, regrowing, and accelerating the healing process of damaged organs and tissues in the body. There are several techniques to this, using a combination of multiple disciplines such as biology, chemistry, material science, and engineering. Tissue engineering is a diverse field, and, currently, there are three methods in which it is currently being done.

Table 8.2: Structural congenital defects of interest in tissue engineering.

Condition	Description	Ref.
Spina bifida	Spina bifida is a condition that affects the spine and is usually apparent at birth. It is a type of neural tube defect (NTD). Spina bifida can happen anywhere along the spine if the neural tube does not close all the way.	[68]
Urogenital defects	Congenital urogenital anomalies are birth defects and inherited conditions that affect the urinary tract (kidneys, ureters, and bladder) or the genital tract (reproductive organs), also called the genitourinary system. Urogenital anomalies can affect urine flow and cause kidney damage that may lead to nephrotic syndrome, acute kidney failure, or chronic kidney failure.	[69]
Gastroschisis	Gastroschisis is a birth defect of the abdominal (belly) wall. The baby's intestines are found outside of the baby's body, exiting through a hole beside the belly button.	[70]
Diaphragmatic hernia	Diaphragmatic hernia is a birth defect, in which there is a hole in the diaphragm (the large muscle that separates the chest from the abdomen). Organs in the abdomen (such as intestines, stomach, and liver) can move through the hole in the diaphragm and upward into a baby's chest.	[71]
Esophageal atresia	Esophageal atresia is a birth defect, in which part of a baby's esophagus (the tube that connects the mouth to the stomach) does not develop properly. It is a birth defect of the swallowing tube (esophagus) that connects the mouth to the stomach.	[72]
Orofacial cleft	Orofacial cleft is a group of conditions that includes cleft lip, cleft palate, or both together. A cleft lip has an opening in the upper lip that may extend into the nose. The opening may be on one side, both sides, or in the middle. A cleft palate occurs when the roof of the mouth contains an opening into the nose. These disorders can result in feeding problems, speech problems, hearing problems, and frequent ear infections.	[73]

Conventional tissue engineering (Figure 8.4) is the most known method of tissue engineering. This includes two approaches: (1) transplantation of a tissue grown on a matrix (artificial or natural) in vitro [74] and (2) in situ regeneration of tissue via implantation of an artificial matrix laden with growth factors to initiate host cell regeneration of the damaged tissue [75].

Cell therapy (Figure 8.5) is the utilization of live cells and their expansion in the host cells for in vivo transplantation [76]. This involves the delivery of cells through systemic injection into the bloodstream [77] or through direct transplantation into a local tissue [78]. The cells used in this application come from either allogenic or autologous sources. This method is also known as stem cell therapy.

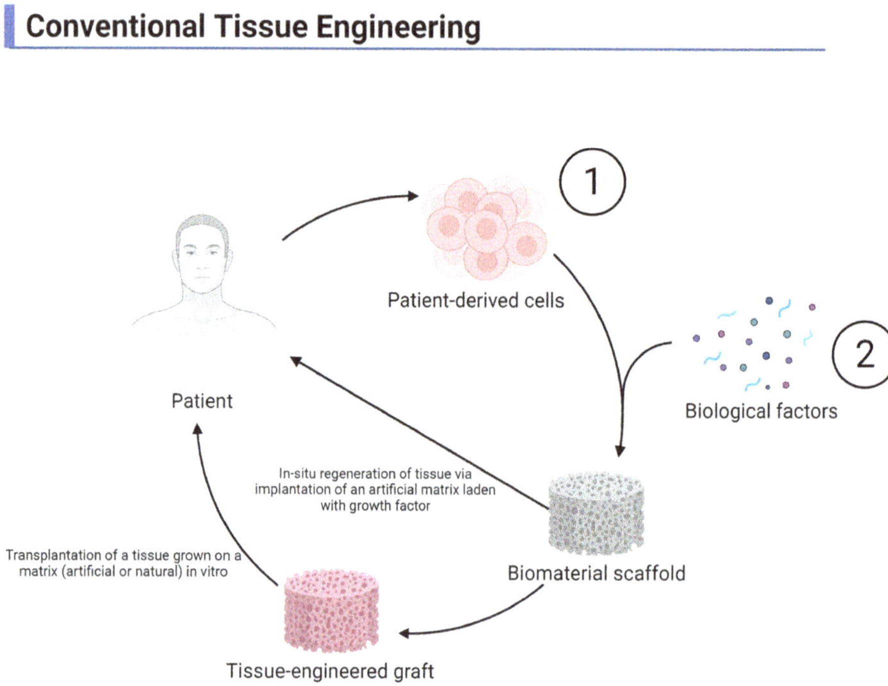

Figure 8.4: Tissue engineering principles using biomaterials. (1) Patient's cells are harvested and seeded on a biomaterial scaffold. The scaffold is then reintroduced inside the body to help regenerate the damaged organ. (2) Biological factors (growth factors) are incorporated in a biomaterial scaffold and reimplanted inside the patient's body. Nascent tissue will then regrow into the scaffold and help regenerate the damaged organ or tissue.

Closed-loop methods (Figure 8.6) involves the use of extracorporeal or implantable devices that house cell-laden polymeric scaffolds within a semipermeable membrane to facilitate gas and nutrient exchange, while preventing the movement of antibodies, pathogens, and immune cells [79]. Examples are artificial pancreas, liver, and kidney devices [80].

8.2.1 Conventional tissue engineering

Growing a functional organ outside the body, to say the least, is a very complex task. To create an organ outside the body, it must first have the right anatomical features and complete functional physiology, and must be accepted by the host's body, post transplantation. However, if successful, this will alleviate the burden on organ transplant waiting lines and could potentially save lives of millions of people, worldwide. As of now, there is significant waiting time for patients who need live organs. The organs that are most in need are the heart, liver, kidneys, and

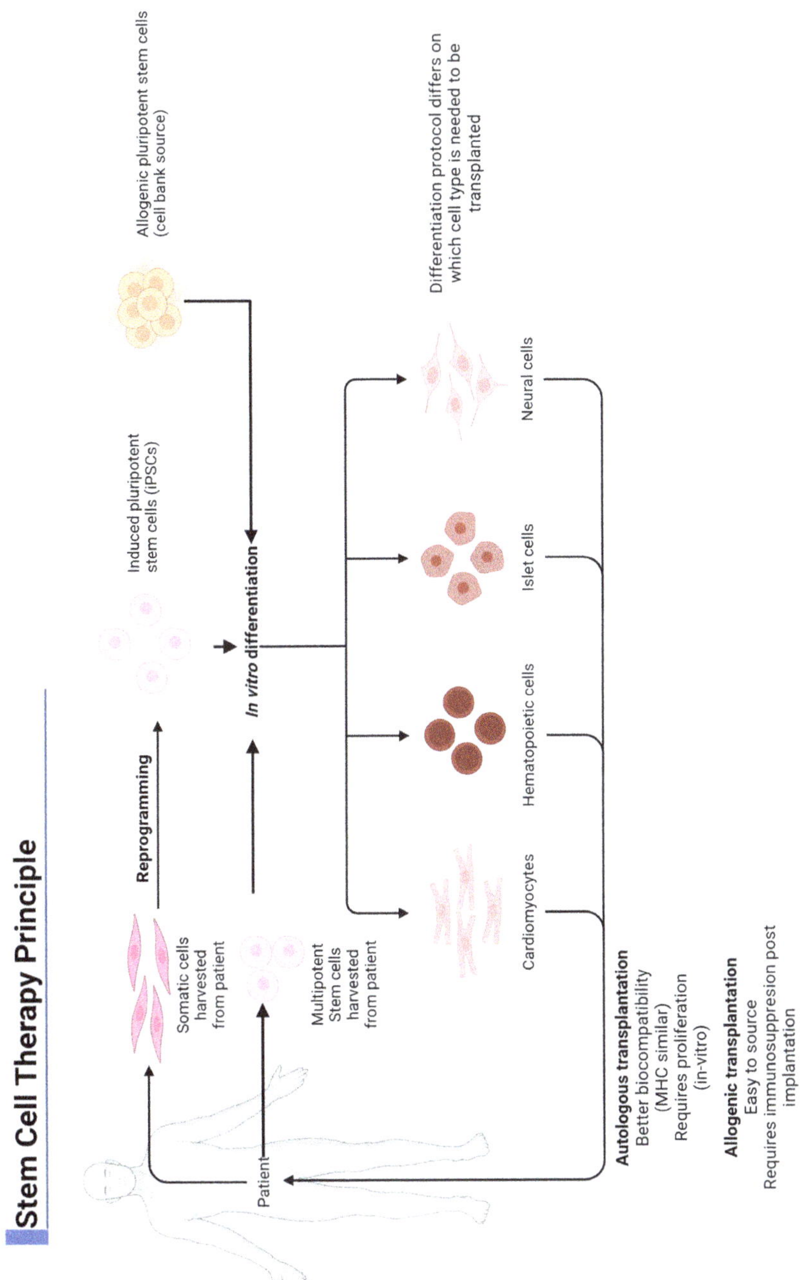

Figure 8.5: Stem cell therapy principles: 1) Somatic cells are reprogrammed to become iPSCs or multipotent stem cells are harvested from the patient (autologous) 2) Pluripotent stem cells are taken from a cell bank source (allogenic). The stem cells are differentiated into the cell line of interest and are then reintroduced into the patient's body to help regenerate the damaged organ or tissue.

Closed-loop Tissue Engineering Methods

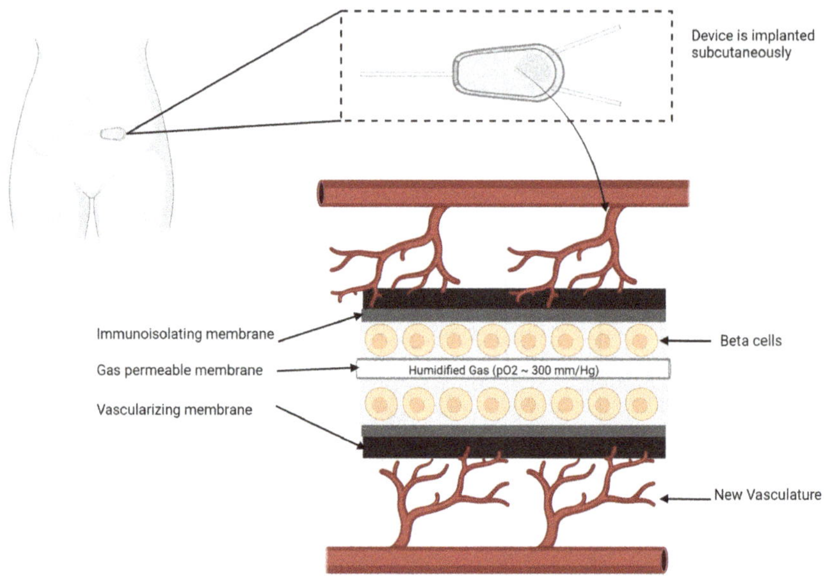

Figure 8.6: Closed-loop tissue engineering methods. A device laden with living cells that will help replace the function of a damaged organ. (Artificial pancreas device) meant to create insulin, in situ. The device is meant to support the cells and protect it from immune response of the body, while serving to deliver vital biochemical substances (e.g., insulin).

cornea. As of now, these organs may only come from donors via cadaver extraction or compatible individuals, making it severely unavailable to most people who need a transplant. People who need these organs may have suffered from several physical injuries, perfusion accidents, or congenital defects.

The first method in conventional tissue engineering involves the in vitro growth of the tissue or organ [81]. In vitro means outside the body, and this method, then, requires multiple components to be realized. The first component in this is the scaffolding material for the cells. Cells organize themselves based on the scaffolding they are growing in. Creating a well-defined structure from scratch is a daunting challenge; hence some researchers are looking into decellularized scaffolds to create organs that can use the host's cells to regrow the organ and remove the need for immunosuppressants, post transplantation, and decrease the risk of organ rejection. The scaffold may come from animals that have organ anatomical features almost analogous to humans. The leading candidates are swine models, as their hearts, kidneys, and liver are, structurally and size wise, analogous to human organs.

8.2.1.1 3D decellularized scaffolds

Decellularization is a process that isolates the extracellular matrix from cells and cellular material, while preserving the 3D ultrastructure of the organ [82, 83]. This can be achieved by the combined application of physical, chemical, and enzymatic methods. In a published protocol on decellularization of mice heart, four major steps are classically needed [84–86]

1. Lysis of the cell membranes
2. Separation of the cellular components from the ECM
3. Solubilization of cytoplasmatic and nuclear components
4. Removal of cell debris

A technique called perfusion-decellularization is commonly used to obtain whole-organ scaffolds as a platform to bioengineer functional organs [87, 88]. Organs are harvested from donors and, then, undergo the process of decellularization, yielding acellular scaffolds composed of extracellular matrix that preserve the macro- and micro-architecture of the original organ. Decellularized scaffolds are then repopulated with regenerative cells (e.g., stem cells) and placed under biomimetic in vitro culture conditions, inside a bioreactor. After this period of culture, the bioengineered organ can be transplanted to supplement or completely replace the function of a failing organ in the recipient. The main artery supplying the organ is cannulated and decellularizing agents (e.g., detergents, enzymes) are delivered through the vasculature under pressure, with the help of a pump system. The decellularizing agent reaches all the tissues with minimal diffusion distance possible, thanks to its distribution throughout the organ's capillary network. Decellularizing agents then provoke the lysis, detachment, or digestion of cellular elements in the tissues. Cellular debris is then removed from the tissue via the venous system, leaving behind the extracellular matrix scaffold composed of collagen, elastin, laminin, fibronectin, glycosaminoglycans, and other elements that preserve key microarchitectural components such as basement membranes.

Cells that are utilized in this perfusion-decellularization process may come from autologous sources, e.g., stem cells coming from the host [89, 90]. However, this process takes several weeks to months to differentiate the stem cells to the specific cell type that is needed to populate the decellularized organ. Introducing these cells in the right areas of the scaffold is also a major task as an ectopic (misplaced) cell population may cause the organ to not function properly. Furthermore, even after using autologous cells to elicit nonimmune response, residual epitope may still induce immunogenic reaction and organ rejection, due to variations in methods of decellularization [91, 92]. Until now, there is still no standardized way to ensure that scaffolds are completely devoid of cell materials and residual epitope that can induce immune response. Lastly, the tissue or organ scaffolds must also pass long-term preservation time (weeks to months) to create a stock of decellularized scaffolds that can be used on demand, in case of rejection or failure to create a functional organ.

8.2.1.2 2D scaffolds

Research in 2D scaffolds is also growing, as researchers can have better design controls in creating and modulating how the cells behave while populating the scaffolding material, in comparison to decellularized scaffolds. The lowest hanging fruit in this area is with the use of 2D scaffolds, primarily in treating 2D damage in the body. These scaffolds are usually simple sheets of polymeric materials applied in assisting regeneration of damaged tissues in the body. These scaffolds are quite simple to make due to their geometrical simplicity in comparison to the 3D architecture of organs and are applied to 2D damages or as support structures in the body, while facilitating healing.

For example, artificial skin grafts used for body burns have utilized biomaterials laden with cells [93–95]. Severe burn victims (third degree) suffer from delayed and prolonged healing of their skin, due to the destruction of all the skin layers. This kind of burn may require grafting from other parts of the body to ensure that the damaged area can regenerate to preserve the integrity of the skin. This is important, as the skin has the primary role of protecting the body from outside pathogens. If this barrier is compromised, then, the body is prone to infection. Clinicians conventionally use autologous skin graft to transfer some healthy skin onto the burnt area to facilitate some degree of healing [96–98]. However, in burn victims that have limited healthy skin to graft, skin expanders are usually implanted under the skin layers to increase the available surface area for grafting [99, 100]. This technique has a few limitations and creating a biomaterial that can act as an alternative for autologous skin graft is a better treatment regimen for patients. Polymers of that are components of the natural ECM of the skin and are good candidates for this application. Gelatine [101], hyaluronic acid [102], and collagen [103]-based electrospun 2D mats can create a natural scaffold for stem cells harvested from the patient. In this strategy, the stem cells are first harvested from the patient and differentiated in vitro into skin epithelial cells. This cell line is then proliferated and transferred into the electrospun scaffold to then populate and help with the nascent cell. Acellular method where polymeric scaffolds are directly placed in the burnt area can also facilitate some degree of healing for burn wounds that still have some remaining epithelial cells on its surface (e.g., second-degree burn). The scaffold, then, can also deliver growth factors [104, 105] that help facilitate the proliferation of new skin cells and formation of new vascularization networks.

Chronic wounds caused by delayed healing secondary to an underlying condition, e.g., diabetes, can also be helped with biomaterial-based skin scaffolds. Compared to burn wounds, chronic wounds are characterized by the following: Alkaline $pH > 7.15$, excessive release of pro-inflammatory cytokines, overproduction of reactive oxygen species (ROS), high level of proteases that disrupts the granulation phase of wound healing, lack of stem cells to differentiate into adult skin epithelium, and persistent infection. In this regard, strategies to heal this type of wounds rest in normalizing these deviations by creating a wound healing material, which can release agents that counteract these characteristics. For example, it has been demonstrated that chronic wound fluid contains high levels of elastase and matrix metalloproteinases (MMPs), including MMP-1, MMP-2, MMP-8, MMP-9, and MMP-13 [106, 107]; these are proteinases that disrupt the creation of collagen [108]. Collagen is an important component for extracellular matrix, which cells use to organize and help close the wound. Here, a clinical material called Promogran™ [109] was developed as a highly absorbent wound dressing made of collagen and oxidized regenerated cellulose, which, on contact with the wound exudate, becomes a soft gel. In turn, the formed gel can make inactive elastase and MMPs as well to bind to GFs, preventing them from degradation, and helping the wound to heal.

In other conditions such as hernia, the muscle layer is weakened and the tissue or an organ such as the bowel exits through the wall of its cavity. Various types of hernias can occur, most commonly involving the abdomen, and specifically the groin [110, 111]. Common causes of hernia are due to overexertion of an already weakened muscle; once it happens, surgical intervention can be done to restore the position of the displaced tissues and organ (colon). In addition to this, muscle reinforcement methods using a 2D polymeric mesh material are done [112]. The materials used for this mesh are polypropylene or copolymers of polyethylene and polypropylene [112, 113]. Biomeshes have also been introduced to create support for post hernia repair surgery [114]; these are made with natural animal-sourced extracellular matrix; however compared to synthetic meshes, these biomeshes have lower tensile strength. The advantages of using biomeshes are reduced bowel adhesion and infection [115]. Approaches utilizing mesh have been able to demonstrate faster return to usual activity, lower rates of persistent pain, shorter hospital stays, and a lower recurrence rate of hernia.

In more advanced applications of 2D scaffolds for tissue engineering such as in cardiology: Post myocardial infarction causes a reduced amount of cardiac capacity [116, 117]. This is usually measured via ejection fraction [118]. This is the amount of blood the heart is capable of pumping into the systemic circulation. In people who have suffered a heart attack, this ejection fraction is significantly reduced, consequently affecting their capability to maintain normal physiological functions and physical activities. This is due to the formation of scarring in the heart muscles. These scar tissues are unable to contract and help with the pumping of blood in the heart. One strategy where 2D scaffolding materials are being investigated upon is with the development of cardiac patches. A cardiac patch is a functioning piece of laboratory-grown heart tissue that can be used in replacing the patient's damaged one, before the entire organ is affected [119, 120]. This patch is directly implanted on top of the infarcted area of the heart, with the goal of integrating it into the myocardial tissue and restoring its normal ejection fraction capability. There are multiple cardiac patch designs utilizing various polymeric materials, but the underlying design consideration is creating a 2D scaffold, whereby cells (cardiomyocytes) can proliferate and mature and organize into a functional, beating scaffold.

The materials used for creating a functional cardiac patch are mostly polymers from naturally derived sources, both animals and plants. Examples are collagen, fibrin, silk fibroin, decellularized ECM, alginate, and cellulose [119, 120]. This is due to the rationale that naturally derived materials have better degradability, compatibility, and immune response in contrast to s (poly (vinyl alcohol) (PVA), poly(lactic-co-glycolic) acid (PLGA), poly-(Lsynthetic polymer-lactic) acid (PLLA) and polyurethanes (PU)). Synthetic polymers, however, have also been seen to synergize naturally occurring polymers, for example, Gelatine methacrylate (GelMa), and have better mechanical stability and ease in fabrication into patches using various methods, in contrast to natural polymers. There are different methods where these materials are being used for the creation of these cardiac patches. Natural materials that are either electrospun or 3D-printed give a natural mimicking scaffold for cardiomyocytes. There are pros and cons in these two methods. The first, is that electrospun fibers generally can give a much similar microenvironment to a local ECM than a 3D-printed scaffold, since electrospinning can create random nanoscale fibrous structures, depending on the parameters used during the process, while 3D printing must use a suitable bio-ink (typically a hydrogel) [121–123] that can support the cells during the printing process and has a microstructural limit that is less like a native ECM [123–126]. However, electrospun nanofibers are limited to a monoculture of cells that can be seeded on it and it requires a two-step process of scaffold fabrication and cell loading, while 3D bioprinting does the fabrication and cell loading at the same time and can use multiple 3D printing heads to allow for multiple cell types to be incorporated into the structure.

The functionality of these cardiac patches is scored based on their ability to provide an adequate maturation and proliferation environment for the cardiomyocytes and subsequently perform a synchronized contractile motion. Cardiomyocytes, unlike skeletal muscle cells, are involuntary and their contractions are controlled by the sympathetic and parasympathetic nervous system [127, 128]. The sympathetic nervous system (SNS) releases the hormones (catecholamines – epinephrine and norepinephrine) to accelerate the heart rate, while the parasympathetic nervous system (PNS) releases the hormone acetylcholine that slows it down. Furthermore, there are microscopic structures that identify cardiac muscles from skeletal muscles, and these are called intercalated discs. These structures connect cardiomyocytes and support synchronized contraction of the cardiac tissue.

In creation of cardiac patches, one must ensure that it can integrate into the native tissue to be able to react synchronously with the sympathetic and parasympathetic nervous system. To do so, neovascularization must happen between the cardiac patch and the tissue [129–131] to allow transfer of cardiac regulating hormones. If not, the patch will be unable to synchronize its contraction in relation to the native tissue, which will lead to arrhythmia (irregular heartbeat). It has been shown that the addition of growth factors such as vascular endothelial growth factor (VEGF), hepatocyte growth factor (HGF), and insulin-like growth factor-1 (IGF-1), into the cardiac patches [132, 133] can regulate the growth, survival, and migration of cardiomyocytes. However, this adds to the complexity of the scaffold as the proper release kinetics as well as stability of these growth factors must be ensured, pre- and post-implantation.

Although cardiac patches have shown significant and promising results in animal models [134], there is still a long way to go, before it can be translated into the clinic. First, for a cardiac patch to be placed, an open-heart surgery is the only means to do so [135]. Open-heart surgeries are highly invasive and come with high risks for patients, who already had previous cardiac accidents and who are not in full cardiac health. Thus, finding a way to implant these cardiac patches through minimally invasive means must be devised. Since this may be considered as a combination product with various fabrication steps that have to be controlled, retaining the viability and functionality of cells remains a block throughout the large-scale production of such therapeutic patches. Although various studies have found that cell retention and engraftment are enhanced to some extent with cardiac patches, the therapeutic efficiency is still far below clinical requirement. This indicates that further understanding on the basic requirements of a cardiac patch is needed.

8.2.1.3 Tubular scaffold constructs

Tubular constructs are types of tissue scaffolding that are specifically applied to tissues that share the same geometric features, such as nerve fibers, vasculature, some types of tendons, canals, and tubes (vagina, trachea, esophagus, etc.). In terms of geometric complexity, tubular scaffolds are just a degree more complex than 2D scaffolds but are less complex than 3D scaffolds. Although they can be considered as 3D, tubular scaffolds can be made from 2D sheets attached from end to end to form a cylinder.

In nerve tissue engineering, nerve conduits are grown from tubular polymeric constructs [136, 137]. Peripheral nerves are usually damaged by trauma, autoimmune diseases, or genetic factors [138]. When left untreated, it may lead to limb desensitization, muscle atrophy, and permanent loss of neurological function in the affected organ. Nerve guidance conduits (NGC) are used in connecting severed peripheral nerve fibers to allow for the retransmission of electrical signals from the central nervous system and to regain sensation and motor control in the affected body part. NGC can be made from natural and synthetic polymers, and both types of materials have their own merits; however, the overarching theme for NGC is that it should allow electrical impulses to be transmitted from the proximal to the distal end of the nerve, while facilitating the regrowth of nascent neuronal cells (Schwann cells). The polymers should be electrically conductive and biocompatible. There is no single class of polymer that can be used for this application; thus mixing of synthetic and natural polymers is being done to exploit the properties of both. Polypyrrole (PPy), for example, is a conductive polymer that shows promise in nerve tissue engineering [139]. Its conductive properties and biocompatibility have been well-documented, but it has very low mechanical properties. It has been blended with PLGA, PCL, Poly (caprolactone fumarate) and natural polymers (e.g., HA) to overcome its mechanical property limitations [140–142]. Furthermore, it increases the biological stability of PPy by reducing the coupling defects in its polymeric backbone.

Another design consideration for NGCs is the hierarchical structure that the nerve must form during regrowth. The alignment of nascent cells is vital in transmitting the electrical signals properly along the nerve [143]. Thus, NGCs must provide this alignment through the micro or nanotopology of the scaffold. Here, electrospinning has shown its value in creating aligned nanofibers to guide the growth of these nascent cells along its strands. Conventional electrospinning set-ups create random nonwoven fiber structures that are usually capable of mimicking ECM micro topologies. Thus, modifications on electrospinning setups are done to create the required fiber alignment. The simplest way it is done is through attaching a rotating collector and increasing its rotational speed to upward of 700 rpm. This high rotational speed allows for the fibers to be deposited on the surface of the collector in an aligned manner. However, this may not be enough, as this will lead to structural limitations and implantation difficulties. Thus, a combination of randomly and aligned nanofibers are developed. Varying the electrical field on the surface of the collector allows for this combined and controlled patterning of aligned and random nanofibers [144]. Here, copper wires are set up horizontally on the collector, and cellophane tapes are set over the fixed copper wires in the vertical and horizontal axis. The direction of the nanofiber alignment is influenced by the way the cellophane tape is attached on to the copper wires. A horizontal cellophane tape attachment yields a vertical nanofiber alignment, and a vertical cellophane tape yields a horizontal nanofiber alignment. The nanofibers that are collected on the cellophane tape are aligned in parallel due to the adjustment of the electric field between the spinning needle and the collector caused by the conduction of the copper line and the insulation of the cellophane tape. The patterns can then be adjusted, depending on the configuration of the copper wire and cellophane tape, subsequently combining the properties of random and aligned nanofibers in a single mat. The nanofiber mat produced in this manner can then be rolled into an NGC, with the aligned segment directing the growth of Schwann cells along its fiber alignment, while the random segment can provide ease of suturability, during implantation.

Vascular grafts are used on damaged or diseased blood vessels when surgeons need to redirect blood flow by replacing the blood vessel [145, 146]. Blood vessels are an important part of the body, as they are responsible for the transport of blood that carries nutrients and gases, and facilitate elimination of waste from the tissues and organs, when pathologies, such as in cases of [147–149]. These vascular diseases may require surgical intervention and repair the damaged vessel. In severe cases where the vessel is too damaged, a graft is placed to restore proper blood flow. There are multiple types of vascular grafts: the first is the autologous native graft [150], where a healthy vessel is harvested from a different part of the body and retransplanted to where the disease is located. This is common in coronary bypass surgeries [151], where a small section of the femoral artery is cut and then reattached to the coronary arteries to bypass or redirect the blood flow into the heart muscle, post-myocardial infarction. In cases of aneurysms such in endovascular, abdominal, and thoracic aneurysm repair [152], synthetic grafts made from polymers, such as polyester [153, 154], expanded-polytetrapolyethylene [155], and polyurethane [156] are used to repair the diseased vessel. Regardless of material, design criteria for this type of graft include its hemocompatibility and mechanical stability. These two properties are the most vital, since the graft should interact with blood components, and therefore not evoke any blood clotting cascade or lysis of red blood cells; the graft should also be mechanically compliant with the variation of blood pressure in the body [157]. One last important consideration is that these grafts are nonnative and do not integrate and respond well with the local vascular system; they merely act as a pipes for restoring blood flow in the affected site of the body. This, in turn, may shorten the lifespan of the graft and cause bleeding complications and thrombogenic events, over time. Researchers are creating vascular grafts that have the capability to integrate with the local vascular system by allowing vascular cells to proliferate into the graft's structure [158]. Here, electrospun fibers are also used to create a suitable transient graft that allows for restoration of blood flow, while facilitating growth of new vascular tissue to replace the graft [159, 160]. The graft, in the beginning, will serve a similar function as a synthetic polymeric graft, but over time, the micro topology and material effects can stimulate the growth of a new vessel. This is like what an NGC tries to achieve but applied onto blood vessels. This type of graft should possess the following characteristics: biocompatible, nonthrombogenic, mechanically durable, and simultaneously degraded along with the native vascular tissue formation. The first three requirements are like what a synthetic graft must possess, but the addition of the degradation rate of the polymer to coincide with the proliferation rate of the nascent vascular tissue can be challenging. But since electrospun fibers mimic native ECM and can be made from various natural and synthetic polymers, adding site specific growth factors can also be done to accelerate the proliferation of vascular tissue to regrow a new vessel, before the mechanical stability of the graft is compromised. Furthermore, it must also be considered that a blood vessel is not just a simple pipe and is composed of layers of distinct features. The arterial wall is composed of three different layers: (i) intima, (ii) media, and (iii) adventitia, which can constrict and dilate in response to the physiological condition of the body [161]. Thus, the graft must allow for the formation of these layers and the hierarchical structure of the vessel. This is done by creating a layer-by-layer approach comprising multiple polymeric materials with varied micro topology to stimulate the growth of various layers of the vessel.

Perhaps one of the most complicated physiological structures that can be made using tubular scaffolds is the trachea. The trachea, or what is commonly called the windpipe, is a hollow tubular structure part of the respiratory tract. The trachea is composed of about 20 rings of tough cartilage. The back part of each ring is made of muscle and connective tissue. Moist, smooth tissue called mucosa lines the inside of the trachea. The trachea widens and lengthens slightly with each breath in, returning to its resting size with each breath out. In cases of genetic anomalies (tracheomalacia), infection (tuberculosis), stricture, or cancer, the trachea can be deformed and needs to be surgically corrected. Advances in tissue engineering have led researchers to create, ex-vivo, a lab-grown trachea that can be transplanted [162, 163]. To create an artificial trachea, the patient is first imaged by a CT scan. This is so that the tissue engineers can make a scaffold that fits perfectly within the patient's body, by mapping out the dimensions needed for growing the scaffold. The scaffold is then made from nano-sized PET (polyethylene terephthalate) fibers using these specifications. It is then seeded with stem cells from the patient's bone marrow by placing it in a bioreactor with transcription factors. These help force the stem cells to differentiate into trachea-specific cells. These cells then grow and divide to produce cartilage around the scaffold, after 2–3 days. Finally, it is surgically implanted by suturing it in the patient's throat and lungs. However, this process has been riddled with controversy, as the engineered trachea that were transplanted to humans had caused deaths to those who received it. There had been reports that the laboratories responsible for creating the engineered tracheas lacked good manufacturing process (GMP) certifications. Furthermore, the trachea caused the patients to accumulate mucus on their lower tracheal tract, evidence that the cilia responsible for the expulsion of mucus are dysfunctional or nonexistent. It shows that tissue engineers must be wary of the complexity of the organ they would like to create, in order not to cause harm to people who will receive the transplants.

8.2.1.4 3D scaffolds

3D scaffold structures are being realized to solve more complex tissue engineering issues. Most pathology that involve the organs in the body either affect the whole organ or a portion of an organ, which is 3D in structure. Hence, to regrow the tissues that are damaged fully within a reasonable timeline and with correct anatomical feature, 3D scaffolds are being developed [164, 165]. There is a plethora of ways in which 3D scaffolds can be made. But an overarching feature in all these scaffolds is that it should allow cell proliferation, differentiation, and penetration into its network to allow for the scaffold to integrate into the local tissue where it is implanted. Like other scaffolding types (2D and tubular), 3D scaffolds can be acellular, that is, pure biomaterial that allows for nascent cells to integrate into the material and guide regeneration, or it can be seeded with cells in vitro and implanted into the site where the damaged organ is located. The 3D scaffold should temporarily withstand the external loads and stresses caused by the formation of the new tissue, while preserving mechanical properties close to that of the surrounding tissue. It was demonstrated that the tissue-specific mechanical characteristic of stiffness could control the differentiation of laden stem cells. Furthermore, the scaffold designs such as sponges, meshes, foams, etc., are able the control biodegradation as a key factor in tissue engineering. The degradation of biomaterials could be surface or bulk. In contrast to bulk degradation that breaks the internal structure of the material, surface degradation maintains the bulk structure. The rate of degradation should match the tissue growth, without separation of toxic byproducts. The degradation of a biomaterial could be achieved by physical, chemical, biological, or combined processes influencing the biocompatibility of the 3D scaffold. For example, incorporating different biodegradable components in the construct triggers hydrolytic degradation, while processes such as enzymatic digestion and cell-driven degradation biologically change the implant material. When the application of a scaffold does not require a complete degradation (for example, in articular cartilage repair), permanent (nondegradable) or semi-permanent scaffolds could be used [165]. When implanted in the body, toxic, immunological, or foreign body responses should not occur, which prove that the scaffold is biocompatible. The surface properties of a scaffold should also be designed in such a way as to facilitate cell attachment, homogeneous distribution, proliferation, and cell-to-cell contacts. The scaffold geometry should maintain a porous or fibrous design and provide high surface-to-volume ratio for cell attachment and tissue development. Nanostructured surfaces demonstrate high surface energy as opposed to polished materials that result in enhanced hydrophilicity and, therefore, improved adhesion of proteins and cell attachment.

Acellular scaffolds are very common in bone tissue engineering [166, 167]. Here, inorganic materials such as beta-TCP or hydroxyapatite (HA), can be combined with other polymeric materials, so that it can initiate the growth of lost bone mass and guide the growth of a defective bone to correct it [168, 169]. These inorganic materials constitute the inorganic components of the bone and by combining it with osteogenic growth factors, it can stimulate the growth of bone tissue and provide structural support to the affected site. Bone fillers is an example that is used in multiple conditions, such as oral maxillofacial reconstruction or other facial reconstruction surgery. NanOss® bone void filler [170, 171], originally from Angstrom Medica, is the first nanotechnology medical device to receive clearance by the US Food and Drug Administration, in 2005. According to that original developing company and Pioneer Surgical Technology Inc. that now markets the material, NanOss® is an innovative structural biomaterial that is highly osteoconductive and remodels over time into human bone, with applications in the sports medicine, trauma, spine, and general orthopedics markets. It is formulated by precipitating nanoparticles of calcium phosphate in aqueous phase, and the resulting white powder is compressed and heated to form a dense, transparent, and nanocrystalline material. It is claimed to be the first material that duplicates the microstructure, composition, and performance of human bone. Bone fillers are injectable and are highly suited for facial reconstruction surgery, where voids must be filled to regain the function and aesthetics of the affected bone. But there are some instances where solid 3D scaffolds are needed to serve as a mechanical reinforcement, especially in load-bearing bone structures. Here, metallic materials, such as magnesium (Mg) and its alloys and other metals are the materials used in bone tissue engineering. Magnesium-based scaffolds are bioresorbable and with suitable mechanical properties and biocompatibility that also allows bone cell activation [172–174]. Magnesium-based alloys are fabricated to be porous, so that bone cells can infiltrate the whole scaffold. Porous Mg has better degradation behavior (slower hydrogen evolution) and slower decrement of compressive yield strength in simulated body fluid (SBF) immersion tests. And, Mg and its alloys can be made with wide ranging mechanical properties: elongation (from 3% to 21.8%) and tensile strength (from 86.8 to 280 MPa), while its elastic modulus (41–45 GPa) makes it closer to that of the bone, compared to other metals. This is important in preventing stress shielding which can initiate bone decomposition around the implant.

Regardless of application, acellular 3D scaffolds must possess the ability to allow infiltration of cells into the scaffolding material, while being able to be degraded or absorbed, once the new tissue has been regrown. This can be a daunting task in some other parts of the body where tissue turnout is slower than the rest of the body, such as in vital organs (e.g., heart and kidneys). That is why, in these more complex applications, live cells are integrated into the scaffolding material due to the nature of the damage of the tissue it intends to regenerate. Here, cell-laden 3D scaffolds are the preferred route of treatment. For example, in post myocardial infarction where a scar tissue has formed and has caused reduced cardiac output, or in cases of kidney ischemia where renal perfusion and filtration is reduced, cell-laden biomaterials can be injected to stimulate the growth of new functional tissue to replace the damaged tissue [175, 176]. Compared to bone tissue engineering, where the focus is to provide structural support with minimal biological functionality, cell-laden 3D scaffolds aim to provide or restore the physiological function of the affected organ.

Incorporating cells together with the biomaterials prior to implantation comes with its own challenges. First is that the cells must be compatible with the biomaterial in terms of number of cells surviving. In injectable cell-laden hydrogels, this means that the cells must be injected together with the biomaterial [176, 177], subjecting the cells to shear stresses that can induce cell death. Furthermore, the biomaterial must have shear thinning properties to alleviate the shear forces that the cells experience during implantation. Shear thinning is a rheological property of a non-Newtonian fluid, meaning that the viscosity of the material is reduced when a shear strain is applied to it and regains its high viscosity or resistance to flow when the force is removed. In most cases, once a polymer has been cross-linked and forms a hydrogel, extruding it from a needle can be cumbersome, as it requires large forces to be able to inject into the tissue. Not only that this can lead to high shear stresses on the cells, it can also result in uneven distribution of dosing in the affected area. Additionally, cell-laden injectable hydrogels must possess porosity to ensure that it can allow cell proliferation into its network. Porosity, pore size, and pore interconnectivity are critical parameters determining the performance of 3D scaffolds in tissue engineering. Pore size affects cell attachment, migration, morphology, and proliferation; porosity and pore structure also have a strong influence on the mechanical properties of the matrix, the supply of nutrients and oxygen, and the removal of waste products. Different cells and tissues have been shown to require scaffolds of different pore size for maximum tissue growth and development: reports of the optimal average pore size range from as small as 5 μm for neovascularization to 400–500 μm for bone and cartilage regeneration. In general, relatively small pores are considered advantageous initially to maximize cell attachment and interactivity, while highly porous materials provide less diffusional resistance for nutrient and gas exchange. As porous scaffolds become increasingly occluded due to tissue deposition and development, the ability to increase the pore size and porosity in situ during cell culture offers significant advantages for improving nutrient, growth factor, and oxygen penetration into the matrix, for faster and better tissue growth. The idea is to slowly degrade the injected scaffold and be replaced by nascent tissue with functioning physiological properties and coordination with the local implantation site. The cells must be able to self-organize and integrate to the local tissue without the biomaterial hindering the process.

Hydrogels are also being used as bioinks in 3D printing [178–180]. This is an emerging technology that aims to produce biological structures that can replace cadaver-based scaffolds (decellularized scaffold) technology. It aims to produce and print organs on demand, using the patient's own stem cells that had been differentiated to cell lines needed to produce the organ. However, creating fine details, large structures, and incorporating cells directly using 3D bioprinting remains a challenge. Cells have difficulty surviving in conditions that are not physiologically similar to those that can be found inside the body. Slight changes in temperature, pH, and diffusivity of gas and nutrients may cause cells to undergo apoptosis or necrosis. Bioinks may, to a certain degree, provide physiologically similar conditions to maintain a healthy population of cells during the printing process, but there are some limitations that can hinder this; first is that printing time usually takes hours before a structure is completed, especially in conventional 3D printing methods that use a top-down printing approach with a nozzle head and a collector bed setup [181–183]. The more complex the geometry being printed, the longer it takes for cells to be in an unfavorable condition, increasing the likelihood of cells dying in the process. Therefore, there is increasing need for faster printing technologies, such as the stereolithography method where a bottom-up approach is used [184–186]. Here, the cells are maintained in a bioink bed, which when exposed to light can create geometric structures faster than the conventional printing head method. However, this would mean using photo cross-linkers that are usually toxic to cells. But, the upside of this method is that it can create complex geometric structures without limitations of the rheological properties of the bioink, since it is a bottom-up approach. Once the structure has solidified, post-cross-linking, the structure is lifted and can be transferred into a bioreactor immediately, increasing the ratio of healthy versus dead population of cells embedded on the bioink. It must also be considered that the whole printing process must be maintained aseptically, as introduction of pathogens into the printed structure may cause the cells to all die off.

Another challenge that 3D bioprinting is facing is the likelihood of the cells not self-organizing and not becoming a functional organ or tissue [187, 188]. The anatomical features of an organ are very complex, since not just one type of cells forms an organ or even a tissue. Printing multiple types of cells in one setting can be cumbersome and almost impossible, even in a perfect bioreactor condition. Fully matured and differentiated cells have different physiological requirements to be met. Hence, doing multicell line printing is difficult, as these requirements per cell line must be met during the printing process. Thus, one strategy is to use stem cells and have different growth factors incorporated in the bioink, so that the scaffold itself can trigger multiple cell line differentiation post printing [189, 190]. However, this is still far from being realized, as there can be multiple triggers that can drive differentiation of stem cells into a particular cell lineage. More basic science is still needed to fully understand what kind of biochemical and physicochemical cues are needed to differentiate stem cells to specific cell lines needed for the printed organ.

8.2.2 Stem cell biomaterial-based tissue engineering

Stem cells are cells that can transform into different cell line in the body [191]. Typically, stem cells are unspecialized and have no specific functions in the body, except that they have the potential to become specialized cells, such as muscle cells, blood cells, and brain cells. Stem cells are also important in repairing damaged parts of the body, as they can be recruited in the injured area and become part of the new tissue. Stem cells are divided into two types: pluripotent and multipotent stem cells. Pluripotent stem cells (pSC) can differentiate into any kind of cells, while multipotent stem cells can transform into limited multiple types. They can be further classified based on where they are derived from. The most pluripotent are the embryonic stem cells (ESC), which, as the word suggests, comes from early stage developing embryos. Since an embryonic tissue has the potential to develop into all the types of organs in the body, the isolated stem cells derived from an embryo possess the ability to become whatever kind of cell it may choose, given the right biochemical signaling. There is also what is called induced pluripotent stem cells (iPS cells), different from embryonic stem cells (ESC cells), which form the inner cell mass of an embryo but also are pluripotent, as they come from somatic or adult cells and, with proper biochemical cues, can revert to being stem cells that can eventually give rise to all the cell types that make up the body. Next, are tissue-specific stem cells (a.k.a. somatic or adult stem cells). Typically, these stem cells can generate different cell types for the specific tissue or organ in which they are localized. Mesenchymal stem cells (MSCs) also known as mesenchymal stromal cells or medicinal signaling cells are multipotent stromal cells that can differentiate into a variety of cell types, including osteoblasts (bone cells), chondrocytes (cartilage cells), myocytes (muscle cells), and adipocytes (fat cells which give rise to marrow adipose tissue).

Advances in our understanding of stem cells give rise to a new field of regenerative medicine in which stem cells are envisioned to replace or regrow damaged or diseased organs and tissue in the body [192, 193]. This is called cell therapy and has been progressing well, in the last two decades of development. There are two ways in which this is envisaged. First is with the use of autologous stem cells, that is, the cells are harvested and retransplanted from the same patient needing the therapy, while the other method is to use allogenic stem cells, in which the cells come from a cell line bank and are differentiated and transplanted to multiple patients who need the therapy. There are some pros and cons in these two types of stem cell therapy. The most obvious is that the immunogenicity of allogenic stem cells are not observed in autologous stem cells, that is, patients who receive allogenic stem cells must undergo a parallel treatment of immunosuppressants for their body to accept the transplanted cells [194–196, while autologous stem cell treatment generally does not require immuno suppressants and has a higher likelihood of transplantation acceptance and success in therapy. However, autologous stem cell therapy requires more patient burden [197, 198], as the stem cells must be harvested and isolated first, typically within the patient's bone marrow, and then further proliferated and differentiated into whatever cell line the therapy requires. These proliferation and differentiation steps are costly and time-consuming, and for patients who are in immediate need of treatment, this delay may be a matter of life and death. In contrast, allogenic stem cell therapy can be an off-the-shelf solution in which the cells are already proliferated and differentiated to meet the patient's need.

8.2.2.1 Role of biomaterials in stem cell differentiation

There is a wide array of potential applications for stem cells in regenerative therapy. People who might benefit from stem cell therapies include those with spinal cord injuries, type 1 diabetes, Parkinson's disease, amyotrophic lateral sclerosis, Alzheimer's disease, heart disease, stroke, burns, cancer, and osteoarthritis. Stem cells may have the potential to be grown to become new tissue for use in transplant and regenerative medicine. Researchers continue to advance the knowledge on stem cells differentiation methods to transplant in regenerative medicine. It must be understood that it is not pure conventional therapy and does not implant undifferentiated stem cells in the body with the hope that it will transform to the local tissue it has been placed in. The stem cells are first proliferated in vitro and exposed to biochemical and physical cues to allow them to differentiate into a specific cell type in the body. This differentiation process is specific to each cell type, and slight variations may lead to low yield of differentiated cells, improper differentiation, or contamination of stem cells into the final product. It must be considered that the final cell product must contain almost pure fully differentiated cells, with no stem cell population included, as it may cause teratogenic effects or, when once implanted, might differentiate into another cell type and cause ectopic cell type in body. For example, if cardiomyocyte cell product is implanted in a patient and a significant number of stem cells have been included in the lot, these stem cells might differentiate into fibroblast cells that can harden the heart muscle and hinder its pumping motion. Since stem cell differentiation is still a growing science, there are still a lot of unknowns and proper control of the differentiation protocol must be done to ensure that the implanted cells are of high quality. These control parameters use biological testing such as genome sequencing, using quantitative polymerase chain reaction (q-PCR) and Western Blot to trace hallmark genes and proteins specific for a cell type and flow cytometry, to determine the population distribution of differentiated cells to undifferentiated or improper differentiated cells.

Biomaterials already have a role in this step of stem cell therapy [199–201]. It has been shown that certain biomaterials, when introduced to a population of stem cells, can guide differentiation into specific cell types. Differentiation of stem cells can be modulated by various signals, including **soluble growth factor signal, cell adhesion signal, mechanical signal, and topological cues** [202, 203]. These signals can be incorporated into biomaterials to induce desirable cellular differentiation. Biomaterials can deliver soluble growth factors locally to stem cells at a controlled rate for a long period. Stem cell adhesion to specific adhesion molecules presented by biomaterials can induce specific differentiation. Mechanical signals and topological cues can be delivered to stem cells seeded onto biomaterial scaffolds.

Delivery of soluble growth factors

Growth factors entrapped inside biomaterials can be released via diffusion or degradation of delivery systems, or a combination of these two mechanisms [204]. There are multiple ways in which growth factors can influence stem cell differentiation. For example, vascular endothelial growth factor (VEGF) released from a hydrogel delivery system [205, 206] can induce vascular differentiation of human ESCs, and insulin-like growth factor-1 and transforming growth factor-β1 (TGF-β1) released from gelatine micro particles can induce chondrogenic differentiation of MSCs [207, 208]. Designing a growth factor protein delivery system using polymers can be a challenge. Growth factors can lose their bioactivity mainly from denaturation of the protein during loading process of the polymer carrier. When exposed to harsh synthesis methods, such as heat, acidic or alkali pH, and exposure to sonication and organic solvents, these growth factor proteins are often denatured and lose their biological activity.

In one example, the use of heparin for scaffolds avoids the use of organic solvents, as heparin has natural affinity to polymers [209, 210]. Heparin can, then, be used for the incorporation of growth factors into polymeric delivery systems. An additional benefit to using heparin in growth factor delivery systems is that it can deliver dual or multiple growth factors with high affinities, given that they have high affinity to it; this includes VEGF, fibroblast growth factor 2 (FGF2), and platelet-derived growth factor (PDGF), which, when used together can enhance tissue regeneration, especially in bone tissue. Another example of using heparin is with heparin-conjugated fibrin. This material has been used to induce differentiation of ESC-derived neuronal progenitor cells into neurons (e.g., oligo dentrocytes) by delivering multiple soluble factors (neurotrophin-3, PDGF, and sonic hedgehog) [211]. Furthermore, heparin can also potentiate the bioactivity of growth factors such as bone morphogenetic protein 2 (BMP-2) [212, 213].

Use of cell adhesion ligands in biomaterial scaffolds

Cells can adhere to biomaterial scaffolds based on interactions between cell surface receptors (i.e., integrin) and cell adhesion peptides (e.g., arginine-glycine-aspartic acid (RGD), an integrin-binding peptide found in fibronectin, laminin, and collagen) [214–216]. With regards to stem cell adhesion, they base their interactions between specific integrin and adhesion peptide, and this can modulate stem cell behavior. Depending on the material of the scaffolds, natural polymers such as collagen contain cell adhesion peptides that allow the direct adhesion of cells, without further modification, in contrast to synthetic polymeric scaffolds, which do not contain cell adhesion peptides and require modification to add these cell adhesion ligands. Thus, cell adherence to synthetic polymers uses conjugated cell adhesion peptides or cell adhesion proteins (e.g., fibronectin, vitronectin, collagen, and laminin) adsorbed on the polymers from serum. Although the adsorption of cell adhesion proteins containing cell adhesion peptides to scaffold surfaces facilitates cell adhesion, this is a highly unreliable process, as desorption of these peptides can happen. Thus, covalent conjugation of cell adhesion peptides to scaffolds would be better, since it is more reproducible and controllable, by adjusting the density of peptide conjugation. This would lead to better stem cell adhesion and modulation in their cell behavior.

Here, RGD peptides covalently conjugated to polyethylene glycol (PEG) diacrylate hydrogels induced osteogenic differentiation of human mesenchymal stem cells (MSCs) [217–219]. The process of osteogenic differentiation was concentration-dependent on higher RGD peptide concentration and showed significantly greater alkaline phosphatase activity of the MSCs. Adding RGD peptide also stimulated chondrogenic differentiation of stem cells. Compared to pure PEG diacrylate hydrogels, RGD peptide-conjugated PEG diacrylate hydrogels enhanced the expression of cartilage-specific markers (i.e., aggrecan and collagen type II) of human embryonic stem cells (ESCs) [220].

Mechanical cell signaling using biomaterials

Cells are highly influenced by mechanical stimuli inside the body and can behave based on the dynamic environment they are subjected into. With this knowledge, mechanical signals can influence stem cell differentiation as well. For example, shear force or cyclic strain induces vascular differentiation of ESCs [221, 222]. Using biomaterials as a cell adhesion substrate, it can also be used to deliver mechanical signals to cells seeded on these biomaterials. Mechanical forces such as stretching, shear, and compression can be applied to these biomaterial scaffolds and be transmitted to the seeded cells via cell surface receptors (i.e., integrin), which serve as mechanical transducers between the biomaterial scaffolds and cytoskeleton of the cell. These mechanical signals are converted into biochemical responses and activate specific genes that modulate stem cell differentiation.

Interestingly, pure mechanical signal alone can induce the differentiation of stem cells or progenitor cells, without the need for soluble growth factors. Here, research has demonstrated that cyclic strain can promote cardiomyogenesis of embryonic stem cell-derived cardiomyocytes (ESCCs). ESCCs cultured on poly(lactide-co-caprolactone), PLCL scaffolds, and subjected to cyclic strain in vitro, upregulated cardiac gene expression compared to unstrained controls. Upon in vivo implantation of ESCCs seeded on elastic scaffolds or nonelastic scaffolds onto the surface of rat's beating heart, cardiac gene expression was observed to be upregulated in the elastic scaffolds in comparison to unstrained control scaffolds [220, 223]. This is evidence that cyclic strain has an effect in inducing cardiomyogenesis in ESC.

Another example is with cells in knee cartilage. These cells are exposed to cyclic compressional forces, in vivo, through constant motion of the joint. Thus, it is theorized that cyclic compression can induce chondrogenic differentiation of stem cells. In a study conducted using hydrogels as a scaffold, bone marrow-derived MSCs were cultured in PEG diacrylate hydrogels, and with application of cyclic compressive strain to the hydrogels, in vitro, have upregulated the gene expression of cartilage-related markers such as Sox-9, type II collagen, and aggrecan, even in the absence of soluble growth factor (TGF-beta1) [218, 224].

Not only can mechanical signals or soluble signals be used to stimulate the differentiation of stem cells, to a higher extent, the combination of both methods can also show synergism. In a research example, either cyclic strain or TGF-β1 alone can stimulate the cardiomyogenic differentiation of stem cells, but combining TGF-β1 and cyclic strain application can upregulate the expression of cardiac-specific markers in MSCs better than TGF-β1 supplementation alone. The results have shown that the combination of cyclic strain and TGF-β1 can increase the expression of troponin-I or cardiac myosin heavy chain by an average of 44.5% of treated MSCs.

Micro and nano topological cues for directing stem cell differentiation

Micro and nano topological cues from biomaterial substrates can also influence the differentiation of stem cells. The advent of micro and nano fabrication techniques has enabled the production of micro and nano patterned culture surfaces and has allowed the study of stem cells behavior in response to these topologies [225, 226]. ESCs can differentiate in various types of tissues or organs by forming embryoid bodies (EBs). (EBs are 3D aggregates of pluripotent stem cells. EBs are differentiations of human embryonic stem cells compromising the three embryonic germ layers). However, the inability to homogeneously direct ESC lineage commitment limits the clinical application of ESCs. One reason for the heterogeneity of ESC lineage commitment is the wide variation in EB size. Controlling EB size can precisely modulate ESC differentiation. Conventional techniques for the formation of EBs, such as hanging drop and suspension cultures, do not allow for precise control of EB size. In a study performed using a microwell array system created by photolithography and plasma etching techniques, uniformly sized EB formation has been achieved [227]. The results have shown that EBs are formed from human culture ESCs at a higher rate (78%) in the microwells, in the size range of 280 to 359 µm. In contrast, only (31%) of EBs formed using normal culture on plastic were of the same size range. Another method for the production of uniformly sized EBs involves microfabricated adhesive stencils [227, 228]. EBs with sizes ranging from 100 to 500 µm in diameter were formed using this method, and EB size was found to influence EB lineage commitment. Smaller (100 µm) EBs exhibited increased expression of ectodermal markers compared to larger (500 µm) EBs, whereas the 500 µm sized EBs showed increased expression of mesodermal and endodermal markers compared to the 100 µm sized EBs.

Enhancing cell delivery and engraftment using biomaterials

Delivery of cells toward the affected area of the body is half the battle in cell therapy. Three stages make up a typical cell therapy procedure: (1) in vitro preparation of cell suspensions; (2) injection procedure; and (3) retention of the administered cells, post-injection. Focusing on one stage only can yield optimized settings that are not favorable to the entire procedure, and therefore it is essential that a systematic investigation considers all three stages to outline optimal transplantation parameters. Cell delivery is a science on its own and comes with its very own challenges. Unlike other drug products [229, 230], cells cannot be delivered intravenously, as it could lead to ectopic cell localization and teratogenicity. Cell delivery is a suite of methods that has two purposes – first to precisely deliver cellular therapeutics to a target site (tissue or organ) and second, to ensure that the delivered cell therapeutics work. Cell delivery is important because cellular therapeutics is fragile, and preservation of their functionality is essential for efficacy [231].

It has been described earlier that injectable hydrogels can be used as a delivery vehicle for cells during transplantation [232, 233]. Not only can they enhance the delivery of cells, these biomaterials can also improve the engraftment of the implanted cells, when compared to simply injecting cell suspension products. Typically, cells are suspended in phosphate buffer solution either before or during transplantation. However, cells can sediment fast in this media and can lead to nonuniform dosing. Using hydrogels can make the cells sediment less as the viscosity of the gel can hold the cells homogenously during implantation. The hydrogel can also act as scaffolding for the cells and can be loaded with growth factors that can enhance vascularization toward the newly implanted hydrogel-cell material.

8.2.3 Biomaterials in closed-loop tissue engineering methods

Device-cell combination products is a way to create a functional organ that does not involve regrowing the diseased organ but aims to create an implantable extracorporeal organ that can function like native organs. This method is called closed-loop tissue engineering [234, 235] as the functional tissues or cells are enclosed inside a device and can interact with the body only through exchanges of nutrients, gasses, and the biochemical signaling via a semi-permeable membrane.

One example of this is the development of artificial pancreas. The pancreas is part of the endocrine system responsible for secreting various hormones, most famously, insulin, which is needed to transport glucose from the blood toward the intracellular space for metabolism. People who have Type I diabetes have a malfunctioning pancreas in which the beta cells that are responsible for insulin production are attacked by the immune cells of the person. The cause of this phenomenon is still unknown, but when this happens, the person, generally, will have to be dependent on synthetic insulin for the rest of his/her life to survive. There are some electronic closed-loop systems that are being used by T1D patients to make insulin administration and glucose monitoring easy and streamlined [236–238]. The closed-loop system incorporates an insulin drug reservoir and delivery with continuous glucose monitoring to track the level of blood glucose in the body in real time and deliver the right amount of insulin, based on that current glucose level reading. However, there are limitations to this technology, as the insulin must be replenished once it is used up, and the CGM monitor needle has to be replaced every two weeks, as the sensitivity of the glucose monitor can be reduced due to the body's immune response toward the needle sensor, not to mention that this system is very bulky and can limit the person's day-to-day activity. Hence, the appeal of creating an implantable closed-loop synthetic tissue-based pancreas is still very big, as patients can have it implanted and not worry about those limitations. One leading device being explored for clinical application is the Procyon cell encapsulation device [239]. It is a small, flat, thin, highly durable, flexible, implantable chamber, which features mitigation of foreign body response, promotes the formation of vascular structures on its surface enabling the rapid diffusion of nutrients to the cells inside, and the rapid absorption of insulin (or other therapeutic molecules) secreted by the encapsulated cells, while providing a barrier protecting them from attacks by the body's immune system, without the need for immunosuppressive drugs. The Procyon technology, designed with practical clinical use as a driver, includes integration of oxygen delivery to the implantable device, which enables tighter packing of cells while maintaining their viability and functionality However, one challenge is that the cells must be able to survive inside the device. It has been found that the beta cells, especially in their first months of encapsulation inside the device, require high amounts of oxygen to survive [240]. This would mean that the implant must be tethered to an external oxygen source to supply oxygen to the cells until they are self-sufficient and able to survive with normal oxygen supply from the newly formed vasculature that may form around the implant. Second, it is still unsure whether the beta cells can respond immediately with a spike in glucose level.

8.3 Conclusions

Tissue and organ regeneration is now being realized with the help of advances in biomaterials and cell biology. Multiple approaches are now being developed to help solve organ scarcity, tissue and organ repair, and treatment for various diseases using cell therapy. There is still need for a complete understanding of materials and cell interaction to optimize the growth and create functionality to the grown tissue. Pure biomaterial implantation to guide the growth of damaged tissue is now being realized in the clinic with relatively good signs of success. This demonstrates that biomaterials can help in regenerating tissues and restore their functionality to a certain extent, in contrast to simply letting the normal healing process occur. This shows positive signs for the future, that a fully functional lab-grown organ, tissue, and closed-loop cell devices can also be developed with the help of biomaterial sciences. However, even though growing small portions of tissue, in vitro, can be done right now, ensuring that it can be transplanted and function as intended remains a challenge. Thus, better understanding of anatomy and physiology can help biomaterial scientists with designing methods that can help realize this grail.

References

[1] Agudo J, Park ES, Rose SA, Alibo E, Sweeney R, Dhainaut M, et al. Quiescent tissue stem cells evade immune surveillance. Immunity. 2018;48(2):271–85. e5.

[2] Yokoi H, Isaji S, Yamagiwa K, Tabata M, Sakurai H, Usui M, et al. Donor outcome and liver regeneration after right-lobe graft donation. Transplant International. 2005;18(8):915–22.

[3] Marcos A, Fisher RA, Ham JM, Shiffman ML, Sanyal AJ, Luketic VA, et al. Liver regeneration and function in donor and recipient after right lobe adult to adult living donor liver transplantation12. Transplantation. 2000;69(7):1375–79.
[4] Fausto N, Riehle KJ. Mechanisms of liver regeneration and their clinical implications. Journal of Hepato-biliary-pancreatic Surgery. 2005;12(3):181–89.
[5] Davies MJ. The oxidative environment and protein damage. Biochimica Et Biophysica Acta (Bba)-proteins and Proteomics. 2005;1703(2):93–109.
[6] Das K, Das S, Dhundasi S. Nickel, its adverse health effects & oxidative stress. Indian Journal of Medical Research. 2008;128(4):412.
[7] Hage A, Hage F, Valdis M, Guo L, Chu MW. The Ross procedure is the optimal solution for young adults with unrepairable aortic valve disease. Annals of Cardiothoracic Surgery. 2021;10(4):454.
[8] Braiman-Wiksman L, Solomonik I, Spira R, Tennenbaum T. Novel insights into wound healing sequence of events. Toxicologic Pathology. 2007;35(6):767–79.
[9] Cañedo-Dorantes L, Cañedo-Ayala M. Skin acute wound healing: A comprehensive review. International Journal of Inflammation. 2019;2019.
[10] Schmidt-Bleek K, Schell H, Schulz N, Hoff P, Perka C, Buttgereit F, et al. Inflammatory phase of bone healing initiates the regenerative healing cascade. Cell and Tissue Research. 2012;347(3):567–73.
[11] Krafts KP. Tissue repair: The hidden drama. Organogenesis. 2010;6(4):225–33.
[12] Giannoudi M, Harwood P. Damage control resuscitation: Lessons learned. European Journal of Trauma and Emergency Surgery. 2016;42(3):273–82.
[13] Periayah MH, Halim AS, Saad AZM. Mechanism action of platelets and crucial blood coagulation pathways in hemostasis. International Journal of Hematology-oncology and Stem Cell Research. 2017;11(4):319.
[14] Broos K, Feys HB, De Meyer SF, Vanhoorelbeke K, Deckmyn H. Platelets at work in primary hemostasis. Blood Reviews. 2011;25(4):155–67.
[15] Engelmann B, Massberg S. Thrombosis as an intravascular effector of innate immunity. Nature Reviews. Immunology. 2013;13(1):34–45.
[16] Russo MV, McGavern DB. Immune surveillance of the CNS following infection and injury. Trends in Immunology. 2015;36(10):637–50.
[17] Alhajj M, Bansal P, Goyal A. Physiology, Granulation Tissue. StatPearls [Internet]. 2020.
[18] Häkkinen L, Larjava H, Koivisto L. Granulation tissue formation and remodeling. Endodontic Topics. 2011;24(1):94–129.
[19] Greaves NS, Ashcroft KJ, Baguneid M, Bayat A. Current understanding of molecular and cellular mechanisms in fibroplasia and angiogenesis during acute wound healing. Journal of Dermatological Science. 2013;72(3):206–17.
[20] Barrientos S, Stojadinovic O, Golinko MS, Brem H, Tomic-Canic M. Growth factors and cytokines in wound healing. Wound Repair and Regeneration. 2008;16(5):585–601.
[21] Franchini A, Ottaviani E. Repair of molluscan tissue injury: Role of PDGF and TGF-β. Tissue & Cell. 2000;32(4):312–21.
[22] Denhardt DT, Noda M. Osteopontin expression and function: Role in bone remodeling. Journal of Cellular Biochemistry. 1998;72(S30–31):92–102.
[23] Hijova E. Matrix metalloproteinases: Their biological functions and clinical implications. Bratislavske Lekarske Listy. 2005;106(3):127–32.
[24] Abraham D, Ponticos M, Nagase H. Connective tissue remodeling: Cross-talk between endothelins and matrix metalloproteinases. Current Vascular Pharmacology. 2005;3 (4):369–79.
[25] Enoch S, Leaper DJ. Basic science of wound healing. Surgery (Oxford). 2008;26(2):31–37.

[26] Ehrlich HP. Collagen Considerations in Scarring and Regenerative Repair. Scarless Wound Healing: CRC Press; 2000. p. 111–25.
[27] Ehrlich HP, Krummel TM. Regulation of wound healing from a connective tissue perspective. Wound Repair and Regeneration. 1996;4(2):203–10.
[28] Greenhalgh DG. Wound healing and diabetes mellitus. Clinics in Plastic Surgery. 2003;30(1):37–45.
[29] Sharma S, Schaper N, Rayman G. Microangiopathy: Is it relevant to wound healing in diabetic foot disease?. Diabetes/metabolism Research and Reviews. 2020;36:e3244.
[30] Markel TA, Lou DC, Pfefferkorn M, Scherer III L, West K, Rouse T, et al. Steroids and poor nutrition are associated with infectious wound complications in children undergoing first stage procedures for ulcerative colitis. Surgery. 2008;144(4):540–47.
[31] Grada A, Phillips TJ. Nutrition and cutaneous wound healing. Clinics in Dermatology. 2021.
[32] Teller P, White TK. The physiology of wound healing: Injury through maturation. Perioperative Nursing Clinics. 2011;6(2):159–70.
[33] Edwards R, Harding KG. Bacteria and wound healing. Current Opinion in Infectious Diseases. 2004;17(2):91–96.
[34] Schreier T, Degen E, Baschong W. Fibroblast migration and proliferation during in vitro wound healing. Research in Experimental Medicine. 1993;193(1):195–205.
[35] Ellis S, Lin EJ, Tartar D. Immunology of wound healing. Current Dermatology Reports. 2018;7 (4):350–58.
[36] Stenn KS, Depalma L. Re-epithelialization. In: Clark RAF, Henson PM, editors. The Molecular and Cellular Biology of Wound Repair. Boston, MA: Springer US; 1988. p. 321–35.
[37] Wu J, Liu Y, Cao Q, Yu T, Zhang J, Liu Q, et al. Growth factors enhanced angiogenesis and osteogenesis on polydopamine coated titanium surface for bone regeneration. Materials & Design. 2020;196:109162.
[38] Smith QT. Collagen metabolism in wound healing. In: Day SB, editor. Trauma: Clinical and Biological Aspects. Boston, MA: Springer US; 1975. p. 31–45.
[39] Sardari K, Kakhki EG, Mohri M. Evaluation of wound contraction and epithelialization after subcutaneous administration of Theranekron® in cows. Comparative Clinical Pathology. 2007;16(3):197–200.
[40] Strihler B. Times, Cells, and Aging. Elsevier. 2012.
[41] Shay JW. Telomeres and aging. Current Opinion in Cell Biology. 2018;52:1–7.
[42] Oeseburg H, de Boer RA, van Gilst WH, van der Harst P. Telomere biology in healthy aging and disease. Pflügers Archiv-European Journal of Physiology. 2010;459(2):259–68.
[43] Prusiner SB. Neurodegenerative diseases and prions. New England Journal of Medicine. 2001;344(20):1516–26.
[44] Bazan NG, Molina MF, Gordon WC. Docosahexaenoic acid signalolipidomics in nutrition: Significance in aging, neuroinflammation, macular degeneration, Alzheimer's, and other neurodegenerative diseases. Annual Review of Nutrition. 2011;31:321–51.
[45] Dequeker J, Aerssens J, Luyten FP. Osteoarthritis and osteoporosis: Clinical and research evidence of inverse relationship. Aging Clinical and Experimental Research. 2003;15(5):426–39.
[46] Dequeker J, Luyten F. The history of osteoarthritis-osteoarthrosis. Annals of the Rheumatic Diseases. 2008;67(1):5–10.
[47] Davidson A, Diamond B. Autoimmune diseases. New England Journal of Medicine. 2001;345(5):340–50.
[48] Cooper GS, Stroehla BC. The epidemiology of autoimmune diseases. Autoimmunity Reviews. 2003;2(3):119–25.

[49] Banchereau J, Pascual V. Type I interferon in systemic lupus erythematosus and other autoimmune diseases. Immunity. 2006;25(3):383–92.
[50] Ramos PS, Criswell LA, Moser KL, Comeau ME, Williams AH, Pajewski NM, et al. A comprehensive analysis of shared loci between systemic lupus erythematosus (SLE) and sixteen autoimmune diseases reveals limited genetic overlap. PLoS Genetics. 2011;7(12): e1002406.
[51] Lam GK, Petri M. Assessment of systemic lupus erythematosus. Clinical and Experimental Rheumatology. 2005;23(5):S120.
[52] Kinsey D, Paul CP, Taylor D, Caricchio R, Kulathinal RJ, Hayes-Conroy A. The whole lupus: Articulating biosocial interplay in systemic lupus erythematosus epidemiology and population disparities. Health & Place. 2018;51:182–88.
[53] Narula N, Dannenberg AJ, Olin JW, Bhatt DL, Johnson KW, Nadkarni G, et al. Pathology of peripheral artery disease in patients with critical limb ischemia. Journal of the American College of Cardiology. 2018;72(18):2152–63.
[54] Clarkson AN, Sutherland BA, Appleton I. The biology and pathology of hypoxia-ischemia: An update. Archivum Immunologiae Et Therapiae Experimentalis. 2005;53(3):213–25.
[55] Khan M, Kwiatkowski P, Rivera BK, Kuppusamy P. Oxygen and oxygenation in stem-cell therapy for myocardial infarction. Life Sciences. 2010;87(9–10):269–74.
[56] Unnithan AKA, Mehta P. Hemorrhagic stroke. StatPearls [Internet]. 2020.
[57] Hui C, Tadi P, Patti L. Ischemic Stroke. 2018.
[58] Gatti GL, Freda N, Giacomina A, Montemagni M, Sisti A. Cleft lip and palate repair. Journal of Craniofacial Surgery. 2017;28(8):1918–24.
[59] Liao Y-F, Mars M. Long-term effects of palate repair on craniofacial morphology in patients with unilateral cleft lip and palate. The Cleft Palate-craniofacial Journal. 2005;42 (6):594–600.
[60] Maranho DA, Nogueira-Barbosa MH, Simão MN, Volpon JB. Ultrasonographic evaluation of Achilles tendon repair after percutaneous sectioning for the correction of congenital clubfoot residual equinus. Journal of Pediatric Orthopaedics. 2009;29(7):804–10.
[61] Shah SR, Wishnew J, Barsness K, Gaines BA, Potoka DA, Gittes GK, et al. Minimally invasive congenital diaphragmatic hernia repair: A 7-year review of one institution's experience. Surgical Endoscopy. 2009;23(6):1265–71.
[62] Lemire RJ. Neural tube defects. Jama. 1988;259(4):558–62.
[63] Frey L, Hauser WA. Epidemiology of neural tube defects. Epilepsia. 2003;44:4–13.
[64] Stoll C, Dott B, Alembik Y, Roth M-P. Associated congenital anomalies among cases with Down syndrome. European Journal of Medical Genetics. 2015;58(12):674–80.
[65] Manning N, Archer N. A study to determine the incidence of structural congenital heart disease in monochorionic twins. Prenatal Diagnosis: Published in Affiliation with the International Society for Prenatal Diagnosis. 2006;26(11):1062–64.
[66] Aksu-Menges E, Eylem CC, Nemutlu E, Gizer M, Korkusuz P, Topaloglu H, et al. Reduced mitochondrial fission and impaired energy metabolism in human primary skeletal muscle cells of Megaconial Congenital Muscular Dystrophy. Scientific Reports. 2021;11(1):1–15.
[67] Ouedraogo M, Baudoux T, Stévigny C, Nortier J, Colet J-M, Efferth T, et al. Review of current and "omics" methods for assessing the toxicity (genotoxicity, teratogenicity and nephrotoxicity) of herbal medicines and mushrooms. Journal of Ethnopharmacology. 2012;140(3):492–512.
[68] Sav A. Pathological Anatomy of Spina Bifida. The Spina Bifida: Management and Outcome. Milano: Springer Milan; 2008. p. 43–57.

[69] Limwongse C. Syndromes and malformations of the urinary tract. In: Avner E, Harmon W, Niaudet P, Yoshikawa N, editors. Pediatric Nephrology: Sixth Completely Revised, Updated and Enlarged Edition. Berlin, Heidelberg: Springer Berlin Heidelberg; 2009. p. 121–56.
[70] Schwartz MZ, Timmapuri SJ. Gastroschisis. In: Puri P, editor. Pediatric Surgery: General Principles and Newborn Surgery. Berlin, Heidelberg: Springer Berlin Heidelberg; 2020. p. 1177–87.
[71] Mehollin-Ray AR. Congenital diaphragmatic hernia. Pediatric Radiology. 2020;50(13):1855–71.
[72] Beasley SW. Anatomy. In: Beasley SW, Myers NA, Auldist AW, editors. Oesophageal Atresia. Boston, MA: Springer US; 1991. p. 45–58.
[73] Orofacial C, Vohr H-W, editor. Encyclopedic Reference of Immunotoxicology. Berlin, Heidelberg: Springer Berlin Heidelberg; 2005. p. 492.
[74] Baiguera S, Jungebluth P, Burns A, Mavilia C, Haag J, De Coppi P, et al. Tissue engineered human tracheas for in vivo implantation. Biomaterials. 2010;31(34):8931–38.
[75] Silva CR, Babo PS, Gulino M, Costa L, Oliveira JM, Silva-Correia J, et al. Injectable and tunable hyaluronic acid hydrogels releasing chemotactic and angiogenic growth factors for endodontic regeneration. Acta Biomaterialia. 2018;77:155–71.
[76] Kim SU, De Vellis J. Stem cell-based cell therapy in neurological diseases: A review. Journal of Neuroscience Research. 2009;87(10):2183–200.
[77] Sizikov AA, Nikitin PI, Nikitin MP. Magnetofection in vivo by nanomagnetic carriers systemically administered into the bloodstream. Pharmaceutics. 2021;13(11):1927.
[78] Thum T, Bauersachs J, Poole-Wilson PA, Volk H-D, Anker SD. The dying stem cell hypothesis: Immune modulation as a novel mechanism for progenitor cell therapy in cardiac muscle. Journal of the American College of Cardiology. 2005;46(10):1799–802.
[79] Rivera FJM, Arciniegas AJR. Additive manufacturing methods: Techniques, materials, and closed-loop control applications. The International Journal of Advanced Manufacturing Technology. 2020;109(1):17–31.
[80] Turksoy K, Bayrak ES, Quinn L, Littlejohn E, Cinar A. Multivariable adaptive closed-loop control of an artificial pancreas without meal and activity announcement. Diabetes Technology & Therapeutics. 2013;15(5):386–400.
[81] Mota C, Camarero-Espinosa S, Baker MB, Wieringa P, Moroni L. Bioprinting: From tissue and organ development to in vitro models. Chemical Reviews. 2020;120(19):10547–607.
[82] Waqanivavalagi SW, Bhat S, Ground MB, Milsom PF, Cornish J. Clinical performance of decellularized heart valves versus standard tissue conduits: A systematic review and meta-analysis. Journal of Cardiothoracic Surgery. 2020;15(1):1–13.
[83] Kasimir M-T, Rieder E, Seebacher G, Silberhumer G, Wolner E, Weigel G, et al. Comparison of different decellularization procedures of porcine heart valves. The International Journal of Artificial Organs. 2003;26(5):421–27.
[84] Lu T-Y, Lin B, Kim J, Sullivan M, Tobita K, Salama G, et al. Repopulation of decellularized mouse heart with human induced pluripotent stem cell-derived cardiovascular progenitor cells. Nature Communications. 2013;4(1):1–11.
[85] Hillebrandt KH, Everwien H, Haep N, Keshi E, Pratschke J, Sauer IM. Strategies based on organ decellularization and recellularization. Transplant International. 2019;32(6):571–85.
[86] Alexanian RA, Mahapatra K, Lang D, Vaidyanathan R, Markandeya YS, Gill RK, et al. Induced cardiac progenitor cells repopulate decellularized mouse heart scaffolds and differentiate to generate cardiac tissue. Biochimica Et Biophysica Acta (Bba)-molecular Cell Research. 2020;1867(3):118559.
[87] Arenas-Herrera J, Ko I, Atala A, Yoo J. Decellularization for whole organ bioengineering. Biomedical Materials. 2013;8(1):014106.

[88] Tapias LF, Ott HC. Decellularized scaffolds as a platform for bioengineered organs. Current Opinion in Organ Transplantation. 2014;19(2):145.
[89] Wang Y, Bao J, Wu Q, Zhou Y, Li Y, Wu X, et al. Method for perfusion decellularization of porcine whole liver and kidney for use as a scaffold for clinical-scale bioengineering engrafts. Xenotransplantation. 2015;22(1):48–61.
[90] Murphy SV, Atala A. Organ engineering–combining stem cells, biomaterials, and bioreactors to produce bioengineered organs for transplantation. Bioessays. 2013;35(3):163–72.
[91] Brown BN, Valentin JE, Stewart-Akers AM, McCabe GP, Badylak SF. Macrophage phenotype and remodeling outcomes in response to biologic scaffolds with and without a cellular component. Biomaterials. 2009;30(8):1482–91.
[92] Edri R, Gal I, Noor N, Harel T, Fleischer S, Adadi N, et al. Personalized hydrogels for engineering diverse fully autologous tissue implants. Advanced Materials. 2019;31(1):1803895.
[93] Kaur A, Midha S, Giri S, Mohanty S. Functional skin grafts: Where biomaterials meet stem cells. Stem Cells International. 2019;2019.
[94] Sheikholeslam M, Wright ME, Jeschke MG, Amini-Nik S. Biomaterials for skin substitutes. Advanced Healthcare Materials. 2018;7(5):1700897.
[95] Chen F-M, Liu X. Advancing biomaterials of human origin for tissue engineering. Progress in Polymer Science. 2016;53:86–168.
[96] Singh M, Nuutila K, Kruse C, Robson MC, Caterson E, Eriksson E. Challenging the conventional therapy: Emerging skin graft techniques for wound healing. Plastic and Reconstructive Surgery. 2015;136(4):524e–30e.
[97] Patterson CW, Stark M, Sharma S, Mundinger GS. Regeneration and expansion of autologous full-thickness skin through a self-propagating autologous skin graft technology. Clinical Case Reports. 2019;7(12):2449–55.
[98] Mcheik JN, Barrault C, Levard G, Morel F, Bernard F-X, Lecron J-C. Epidermal healing in burns: Autologous keratinocyte transplantation as a standard procedure: Update and perspective. Plastic and Reconstructive Surgery Global Open. 2014;2:9.
[99] Davis C, Boyd C, de Acosta Andino DAM, Kumbla PA, Sanchez RJ, Kurapati S, et al. Dermal autografts in breast reconstruction: A review of past and current trends. Annals of Plastic Surgery. 2020;84(5):618–22.
[100] Baluch N, Nagata S, Park C, Wilkes GH, Reinisch J, Kasrai L, et al. Auricular reconstruction for microtia: A review of available methods. Plastic Surgery. 2014;22(1):39–43.
[101] Kim JW, Kim MJ, Ki CS, Kim HJ, Park YH. Fabrication of bi-layer scaffold of keratin nanofiber and gelatin-methacrylate hydrogel: Implications for skin graft. International Journal of Biological Macromolecules. 2017;105:541–48.
[102] Chanda A, Adhikari J, Ghosh A, Chowdhury SR, Thomas S, Datta P, et al. Electrospun chitosan/polycaprolactone-hyaluronic acid bilayered scaffold for potential wound healing applications. International Journal of Biological Macromolecules. 2018;116:774–85.
[103] Hernández-Rangel A, Martin-Martinez ES. Collagen based electrospun materials for skin wounds treatment. Journal of Biomedical Materials Research. Part A. 2021.
[104] Lai H-J, Kuan C-H, Wu H-C, Tsai J-C, Chen T-M, Hsieh D-J, et al. Tailored design of electrospun composite nanofibers with staged release of multiple angiogenic growth factors for chronic wound healing. Acta Biomaterialia. 2014;10(10):4156–66.
[105] Sundaramurthi D, Krishnan UM, Sethuraman S. Electrospun nanofibers as scaffolds for skin tissue engineering. Polymer Reviews. 2014;54(2):348–76.
[106] Wiegand C, Schönfelder U, Abel M, Ruth P, Kaatz M, Hipler U-C. Protease and pro-inflammatory cytokine concentrations are elevated in chronic compared to acute wounds and

can be modulated by collagen type I in vitro. Archives of Dermatological Research. 2010;302(6):419–28.

[107] Castruita-de la Rosa C, Garza-Veloz I, Cardenas-Vargas E, Castañeda-Miranda R, Solis-Sanchez LO, Ortiz-Rodriguez JM, et al. Biological activity and implications of the metalloproteinases in diabetic foot ulcers. The Role of Matrix Metalloproteinase in Human Body Pathologies. 2017:169.

[108] Vincenti MP, Brinckerhoff CE. Transcriptional regulation of collagenase (MMP-1, MMP-13) genes in arthritis: Integration of complex signaling pathways for the recruitment of gene-specific transcription factors. Arthritis Research & Therapy. 2002;4(3):1–8.

[109] Cullen B, Smith R, Mcculloch E, Silcock D, Morrison L. Mechanism of action of promogran, a protease modulating matrix, for the treatment of diabetic foot ulcers. Wound Repair and Regeneration. 2002;10(1):16–25.

[110] Morales-Cárdenas A, Ploneda-Valencia CF, Sainz-Escárrega VH, Hernández-Campos AC, Navarro-Muniz E, López-Lizarraga CR, et al. Amyand hernia: Case report and review of the literature. Annals of Medicine and Surgery. 2015;4(2):113–15.

[111] Jansen PL, Mertens P, Klinge U, Schumpelick V. The biology of hernia formation. Surgery. 2004;136(1):1–4.

[112] See CW, Kim T, Zhu D. Hernia mesh and hernia repair: A review. Engineered Regeneration. 2020;1:19–33.

[113] Shankaran V, Weber DJ, Lawrence Reed I R, Luchette FA. A review of available prosthetics for ventral hernia repair. Annals of Surgery. 2011;253(1):16–26.

[114] Novitsky YW. Biology of biological meshes used in hernia repair. Surgical Clinics. 2013;93 (5):1211–15.

[115] Cevasco M, Itani KM. Ventral hernia repair with synthetic, composite, and biologic mesh: Characteristics, indications, and infection profile. Surgical Infections. 2012;13(4):209–15.

[116] Thygesen K, Alpert JS, White HD. Infarction JEAAWTFftRoM. Universal definition of myocardial infarction. Journal of the American College of Cardiology. 2007;50(22):2173–95.

[117] Anderson JL, Morrow DA. Acute myocardial infarction. New England Journal of Medicine. 2017;376(21):2053–64.

[118] Gerber Y, Weston SA, Berardi C, McNallan SM, Jiang R, Redfield MM, et al. Contemporary trends in heart failure with reduced and preserved ejection fraction after myocardial infarction: A community study. American Journal of Epidemiology. 2013;178(8):1272–80.

[119] Song Y, Wang H, Yue F, Lv Q, Cai B, Dong N, et al. Silk-based biomaterials for cardiac tissue engineering. Advanced Healthcare Materials. 2020;9(23):2000735.

[120] Shafiee A, Atala A. Tissue engineering: Toward a new era of medicine. Annual Review of Medicine. 2017;68:29–40.

[121] Fioretta ES, Motta SE, Lintas V, Loerakker S, Parker KK, Baaijens FP, et al. Next-generation tissue-engineered heart valves with repair, remodelling and regeneration capacity. Nature Reviews. Cardiology. 2021;18(2):92–116.

[122] Ramiah P, du Toit LC, Choonara YE, Kondiah PP, Pillay V. Hydrogel-based bioinks for 3D bioprinting in tissue regeneration. Frontiers in Materials. 2020;7:76.

[123] Kumar S, Tharayil A, Thomas S. 3D bioprinting of nature-inspired hydrogel inks based on synthetic polymers. ACS Applied Polymer Materials. 2021;3(8):3685–701.

[124] Panwar A, Tan LP. Current status of bioinks for micro-extrusion-based 3D bioprinting. Molecules. 2016;21(6):685.

[125] Gopinathan J, Noh I. Recent trends in bioinks for 3D printing. Biomaterials Research. 2018;22(1):1–15.

[126] Yin J, Yan M, Wang Y, Fu J, Suo H. 3D bioprinting of low-concentration cell-laden gelatin methacrylate (GelMA) bioinks with a two-step cross-linking strategy. ACS Applied Materials & Interfaces. 2018;10(8):6849–57.

[127] Recordati G. A thermodynamic model of the sympathetic and parasympathetic nervous systems. Autonomic Neuroscience. 2003;103(1–2): 1–12.

[128] Takayama Y, Kushige H, Akagi Y, Suzuki Y, Kumagai Y, Kida YS. Selective induction of human autonomic neurons enables precise control of cardiomyocyte beating. Scientific Reports. 2020;10(1):1–13.

[129] Vunjak-Novakovic G, Tandon N, Godier A, Maidhof R, Marsano A, Martens TP, et al. Challenges in cardiac tissue engineering. Tissue Engineering. Part B, Reviews. 2010;16(2):169–87.

[130] Miyagi Y, Chiu LL, Cimini M, Weisel RD, Radisic M, Li R-K. Biodegradable collagen patch with covalently immobilized VEGF for myocardial repair. Biomaterials. 2011;32(5):1280–90.

[131] Su T, Huang K, Daniele MA, Hensley MT, Young AT, Tang J, et al. Cardiac stem cell patch integrated with microengineered blood vessels promotes cardiomyocyte proliferation and neovascularization after acute myocardial infarction. ACS Applied Materials & Interfaces. 2018;10(39):33088–96.

[132] O'Neill HS, O'Sullivan J, Porteous N, Ruiz-Hernandez E, Kelly HM, O'Brien FJ, et al. A collagen cardiac patch incorporating alginate microparticles permits the controlled release of hepatocyte growth factor and insulin-like growth factor-1 to enhance cardiac stem cell migration and proliferation. Journal of Tissue Engineering and Regenerative Medicine. 2018;12(1):e384–e94.

[133] Báez-Díaz C, Blanco-Blázquez V, Sánchez-Margallo F-M, Bayes-Genis A, González I, Abad A, et al. Microencapsulated insulin-like growth factor-1 therapy improves cardiac function and reduces fibrosis in a porcine acute myocardial infarction model. Scientific Reports. 2020;10(1):1–11.

[134] Orton EC, Mama K, Hellyer P, Hackett TB. Open surgi repair of tetralogy of Fallot in dogs. Journal of the American Veterinary Medical Association. 2001;219(8):1089–93.

[135] Cui J, Li J, Mathison M, Tondato F, Mulkey SP, Micko C, et al. A clinically relevant large-animal model for evaluation of tissue-engineered cardiac surgical patch materials. Cardiovascular Revascularization Medicine. 2005;6(3):113–20.

[136] Yucel D, Kose GT, Hasirci V. Polyester based nerve guidance conduit design. Biomaterials. 2010;31(7):1596–603.

[137] Kim Y-T, Haftel VK, Kumar S, Bellamkonda RV. The role of aligned polymer fiber-based constructs in the bridging of long peripheral nerve gaps. Biomaterials. 2008;29(21):3117–27.

[138] Stucky CL, Mikesell AR. Cutaneous pain in disorders affecting peripheral nerves. Neuroscience Letters. 2021;765:136233.

[139] Xu H, Holzwarth JM, Yan Y, Xu P, Zheng H, Yin Y, et al. Conductive PPY/PDLLA conduit for peripheral nerve regeneration. Biomaterials. 2014;35(1):225–35.

[140] Boni R, Ali A, Shavandi A, Clarkson AN. Current and novel polymeric biomaterials for neural tissue engineering. Journal of Biomedical Science. 2018;25(1):1–21.

[141] Sánchez-Pech JC, Rosales-Ibáñes R, Cauich-Rodriguez JV, Carrillo-Escalante HJ, Rodríguez-Navarrete A, Avila-Ortega A, et al. Design, synthesis, characterization, and cytotoxicity of PCL/PLGA scaffolds through plasma treatment in the presence of pyrrole for possible use in urethral tissue engineering. Journal of Biomaterials Applications. 2020;34(6):840–50.

[142] Guo B, Ma PX. Conducting polymers for tissue engineering. Biomacromolecules. 2018;19(6):1764–82.

[143] Li Y, Xiao Y, Liu C. The horizon of materiobiology: A perspective on material-guided cell behaviors and tissue engineering. Chemical Reviews. 2017;117(5):4376–421.

[144] Kim JI, Hwang TI, Aguilar LE, Park CH, Kim CS. A controlled design of aligned and random nanofibers for 3D bi-functionalized nerve conduits fabricated via a novel electrospinning set-up. Scientific Reports. 2016;6(1):1–12.

[145] Nemeno-Guanzon JG, Lee S, Berg JR, Jo YH, Yeo JE, Nam BM, et al. Trends in tissue engineering for blood vessels. Journal of Biomedicine & Biotechnology. 2012;2012.

[146] Wang X, Lin P, Yao Q, Chen C. Development of small-diameter vascular grafts. World Journal of Surgery. 2007;31(4):682–89.

[147] Trollope AF, Golledge J. Angiopoietins, abdominal aortic aneurysm and atherosclerosis. Atherosclerosis. 2011;214(2):237–43.

[148] Stary HC, Chandler AB, Dinsmore RE, Fuster V, Glagov S, Insull W, et al. A definition of advanced types of atherosclerotic lesions and a histological classification of atherosclerosis: A report from the Committee on Vascular Lesions of the Council on Arteriosclerosis, American Heart Association. Circulation. 1995;92(5):1355–74.

[149] Yokobori AT, Ichiki M, Tomono Y, Tanaka D, Kanno T, Sugiura R, et al. Frequency response of blood vessel wall with atherosclerosis and aneurysm. Bio-medical Materials and Engineering. 2011;21(3):171–77.

[150] Watanabe T, Kanda K, Ishibashi-Ueda H, Yaku H, Nakayama Y. Autologous small-caliber "biotube" vascular grafts with argatroban loading: A histomorphological examination after implantation to rabbits. Journal of Biomedical Materials Research Part B. 2010;92(1):236–42.

[151] Pevni D, Ben-Gal Y, Mohr R, Ganiel A, Paz Y, Kramer A, et al. One or two internal thoracic grafts? Long-term follow-up of 957 off-pump coronary bypass surgeries. The Annals of Thoracic Surgery. 2017;104(1):70–77.

[152] Murphy EH, Szeto WY, Herdrich BJ, Jackson BM, Wang GJ, Bavaria JE, et al. The management of endograft infections following endovascular thoracic and abdominal aneurysm repair. Journal of Vascular Surgery. 2013;58(5):1179–85.

[153] Manju S, Muraleedharan CV, Rajeev A, Jayakrishnan A, Joseph R. Evaluation of alginate dialdehyde cross-linked gelatin hydrogel as a biodegradable sealant for polyester vascular graft. Journal of Biomedical Materials Research. Part B, Applied Biomaterials. 2011;98(1):139–49.

[154] Ricco J-B, Assadian A, Schneider F, Assadian O. In vitro evaluation of the antimicrobial efficacy of a new silver-triclosan vs a silver collagen-coated polyester vascular graft against methicillin-resistant Staphylococcus aureus. Journal of Vascular Surgery. 2012;55(3):823–29.

[155] Yan S, Li Y, Jiang Y-C, Xu Y, Wang D, Zhang X, et al. Expanded polytetrafluoroethylene/silk fibroin/salicin vascular graft fabrication for improved endothelialization and anticoagulation. Applied Surface Science. 2021;542:148610.

[156] Asadpour S, Yeganeh H, Ai J, Ghanbari H. A novel polyurethane modified with biomacromolecules for small-diameter vascular graft applications. Journal of Materials Science. 2018;53(14):9913–27.

[157] Mokhtari N, Zargar Kharazi A. Blood compatibility and cell response improvement of poly glycerol sebacate/poly lactic acid scaffold for vascular graft applications. Journal of Biomedical Materials Research. Part A. 2021;109(12):2673–84.

[158] Naito Y, Shinoka T, Duncan D, Hibino N, Solomon D, Cleary M, et al. Vascular tissue engineering: Towards the next generation vascular grafts. Advanced Drug Delivery Reviews. 2011;63(4–5):312–23.

[159] Sell SA, McClure MJ, Garg K, Wolfe PS, Bowlin GL. Electrospinning of collagen/biopolymers for regenerative medicine and cardiovascular tissue engineering. Advanced Drug Delivery Reviews. 2009;61(12):1007–19.

[160] Zhang Y, Li X, Guex A, Liu S, Müller E, Malini RI, et al. A compliant and biomimetic three-layered vascular graft for small blood vessels. Biofabrication. 2017;9(2):025010.

[161] Levy BI, Tedgui A. Biology of the Arterial Wall: Springer Science & Business Media; 2007.
[162] Mansfield EG, Greene VK, Auguste DT. Patterned, tubular scaffolds mimic longitudinal and radial mechanics of the neonatal trachea. Acta Biomaterialia. 2016;33:176–82.
[163] Hsieh C-T, Liao C-Y, Dai N-T, Tseng C-S, Yen BL, Hsu S-H. 3D printing of tubular scaffolds with elasticity and complex structure from multiple waterborne polyurethanes for tracheal tissue engineering. Applied Materials Today. 2018;12:330–41.
[164] Wen Y, Xun S, Haoye M, Baichuan S, Peng C, Xuejian L, et al. 3D printed porous ceramic scaffolds for bone tissue engineering: A review. Biomaterials Science. 2017;5(9):1690–98.
[165] Nikolova MP, Chavali MS. Recent advances in biomaterials for 3D scaffolds: A review. Bioactive Materials. 2019;4:271–92.
[166] Benesch J, Mano JF, Reis RL. Proteins and their peptide motifs in acellular apatite mineralization of scaffolds for tissue engineering. Tissue Engineering. Part B, Reviews. 2008;14(4):433–45.
[167] Guda T, Appleford M, Oh S, Ong J. A cellular perspective to bioceramic scaffolds for bone tissue engineering: The state of the art. Current Topics in Medicinal Chemistry. 2008;8(4):290–99.
[168] Bao C, Chong MS, Qin L, Fan Y, Teo EY, Sandikin D, et al. Effects of tricalcium phosphate in polycaprolactone scaffold for mesenchymal stem cell-based bone tissue engineering. Materials Technology. 2019;34(6):361–67.
[169] Yanoso-Scholl L, Jacobson JA, Bradica G, Lerner AL, O'Keefe RJ, Schwarz EM, et al. Evaluation of dense polylactic acid/beta-tricalcium phosphate scaffolds for bone tissue engineering. Journal of Biomedical Materials Research. Part A. 2010;95(3):717–26.
[170] Epstein NE. High posterior cervical fusion rates with iliac autograft and Nanoss/bone marrow aspirate. Surgical Neurology International. 2017;8.
[171] Walsh W, Christou C, Low A, Yu Y, Oliver R, Bertollo N, et al. Bone graft materials: A comparison of NanOss Bioactive 3d and VitOss BA in a challenging model. Orthopaedic Proceedings: The British Editorial Society of Bone & Joint Surgery. 2013;359.
[172] Yazdimamaghani M, Razavi M, Vashaee D, Moharamzadeh K, Boccaccini AR, Tayebi L. Porous magnesium-based scaffolds for tissue engineering. Materials Science and Engineering: C. 2017;71:1253–66.
[173] Adhikari U, Rijal NP, Khanal S, Pai D, Sankar J, Bhattarai N. Magnesium incorporated chitosan based scaffolds for tissue engineering applications. Bioactive Materials. 2016;1(2):132–39.
[174] Bakhsheshi-Rad HR, Dayaghi E, Ismail AF, Madzlan A, Akhavan Farid A, Xiongbiao C. Synthesis and in-vitro characterization of biodegradable porous magnesium-based scaffolds containing silver for bone tissue engineering. Transactions of Nonferrous Metals Society of China. 2019;29(5):984–96.
[175] Yang J, Zhang YS, Yue K, Khademhosseini A. Cell-laden hydrogels for osteochondral and cartilage tissue engineering. Acta Biomaterialia. 2017;57:1–25.
[176] Zarrintaj P, Khodadadi Yazdi M, Youssefi Azarfam M, Zare M, Ramsey JD, Seidi F, et al. Injectable cell-laden hydrogels for tissue engineering: Recent advances and future opportunities. Tissue Engineering Part A. 2021;27(11–12):821–43.
[177] Nguyen TP, Li F, Shrestha S, Tuan RS, Thissen H, Forsythe JS, et al. Cell-laden injectable microgels: Current status and future prospects for cartilage regeneration. Biomaterials. 2021;279:121214.
[178] Pataky K, Braschler T, Negro A, Renaud P, Lutolf MP, Brugger J. Microdrop printing of hydrogel bioinks into 3D tissue-like geometries. Advanced Materials. 2012;24(3):391–96.
[179] Chimene D, Lennox KK, Kaunas RR, Gaharwar AK. Advanced bioinks for 3D printing: A materials science perspective. Annals of Biomedical Engineering. 2016;44(6):2090–102.

[180] Gao F, Ruan C, Liu W. High-strength hydrogel-based bioinks. Materials Chemistry Frontiers. 2019;3(9):1736–46.

[181] Anupama Sekar J, Athira R, Lakshmi T, Velayudhan S, Bhatt A, Anil Kumar P, et al. 3D bioprinting in tissue engineering and regenerative medicine: Current landscape and future prospects. Biomaterials in Tissue Engineering and Regenerative Medicine: Springer. 2021;561–80.

[182] Raphael B, Khalil T, Workman VL, Smith A, Brown CP, Streuli C, et al. 3D cell bioprinting of self-assembling peptide-based hydrogels. Materials Letters. 2017;190:103–06.

[183] Katakam P, Dey B, Assaleh FH, Hwisa NT, Adiki SK, Chandu BR, et al. Top-down and bottom-up approaches in 3D printing technologies for drug delivery challenges. Critical Reviews™ in Therapeutic Drug Carrier Systems. 2015;32:1.

[184] Liu X, Yue T, Kojima M, Huang Q, Arai T. Bio-assembling and bioprinting for engineering microvessels from the bottom up. International Journal of Bioprinting. 2021;7:3.

[185] Nie M, Takeuchi S. Bottom-up biofabrication using microfluidic techniques. Biofabrication. 2018;10(4):044103.

[186] Gaspar VM, Lavrador P, Borges J, Oliveira MB, Mano JF. Advanced bottom-up engineering of living architectures. Advanced Materials. 2020;32(6):1903975.

[187] Benmeridja L, De Moor L, De Maere E, Vanlauwe F, Ryx M, Tytgat L, et al. High-throughput fabrication of vascularized adipose microtissues for 3D bioprinting. Journal of Tissue Engineering and Regenerative Medicine. 2020;14(6):840–54.

[188] Vijayavenkataraman S, Yan W-C, Lu WF, Wang C-H, Fuh JYH. 3D bioprinting of tissues and organs for regenerative medicine. Advanced Drug Delivery Reviews. 2018;132:296–332.

[189] Heid S, Boccaccini AR. Advancing bioinks for 3D bioprinting using reactive fillers: A review. Acta Biomaterialia. 2020.

[190] Bittner SM, Guo JL, Mikos AG. Spatiotemporal control of growth factors in three-dimensional printed scaffolds. Bioprinting. 2018;12:e00032.

[191] Alison MR, Poulsom R, Forbes S, Wright NA. An introduction to stem cells. The Journal of Pathology: A Journal of the Pathological Society of Great Britain and Ireland. 2002;197 (4):419–23.

[192] Mahla RS. Stem cells applications in regenerative medicine and disease therapeutics. International Journal of Cell Biology. 2016;2016.

[193] Kolios G, Moodley Y. Introduction to stem cells and regenerative medicine. Respiration. 2013;85(1):3–10.

[194] Vantyghem M-C, de Koning EJ, Pattou F, Rickels MR. Advances in β-cell replacement therapy for the treatment of type 1 diabetes. The Lancet. 2019;394(10205):1274–85.

[195] Liang J, Zhang H, Kong W, Deng W, Wang D, Feng X, et al. Safety analysis in patients with autoimmune disease receiving allogeneic mesenchymal stem cells infusion: A long-term retrospective study. Stem Cell Research & Therapy. 2018;9(1):1–10.

[196] Uccelli A, Pistoia V, Moretta L. Mesenchymal stem cells: A new strategy for immunosuppression?. Trends in Immunology. 2007;28(5):219–26.

[197] Anderson K, Giralt S, Mendoza TR, Brown J, Neumann J, Mobley G, et al. Symptom burden in patients undergoing autologous stem-cell transplantation. Bone Marrow Transplantation. 2007;39(12):759–66.

[198] Wang XS, Shi Q, Shah ND, Heijnen CJ, Cohen EN, Reuben JM, et al. Inflammatory markers and development of symptom burden in patients with multiple myeloma during autologous stem cell transplantation. Clinical Cancer Research. 2014;20(5):1366–74.

[199] Elisseeff J, Ferran A, Hwang S, Varghese S, Zhang Z. The role of biomaterials in stem cell differentiation: Applications in the musculoskeletal system. Stem Cells and Development. 2006;15(3):295–303.

[200] Zhao X, Cui K, Li Z. The role of biomaterials in stem cell-based regenerative medicine. Future Medicinal Chemistry. 2019;11(14):1777–90.
[201] Xu Y, Chen C, Hellwarth PB, Bao X. Biomaterials for stem cell engineering and biomanufacturing. Bioactive Materials. 2019;4:366–79.
[202] Viswanathan P, Ondeck MG, Chirasatitsin S, Ngamkham K, Reilly GC, Engler AJ, et al. 3D surface topology guides stem cell adhesion and differentiation. Biomaterials. 2015;52:140–47.
[203] Chen W, Han S, Qian W, Weng S, Yang H, Sun Y, et al. Nanotopography regulates motor neuron differentiation of human pluripotent stem cells. Nanoscale. 2018;10(7):3556–65.
[204] Cabanas-Danés J, Huskens J, Jonkheijm P. Chemical strategies for the presentation and delivery of growth factors. Journal of Materials Chemistry B. 2014;2(17):2381–94.
[205] Tabata Y, Miyao M, Ozeki M, Ikada Y. Controlled release of vascular endothelial growth factor by use of collagen hydrogels. Journal of Biomaterials Science. 2000;11(9):915–30.
[206] Oliviero O, Ventre M, Netti P. Functional porous hydrogels to study angiogenesis under the effect of controlled release of vascular endothelial growth factor. Acta Biomaterialia. 2012;8(9):3294–301.
[207] Solorio LD, Vieregge EL, Dhami CD, Dang PN, Alsberg E. Engineered cartilage via self-assembled hMSC sheets with incorporated biodegradable gelatin microspheres releasing transforming growth factor-β1. Journal of Controlled Release. 2012;158(2):224–32.
[208] Kudva AK, Dikina AD, Luyten FP, Alsberg E, Patterson J. Gelatin microspheres releasing transforming growth factor drive in vitro chondrogenesis of human periosteum derived cells in micromass culture. Acta Biomaterialia. 2019;90:287–99.
[209] Jeon O, Song SJ, Kang S-W, Putnam AJ, Kim B-S. Enhancement of ectopic bone formation by bone morphogenetic protein-2 released from a heparin-conjugated poly (L-lactic-co-glycolic acid) scaffold. Biomaterials. 2007;28(17):2763–71.
[210] Liang Y, Kiick KL. Heparin-functionalized polymeric biomaterials in tissue engineering and drug delivery applications. Acta Biomaterialia. 2014;10(4):1588–600.
[211] Sakiyama-Elbert SE. Incorporation of heparin into biomaterials. Acta Biomaterialia. 2014;10 (4):1581–87.
[212] Hettiaratchi MH, Miller T, Temenoff JS, Guldberg RE, McDevitt TC. Heparin microparticle effects on presentation and bioactivity of bone morphogenetic protein-2. Biomaterials. 2014;35(25):7228–38.
[213] Kim S, Cui Z-K, Kim PJ, Jung LY, Lee M. Design of hydrogels to stabilize and enhance bone morphogenetic protein activity by heparin mimetics. Acta Biomaterialia. 2018;72:45–54.
[214] Lee TT, García JR, Paez JI, Singh A, Phelps EA, Weis S, et al. Light-triggered in vivo activation of adhesive peptides regulates cell adhesion, inflammation and vascularization of biomaterials. Nature Materials. 2015;14(3):352–60.
[215] Liu Q, Zheng S, Ye K, He J, Shen Y, Cui S, et al. Cell migration regulated by RGD nanospacing and enhanced under moderate cell adhesion on biomaterials. Biomaterials. 2020;263:120327.
[216] Karimi F, O'Connor AJ, Qiao GG, Heath DE. Integrin clustering matters: A review of biomaterials functionalized with multivalent integrin-binding ligands to improve cell adhesion, migration, differentiation, angiogenesis, and biomedical device integration. Advanced Healthcare Materials. 2018;7(12):1701324.
[217] Yang F, Williams CG, Wang D-A, Lee H, Manson PN, Elisseeff J. The effect of incorporating RGD adhesive peptide in polyethylene glycol diacrylate hydrogel on osteogenesis of bone marrow stromal cells. Biomaterials. 2005;26(30):5991–98.

[218] Steinmetz NJ, Bryant SJ. The effects of intermittent dynamic loading on chondrogenic and osteogenic differentiation of human marrow stromal cells encapsulated in RGD-modified poly (ethylene glycol) hydrogels. Acta Biomaterialia. 2011;7(11):3829–40.
[219] Chun C, Lim HJ, Hong K-Y, Park K-H, Song S-C. The use of injectable, thermosensitive poly (organophosphazene)–RGD conjugates for the enhancement of mesenchymal stem cell osteogenic differentiation. Biomaterials. 2009;30(31):6295–308.
[220] Jang H-K, Kim B-S. Modulation of stem cell differentiation with biomaterials. International Journal of Stem Cells. 2010;3(2):80.
[221] Huang Y, Qian J-Y, Cheng H, Li X-M. Effects of shear stress on differentiation of stem cells into endothelial cells. World Journal of Stem Cells. 2021;13(7):894.
[222] Potter CM, Lao KH, Zeng L, Xu Q. Role of biomechanical forces in stem cell vascular lineage differentiation. Arteriosclerosis, Thrombosis, and Vascular Biology. 2014;34(10):2184–90.
[223] Gwak SJ, Bhang SH, Kim IK, Kim SS, Cho SW, Jeon O, et al. The effect of cyclic strain on embryonic stem cell-derived cardiomyocytes. Biomaterials. 2008;29(7):844–56.
[224] Terraciano V, Hwang N, Moroni L, Park HB, Zhang Z, Mizrahi J, et al. Differential response of adult and embryonic mesenchymal progenitor cells to mechanical compression in hydrogels. Stem Cells. 2007;25(11):2730–38.
[225] Kang KB, Lawrence BD, Gao XR, Guaiquil VH, Liu A, Rosenblatt MI. The effect of micro-and nanoscale surface topographies on silk on human corneal limbal epithelial cell differentiation. Scientific Reports. 2019;9(1):1–8.
[226] Zhang Y, Gordon A, Qian W, Chen W. Engineering nanoscale stem cell niche: Direct stem cell behavior at cell–matrix interface. Advanced Healthcare Materials. 2015;4(13):1900–14.
[227] Choi YY, Chung BG, Lee DH, Khademhosseini A, Kim J-H, Lee S-H. Controlled-size embryoid body formation in concave microwell arrays. Biomaterials. 2010;31(15):4296–303.
[228] Zorlutuna P, Annabi N, Camci-Unal G, Nikkhah M, Cha JM, Nichol JW, et al. Microfabricated biomaterials for engineering 3D tissues. Advanced Materials. 2012;24(14):1782–804.
[229] Mitrečić D, Nicaise C, Gajović S, Pochet R. Distribution, differentiation, and survival of intravenously administered neural stem cells in a rat model of amyotrophic lateral sclerosis. Cell Transplantation. 2010;19(5):537–48.
[230] Epstein SE, Luger D, Lipinski MJ. Paracrine-mediated systemic anti-inflammatory activity of intravenously administered mesenchymal stem cells: A transformative strategy for cardiac stem cell therapeutics. Circulation Research. 2017;121(9):1044–46.
[231] Pandey AC, Lancaster JJ, Harris DT, Goldman S, Juneman E. Cellular therapeutics for heart failure: Focus on mesenchymal stem cells. Stem Cells International. 2017;2017.
[232] Alagarsamy KN, Yan W, Srivastava A, Desiderio V, Dhingra S. Application of injectable hydrogels for cardiac stem cell therapy and tissue engineering. Reviews in Cardiovascular Medicine. 2019;20(4):221–30.
[233] Cai L, Dewi RE, Heilshorn SC. Injectable hydrogels with in situ double network formation enhance retention of transplanted stem cells. Advanced Functional Materials. 2015;25(9):1344–51.
[234] Sodian R, Lemke T, Fritsche C, Hoerstrup SP, Fu P, Potapov EV, et al. Tissue-engineering bioreactors: A new combined cell-seeding and perfusion system for vascular tissue engineering. Tissue Engineering. 2002;8(5):863–70.
[235] Zhang H, Zhou L, Zhang W. Control of scaffold degradation in tissue engineering: A review. Tissue Engineering. Part B. 2014;20(5):492–502.
[236] Cappon G, Vettoretti M, Sparacino G, Facchinetti A. Continuous glucose monitoring sensors for diabetes management: A review of technologies and applications. Diabetes & Metabolism Journal. 2019;43(4):383–97.

[237] Galindo RJ, Aleppo G. Continuous glucose monitoring: The achievement of 100 years of innovation in diabetes technology. Diabetes Research and Clinical Practice. 2020.
[238] Boyce E. Knowledge is power? How continuous blood glucose monitoring systems are changing the management of type 1 diabetes mellitus. Pediatric Nursing. 2020;46(4):179–95.
[239] Goswami D, Domingo-Lopez DA, Ward NA, Millman JR, Duffy GP, Dolan EB, et al. Design considerations for macroencapsulation devices for stem cell derived islets for the treatment of type 1 diabetes. Advanced Science. 2021;8(16):2100820.
[240] Papas KK, De Leon H, Suszynski TM, Johnson RC. Oxygenation strategies for encapsulated islet and beta cell transplants. Advanced Drug Delivery Reviews. 2019;139:139–56.

Concluding remarks

This book is a testament for the decades of hard work of numerous scientists from various fields. Their works have contributed to the advent of biomaterial science and its advancement toward clinical translation. May this book serve as a guide for future researchers of biomaterials to enable them to design rationale and functional biomaterials that can help solve challenges in the biomedical field and the clinic. Ultimately, designing an effective biomaterial can reach millions of people and help alleviate their medical maladies.

Giving perspective on anatomy and physiology allows for a systematic way of thinking for biomaterial scientists when designing biomaterials. As complex as anatomy and physiology can be, this book tried to distill relevant knowledge within the field and give people coming from non-biology or clinical background a working knowledge within those topics. The role of biomaterials in various biomedical fields had been discussed in this book and had detailed the importance of anatomy and physiological concepts in those specific applications. Anatomy is the architecture and roadmap for biomaterial scientists as it gives information to the complexity level of biological structures in the body. Physiology is the way these structures normally function and give information on how biomaterials should behave in a way they are not negatively disruptive but only to restore and maintain homeostasis or alleviate and improve a physiological malady. Equipped with this knowledge, it allows biomaterial scientists to create functionalities that fit within the physiological and anatomical bounds of the human body, which is vital and must be part of every design considerations of all biomedical devices, drug delivery systems, and implants.

Most importantly, learning the relevant concepts about the human body allows biomaterial scientists to assess the risks and benefits of biomaterials they are designing and as much as possible create biomaterials that can avoid detrimental effects to the patient. Although no biomaterial is completely risk-free and functionality is the arm that pushes the field into further heights, biomaterialists should always remember that no amount of functionality is greater than patient's safety. Safety and reliability should go hand in hand with the advancement of the field. Hence, knowing and understanding the human body must be a requirement in the field of biomaterials, because we must first understand our nature before improving it.

Ludwig Erik Aguilar, Ph.D.
Author

https://doi.org/10.1515/9783110655377-009

Index

https://doi.org/10.1515/9783110655377-010

www.ingramcontent.com/pod-product-compliance
Lightning Source LLC
Chambersburg PA
CBHW061153240926
51846CB00006B/997